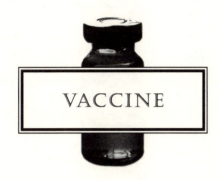

VACCINE

VACCINE

THE CONTROVERSIAL STORY OF
MEDICINE'S GREATEST LIFESAVER

Arthur Allen

W. W. NORTON & COMPANY

NEW YORK   LONDON

For information about permission to reproduce selections from this book, write to
Permissions, W. W. Norton & Company, Inc., 500 Fifth Avenue,
New York, NY 10110

Manufacturing by Maple-Vail Book Manufacturing Group
Book design by Dana Sloan
Production Manager: Amanda Morrison

Library of Congress Cataloging-in-Publication Data

Allen, Arthur, 1959–
   Vaccine : the controversial story of medicine's greatest lifesaver / Arthur Allen.
   1st. ed.
        p. ; cm.
   Includes bibiliographical references and index.
   ISBN-13: 978-0-393-05911-3 (hardcover)
   ISBN-10: 0-393-05911-1 (hardcover)
1. Vaccination—History. 2. Communicable diseases—Prevention—History.
I. Title.
[DNLM: 1: Vaccination—history—United States. 2. Vaccines—
history—United States. 3. Dissent and Disputes—history—United States.
4. Health Policy—history—United States. 5. History, Modern 1601– —United
States. 6. Vaccination—adverse effects—United States. 7. Vaccines—adverse
effects—United States. QW 805 A425v 2007]
   RA638.A45 2007
   614.4'70973—dc22

                                                              2006019480

W. W. Norton & Company, Inc., 500 Fifth Avenue, New York, N.Y. 10110
www.wwnorton.com

W. W. Norton & Company, Ltd., Castle House, 75/76 Wells Street, London W1T 3QT

2 3 4 5 6 7 8 9 0

TO MARGARET, ISAAC, AND LUCY—MY LIFE

&

TO MY PARENTS,
WHO TOOK SOME ADVICE
AND IGNORED THE REST

CONTENTS

Introduction: Vaccination and Politics   11

Part One: ORIGINS

1   Experimenting on the Neighbors with Cotton Mather   25

2   The Peculiar History of Vaccinia   46

3   Vaccine Wars: Smallpox at the Turn
of the Twentieth Century   70

Part Two: GOLDEN AGE

4   War Is Good for Babies   115

5   The Great American Fight Against Polio   160

6   Battling Measles, Remodeling Society   215

Part Three: CONTROVERSY

7   DTP and the Vaccine Safety Movement   251

8   No Good Deed Goes Unpunished   294

9   People Who Prefer Whooping Cough   327

10   Vaccines and Autism?   371

Epilogue: Our Best Shots   421

Acknowledgments   443

Notes   445

Index   499

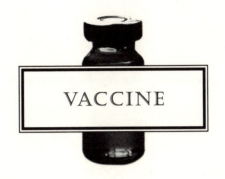

VACCINE

INTRODUCTION

—

VACCINATION AND POLITICS

On December 21, 2002, President George W. Bush rolled up his sleeve, presented a deltoid muscle on his left arm, and was pricked 15 times with a tiny, bifurcated needle, whose prongs held between them a droplet of vaccinia virus derived from an infected calf. The Commander in Chief's vaccination was the keystone of a public health campaign to immunize 10 million police and health workers against smallpox by the fall of 2003, preparing the nation for a terrorist germ warfare attack. "We believe that regimes hostile to the United States may possess this virus," Bush said. He didn't name Iraq, but two of his highest aides did, under cover of anonymity.[1]

The president's vaccination was a highly politicized public health gesture, a symbolic act demonstrating that Saddam Hussein's capabilities and plans were real and evil enough to justify aggression against his regime. The last samples of smallpox virus in the world were supposedly kept in locked freezers at the CDC in Atlanta and in Novosibirsk, Russia. But the Soviets, we had learned from defectors, had secretly produced 20 tons of smallpox, and with the Soviet biowarfare complex dispersed to the four winds by economic collapse, no one was entirely sure that a cynical Russian expatriate might not be fiddling with some of that virus in a laboratory in Baghdad.[2] The government had no evidence whatsoever, it would eventually be learned, that Saddam had acquired smallpox. But in those feverish days the mere suggestion that he might have done so provided a peculiar, circular justification for smallpox vaccination, and for war. If we vaccinated our president, surely we were doing it because Saddam wanted to deploy smallpox against us. If Saddam was indeed evil enough to start an epidemic of this dread disease, and armed with the power to do it, then his overthrow was an urgent necessity. If we invaded Iraq, then he might use smallpox as a terror weapon. This was how we came, at the beginning of the twenty-first century, to vaccinate our president against an extinct disease.

Americans had not been routinely vaccinated against smallpox for 30 years, and the vaccine had not been updated in any way. The inoculation Bush received was little different in its technique from that first described by Edward Jenner, an English country doctor, in 1796. Vaccinia, which Jenner found growing on cows, was the "shot" that put the *vacca* in vaccination; Louis Pasteur gave the name *vaccines* to his concoctions against rabies and fowl cholera a century later in tribute to Jenner. Smallpox vaccination was one of the first successful medical interventions; because of it, smallpox became the first and only contagious disease ever eradicated. In a rare case of cold war cooperation, the Soviet and U.S. governments worked together through the U.N. World Health Organization marshalling thousands of medical personnel to chase down the deadly virus and administer the knockout blows of vaccine that quelled smallpox. Its defeat, in 1980, was the culmination of an era of passionate medical idealism; campaigns to eliminate other infectious diseases—malaria, polio, measles, and hepatitis B—came on its heels.

Many of the scientists and doctors who cornered the smallpox virus in remote African and Indian villages now occupied senior positions in public health in America, and a few sat on the Centers for Disease Control's advisory committee for immunization practices, which the White House asked in 2002 to decide how to proceed on smallpox vaccination. The public health administration was profoundly ambivalent on the subject. Evidence of weaponized Iraqi smallpox was scanty; John Modlin, the Dartmouth pediatrician who headed the advisory committee, was given a CIA briefing on the evidence and later said he learned no more than had already been published in *The Washington Post*.[3] There were public health officials who recognized that the vaccination enterprise had long relied upon military and other social mobilizations. Without fear, history had shown, it was difficult to get people vaccinated. Many of the president's advisors, including Vice President Dick Cheney, his chief aide Irwin "Scooter" Libby, and Marvin Olasky, the author of Bush's "compassionate conservatism" philosophy seemed to regard the return of smallpox with *schadenfreude*, almost as if they were hoping for an occurrence of the scourge that would validate their worldview. The threat, according to Olasky in a *Washington Times* newspaper column, revealed the "ignorance" of the Centers for Disease Control (CDC) and the World Health Organization (WHO) campaign to eradicate the disease. "Because liberalism dominates American and European culture, we stopped inoculating against smallpox, and now we are more vulnerable to it than at any time over the past two centuries," Olasky wrote.[4] The majority of physicians, even the specialists, were part of that liberal culture and were skeptical of the smallpox campaign. According to one insider, the White House had to "kick, punch and stomp" the CDC panel into accepting its plan. Another described the panel as "sheep"; faced with a risk they couldn't properly evaluate, the panel voted 8–1 (Paul Offit, a pediatric immunologist from the University of Pennsylvania, was the sole dissident) to recommend the vaccination of 500,000 hospital workers, police officers, and firefighters in the first month of 2003, and 10 million others by the end of the summer.

Within a year of its portentous start the smallpox vaccination campaign had sputtered out. County and state public health officials complained of the costs and liability risks posed by the vaccine. Many said their time and money would be better spent encouraging vaccination against the flu, a quantifiable killer. (The administration did not turn its full attention to the threat of pandemic flu until three years later, when hurricanes Katrina and Rita finally forced it to reckon with the likelihood—and political impact—of natural disasters.) The nurses' unions and scores of hospitals were refusing to join the smallpox vaccination campaign, following the precedent set by 6,000 Washington and New York mail sorters who balked at getting prophylactic vaccination in the aftermath of the anthrax mailings of October 2001. The postal workers had heard about reports of dangerous side effects from the Pentagon's anthrax vaccination of troops that began in 1998; rather than risk the autoimmune and neurological disorders claimed by some military veterans, the postal employees preferred to face the small risk that residual anthrax spores in their bodies would germinate after antibiotics had worn off.

In a nation ambiguously mobilized for warfare, smallpox vaccination had become thoroughly tangled in politics. So, as we'll see, had vaccination in general. The way in which smallpox vaccination returned showed something about the broader view of public health and how it had changed in the last half-century. There was a time not so long ago when nearly all Americans, grateful for the defeat of polio by Jonas Salk's famous shots, eagerly embraced vaccination. But the nurses and mail sorters were not the only vaccine resisters of the new millenium—far from it. Many parents of small children were concluding that vaccination was not an immutable part of life, a wholesome and unavoidable rite of passage, but rather a medical procedure that clashed with their concepts of personal autonomy, informed consent, and acceptable risk.

I became curious about vaccination shortly after my son was born in 1996, when I heard that an older vaccine against whooping cough—at the time I hardly knew what whooping cough was—had been replaced by a safer version. As a parent, I was a little surprised that any vaccine given to millions of babies had been considered dangerous enough to be replaced; vaccines seemed as simple and as unassailably good for you as mother's milk. The possibility that they weren't whetted my journalistic instincts. As I looked into the history of the whole-cell pertussis vaccine I came across a still-unfinished debate about the relative dangers of vaccination in general. Gradually, as I examined vaccines in a series of articles for *The New Republic*, *The Washington Post Magazine*, *The New York Times Magazine*, and *Salon*, it seemed to me that the best way to gain an understanding of why our children were vaccinated against particular diseases—and why some people were challenging these choices—was to delve into the history of vaccines.

WHAT IS A VACCINE? Under the definition I'll be using in this book, a vaccine is a substance that introduces a whole or partial version of a pathogenic microorganism into the body in order to train the immune system to defend itself when the organism threatens to cause an infection through natural means. A vaccine works by stimulating the immune system to create antibodies and immune cells that recognize the pathogen and are thus prepared to battle it when it presents itself at the portals of the body. From a scientific perspective, vaccination has evolved from a purely empirical procedure into a technology that benefits from our growing ability to understand and manipulate germs and the immune system. Although vaccination works by protecting an individual, it rarely works perfectly to protect all individuals who receive it, and some individuals cannot be vaccinated. Thus, vaccination campaigns that seek to protect large populations require cultural as well as scientific innovation. Unlike cancer pills, asthma sprays, or insulin shots, vaccines prevent rather than actively fight illness. Like these drugs, vaccines carry some measure of risk to the patient. To convince people to take this slight risk in the interest of fighting a disease that is not currently harming them, or their children, doctors and public health officials must try to shape how people think about infectious disease. Whatever the idealistic objectives of a vaccine, the authority of its administrator is key. The history of vaccines as a disease-fighting measure is thus partly the story of how authorities imposed a medical procedure on people, and how people responded.

In America, vaccination is the first act the state requires of a person; without it, or a legal exemption, a kid can't even get into nursery school. But while vaccination seems anodyne to most people, it has often been dogged by controversy. Over time, vaccines have become a victim of their own success: since the great majority of children in this country no longer come down with the terrible infectious diseases that shots protect against, the other diseases of childhood—autism, juvenile diabetes, ADHD, asthma—have become more visible; the prevalence of some of these diseases is even on the rise. A recent survey published in *The Journal of the American Medical Association* found that a quarter of all American parents are now reluctant to have their children vaccinated.[5] The social history of vaccines, which is to say their use and acceptance in America, does not always parallel the scientific history. While vaccination seems to be more efficient and safer than ever before, public ambivalence about the practice has rarely been higher.

In the first 200 or so years of vaccination's history in America, it could only prevent smallpox. Starting in the early 1900s and especially after

World War II, the pace of development accelerated and vaccines conquered a plethora of major diseases. Optimism about their potential made the postwar a Golden Age of public acceptance for vaccination, although there were plenty of troubles along the way. It was only in the 1970s, when most vaccine-preventable diseases had been soundly beaten back, that the angst began: first there were claims that vaccines were unnecessary and dangerous, and that some caused brain damage; later came the hypothesis that they were causing the increase in chronic illness. While only a small minority of American parents openly shunned vaccination, a 30-year groundswell of dissent has led us to a point where vaccination might be said to be in crisis. Parents armed with Internet educations, manufacturers wary of legal liability, a cost-cutting government obsessed with germ warfare—all have made an already dissonant debate over vaccines harsher. In 2006, despite emerging microbial threats such as West Nile Virus, SARS, AIDS, and avian flu, the public was almost as skeptical of vaccinating its children as it had been a century earlier.

But if science had improved vaccines, public health had learned to be more circumspect in their use. As the smallpox vaccination campaign got under way in 2003, it was accompanied by words of caution. A few old doctors could recall a previous bioterrorism scare, in 1942, during the course of which the military had accidentally infected 300,000 men with hepatitis B, killing more than 100 of them with a contaminated yellow fever vaccine. One of the infected men was Harold Hinman, a public health service doctor en route to El Salvador. The physician's son, Alan Hinman, went on to become a pillar of the postwar U.S. vaccine establishment, with a reputation built partly on his creation of the first vaccine-safety monitoring system. Other admonitory voices recalled the 1976 swine flu fiasco, in which 40 million Americans were vaccinated to prevent a flu pandemic that never came and after which the government paid $100 million to people who claimed the vaccine had caused an autoimmune disease. And then there was the last mass smallpox vaccination campaign, in 1947, when 6 million New Yorkers were vaccinated after a traveling businessman brought a case from Mexico to Manhattan. During that campaign, the smallpox vaccine had caused far more casualties than the disease had.

In 2003, the leaders of the public health service drew lessons from this

thorny past. They held public forums to discuss the vaccination campaign, screened out people whom the vaccine might harm, and established a surveillance system to detect adverse reactions. The result of all
of these elaborate precautions was a vaccination campaign that never got
off the ground—which was probably a good thing. Doctors and nurses
face lethal bugs in their hospitals every day and tend not to be credulous
types. By 2005 fewer than 40,000 people had chosen to be vaccinated
against smallpox. The Bush administration had seemingly distorted the
truth and manipulated public fears to achieve its goals. Just as the failure
to find weapons of mass destruction in Iraq had cast doubt on the war's
necessity and soured the public on its harsh costs, the CDC's smallpox
vaccination requirements had exacted a price in trust. It was clear that the
CDC had endorsed a policy many of its wiser heads did not believe in.
Because the CDC could not offer a solid justification for the figure of 10
million vaccinees, it suffered the first major crisis of credibility in its history. Many professionals in state health departments felt that the vaccination campaign was not based on one of the sober risk assessments for
which the agency was justifiably known, but on an attempt to build
national consensus for war by stoking fear. In that sense it was a perver-
sion of the social contract that had always been at the heart of vaccination
programs in America.

When I vaccinate my child against tetanus, which grows in the soil,
shelters in rusty nails, and does not spread from child to child, I am protecting my child alone. But I also vaccinate her against measles, polio,
whooping cough, chickenpox, mumps, meningitis, hepatitis, diphtheria,
and pneumonia. In doing so I help eliminate the safe haven from which
these organisms might launch an attack on somebody else's child, on a
teenager whose immunity has waned, or on an adult who was never vaccinated. Public health means recognizing that complete personal responsibility is unattainable. Within limits, we must all help look after one
another—that is our social contract.

In telling the story of vaccination this book makes an assessment that
is as fair as I can make it, based on the available evidence. I am neither a
scientist nor someone with personal experience of a severe vaccine reaction (you will read about both in these pages). I am neither a parent like
George Mead of Oregon, propelled into antivaccine activism by dismay

over his child's decline into autism, nor a vaccinologist like Paul Offit, driven by a belief in the lifesaving imperative of vaccines. I have tried to steer clear of the defensive posture of some public health figures who equate criticism with disloyalty or ignorance. Scientists try to base their decisions on the demonstrated facts, but even the best scientists are most confident of the facts established in their own laboratories.

I do, however, bring personal agendas to this book. First, I have a healthy respect for germs and the doctors who treat them—especially since nearly dying from a *Streptococcus pyogenes* infection that snuck up on me out of nowhere, it seemed, when my children were 4 and $1\frac{1}{2}$. Second, I grew up in a middle-class Jewish home in Cincinnati in which pediatricians and other doctors were both close friends and respected authorities. I also had an older sister, a nurse and yoga practitioner, whose alternative approaches to health challenged, puzzled, and sometimes angered the rest of us. I struggle to this day to understand whether these beliefs helped or hurt my sister, who died of breast cancer in 2003. She blamed many of her problems on environmental toxins and "toxic" childhood experiences. When I hear the parents of suffering children insist that the "vaccine machine" inflicted their pain I can't help but think that anger is a stage of reckoning with a profoundly damaged child. Anger has its purposes, after all. It can motivate and transform despair, whether or not its target is misplaced.

My discussions with parents skeptical of vaccination, I think, have helped clarify for me a certain insight into the philosophical underpinnings of health beliefs. Scientists, and scientific thinkers in general, make different assessments of risk than do most of the rest of us. The scientist is accustomed to reducing a problem to a set of solvable variables. Thus, the vaccine developer, or public health scientist, sees a disease that kills and sickens a certain number of people each year. She sees a vaccine that has the potential to combat that disease. At that point she becomes convinced that people should be given the vaccine, as long as it does not sicken people too often and can be produced inexpensively enough for health insurers to pay for it; as long as it does not counteract the effects of other vaccines or overly complicate vaccine schedules. This is the basic utilitarian equation of vaccines. But the rest of us carry around other, sometimes tangled and contradictory, often emotionally laden, criteria.

Unlike the scientist (or the scientific ideal anyway), few of us can escape generalizing from personal experience. If our child is harmed by a vaccine, we will tend to believe other parents who say their child was harmed by a vaccine. Many of us are likely to have an instinctual aversion to taking a drug unless we can be convinced it is absolutely worthwhile. Unlike the scientist, we do not quantify the risk on a population-wide scale. To do so would seem profoundly inhuman even to the most altruistic among us. Instead we ask, Why bother? or What's in it for me? In the case of a pediatric vaccine, we may think, Things are going along pretty well for my child at the moment, but you never know what the future will hold—so why take a chance? Whether we are preparing for terrorism or planning vaccine schedules, we are forced to reckon with the fact that people assess risks in different ways. This is right and necessary. The day when all our decisions are based on equations fed into computers would be a sad day indeed. Still, it's worth doing the equations, and everyone who wishes to share in the decision-making must recognize science for what it can tell us.

This book deals with preventive vaccines against infectious diseases. For reasons of space and thematic unity, the book does not delve into the emerging field of therapeutic vaccines, which seek to prime the immune system to fight things like cancer, asthma, diabetes, and cocaine addiction. In general, when I speak of vaccines I am referring to substances that provoke *active* immunity, meaning that they stimulate the patient's immune system to create antibodies and to train certain immune cells to recognize a particular virus, bacteria, or parasite. I will refer to immunizing substances that provide *passive* immunity as gamma globulins, IGG (immune gamma globulin), or serums. These are therapeutic substances, antibodies to a particular germ, harvested from human or animal blood, that are injected or infused into a sick patient. In a few historical circumstances, passive immunogens have also been administered preventively on a large scale. For example, children in the early part of the twentieth century were given diphtheria antitoxin, produced in the blood of horses, to protect them from bacteria circulating in their families or communities. In contemporary practice, a patient hospitalized with a deadly infection may receive intravenous gamma globulins that are essentially a more sophisticated and safer version of the horse serums of yesteryear.

Many new-generation vaccines with complex designs are being tested as this book is being written. But in general, the vaccines against bacteria, viruses, and, to a lesser extent, parasites discussed here can be classified into several groups:

- *Live-virus vaccines:* Examples include the vaccines that protect against smallpox, measles, mumps, and rubella, as well as the oral polio vaccine. The live-virus vaccine is a disease virus that has been modified—by exposure to animal tissues, or, most recently, through genetic manipulation—so as to not cause disease while still causing an infection that generates immunity. Some newer live-virus vaccines against rotavirus, dengue fever, and other diseases are viral chimeras; scientists start with a particular virus' genetic backbone, then modify it by adding genes from other viruses.
- *Killed-virus vaccines:* Examples include the Salk polio vaccine and most flu vaccines. These are viruses that have been inactivated through exposure to heat or to chemicals such as formaldehyde.
- *Live-bacteria vaccines:* The best-known example is the Bacille Calmette-Guerin (BCG), a vaccine that has been used for decades, with limited efficacy, to combat tuberculosis. To make BCG, tuberculosis bacteria from cows were modified in culture to provide immunity without disease. New, genetically modified live-virus vaccines against tuberculosis and typhoid fever are being developed.
- *Killed-bacteria vaccines:* These vaccines include the original pertussis, cholera, and typhoid fever vaccines. Made of entire bacteria that are cultured and then killed with chemicals, most of these vaccines are considered crude and are being phased out of use in the developed world, usually replaced by bacterial subunit vaccines.
- *Bacterial subunit vaccines:* Most bacterial vaccines in use in the United States at present are what is known as subunit vaccines. They consist of bacterial proteins and sugars modified in a variety of ways. The common diphtheria and tetanus vaccines are made by chemically altering toxins from the bacteria; whooping cough vaccines by purifying certain pertussis proteins. Polysaccharide vaccines, including certain typhoid, haemophilus and pneumonia vaccines, are made from the purified components of bacterial cell walls. Newer vaccines against diseases like

Haemophilus influenzae type b (known as Hib) meningitis, and *Streptococcus pneumoniae*, were made by chemically fusing bacterial proteins with polysaccharide cell-wall components.

Be it viral or bacterial, live or killed, a vaccine's success as a public health measure relies on three legs of support: (1) the public, which must be confident of the safety and worth of the procedure; (2) manufacturers, who seek to generate profits by making vaccines; and (3) government and public health professionals, who negotiate with the others to further population-wide health goals. As you will see throughout this book, none of these legs is entirely stable.

PART ONE

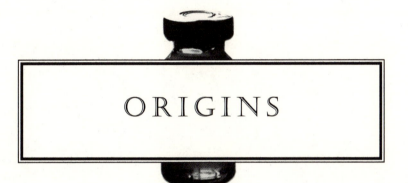

ORIGINS

CHAPTER 1

—

EXPERIMENTING ON THE NEIGHBORS WITH COTTON MATHER

Pray, Papa! Pray to God to bless us, for we are inoculated.

—FRANCES BOSCAWEN, 1755[1]

"THE GRIEVOUS CALAMITY of the smallpox has now entered the town."

Now *there* was a sentence to spark terror in every heart. Fear and excitement probably mingled in the mind of the Reverend Cotton Mather as he wrote it, on May 26, 1721. As it usually did in colonial Boston, the pox had entered from the sea. From the windows of his house on Ship Street by the harbor, Mather could have seen the vessel that brought it ashore. The *HMS Seahorse*, docked at Long Wharf, had sailed past quarantine at Spectacle Island on April 21 after its voyage from the British Caribbean. Sick slaves were found on board, and now the disease was well entrenched around 12,000-souled Boston.

The pious response to a smallpox epidemic was to accept its divine judgment as an opportunity for repentance. The practical response was to head for the hills. But the next few sentences in Mather's diary signaled the start of something new: "The practice of conveying and suffering the Smallpox by inoculation has never been used in America or indeed in our Nation," he wrote. "But how many lives might be saved by it if it were practiced? I will procure a council of physicians and lay the matter before them." History had chosen Mather to introduce vaccination to America, although he was setting in motion the sort of changes that would some-day crumble his God-fearing tower of belief.[2]

There were many faces to Mather: He was a man of God, a politician, a preacher, a grieving father and multiple widower. At times he was play-ful and lively, even with religious foes; there was scarcely anyone more interesting to talk to in New England. Other times he condemned his enemies to eternal damnation.

Bostonians, for the most part, did not wish to accept Cotton Mather's deliverance. Mather was the grandson of Richard Mather, a Puritan founder of the Massachusetts colony, and the son of Increase Mather, prominent author, politician, former president of Harvard University, and octogenarian hypochondriac. Mather the younger was 58 years old, and the flock was leaving his church, tired of his incessant, foolish poli-ticking, his thundering condemnation of political enemies, his manic-depressive wife and self-pitying sermons. They could not pardon his endorsement of the Salem witch trials of 1692, even though Mather, ever the political animal, eventually came round to condemning the trials. In small-town Boston the people rolled their eyes at Mather and his dire proclamations. Mather had become a tragicomic figure.

Yet for all his conviction of the absolute rule of an angry God, Mather was among the most learned, scientifically curious men of his day, a member of the prestigious Royal Society in London, led by Sir Isaac Newton. He owned America's largest library, with about 8,000 tomes, and was its most prolific writer, author of 400 works, with a writer's curiosity and vanity. Embracing inoculation, he was probably animated in equal parts by political ambition, hopes for salvation, and sheer inquisitiveness.

A grim conviction of God's punishing grip on humankind showed in Mather's earlier writings about smallpox. This disease he saw approach-ing Boston like "an angel with flaming sword" was caused, in one of the common views of the age, by an "innate seed" that sprouted from the body in reaction to the mysterious airs and sparks of a shifting environ-ment. Smallpox was intrinsic to civilization. For a mind like Mather's, it was intrinsic to Mankind's state of sin. When the spark fired the smallpox within you, Mather wrote in a seventeenth-century medical treatise, *The Angel of Bethesda*, the only path was to bow before the terrible heavens and prepare for the worst: "There is a poison within thee, the poison of an evil heart which departs from the living God. . . . All of the watery pustules which now fill thy skin are but little emblems of the errors which thy life has been filled withal. Make thy lamentation, Lord, from the sole of the foot, even to the head, there is no soundness in me; only putrefying sores."[3] He was firm in his conviction, ecstatic in tidings of woe, yet it would be Mather who did more than any of his confederates to bring medical innovation to eighteenth-century America. It was he who overturned the notion of hopelessness before the Providence of smallpox.

In this, the first century of vaccination, the questions about the efficacy of the procedure were often secondary to doubts about the morality of using it to intercede against an illness. Among medical interventions, vaccination was uniquely vulnerable to accusations of sin. There was no religious injunction against a doctor's use of bleeding, for example, or the application of massive doses of toxic minerals such as mercury, cadmium, and sulfur. But these were treatments of disease. To sicken oneself as a way of preventing God from sickening you—that was a terrifying spiritual risk, an act of supreme arrogance (as well as a potential violation of the Hippocratic oath to first, do no harm).

Inoculation, or variolation, as it has come to be known, was a procedure in which smallpox virus, usually procured directly from the pustule of a sick patient, was scratched into a person's skin with a knife or other sharp object. The word *vaccine* was not coined until 1800 by Richard Dunning, a physician in Plymouth, England, to describe the invention by his friend Edward Jenner, who discovered that a virus growing on cows could protect people from smallpox. But although many histories credit Jenner as the first vaccinator, variolation had been practiced in parts of Asia for up to a millenium earlier. In medical terms, variolation was also a form of vaccination, and it was also a tool of public health.

The Boston smallpox epidemic of 1721 marked a milestone in the story of medicine and public health, and it was also a chapter in the battle between secular and sacred authority. Ironically, Mather and his clerical allies used their churchly power to push through a scientific innovation in the face of strong resistance from the medical profession. Over time

Mather's campaign encouraged further improvements in medical science and growing faith in the efficacy of doctoring. The ability of medical interventions to save people and prolong lives would gradually alter humankind's vision of its place in the world. It would weaken the religious conviction of illness as sin that Mather so fervently expounded.

Boston, 1721

It is difficult to imagine a world in which God's awesome power was an undeniable fact for everyone, a world in which loved ones were swept away by diseases whose sudden appearance was as mysterious as their departure. Disease, like a spiritual journey, was a passive experience. Disease states, like visions, were visited upon people. The prevalence of death, especially among children, was a state of things that molded a stout heart or a crazed mind. Mather himself lost 13 children. "A dead child is a sign no more surprising than a broken pitcher or a blasted flower," he once preached. His dour father had warned that "great affections breed great afflictions," yet for Mather, "the dying of a child is like the tearing of a limb from us."

Smallpox before vaccination may have provided an excellent opportunity to bear witness to God, but this is not to say that Bostonians made no effort to save their skins. Although germ theory in its present form would not be accepted for another 150 years, smallpox had been around long enough in Europe for people to realize that it spread from person to person. It was understood that quarantine, flight, and prior exposure were means of avoiding the contagion of smallpox, though who or what caused the disease was less clear. One influential idea, attributed to the seventeenth-century physician Thomas Sydenham, was that diseases returned when the "epidemic constitution," a vaguely defined combination of weather and other environmental conditions, was propitious. In 1721, perhaps half of Boston was immune to smallpox, since the last epidemic had struck in 1702. Even as Mather took note of the impending catastrophe, about a thousand townspeople—a good portion, no doubt, of those who had the means to do so—were packing their belongings to flee for the countryside. The council of physicians that Mather proposed to consider variation never occurred. Instead, his letter notified doctors that this erratic, headstrong preacher was up to something, mobilizing them against his plans. Mather seems to have been loathed by many of the artisans, merchants, and professionals of Boston; they had little patience for his self-promoting politicking. That antipathy, along with the shocking novelty of the method he proposed, led to an enormous outcry. He was, however, able to convince a single physician, a member of his congregation named Zabdiel Boylston, to begin variolating patients. On June 26, 1721, Boylston—whose wife and other young children had been shipped out to the country to avoid the epidemic— performed the operation, using a toothpick and quill for the first time on three patients: his 6-year-old son, Tommy, his slave Jack, and Jack's $2\frac{1}{2}$- year-old son Jackie.[4] The operation apparently took place at Boylston's modest house in Brookline; Tommy took sick on the seventh day. After three days of fevers that "did considerably terrify" his father, the child and his experiment-mates came out the other side, "no longer gravely ill."[5]

Subsequently, Boylston carried out most variolations at his office on Dock Square, a site now occupied by Sam Adams Park, adjacent to Faneuil Hall. At the time this was the water's edge in the heart of Boston Harbor. As the epidemic continued through the fall and into the next year, Boylston variolated a total of 248 patients. Many of the first were dependents of Mather's and loyal fellow ministers; male heads of household seemed most comfortable with the procedure after their slaves and children had survived it. But after only a few months of good results, variolation had become popular enough among the elite that several Harvard students underwent it. At the peak of illness, smallpox patients were covered with thousands of boils. Boylston would harvest pus by puncturing a boil with a lancet and squeezing the fluid into a stoppered glass jar. To variolate a new patient, Boylston cut a slit into the arm or leg and liberally dabbed on the fluid. Throughout the terrifying ordeal of fever and convalescence following variolation, Boylston's patients stayed in bed in darkened rooms. Relatives and friends paid visits, "praising God all the while," as a contemporary stated. It is no wonder that deliberate infection with this deadly disease was a repugnant, horrifying concept to those who had lived through a smallpox epidemic, which is to say, most adults. Belief in its success was a matter of faith given the paucity of evidence available even to a well-educated Bostonian. Perhaps only someone as self-righteous as Mather could have had the nerve to introduce it.

Boylston first gave public notice of his variolations in a letter to the *New England Gazette* on July 17. At a meeting four days later, all 12 of the town's other physicians voted to urge Boylston to cease and desist. An immigrant French surgeon, Laurence Dalhonde, alleged that variolation was unsafe, having killed or seriously injured sailors upon whom it was tried at Cremona, Italy, in 1696. Boylston invited the other doctors to visit the seven patients he had variolated up to then, but he was shunned. The town's selectmen ordered Boylston to stop, but he ignored them and continued to variolate under the protection of the Mathers and their allies. As smallpox whipped through the town, Boylston's variolations, rather than the pox itself, became the principle fixation of the popular anguish. As "widows multiply," Boston became a "dismal picture and emblem of hell, fire and darkness filling it and a lying spirit reigning there," Mather wrote.[6] The "deluded" townspeople responded to Boylston with "a Satanic fury. They rave, they rail, they blaspheme" against the "devilish invention" that was "against divine Providence."[7] Himself but a recent convert to this strange practice, Mather thought he knew who was behind the opposition: "the devil in a great rage at being cheated of a great feast of death."

Sixteen-year-old Benjamin Franklin, who would become a champion of variolation years later in Philadelphia, helped to publish many of the attacks on it in his brother James's newspaper, the *New England Courant*. The Franklins, supporters of more economic autonomy for Massachusetts, opposed Mather and the clergy because of their close ties to the governor and the crown. As public alarm increased, Boylston found he was not safe on the streets and for a while went underground, variolating patients wherever he could find a safe house. The town's doctors and merchants heaped invective on Mather and Boylston in the press and in pamphlets, and the ministers gave as good as they got. Reading these pamphlets, many of them preserved at the National Library of Medicine, a student of contemporary vaccination controversies can't help but feel a sense of recognition. Lurking behind the disagreement over the benefits and risks of a medical procedure were political animosity, professional competition and insecurity, the discomfort of indecision, and the fear of death.

At the heart of the dispute was a great irony: organized opposition to Mather was led by the doctors of Boston, in particular the most highly trained physician among them, an arrogant Scotsman named William Douglass. The Congregationalist clergy, meanwhile, accustomed to speaking for an angry, merciless God, found themselves in the position of defending a practice on the grounds that its lifegiving properties could only have been put in their hands by a gentle, loving deity. "We who have stood by and seen [Boylston's] tenderness, courage and skill in the hazardous operation cannot enough value the Man and give praise to God," the preachers said in one letter to the *New England Gazette*. As for Douglass, who had emigrated to Boston in 1715, he was "a man of good learning but mischievously given to criticism," a fellow Scot wrote some years later.[8] Unlike most American colonial physicians, who had learned their trade by apprenticeship, Douglass was the product of three European medical schools, the only diploma-carrying MD in Boston. The pedigree may not have guaranteed much power to heal, given the dangerous excesses of early-eighteenth-century medicine, but it did give Douglass a halo of superior authority. Within a few years, variolation would catch on and Douglass himself would be a leading variolator during the next Boston epidemic, in 1730. His attack on Mather and Boylston was in the first instance a defense of his professional standing—a staking out of turf.

Upon arriving in Boston from Europe, Douglass had attempted to cultivate Mather by lending him books and journals from the mother country. Indeed, Mather found the first published accounts of variolation in copies of the *Philosophical Transactions*, a publication of the British Royal Society, that he borrowed from Douglass.[9] When it became clear that his journals had inspired Mather's experiment, Douglass furiously demanded them back. These articles were nothing more than "virtuoso amusements" for dilettantes, he wrote. They were of no medical importance, and allowing Mather and Boylston to make use of them had been like placing "dangerous edge tools in the hands of fools."[10] Douglass was enraged by the pious support Boylston marshalled in the clergy, who supported him in a broadside published a few days after the town selectmen had ordered an end to variolation. For Douglass, this was unwarranted meddling in civil matters by the unqualified ministers of the soul. It was

all so amateurish—nothing of the sort would have happened back in civilized Europe, he believed. Boylston, an "illiterate Quackish character,"
had been "rash and heedless" to variolate without the town's permission,
and the ministers were operating outside their area of expertise by
bestowing this "ignorant cutter for the stone"—a derogatory reference to
the surgical profession—with "all the fulsome common place of quack
advertisements."[11] In Britain's medical hierarchy, which did not exist in
the colonies, surgeons were well down the ladder from physicians like
Douglass. The variolation campaign, he wrote, was an "infatuation" typical of Puritan ministers like Mather, "a man of whim and credulity" who
took the smallpox epidemic as "a fit opportunity to make experiments on
his neighbors."[12]

Douglass brought out the self-righteous prig in Mather, who enjoyed
how the controversy isolated and scoured him.[13] When a firebomb was
tossed through a window of Mather's house, he responded with hysterical
elation. The bomb, which failed to explode, came attached to a note that
declared, *Mather, you dog, Damn you, I'll inoculate you with this.* For
Mather, who throughout his life had described each toothache, chest
pain, and fever with the dire conviction of his impending doom, the fire-
bombing refurbished his self-image.[14] He was "overcome with unutter-
able joy at the prospect of approaching martyrdom."[15] It was thrilling to
be targeted "for no other cause pretended but this: that I have saved the
lives of dying people."

For all the political intrigue, the substance of Douglass's response to
variolation was reasonable for a physician of the time. As a means of stopping smallpox, variolation was untested and could spread other diseases
from patient to patient. Mather's belief in invisible smallpox germs was
almost on a par with the "wonders of the invisible world" to which he had
attributed the bewitching of Salem's children. What's more, the variolated patients were not being properly quarantined, and their free circulation around the town, Douglass warned, made them a menace to
society. "If a man should willfully throw a bomb into a town . . . ought he
not to die?" Douglass asked. "So if a man should willfully bring infection
from a person sick of a deadly and contagious disease into a place of
health—is not the mischief as great?"[16]

In a number of ways, the reactions to variolation set the pattern for future generations of vaccination skeptics. The variolation debate triggered the first American dispute over medical statistics, for example. In 1726, Boylston published an account of the affair in which he listed each of the 248 variolations he practiced—certainly one of the earliest lengthy medical case series. Boylston reported that 6 of the patients—about 3 percent—had died under variolation. By comparison, 844, or 14 percent of the 5,980 townspeople who took ill of smallpox "in the natural way" had died—a grim tithing. These early case histories were met in turn with some of the earliest charges of scientific fraud. In an account of the events published in 1730, Douglass charged[17] that Boylston's account had suppressed deaths and injuries among his patients, and left out some variolations—"20 or 30 were concealed from us, many of them without a doubt in their silent graves."[18] This sort of suspicion must have been widespread in the colonies, because following the Boston smallpox epidemic of 1753, Benjamin Franklin described the organization of a "strict and impartial inquiry," in which constables from each ward of the city went forth, accompanied by partisans for and against variolation, to collect information on the morbidity and mortality of "natural" and variolated smallpox cases in each household.

Boylston's account of the campaign made it appear that variolation was safer than getting smallpox the normal way. But whatever they made of the risk/benefit equation, Bostonians were still, on balance, reluctant to variolate. What was the sense of putting one's child or spouse or even one's self at risk from a deadly disease intentionally, even if the long-term chances of survival were thus improved? To be sure, any medical procedure in this period was perilous. The cutting edge, as it were, of eighteenth-century medicine was the use of copious bleeding, the opening of the veins with a lancet to drain off a few quarts of blood, thus to be rid of "evil humors." Other popular treatments included mercury-induced purges, antimony sweats, and vomiting induced by ipecac and other vile concoctions. The little blue mercury pills might give you diarrhea, cause you to salivate uncontrollably, and loosen each tooth from your head, but such were the hardships of a cure. No healthy person, though, had ever intentionally sought out a disease as a preventive measure.

Critics also charged that variolation caused hidden long-term health problems—another recurring theme among antivaccinists. " 'Two or

three years hence you'll see the dreadful effects of this wicked practice . . .
you'll see what happens to the people that are under it,' "critics said. In
Mather's time, such beliefs reflected the depths of fear of a new procedure that cheated death. "Notwithstanding the uncontroverted success of
inoculation, it does not seem to make the progress among the common
people of America which at first was expected," Franklin wrote. "Scruples
of conscience weigh upon many concerning the lawfulness of the practice, and if a relative objects, parents are hesitant to go ahead lest in case
of a disastrous event, perpetual blame should follow."[19]

"Scruples of conscience" come through clearly in a November 2, 1721,
letter from Adam Winthrop, a Massachusetts landowner and councilman
(and hanging judge during the witch trials), to his son, who was in Boston
at the time and wrestling with the decision. To variolate, Winthrop wrote,
was to test fate, for God could kill with natural smallpox or with variolation. He warned his son that placing hopes in variolation would "look like
placing your refuge in ways and creatures rather than God." Death by
infection was distributed by God and thus, if sorrowful, at least nothing to
be regretted; a natural death would at least give "the refreshment to think
this is God's doing." If he variolated, Winthrop warned his son, he would
face the final reckoning with the thought that he had brought it upon
himself. "I should have less distress in burying many children by the
absolute acts of God's providence," Winthrop wrote, "than in being the
means of burying one by my own act and deed."[20]

The Rev. William Cooper, a neighbor of Boylston, attempted to
answer the critics in a pamphlet titled, *Letter to a Friend in the Country,
attempting a solution of the scruples and objectives of a conscientious or religious
nature commonly made against the new way of receiving the smallpox*. Cooper's
broadside was a sort of spiritual catechism of variolation, but it reads a bit
like a question-and-answer page on a CDC website, with the difference
that Cooper is seeking to relieve moral rather than health-related doubts.
Point by point, he soothes the pious individual's concerns about variolation. Against the claim that it is "not lawful for me to make myself sick
when I am well," Cooper argues that while "to bring sickness upon oneself for its own sake is what no man in his right wits would do" to "make
my self sick in such a way as to probably serve my health and save my life
. . . is certainly fitting and reasonable." As for whether it is "not right to

trust in God" Cooper responds that there can be no harm to letting the fuel of smallpox in through an arm rather than the lungs so that "the fuel will burn with less fierceness and consequently danger." Yes, smallpox is God's work, Cooper concludes, yet if God "shows us the way to escape the extremity and destruction, at least, if not the touch of it, should we not accept this gift with adoring thankfulness?"[21]

Some scholars of colonial America believe that Mather's variolation drive was opportunistic—that he adopted the cause primarily to raise the political capital he had lost through his careless endorsement of the Salem trials.[22] But even if inoculation was a political opportunity it must also have carried special existential weight for Mather. A few more epidemics and the City on the Hill that he hoped to build in Boston would no longer be tenable.

This same calculus rang true on a personal level. Mather's firstborn child, Abigail, died after convulsions at five months of age. His first-born son died six years later, four days after his birth, of unknown causes. Other children died of fever and worms, of diarrhea and birth defects. Two toddler daughters fell into the fireplace and burned themselves; a third set fire to herself with a candle. At least two of his children had gotten smallpox during an epidemic in 1702, though both survived. In 1713, as Mather was scribbling a treatise on measles, that disease swept through his household, killing his second wife and three children. As the smallpox epidemic of 1721 raged, Mather's daughter Abigail (named for two earlier, dead Abigails) was suffering through a pregnancy that would ultimately kill her; his daughter Hannah was sick and his sister's family had smallpox. His son Increase, a dreamy rambler with little interest in God, was keeping company with rowdy students. "It was a time of unspeakable trouble and anguish,"[23] he wrote in his journal. Mather's hopes for the future lay with Samuel, or Sammy, his youngest boy, a 16-year-old Harvard student. The moment of truth, for Mather, came in August, when Sammy's chambermaid at Harvard was afflicted with smallpox. Sammy returned home, demanding to be variolated, and Mather had to decide what to do. In his diary, he struggled mightily with the decision, which, characteristically, carried both emotional and political overtones, for if Sammy died under variolation, it would drive a stake into Mather's reputation. The elder Increase eventually convinced Mather to have

Samuel variolated in secret, a fortuitous decision. Sammy nearly died of smallpox—he may have contracted the disease prior to the incisions—but he survived with no permanent effects.[24]

Mather may have seen variolation as a trial that bestowed upon its survivors a special mark of God's providence for the colonists.[25] Just as a sinner could prepare through prayer to meet his maker, he could also prepare by variolation for the trial of smallpox. Painful and risky, variolation offered the sinner a trial by fire—plunging him into a terrifying experience whose perils were unknown and therefore dependent upon God's will. The danger made survival all the more miraculous. People wanted to live, and ultimately they would variolate in large numbers.

Turks, Blacks, and Women

Mather was an avid observer of natural phenomena and a collector of "curiosities," which he periodically described in letters to the Royal Society. He liked to dabble in theories of the body, having studied medicine when, as a youth with a stutter, he feared that handicap would hold him back from the ministry. He would develop a relatively sophisticated theory of why variolation produced a smallpox less damaging than the "natural" kind. When a person breathed in smallpox, Mather wrote in his account of the Boston epidemic, the "venemous miasms"—which he also described as "animalcules," in a prescient hint at germ theory—went directly to the lung, heart, and bowels. Once "the enemy got into the very center of the citadel . . . the invaded party must be very strong indeed if he can struggle with him and after all entirely expel and conquer him." But in variolation, the miasmas approach "by the outworks of the citadel, and at a considerable distance from it. The enemy, 'tis true, gets in so far as to make some spoil. . . . but the vital powers are kept so clear from his assault that they can manage the combat bravely."[26] As Mather theorized, the peripheral smallpox infection in the skin of the arm or leg—the "outworks of the citadel"—apparently allowed the immune system to prepare an antibody response before the virus had attacked the principal organs of the body.

Another area in which Mather showed a greater degree of enlightenment than his foes was his curiosity about other cultures. To the extent that

Westerners were aware of variolation before the early eighteenth century, it was as exotic folk custom, an Oriental or African practice of witch doctors and crones and therefore inimical to racist snobs like Douglass. Mather was not particularly enlightened on race matters, but he did feel that the origins of the practice were not sufficient grounds to reject it. Coffee, tobacco, and silk also came from heathens, he pointed out. Before reading the two articles about variolation published in the Society's *Philosophical Treatises* in 1713 and 1716, Mather had already heard about the practice from a Libyan-born slave, Onesimus. After receiving Onesimus as a gift in 1706, Mather asked him whether he had had the smallpox, since slaves were expected to care for sick children. "Yes and no," Mather quoted Onesimus as responding. The slave showed his master a scar that remained from a childhood variolation in the land of his birth. Onesimus, whom Mather had since sold for misbehavior, was a "pretty clever fellow," he recalled. After reading the *Philosophical Treatises* articles on variolation, Mather conducted an informal poll of slaves in the town and found half a dozen who had been variolated in Africa. Their stories matched up, he said; "the more plainly, brokenly and blunderingly and like Idiots they tell the Story, it will be with reasonable men all the more credible."[27]

In contrast to Mather's patronizing attitude toward these blacks, Douglass was downright hateful. "There is not a race of men on earth who are more false liars, and their accounts of what was done in their country were never depended upon until now," he retorted.[28] Modern day accounts have shown that the slaves who recounted their variolations were neither liars nor idiots; indeed, variolation as they described it continued to be practiced in parts of Africa into the 1970s. And if Mather's "credulity" was self-serving it also contained a seed of genuine curiosity and open-mindedness. This spirit linked Mather to the variolation camp in England, where events were progressing on a parallel track.

In England, the cultural spadework for variolation was conducted by a very modern woman, Lady Mary Wortley Montagu, a brilliant bluestocking whose barrister husband was, for a time, ambassador to Turkey. In 1718, three years after she had suffered a terrible bout of smallpox that almost killed her and permanently marred her beauty, Montagu had her six-year-old son Edward variolated in Constantinople, where she had observed old Greek women performing it with success.

Upon returning to England, Montagu discussed her discovery widely, and when smallpox struck London in 1721, a month before it arrived in Boston, Montagu had her $2\frac{1}{2}$-year-old daughter variolated by Charles Maitland, a court surgeon. The variolation of Mary Montagu on April 21, 1721, was certainly unknown to Mather; presumably it might have dissuaded him from variolating Bostonians, because Montagu was very much not in Mather's political camp. Careless in her love affairs and finances and rebellious in her perspective on the authorities, Montagu was like Mather only in her capacity for objective thought. In the Turkish practice she saw promise where others saw danger and darkness. The Ottomans had embraced variolation to guarantee the beauty and integrity of their harems. Circassian baby girls kidnapped from the North Caucasas were variolated, and those who survived with no scarring might enter the harem. Eventually, variolation caught on as a general hygiene technique and "people take it here as lightly as they take the waters with us," Montagu wrote a friend. Her connections in British society—she was a friend of Princess Caroline and had married into a family of great wealth—helped gain the practice a purchase in the upper echelons, even after the son of the Earl of Sunderland died following variolation in 1722. Montagu's interest encouraged a group of royal physicians to experiment with the procedure and to publish papers on its success. From the beginning, several of the more illustrious British doctors, including Hans Sloane, president of the Royal Society, embraced variolation, while leading clergy members condemned it. In England, Boston's politics of variolation were reversed.

Montagu hated doctors and trembled with indignation at the thought of their taking control of variolation. The practice she had observed in Constantinople was the very essence of simplicity. "The old woman comes with a nutshell full of the matter of the best sort of smallpox and asks what veins you please to have open'd," she wrote to a friend in 1717. Then the woman opened a cut and "puts into the vein as much matter as can lie upon the head of her needle."[29] Other contemporary accounts speak of a mere scratching of the skin. The salient point here is that no large cut was opened, no medicines used to prepare or accompany variolation. Montagu hated being bled, hated the "Physick," or medicines, that her doctors doled out, hated them for having done nothing to save her face from smallpox.[30] In the early eighteenth century, people unlucky enough to have physicians treating them for smallpox were either confined under thick blankets in a sealed, overheated room, or encased between wet sheets with the windows open, even in winter. The resulting sweats or shivers were supposed to get the patient to spew out all possible "morbific matter"—as much of the "innate seed" of smallpox that had always been inside them—as was possible, hopefully without dying in the process.

The court doctors who clustered around Montagu's house after Mary's variolation were so angry with her she feared they would hurt the girl.[31] As variolation spread through the upper classes of Britain, Montagu gradually excused herself from the limelight, since it was clear no one would listen to a woman who promoted the ways of heathens. In a typical gob of bile spat her way, the physician William Wagstaffe dismissed variolation as the "practice of ignorant women," possibly designed by Britain's enemies as "an artful way of depopulating the country." Still, in letters to friends from the time, and in an article she pseudonymously published under the name "A Turkish merchant," Montagu expressed concern about the direction in which variolation was headed. She had helped spread the practice in Britain, she wrote, for the practice was "too important to be left to the doctors" and the "miserable gashes" they inflicted on patients.[32]

Yet doctors wasted little time in burdening it with all their dangerous paraphernalia. Just as Mather may have seen variolation as a preparation for divine trial, the doctors saw the need to "prepare" their patients with trials of their own devising. Before long the procedure had been transformed into a three-week ordeal of bleedings, purgings, sweats, and diets that padded the wallets of the doctors. In his account, Boylston describes making a "pea-sized incision" with a lancet. But slashing and bleeding were intrinsic to the medical thinking of the time, which held that curing illness was a matter of balancing the four humors, often by cutting open the body to let one or the other of them out. The seventeenth and eighteenth centuries introduced various poisons that hastened the process. Mercury induced drooling and diarrhea; antimony provoked the sweats; and sulfur antagonized the side effects of antimony and mercury. These and other chemicals and herbs became part of elaborate "preparation" for variolation, championed by the top physicians in Britain and America. Dr. Richard Mead, who attended Montagu's smallpox illness, wrote that variolation was beneficial because it gave the physician a chance to "drain away, where necessary, some blood, and gently purge the humors" from the patient, and thus "obviate the violence of the approaching fever."[33] Having no immunological insight into how variolation worked, physicians naturally saw its whole point as *preparing* for smallpox, using the techniques at hand.

Dr. Adam Thomson, an Edinburgh-trained Scotsman, brought purge-based variolation to America. Thomson, who lived in Maryland for several years before moving to Philadelphia in 1748, created a two-week course of "cooling regimes" that included a light diet, combinations of antimony and mercury, bleeding and purges.[34] Thomson said he based his practice on a "hint" from the Dutch physician Hermann Boerhaave that mercury might destroy smallpox "animalcules." Thomson's method was probably institutionalized thanks to the influence of Benjamin Rush, founder of the first American medical college at the University of Pennsylvania. Rush, born in Philadelphia but trained at Edinburgh, championed the robust, Scottish school of medicine. He blamed most ills on "hypertension"—too much blood—which he remedied with predictable techniques that were particularly inefficient and frightful during Philadelphia's great yellow fever plague of 1793. Like his countryman William Douglass, Thomson was inclined to put the Montagus and Mathers of the world in their place. He expressed surprise at the fact that more people had not died of variolation given the fact that "almost anyone who knows how to handle a Lancet" had tried a hand at it. Thomson would not reveal the exact formula of his procedure, claiming that in the hands of unskilled ignoramuses, its reputation would be marred.[35] The added frisson of mysterious knowledge enabled his followers to make strong, probably unfounded claims for his success.

It was later revealed that Thomson's recipe involved a first dose of 1.3 grams of mercurous chloride, followed by 3 grams of pulverized cornachin—a retch-inducing mixture of antimony and cream of tartar. The same doses followed for four days and nights. As early as 1767, when mercury and antimony were still in use, but in gentler proportions, Thomson's recipe shocked doctors. "His fibers would be like Behemoth, his nerves like brass, who would venture to subject himself to such a treatment," one commentator wrote.[36] That much calomel would cause the patient to drool copiously and might make his teeth fall out, although the morning purges may have flushed much of the mercury from the system before it filtered into the blood and caused brain damage.

In 1757, with the world ripe for a less punishing method, English surgeon Robert Sutton and his son Daniel began advertising a cheaper, simpler form of variolation, and within eight years they had set up a franchise with 200 practitioners around the world. The Suttons were humble surgeons; the secret to their success was the minimal use of purging and vomiting. Rather than lancets they used thin needles, scratching the pox into the skin so lightly that the patient was "not sensible of the points touching him."[37] In other words, the Suttons went back to the Turkish system—with a few bells and whistles. Music and storytelling were part of the package; at the Suttons's establishment, the Rt. Rev. Robert Houlton was hired at 200 pounds a year to preach a sermon to the variolated as they lay feverishly in bed.

These simplifications took longer to catch on in the colonies. Philadelphia was the colonial variolation center up until the Revolution. With Rush leading the city's physicians, Physick and bleeding would have their day, and the Philadelphia system had broad influence. Inoculation meant three weeks of daily vomiting, purges, sweats, and fevers. Peter Thatcher, a 12-year-old Boston boy inoculated on Saturday, April 14, 1764, described the ordeal thus: "Sunday we took a powder in the morning that work'd me nine times. . . . Monday morning took two pills which worked me twice. . . . Tuesday took a powder that work'd me ten times. . . . Thursday took a powder that worked me 5 times up and once down. . . . Saturday, April 21 took a powder which worked me 4 times down and once up. Felt very sick in the Morning and did not get up till 10 o'clock. I have some Pock come out today. At night took Brimstone [sulfur] in Milk for a sore mouth." On Wednesday, May 2, nearly three weeks after his operation, Thatcher was recovered and in a penitent mood. "Thus, through the Mercy of God, I have been preserved through the Distemper of the Small Pox which formerly was so fatal to many Thousands. . . . Many and heinous have been my sins, but I hope they will be washed away."[38]

That same year, future president John Adams, who was Zabdiel

Boylston's nephew, wrote a letter to his fiancee Abigail Smith, describing his own variolation: "A long and total abstinence from everything in nature that has any taste, two heavy vomits, one heavy Cathartick, four and twenty mercurial and antimonial pills and three weeks close confinement in a house. . . ." Adams was fed on milk and mercury throughout the ordeal, and "every tooth in my head became so loose I believe I could have pulled them all out with my thumb and finger. By such means they conquered the Small Pox, which I had very lightly, but they rendered me incapable of speaking or eating in my old age, in short they brought me into the same situation with my friend Washington, who attributed his misfortune to cracking walnuts in his youth."[39]

During the Revolutionary War, when a smallpox epidemic flattened the Revolutionary army at the gates of Quebec, George Washington was forced to implement mass variolation to give his troops the same protection enjoyed by the British military. Prior to that order, however, variolation was banned in many of the colonies and rarely practiced elsewhere—except in Philadelphia, which was smallpox hell or smallpox Mecca, depending on how one looked at it. Pennsylvania was the only state in the newly minted union that put no controls on inoculation. You could ride into town, be inoculated, and freely walk the streets until you were too sick to stand anymore. In Virginia, variolation was illegal throughout the colonial period, so Thomas Jefferson came to Philadelphia from his home outside Williamsburg to be variolated in 1766. Martha Washington got hers done in 1776, as did Indian Chief Little Turtle. Benjamin Franklin, whose 4-year-old son Francis died of smallpox in 1736, had his $2\frac{1}{2}$-year-old daughter Sarah variolated a decade later.

Despite these examples, variolation was quite unpopular and feared in many places. In Norfolk, Virginia, when a group of loyalist merchants decided to inoculate themselves and some of their new slaves in 1768, pro-independence ruffians rioted at a plantation where the procedure was taking place. The mob chased the sick patients out of the house and through a thunderstorm to the town's smallpox hospital. Thomas Jefferson took the variolators on as legal clients, but a fire at his house destroyed his notes to the case, which never came to trial.[40]

Even Philadelphia ordered a halt to variolations during the first Continental Congress, in 1764, after delegates refused to enter the city

because of the prevalence of smallpox and the authorities' refusal to quarantine those who had been variolated. Soon, though, the practice
resumed. In 1776, during the second Congress, Rhode Island Governor
Samuel Ward caught smallpox and died.[41] Adams and other Founding
Fathers had urged Ward to be inoculated, but to no avail.[42] With her husband still in Philadelphia and smallpox encroaching upon Boston, Abigail
Adams decided to inoculate herself and her three children, and after
three tries was successful. On July 18, while waiting for the symptoms to
hit, she joined a throng cheering the new Declaration of Independence at
the foot of the statehouse in Boston.[43]

Clearly, variolation was a prudent step for an individual—particularly
someone living in Philadelphia, Boston, or any other city, where crossing
paths with smallpox was inevitable—but it was not always good for public
health. If variolated patients were not carefully quarantined for two
weeks, they were liable to spread smallpox to anyone who had not had it.
In the interests of the public good, authorities sometimes threatened to
curtail variolation, while other times they demanded it. Variolation gradually increased in both Philadelphia and Boston, although the methods
and means were not consistent.

Philadelphia, a seaport and way station with a steady influx of immigrants and travelers moving up and down the eastern seaboard, might
have had difficulty controlling variolation even had it tried to do so—10
epidemics swept the town during the century.[44] Still, after 1760 smallpox
was no longer the leading cause of death, and the city's overall death rate
fell for the first time in the century. Public health officials in Puritan
Boston were far stricter than their Quaker brethren. Following the battle
between Mather and Douglass, variolation was generally banned in
Boston except at epidemic time, during which it was widely employed.
Over the century, an informal rule became institutionalized: Before 20
households had reported smallpox, variolations were prohibited out of
fear they would spread smallpox.[45] The city fathers imposed a strict quarantine and posted guards outside the homes of anyone sick with smallpox. After 20 or more houses were affected, however, an epidemic was
declared and everyone vulnerable to the disease encouraged to seek variolation. Writing of Boston's 1753 smallpox epidemic, Franklin noted that
variolation "spread the infection likewise more speedily among those

who did not choose [it], so that in a few months the distemper went through the town and was extinct; and the trade of the town suffered only a short interruption."[47] Variolation accounted for a growing percentage of smallpox cases in Boston through the century: In 1721, 2 percent of all smallpox cases were by variolation; in 1730, 10 percent. By 1778, only 59 of the 2,243 smallpox cases, or 3 percent, were fatal; nearly all the smallpox that year was variolated. Boston's last three smallpox epidemics of the eighteenth century did not cause the overall death rate to increase.

The sociology of vaccination in Boston also changed during this time. In Mather's era, the wealthy had urged variolation while the poor hated and feared it (and couldn't afford it). By the end of the century, Boston merchants, who variolated their families privately, opposed mass variolations because they brought the town's business to a halt, if temporarily.[48] The sting had been taken out of variolation. The key, though, was that Bostonians—in contrast to Philadelphians—had allowed themselves to be corseted for the public good. A sense of community instructed the city's attitude toward smallpox. This became a point of pride. "In New England, the most democratic region on the face of the earth," wrote Benjamin Waterhouse, who would later introduce Jenner's vaccine in America, "the people have voluntarily submitted to more restrictions and abridgements of liberty to secure themselves against that terrific scourge, than any absolute monarch could have enforced."[49]

As "buying the pox" became an inevitable life passage, the rich found ways to make it more pleasant. Jugglers and clowns entertained the guests at the grand variolation hospital run by William Aspinwall (1743–1823) in Brookline, Massachusetts. When the patients were feeling more fit, fife-and-drum corps led them in outdoor marches. Feverish, bedridden patients participated in sing-alongs, and their friends and family were permitted to visit with the inoculees until it was judged that they had become contagious. When they were ready to go home after their three weeks in the hospital, the patients and their clothes were hosed down with sulfur. The procedure itself could be painful and unpleasant, but then so, in their way, were the high colonics and steam baths one came to expect in the *Kurorten* of Europe.

By the time Edward Jenner introduced his vaccine, variolation had become remarkably effective and fairly safe. "Over the course of a century," as a scholar of colonial Philadelphia wrote, "inoculation had transformed smallpox from the dreaded scourge known as the 'speckled monster' to a guest encouraged to visit the family home."[50] Variolation contributed to the growth of Britain's population during the century, though there is scholarly debate over whether smallpox morbidity and mortality fell during the period. England's population, which had grown from 4.9 to 5.2 million from 1681 to 1731, grew to 6.4 million in 1771 and 8.7 million by 1801.[51] In what became the United States, the population grew from 629,000 to 5.3 million between 1730 and 1800, but the sources of growth are too diverse to measure the impact of variolation.[52]

By the late eighteenth century, people concerned about smallpox were aware of both the risks and benefits of variolation. In great cities such as London and Paris there was little hope of generalizing the practice because the populations were too fluid, poor, and resistant to the centralizing authority of public health. However, efforts were being made to variolate large populations when smallpox appeared. Inoculation hospitals and benevolent societies were set up to inoculate the poor. The "general inoculations," as large-scale vaccinations during epidemics were known, were controversial and had limited success, since they were held only on an ad hoc basis.

In the late 1770s, several British physicians proposed the creation of a Society for General Inoculation—a precursor of the National Health Service that would variolate the poor for free. John Haygarth, a philanthropist who helped introduce such a society in Chester, argued that variolation had been, overall, harmful to the poor. The rich procured it for their families, then became indifferent to fighting smallpox.[53] Haygarth argued that the time was ripe for general inoculation in Britain. "No objection can arise from the danger of propagating the infection because, if the inoculation be general, no subjects liable to infection would remain," he stated in a 1793 treatise. "There would be much less hazard than at present from inoculation, which is practiced at various times every year, and sometimes secretly."[54] In other words, Haygarth was proposing that the world follow Boston's example—an egalitarian system of public health, at least as far as smallpox was concerned. If it did, he believed, smallpox could be eradicated. His dream would be realized, but it would take another 200 years.

—

CHAPTER 2

THE PECULIAR HISTORY
OF VACCINIA

But the stumbling strides of microbe hunters are not made by any perfect
logic, and that is the reason I may write a grotesque but not perfect story
of their deeds.

— PAUL DE KRUIF, *Microbe Hunters*

FOR MOST LITERATE PEOPLE, the history of vaccination begins not with
Mather and Montagu but with Edward Jenner, the country doctor from
Gloucestershire who found, growing on cows, a nearly harmless virus
that protected people from smallpox. Jenner's vaccine was safer, more
reliable, and more durable than variolation, and it is still the only vaccine
to have eliminated its reason for being—in 1980, when the World Health
Organization (WHO) declared the disease extinct. For nearly a century
and a half, smallpox was the only vaccine routinely administered, and it
saved millions of lives. But the controversy that marked the return of the
vaccine, amid bioterrorism hysteria in 2002, was only the latest twist in
the remarkable, mysterious life of vaccinia, the virus that is the active
ingredient of the smallpox vaccine.

The shorthand version of Jenner's accomplishment is this: he confirmed through experiment a country legend holding that cowpox, a mild
disease of cows, caused a mild infection in people that thenceforth protected them from smallpox. Over a period of years Jenner tested that idea,

published his results, and became one of the world's first celebrities. The man who brought us the vaccine was a quintessential Englishman, a deceptively obsessive country practitioner most at home studying animals.

The son of a minister and the eighth of nine children, Jenner was raised from age 5 by an older brother after their parents died. As a boy he roamed the fields of his native Berkeley, in Gloucestershire, and in 1763 he became the 14-year-old apprentice to Daniel Ludlow, resident surgeon in the colorfully named village of Chipping Sodbury. Jenner first got onto the cowpox idea in 1770, while flirting with a milkmaid at Ludlow's establishment. Variolation was one of the chief money earners for a surgeon, and Jenner was surprised that the girl had never sought the procedure. He was perhaps even more surprised by her claim that she'd been protected by cowpox. Folklore, Jenner would learn, supported her conviction. Cowpox was not a common disease, but milkmaids were prone to it. And milkmaiding was one of the few vocations, so it was said, that could protect a girl's fair skin. If she managed to avoid smallpox by the time she started hanging about the stable, there was some chance she would remain unblemished for life. There was even a country rhyme that attested to this:

> *Where are you going, my pretty maid?*
> *I'm going a-milking, sir, she said.*
> *What is your father, my pretty maid?*
> *My Father's a farmer, she said.*
> *What is your fortune, my pretty maid?*
> *My face is my fortune, she said.*[1]

Jenner trained under the venerable London surgeon John Hunter, who was a friend of Captain Cook's. When Cook and his *Endeavor* returned from the South Seas in 1771, Jenner organized the expedition's collections of natural history from the islands, earning him an invitation to return to the Pacific with Cook. But Jenner refused, and refused, too, Hunter's offer of a partnership amid the filth and fascination of London's diseased streets. He went home to be close to his dear brother and his chaotic collections of fossils, anatomical preparations, and papers.[2]

As Berkeley's doctor, Jenner was perfectly adequate by the standards of the day, and what's more he charged his patients less the further they had come to see him. Jenner was on the shortish side, well-fed, and dapper. He dressed, for house calls, in a blue coat with yellow buttons and well-polished jockey boots, and he carried a whip with a silver handle. In his free time Jenner collected specimens, recorded observations of nature, and conducted research befitting a gentleman farmer. He grew giant cucumbers, toyed with hot air balloons, and was made a Fellow of the Royal Society in 1789 for his celebrated observations on the predatory conduct of the cuckoo bird.

Jenner's pastoralism was not incidental to the development of vaccine. Indeed, it was his dual character—empiricist and nature's son—that enabled the famous discovery. The folkways about cowpox's effects on smallpox might have interested Jenner no more than his investigations into the fertilizing qualities of human blood were it not for the fact that he abhorred variolation as the result of personal experience. At age 8, Jenner had been sent by his brother to a barn where he was variolated with a large number of other children and adults. It was a wretched experience. "Preparation," according to one account of Jenner's ordeal, "lasted six weeks. He was bled to ascertain whether the blood was fine; he was purged repeatedly till he became emaciated and feeble; he was kept on a very low diet . . . and dosed with a diet drink to sweeten the blood. After this barbarism of human veterinary practice, he was removed to one of the usual inoculation stables and haltered up with others in a terrible state of disease."[3] The traumatic stay left Jenner morbidly sensitized to certain sounds that registered in his feverish brain amidst that scrum of baleful, foul-smelling humanity. "The horrible click of a spoon, knife, or fork falling upon a plate gives my brain a kind of death blow," he said in 1821.[4]

By 1778, the year the smallpox returned to Berkeley, Jenner had become obsessed with the milkmaid legend. He had, in fact, become a bit of a cowpox bore, peppering every surgeon, physician, and farmer he met with questions about the disease. Cowpox was too uncommon for Jenner to collect much in the way of statistics, but his notebooks slowly filled with anecdotes. Doctors did not need Institutional Review Boards to experiment in those days and, by 1796, Jenner had variolated more than 15 farm folk—men and women who'd had the cowpox earlier in life—to test his thesis. None, by his account, became infected with smallpox. In May 1796,

Jenner entered the homestretch of his epochal study. Until then he had only observed the health effects of cowpox passively; now he decided to intentionally spread the disease—from a cowmaid to a child, and from that child to others. As luck would have it, Sarah Nelmes, a milkmaid, had been infected from a cow at a farm near Jenner's home. On May 14, 1796, using a quill and a lancet, Jenner cut two small slits into the arm of a boy named James Phipps and dabbed in the liquid from Nelmes's sore—the liquid that would come to be known as cowpox lymph. Phipps developed the symptoms of cowpox—inflamed joints and red ulcerations at the site of the vaccination, with fever and pain in the groin and limbs a few days later—"troublesome symptoms," Jenner would write later, but trifling compared to smallpox. On July 1, with a lancet kept separate from his cowpox collection, Jenner variolated Phipps with smallpox and waited for a week for symptoms to appear. Nothing happened; cowpox, apparently, had given the boy immunity from the more fearsome disease.

In March 1798, cowpox reappeared at Berkeley. William Summers, a five-year-old boy, was inoculated with matter from the nipple of an infected cow, and the stuff from Summers's sores was used to infect a five-generation chain of patients that included Jenner's 11-month-old son, Robert. With that series completed, and each of the vaccinated children gaining protection, Jenner was ready to publish his treatise. Although an early draft was rejected by the Royal Society, *An Inquiry Into the Causes and Effects of the Variolae Vaccinae, A Disease Discovered in Some of the Western Counties of England, Particularly Gloucestershire, and Known by the Name of The Cow Pox* became an instant hit.[5] "What renders the cowpox virus so extremely singular," Jenner wrote, "is that the person who has been thus affected is forever after secure from the infection of the smallpox; neither exposure to the variolous effluvia, nor the insertion of the matter into the skin, producing this distemper." He was eager to have his technique replace variolation. "I have never known fatal effects arise from the cowpox . . . and as it clearly appears that this disease leaves the constitution in a state of perfect security from the infection of the smallpox, may we not infer that a mode of inoculation may be introduced preferable to that at present adopted?"

Henry Cline, the first London surgeon to vaccinate, wrote to Jenner that August: "I think substituting the cow-pox poison for the smallpox promises to be one of the greatest improvements that has ever been made in medicine."[6] By 1801, about 100,000 people in Europe had been vaccinated. Vaccine quickly spread around the world, with millions vaccinated before a decade was out. The Royal Jennerian Society was created in 1803 to support vaccination; Jenner's bust was placed in the Royal Academy and the empress of Russia named the first vaccinated Russian child Vaccinof. Even Napoleon, in the midst of a bitter war with Britain, immediately freed two prisoners at Jenner's request in 1805: "Ah, Jenner," he is said to have responded, "I can withhold nothing from that man." The emperor's troops were by then under the protection of *le vaccin jennerien*.[7]

Vaccine Spreads to the New World

One of the first nations to get access to vaccine was the United States, where Benjamin Waterhouse, Harvard's first professor of pharmacy, obtained it through his friendship with John Lettsom, a prominent London physician. Other American doctors were testing material they had obtained from their correspondents in England, but Waterhouse got the first "take"—an angry pustule that appeared at the abraded site—when he vaccinated his son Daniel on July 8, 1800. After vaccinating several others in his household, Waterhouse convinced William Aspinwall to variolate them all at his hospital in Brookline on August 2. The test was successful. Waterhouse, however, proved a dubious patron of vaccination in America, botching his self-appointed role through a mixture of avarice and incompetence. Waterhouse's first blunder would perhaps seem natural in today's patent-crazy biomedical community: he tried to extract generous terms for himself from physicians in exchange for sharing the material. In a September 1800 proposal sent to Dr. Lyman Spalding of Portsmouth, New Hampshire, Waterhouse demanded exclusive rights to supply the vaccine—plus a quarter of Spalding's fees. Within a few months, the Waterhouse monopoly had broken down as other doctors found their own suppliers in Britain. There was no natural cowpox in America, an absence that posed an increasingly serious problem as the century went on.

As demand outstripped supply for effective vaccine matter, alert businessmen began trafficking in spurious stuff; farmers and kids vaccinated each other from their sores. The more fake or failed lymph failed to protect people against smallpox, the more cowpox fell into ill repute. "While the deluded patient vainly supposed himself secured from the attacks of the smallpox, his imaginary safety leads him into situations where his life is endangered," Waterhouse wrote, leaving vaccination open to doubt or "contempt."[8] The worst failure occurred at Marblehead, Massachusetts, which suffered a disastrous smallpox outbreak during a vaccination campaign. Using "virus" that his seafaring son had taken from a pustule on the arm of a sailor vaccinated in London, Dr. Elisha Story vaccinated his daughter on October 2, 1800. Two weeks later she broke out with what Story believed was cowpox but was actually smallpox. Waterhouse was meanwhile supplying Story with vaccine that was either contaminated or inactive. The epidemic eventually sickened 1,000 people, killing 68 of them.

It was mainly thanks to President Thomas Jefferson that the reputation of vaccine was saved. During the winter of 1800–1801, none of the Boston physicians succeeded in preserving their vaccine supplies, but fresh virus arrived in March 1801, and Waterhouse almost immediately sent some to Jefferson, asking him to distribute it. The first three batches Jefferson received were ineffective but, in August, using a container designed according to Jefferson's specifications, Waterhouse managed to ship an active supply to Monticello. Jefferson quickly vaccinated 30 people there, including the last members of the Mohican tribe. He distributed the vaccine to doctors in the south and as far north as Philadelphia, where Dr. John Redmond Coxe began vaccinating in November 1801.

Back in Boston, Waterhouse continued to have trouble convincing people of the value of his product. On November 1, Aspinwall ran an ad in a local newspaper claiming that five of six people vaccinated with cowpox at his hospital had come down with smallpox after being subsequently variolated. Finally, Boston's health authorities helped Waterhouse arrange a definitive experiment. On August 16, 1802, 19 young volunteers were vaccinated at the Board of Health office and then variolated in quarantine at Noddle's Island in the harbor. Two controls were also variolated to prove that the smallpox matter was active. The vaccinated youths were all protected, and the successful results were published in December with supporting statements from 11 doctors. Waterhouse had described Jenner's discovery with this headline in a Boston newspaper: "Something New in the Medical Line." The Noddle's Island experiment proved that. It "fixed forever," Waterhouse wrote, "the practce of the new inoculation in Massachusetts."[9]

Jenner had done an experiment and proved a hypothesis, and a grateful world adopted the practice. The problem was, neither Jenner nor anyone else really understood what cowpox was. Over the years, the mystery of cowpox deepened, as the stuff, which eventually would become known as "vaccinia," was transformed through use, passed through thousands of human bodies, shipped from place to place over bumpy roads and rough seas, in containers that only imperfectly controlled its mutations and virulence. People were vaccinated at the pharmacy, "arm to arm" or "fresh from the cow," with lymph that was scratched, pinched, sliced, and kneaded into the skin with lancets, quills, ivory needles, butter knives, and dirty fingernails. They were vaccinated with stuff that had been dried on stone, preserved in tiny glass tubes, stoppered in jars after being squeezed from pimples, or scraped with potato spoons from the belly of a cow. If the patient was lucky, the vaccine took, causing an unsightly but small sore that healed after a light illness that might offer lifelong protection from smallpox.

But there were many unlucky paths from smallpox vaccination. Sometimes the procedure left you with no mark at all, because the vaccinia virus was all dead and had no punch, or your immune system didn't respond to the virus. Sometimes, instead of a small scar, you got fever and an excrutiatingly painful red arm, swollen to the size of your thigh, that kept you out of work for two weeks. From time to time, vaccination killed—because of contamination with streptococcal and staphylococcal bacteria, syphilis, tetanus, or tuberculosis. The pure virus itself could be enough to kill you if your immune system was weakened. Sometimes children went into shock after vaccination, or suffered convulsions. Smallpox vaccine, and the immunizing substances that followed, were all new to the human body. Vaccination was unnatural. It was progress.

Jenner's invention was ahead of its time. It came packaged in a corpus of untrustworthy medical knowledge. Vaccination, as case after case would show, was an empirical procedure. Lacking the knowledge of how it worked, medicine was unable to provide safeguards to its failures. Just as variolation was spoiled by bloodletting and mercury-soaked treatment plans, vaccination relied upon an inadequate infrastructure of cowsheds and lancets laden with bacteria. Live vaccines died or grew bacteria in the unrefrigerated heat. There was no way to test the efficacy of a vaccine except to expose a group of vaccinated people to smallpox.

There were also no licensing boards to test the vaccine or monitor its use. To this day, molecular biologists remain uncertain about just what Jenner's vaccine was. Yes, smallpox protection followed the milkmaid's exposure to cows with certain ulcerations on their teats. But the source of those ulcerations, and the meaning of that protection, lies shrouded in the complex history of civilization's dance with smallpox. Smallpox vaccine, for most of its effective life, remained an empirical remedy—a potion that worked, but as if by magic.

Holy Cow or Horse Heels?

Jenner's 1798 *Inquiry* contained a number of the author's drawings of the manifestations of cowpox disease. One of these is particularly evocative. It depicts the cowmaid Nelmes's hand marred by boils—one each on the wrist, the flesh between thumb and index finger, and the first joint of the index finger. Her gaunt outstretched fingers are reminiscent of Adam's hand reaching for God's on the ceiling of the Sistine Chapel. They suggest the mysterious origins of vaccine.

Vaccination represented a reversal of a centuries-long tide in the war between people and germs. Before Jenner, as historian William McNeill showed in his wonderful 1977 book *Plagues and Peoples*, the history of civilization was the history of epidemics. In the prehistoric societies embodied in the biblical Eden, pathogens failed to thrive because the infectious outbreaks they caused tended to burn themselves out in small, isolated communities. The germs whose infections have killed over the past millenium—from smallpox to measles to tuberculosis to AIDS—are relative newcomers to *Homo sapiens*. Most leapt to us from animals during our postnomadic history (they continue to do so). When large groups started living in and traveling between cities, humans became ideal hosts for these germs and epidemics spread.

Edward Jenner added a peculiar twist to this plot line when he noticed that one of these leaping germs was of benefit to humankind. Benjamin

Jesty, a British farmer, had rendered his wife immune to smallpox by scratching a bit of cow pus into her arm in 1756. But in making systematic *use* of knowledge built upon folk wisdom, Jenner became the first biotechnologist. That was indeed something new in the medical line. For centuries, pathogens coevolved with human society. Now, society was consciously harnessing evolution. Well before anyone understood that microorganisms were what brewed beer, made wine, baked bread, and helped digest food, Jenner put cowpox to work for humanity.

That said, Jenner's understanding of vaccinia was incomplete and shifting. He opened his landmark paper by remarking that his discovery reversed the order of prey that had developed between germs and civilized people. He didn't fully develop this idea, but rather meandered through it in the manner of a gentlemanly stroll, stopping now and then to sniff an exquisite rose or spicebush. "The deviation of man from the stage in which he was originally placed by nature seems to have proved to him a prolific source of diseases," opens his *Inquiry*. "From the love of splendour, from the indulgences of luxury, and from his fondness for amusement he has familiarized himself with a great number of animals which may not originally have been intended for his associates." The dog, the wolf's descendant, was "now pillowed in the lady's lap." Cats were "equally domesticated and caressed," as were farm animals that responded to human masters. These animals carried diseases, and one disease in particular had caught Jenner's attention. Grease, a disease of horses, "bears so strong a resemblance to the smallpox that I think it highly probable it may be the source of the disease." Here, it seemed Jenner would begin a disquisition on the origins of smallpox, but he quickly veered off into a related but different theory. While grease never caused illness in the men who handled horses, Jenner wrote, it was transformed into cowpox when the horsemen handled cows. Jenner believed that grease became cowpox, in the process increasing its activity, "as it rarely happens that the horse affects his dresser with sores, and as rarely that a milkmaid escapes the infection when she milks infected cows."[10]

For all the remarkable insight Jenner showed into the role of domestic animals in disease, he was also introducing a confused theory about the origins of his vaccine. The virus that Jenner and his successors used to prevent smallpox may have started out as an infection native to cows, horses, or even sheep, but if it did, the original virus is now extinct. DNA studies show that cowpox, smallpox, horsepox, and vaccinia are four distinct and seemingly stable viruses. The disease scientists now call cowpox is carried mainly by wild rodents, and sometimes spreads to cats and through cats to humans, though rarely. It is almost never seen in cows. Nor is vaccinia the same as "grease," if from "grease" we are to understand the disease veterinarians call horsepox. That leaves two large, unresolved possibilities: vaccinia is either a virus that has since gone extinct in the wild or it is a man-made virus.

A third, intriguing possibility was kicked around for years: the idea that the virus Jenner found on cows originated as a case of smallpox that milkmaids had given to cows. Perhaps, some speculated, it was one of the many highly attenuated, or weakened forms, of smallpox that decades of variolation had produced in England. Could it be that a milkmaid, itchy with inoculated smallpox, had scratched a sore and then in wiping the lymph on an udder, inadvertently transferred smallpox virus to her cow, and from that cow to the whole race of cows? Peter Razzell, a British medical anthropologist, presented the hypothesis that vaccinia was merely an attenuated form of smallpox in two heavily researched books in 1977.[11] The smallpox used in Suttonian variolation, Razzell believed, was already relatively harmless because it had been weakened by passage through thousands of human arms. There were historical accounts of cows that developed "cowpox" after being exposed to smallpox, he noted. Because cowpox was hard to find and difficult to preserve, most vaccinations in the nineteenth century used pus or dried scabs—"crusts"—from a vaccinated person, rather than cows. Razzell argued that most of this so-called arm-to-arm vaccination was probably spreading mild smallpox rather than "cowpox."

Although it is possible that modified smallpox may have been used occasionally as vaccine, Razzell's thesis that vaccinia originated as smallpox has not stood the test of genetic parsing.[12] Derrick Baxby, a British microbiologist who studied the question closely, has concluded that vaccinia is distinct from all forms of smallpox—and from all other wild poxviruses.[13] Vaccinia might be the descendant of a virus that has since vanished, or it might have resulted from genetic reassortments that occurred during its many passages through man and beasts.[14]

What was clear was that the vaccine that started out in Gloster the cow over time strayed further and further from its original state, until defining just what it consisted of was almost impossible. Vaccine was more of a process than a product.

The vagueness of its origins was one of the factors that led to skepticism about Jenner's vaccine. For no sooner had he brought his creation into the world than up sprang a new group to oppose it—the antivaccinationists, or antivaccinists, terms I use interchangeably in this book. Unlike any other medical procedure, vaccination would prove capable of provoking organized opposition, even social movements against it.

"The Injection of the Morbific Matter"

Perhaps the leading vaccination opponent of Jenner's time was the clergyman-turned-economist Malthus, who aired his views in his treatises on excess population. In his 1806 book *Population*, Malthus opposed vaccination on the grounds that smallpox was "one of those channels, and a very broad one, which Nature has opened for the last thousand years to keep the population down."[15] Tory politicians generally opposed funding compulsory vaccination, their stinginess grounded in reluctance to enhance the life chances of the undeserving poor. A century later, a few American eugenicists opposed public health campaigns against smallpox, diphtheria, and other diseases for similar reasons—because they would preserve too many "unfit" children.[16]

The morally defensible objections to vaccination, on the other hand, were based either upon religious or health concerns, which were not always easily distinguishable. Clergymen argued that vaccination interfered with God's plans or cast doubt upon his omnipotence. Other, more modern cults and denominations focused on the perceived biblical injunction against pollution of the blood. The religious arguments against vaccine were joined by professional variolators and unschooled or unorthodox healers. Among Jenner's earliest foes was John Birch, "surgeon extraordinary" to the Prince of Wales who "treated" smallpox with electric current. Another, Chelsea Hospital physician Benjamin Moseley, had lived in Jamaica for several years and fancied himself an expert on tropical diseases.

And then there were murky, irrational panics over vaccination, com-

pounded by fear of the unknown, hysterical repugnance, and an instinctual mistrust of doctors, which would persist throughout two centuries of
antivaccination activity. Such fears were exemplified in 1802 by James
Gillray's etching, "The Cow Pock—or—the Wonderful Effects of the
New Inoculation!"—in which a wigged physician calmly slices into the
arm of a horrified patient while cows sprout from the bodies of earlier
victims. On the frontspiece of an 1806 book, Jenner was pictured with tail
and hoofs, feeding baskets full of infants to a hideous monster.[17] Critics
often resorted to spuriously simple logic. Vaccination was "the injection
of the morbific matter of a diseased animal into a healthy child," as one
put it. How could that possibly be a good thing? As vaccination became
an accepted practice, antivaccinists jeered at those who accepted it. "That
a people would be found to contaminate their offspring with a poison
taken from the brute creation . . . will stand among the incredible tales of
some future Pliny," Moseley wrote.[18]

As critics exaggerated the failures of vaccine, Jenner and his supporters fiercely defended the sanctity of his approach. In his second vaccination pamphlet, published in 1799, Jenner was on the defensive against
critics who had noted that vaccination sometimes failed, wore off, or
spread smallpox or other diseases. The only reason anyone could fail to
get good results, he wrote, was through "spurious" technique. As even
friends of vaccination began to question whether revaccination was not
necessary in some cases, Jenner dug in his heels. He became obsessed
with his opponents, describing Moseley as a "mad animal," and devoting
some of the 30,000 pounds Parliament awarded him to publishing ghost-
written attacks on his foes. By the third edition of the *Inquiry*, in 1800,
Jenner reported that "the feeble efforts of a few individuals to depreciate
the new practice are sinking fast into contempt." In fact, the criticism was
just getting started.[19]

Wholly Cow

In a passage of a later edition of his *Inquiry* that suggests he'd observed
more failed cowpox infections than he let on, Jenner noted that it was
vital that vaccinators procure their lymph during a phase of the infection
when the infected cow, or human as the case happened to be, was produc-

ing an oozing, nearly clear liquid rather than yellow pus. Arm-to-arm vaccination required the presence of a recently vaccinated person with a ripe vesicle. Jenner conveyed a sense of the precise timing required in an 1807 letter to Swiss-born physician Alexander Marcet, who wished to send a girl for vaccination. Instructing the girl's mother to "bring the young lady to Bedford Place on Thursday 12 o'clock," Jenner promised that "a little one shall be ready to meet her with a Vaccine Pock in the highest state of perfection."[20] The best guarantee that lymph was fresh was to have it staring you in the face.

Over the years and decades, the original Jennerian virus weakened considerably. According to one account, by 1836 the most widely distributed strain had been passed through 2,000 people.[21] When a community ran out of people to vaccinate, it ran out of arms from which to harvest vaccine. Sometimes doctors mailed each other supplies of dried vaccine scabs or dried-out threads that had been passed through a pustule. (As I sat reading Civil War–era letters at the Virginia Historical Society in Richmond one afternoon a few years ago, a dessicated vaccine scab fell from a folded paper into my lap.) This material didn't reliably maintain its potency. Some establishments depended upon poor neighborhoods to supply children who, for a price, could be vaccinated one after another in an endless chain. In 1803, the Spanish crown put 22 orphans aboard a ship bound for the New World to keep alive vaccine virus through successive passages during the trip.[22] Waterhouse, the first American vaccinator, paid neighbor children to become biological warehouses for the vaccine.[23]

Since the germ theory was unproven through most of the nineteenth century, and no one knew the active ingredient of vaccine, purity was an impossibility—how could you purify what you couldn't define? There was no way to know that an unregulated product was actual vaccine, rather than an inadequate or fraudulent replacement. If the vaccine was old it could fail. It could fail if the vaccination was executed poorly by a doctor who extracted lymph at the wrong stage, or stored it too long or in the wrong place, or did not scratch it into the patient's skin in the proper way. Moreover the procedure was risky. At Rivalta, Italy, in one famous case in the 1860s, 44 of 63 children vaccinated from the pustule of a vaccinated child developed syphilis, and many died.[24] It was not uncommon for children to suffer terribly swollen arms or sepsis following contaminated injections.

Over time, experts built up a body of empirical data on correct vaccine use: It was necessary to harvest the vaccine material not before the eighth day after vaccination, but not after the ninth; fever should last three days but not more, and so on. But the data conflicted and the definition of a take varied widely. Some believed vaccination only worked when it produced fever and a swollen arm. Into the 1870s, some frustrated officials argued that the vaccine's impact varied in relation to a given locality's "epidemic constitution."

First Steps to Regulate Vaccine

The earliest attempts to create reliable sources of vaccine followed not long after Jenner's discovery, but it would be decades before they achieved any real success. In March 1812, in the middle of a smallpox epidemic, Dr. James Smith, who had run a Baltimore vaccination clinic since 1802, wrote to Congress and the Maryland legislature seeking help to deliver vaccine. Congress granted Smith part of his wish by making him the U.S. Vaccine Agent and awarding him a franking privilege that allowed his vaccine to be mailed for free. Maryland promised a lottery to raise money for Smith's vaccination program, but the fund was diverted to the construction of the Washington Monument that stands atop Charles Street in downtown Baltimore. In 1822, such regulation of vaccine that existed was returned to the states, where it would remain until 1902.[25]

Vaccination gained a generally positive reputation among U.S. physicians in the first half of the nineteenth century, but the public showed little enthusiasm. The largely rural nation did not suffer great smallpox epidemics, though smallpox prowled the cities. Between 1849 and 1853, for example, 2,396 New Yorkers died of smallpox, with two-thirds of the deaths in children aged 5 or younger, as Dr. Joseph C. Hutchison of the Brooklyn City Dispensary lamented. Hutchison concluded that failure to vaccinate was the prime cause of the epidemic, but hinted that some percentage of the deaths—as well as the public's failure to submit to vaccination—could be laid on the product itself. The vaccine in use in Brooklyn, Hutchison said, was at "the 180th remove from the cow"—and considered fresh at that. It had been furnished by the friend of a friend who knew a physician in Bristol, who had passaged some well-"humanized" vaccine back through a cow.

It took two great wars of the late nineteenth century—the Franco-Prussian conflict and the American Civil War—to hammer home the necessity of a good vaccine supply. The Franco-Prussian War, like many in the future, spread disease, in the process throwing into relief the relative health of large, systematically vaccinated armies and the patchily protected civilian populations from which they were recruited. A case in point: All German conscripts were vaccinated at least once, while the French army had no compulsory vaccination. Of the 800,000-man German army, 8,463 caught smallpox during the war, and of those who fell ill, 459, or 5.4 percent, died. The French army of more than a million men, of whom 700,000 were taken prisoner, reportedly had more than 125,000 infected during the epidemic, of whom about 23,500 died—a fatality rate of 19 percent. That was nearly as many men (28,208) as died fighting the Germans.

Many governments interpreted such statistics as evidence of the need to create or enforce compulsory vaccination laws. In England, where the first compulsory vaccination law took effect in 1840, Parliament gave it teeth in 1871 with a Vaccination Act that authorized local vaccine inspectors to confiscate the goods of poor families who refused the procedure. Germany enacted a law in 1874 requiring all German children to be vaccinated by age 2 and revaccinated around age 12.[26]

In the United States, where the epidemic of the 1870s also struck with considerable deadliness in the cities, there were few moves to compulsory vaccination until much later. Boston had become the first city with a mandatory vaccination law in 1827; Massachusetts the first state to compel vaccination in 1855. New York enabled local school districts to exclude unvaccinated children in 1860, but few outside New York City enforced the law. Even in Manhattan, enforcement was spotty. School officials stubbornly opposed exclusion because so many children were unvaccinated that it could empty classrooms. In fact, it was the spread of public education that enabled the spread of compulsory vaccination laws.[27] But while it was feasible to vaccinate an army with whatever product was at hand, Americans resisted the mandate to vaccinate with an uncertain product. As vaccination laws expanded toward the end of the nineteenth century, so did organized antivaccination resistance.

The solution, as in Europe, was to create vaccine farms, where cattle could be continuously infected with virus harvested "fresh from the cow." To begin this process, vaccine providers needed a source of spontaneous cowpox of the type that Jenner had studied so carefully. But the disease was practically unknown in most of the world and rare in others. The most famous case occurred in a cow near the Loire Valley town of Beaugency in 1866. Although the French stores of Beaugency virus were destroyed during the Franco-Prussian War, lymph from that famous *vache* would travel around the world, becoming the seed virus for the vaccines of many countries and cities for many years to come. The New York City Dispensary, a municipal agency, was the first American institution to procure Beaugency lymph.[28] With physicians pressing the dispensary for vaccine in the winter of 1871, it began inoculating calves with Beaugency virus and by January 15, 1872, 60 calves were serving as vaccine factories.

While the dispensary found Beaugency lymph no more efficacious than vaccine derived from a cow that had been freshly inoculated with "humanized" lymph, it had several advantages. First, it eliminated vaccination as a possible source of syphilis infection—cows didn't get syphilis. Some physicians felt that the bovine virus also produced a stronger take, indicated by a longer-lived vesicle on the patient's arm. Finally, animal vaccination furnished a constant supply of vaccine. "The caprice of a calf we may overrule, whereas human prejudice often proves insurmountable," wrote the dispensary's physician, Frank P. Foster.[29] Henry Austin Martin, a private physician in Boston, was another leading proponent of effective bovine-processed vaccines. As a Union officer during the Civil War, Martin had seen towns and regiments laid to waste by smallpox and was filled with anger by the inefficacious or contaminated wartime vaccine, most of it human scabs sent through the mail. After the war, Martin turned his energies to promoting vaccination in the American Medical Association, while producing vaccine with his own herd of cows.[30] Benjamin Lee, later to be secretary of the Pennsylvania Board of Health, argued in favor of the cow vaccine in 1871 in the *Philadelphia Medical Times*, the city's weekly magazine of medical news, commentary, and gossip. At Philadelphia's Municipal Hospital, where smallpox patients were quarantined and treated, however, presiding physician Dr. William M. Welch still preferred to use "recent human crusts" during an 1880–81 epidemic.

The public's confidence in direct-from-cow vaccination was nowhere more vividly exhibited than in Paris. The French had a soft spot for Jenner, one of those rare foreigners—a Jerry Lewis of the nineteenth century—whom their nation showered with love. Pasteur would do Jenner the ultimate honor by naming his chicken cholera and subsequent attenuated viral concoctions "vaccines"—thus assuring that Jenner's genius would remain imprinted on our language. French vaccine distributors developed a remarkably open and direct system of bovine vaccination that, if nothing else, was entirely transparent. Upon learning that smallpox had broken out in a given neighborhood, Drs. Chambon and St. Yves Menard, the official Parisian vaccinators, would immediately cart a vaccinia-infected cow to the scene. When they reached the affected neighborhood in the Marais, say, or Belleville, the doctors set up a vaccination station in the courtyard of an apartment house or in the street in front of it—a peculiar amalgam of the rustic and cutting-edge scientific, with all the street urchins watching.

"Occasionally, from want of space in the porter's lodge or other reasons, the calf, after being removed from the van, is allowed to remain in the street, its halter being held by an attendant, while another assistant, taking his seat on a camp stool, proceeds to collect lymph from the inoculated area of the animal's side and abdomen with the aid of clamp forceps and lancet," a British doctor reported from the scene. "Infants or adults are brought out into the street, and the extraordinary spectacle may be witnessed of vaccinations being carried out by the medical staff surrounded by an interested crowd of sightseers." During epidemics, people who might have objected to vaccination "from an unknown source, find all their objections on this score removed when they actually see the calf which serves as vaccinifer."[31] The sight of one of these placid, rather clean farm animals seems to have reassured Parisians, for whom *c'est vache* is an expression of endorsement. In one illustration from *Le petit journal* in 1905, a furry-faced cow with an expression of stoic contemplation submits to a mustachioed technician, who tweaks lymph from a shaven teat and hands it on a lancet to a doctor in a top hat.[32]

This remarkable practice appears not to have caught on as a public spectacle in the United States, although the Chambon-Menard virus was imported and is believed by some to be the source of the New York Board of Health strain of vaccine that is still in use. But vaccine stables were based on the European model, and included private and municipal farms. The New York health board, which established the nation's first bacteriological laboratory under Dr. William Prudden in 1886, had earlier established a vaccine stable at 300 Mott Street in the city.[33] When the cows provided more vaccine than the city needed, it sold the excess to other cities or the federal government.[34] In Boston, meanwhile, the bacteriologist Theobald Smith set up a vaccine and sera-producing laboratory in the late nineteenth century. Philadelphia never made its own smallpox vaccine, relying on various large companies, and occasionally New York City, for its supply.

The sheer physical challenges of vaccine production were impressive—cumbersome, riddled with uncertainies, risky by today's standards. Most of the establishments were located next to a central livestock market, providing access to fresh young calves that were isolated and fed good grains and clean water for a few weeks. If the animal appeared healthy and tested free of tuberculosis, it could proceed to the vaccination table. The vaccinators first strapped the calf to a gurney with two of its legs in the air. A patch of the confined animal's abdomen was shaved, a dozen incisions scratched into the animal's skin—not quite deeply enough to draw blood—and vaccine seed virus was smeared into the wounds. The animal was then led into a narrow stall, straitjacketed, and muzzled to keep it from licking its wounds, its tail placed in a splint to keep from flicking them. In a further effort at sterility, the finer establishments employed a stable boy to scoop up each cow pie in the moment of its emanation.[35] Ten days later, if all went well, the cow's underside would have broken out in sores filled with a clear fluid, meaning that the virus "lymph" was ready to be harvested—which was usually done by scraping it out with a rough spoon.

Thus used, the cows were then sold for slaughter. In Paris, each brought about 100 francs, a 30-franc loss for the vaccine institute. In Berlin, the chief rabbi paid a premium for vaccinated calf meat because of the fine eating the animal had done, according to one account. Eventually, when vaccination on an industrial scale became the rule in American cities, vaccine producers cleared a good profit selling their animals to stockyards.[36]

While the return to the cow had eliminated some of the worries of vaccination, many remained. Through the 1890s, processed lymph was generally stored in glass-stoppered jars. Alternatively, ivory needles were dipped into it and wrapped in metal foil for storage or shipping. But even before bacteriologists were able to examine these products, it had become clear that the vaccine in them tended either to lose its potency or become contaminated, or both. Under these circumstances, the public generally avoided vaccination except under the immediate threat of an epidemic.

The biggest step taken to address this concern was also a European import. Workers in the laboratory of Robert Koch in Germany in the 1880s had discovered that by storing the lymph in glycerine it was possible to kill streptococci, tetanus, and other pathogens while preserving the viability of the virus. Thinning the virus in a glycerinated solution also made it last longer and go further. A single cow could now yield up to 6,000 vaccinations, compared with 200 to 300 doses per cow with unglycerinated vaccine. Toward the end of the nineteenth century, vaccine farms sprang up across America, most of them selling glycerinated calf lymph. The new vaccine was hailed as a modern, safe product, although as we shall see, this innovation had unintended consequences of its own.

Britain's Antivaccine Movement

Under the 1871 Vaccination Act, Britain's poor had to submit to vaccination or face fines, confiscation of property, or the workhouse. The law and controversy over mandatory vaccination drew in a wide array of politicians, activists, writers, and scientists as it snowballed into the largest revolt against a medical practice of all time. The national Anti-Vaccination League, formed in response to the 1871 law, was thoroughly middle-class in its composition, but there were also frequent popular uprisings against the practice. Arm-to-arm vaccination was the preferred method of the state for most of the late nineteenth century, and parents were fined up to 20 shillings if they refused to allow lymph from their children to be harvested. Public vaccinators, often paid piece rate for their work, were labeled "baby hunters," and protestors pelted them with eggs and fruit.[37] Working people were anguished at the thought of mingling the blood of their children with that of common paupers. Yet they often felt that they had no choice but to "barter their bodies" in return for work.

In 1889, the Royal Commission on Vaccination was created to determine whether vaccine was safe and effective and whether compulsory vaccination was a sound idea. The commission's activities came at a crucial time in the history of Western medicine. Bacteriology, which had gotten its start in the laboratories of Louis Pasteur and Robert Koch in the 1860s, had made remarkable strides, isolating many important pathogens. The public was led to believe that these discoveries would quickly lead to preventive treatments. "Isolate, attenuate, vaccinate," was Pasteur's famously simple formulation; he meant that it would be a matter of simple hard work to weaken each pathogen and thereby craft life-saving vaccines. In 1884, Pasteur created a worldwide sensation by treating two boys severely bitten by rabid dogs. With this first successful vaccine for humans since smallpox, physicians and public health officials proclaimed a new era in the war against infectious disease.

However, the advances in bacteriology made the inadequacies of smallpox vaccination something of an embarrassment. Two outbreaks of foot-and-mouth disease in cattle were linked to vaccine cows, underscoring the problem. An epidemic that broke out in 1902 in Massachusetts forced the government to order the slaughter of 4,316 animals. A 1908 outbreak was traced to a lot of vaccinia virus that the Mulford company imported from Japan. Mulford's competitor, Parke-Davis, used the Mulford vaccine to inoculate its own cows, which it sold to stockyards where they were mixed with other cattle.

Governments that wanted to make public health part of their expanding powers of control and reform needed to come to grips with its problems. By the turn of the twentieth century, the public was starting to demand clean water, pure air, and decent housing, and the antivaccinators turned this feeling to their advantage. Bacteria were in the water and the air and the streets, they argued—clean them up, and you won't need to vaccinate us. British society was deeply split over the merits of vaccination. H. Rider Haggard's popular novel *Doctor Therne*, a tragicomic farce about antivaccinators and dedicated to the Royal Jennerian Society, spoofs how one-sided the movement was.

The novel tells the story of Dr. Therne, a promising young physician in a small city who is led by circumstance to accept an alliance with the local antivaccination league, led by a florid draper and his hypochondriac wife. Eventually Therne gains a fortune and a seat in Parliament through his opportunism. Therne's description of his rhetorical technique will sound familiar to any public health official who has tried to debunk anti-vaccine conspiracies:

> *I knew my subject thoroughly and understood what point to dwell upon and what to gloze over, how to twist and turn the statistics, and how to marshal my facts in such fashion as would make it very difficult to expose their fallacy. Then when I had done with the general arguments, I went on to particular cases, describing as a doctor can do the most dreadful which had ever come under my notice, with such power and pathos that women in the audience burst into tears.*

People ate up his arguments, Therne said, and "having never seen a case of smallpox, they suppose that the whole race is being poisoned by wicked doctors for their own gain. Hence their fierce energy and heart-felt indignation."[38]

In the book's climax, a smallpox epidemic sweeps through the district and kills Therne's beloved daughter. That same evening, while Therne is speaking at a re-election campaign rally, his daughter's fiancée, a doctor in the mold of the narrator's young, idealistic self, climbs onto the stage, rips off Therne's shirt, and reveals his shameful secret—a recent vaccination scar. "Murderer of your own child," yells the young doctor, "I reveal that which you hide!" Therne, whose career has been built upon the myth that vaccination has no value, is shown to be a hypocrite.

Despite Haggard's fervid book, fiction alone could not dismiss the fact that tainted vaccines provided an excellent reason for shunning manda-tory vaccination. Some of London's finest writers put themselves to work excoriating the system. Lord Alfred Russel Wallace, the naturalist who helped discover evolution, employed a series of arguments that would become the model for antivaccine arguments over decades to come.[39] Wallace led the reader down a path of half-truths, each of which appeared to steal away a bit of the rationale for vaccination. At the end of the path, if you traveled without blinking, was the rejection of smallpox vaccination. He started with a competing philosophy of health. The medical profession, he argued, exaggerated the gravity of smallpox, which wasn't such a bad disease when contracted by an otherwise healthy person. Smallpox, in Wallace's view, was like all other "filth diseases"—it would disappear when nations and cities did away with "foul air and water, decaying organic matter, overcrowding and other unwholesome surroundings."[40] Vaccination's failures, Wallace went on, had been obscured by the fact that public vaccinators did not care for vaccinated patients who subsequently became ill.[41] Further, there had been no controlled experiments comparing populations of vaccinated and unvaccinated people—true enough. In short, Wallace argued, vaccination did not protect against smallpox and weakened, rather than strengthened, the constitution. For proof one need only examine the poor health of the city of London, where vaccination was widespread and the authorities, he claimed, concealed the death and destruction it caused.[42] Vaccine, he concluded, actually *caused* smallpox. Witness the prevalence of smallpox in areas where authorities vaccinated the most. In Ireland there was less smallpox than in Scotland, although Ireland was undervaccinated and Scotland among the most vaccinated areas of the UK. Leicester, which did away with compulsory vaccination in 1873, had only one smallpox death per 10,000 population in 1894, while heavily vaccinated Birmingham suffered 63 cases and 5 deaths per 10,000.

Wallace's claims were eloquently argued, but they ignored inconvenient facts. Vaccination was of course most frequent in areas where smallpox was greatest, because people generally did not vaccinate until an epidemic threatened. And while it was true that overcrowding and poor general health contributed to the spread and mortality of smallpox, there was an important caveat: while smallpox fatality had diminished over the nineteenth century, the death rate from other infectious diseases had risen. It was true that Leicester for a time controlled smallpox without vaccination, but only through rigorous isolation and quarantine practices.[43]

Wallace typified the scientists who would battle vaccination over the years—mavericks who had made their names by overturning established theories, and who as a result identified strongly with antiestablishment points of view. The writer Michael Shermer calls this the "heretic-personality" type. Wallace was also a dabbler in spiritualism and other controversial beliefs; he strayed into bad science, in Schermer's view, because of a personality flaw that made him a little too open-minded. The eighth of nine children, the son of a disgraced small-town lawyer, Wallace had drifted through his early years learning various trades. He was never fully accepted in the aristocratic class to which Darwin belonged.[44] Having dismissed the authorities in his own field, Wallace assumed that the dons of public health were just as unreliable.

George Bernard Shaw was another famous heretic who joined the antivaccination side, though Shaw's attacks were closer to the mark, drawing power from the genuine problems of vaccination and medicine's defensive refusal to admit them. Shaw's most important writing on the subject is in the preface to *The Doctor's Dilemma*, a satirical play about medicine whose protagonist is modeled on Shaw's friend Almroth Wright, the creator of the first typhoid fever vaccine.[45] Shaw was quite rightly skeptical of medical science, which had not yet turned the corner past which the average patient received a net benefit by consulting with the average physician.[46] Medicine was still "very imperfectly differentiated from common cure-mongering witchcraft," he wrote, and people only went to doctors under "the old rule that if you cannot have what you believe in you must believe in what you have."[47] Like many a Brit, Shaw was also sentimental about animals. Experiments on beasts were cruel, morally groundless—"you may not torture my dog, but you may torture dogs"—and largely a waste of time—"burning down London to test a patent fire extinguisher." He felt that vaccination had become a cult, with doctors circling the wagons to defend it despite all flaws. "The Radicals who used to advocate as an indispensable social reform the strangling of the last king with the entrails of the last priest, substitute compulsory vaccination for compulsory baptism without a murmur." And like Wallace, Shaw argued that vaccination supporters "steal credit" from sanitary reforms that had diminished the threat of cholera, typhus, plague, and to a lesser extent tuberculosis. As for Pasteur's rabies vaccine, "the vaccinated people mostly survived, but so do most victims of dog bites."

In 1898, Parliament, following the Royal Commission's recommendation, passed a law allowing "conscientious objectors" to avoid vaccination. The Anti-Vaccination League made wide use of the new clause, signing up objectors with door-to-door campaigns. Vaccination rates, which stood at 80 percent in 1898, fell to 50 percent in 1914 and 18 percent in 1948. That year Britain ended compulsory vaccination. When smallpox broke out, the authorities vaccinated contacts of the patients and quarantined those who would not be vaccinated. This was surprisingly effective. By 1960, four times as many Brits were dying of vaccination side effects than of smallpox.[48]

Having struggled for half a century with compulsory vaccination, Britain made peace with the antivaccinators by essentially surrendering to them. In the United States, compulsory vaccination was only beginning, and so was the struggle over it. There was no federal vaccination law, but as the public health movement grew, state laws tightened, and many cities began excluding unvaccinated pupils from schools. These laws and practices galvanized the previously passive resistance to immunization. The more the public resisted, the more stridently the newly empowered public health officials defended the vaccine. The smarter among them understood the need for improvement in smallpox vaccination. Vaccines were unreliably available, of uncertain origin, and difficult to make safe. They did not always offer good protection. But medicine was not powerful enough to be self-critical, so it persisted in its blinkered unanimity: whatever the dangers and drawbacks of vaccinating, it had to be done, unquestioningly. With a single voice, public health cried, "Vaccinate! Vaccinate!"

CHAPTER 3

—

VACCINE WARS: SMALLPOX AT THE TURN OF THE TWENTIETH CENTURY

To day in all its dimpled bloom
The rosy darling crows with glee;
To morrow in a darkened room
A pallid , wailing infant see,
Whose every vein from head to heel
Ferments with poison from my steel.

— ALEXANDER HOPE HUME[1]

IN NOVEMBER 1901, East Coast newspapers carried reports of a disturbing outbreak of tetanus, the excruciating disease also known as lockjaw. In each of 90 or more cases, a horrible illness and death had followed a few weeks after vaccination against smallpox. A letter in *The New York Times* captured the dilemma: "It is not surprising that people should refuse to submit to vaccination," it noted, "unless they can be given the positive assurance that no bad effects will follow its administration. Since the authorities, in order to protect the majority, make vaccination compulsory, they should also afford entire protection from harmful results to every law-abiding individual."[2]

It was an understandable request, but medical science and public health would never have the power to "afford entire protection" to people undergoing vaccination, or any medical procedure for that matter, and much less so in 1901. Three decades into the Pasteurian era of sanitary science and germ awareness, Jenner's vaccine for all was an accepted dogma of medicine. In America, it was now backed up by truancy laws that required children to go to school and laws that required them to show a scar or certificate of vaccination to attend. Smallpox, it was well known, was highly contagious and deadly, and could only be prevented with vaccination. The champions of this dogma gritted their teeth in the face of the outbreaks of tetanus and toxic shock, badly infected, swollen arms, and occasional cases of encephalitis or eczema vaccinata—all caused by vaccination. In public, at least, doctors and scientists kept up an iron front. The notion of informed consent—the idea that one could advocate a procedure while acknowledging its risks—was 50 years away. Yet ordinary people made their own risk/benefit equations and generally avoided the risk when they could. One survey found only 21 percent of urban and 7 percent of rural U.S. preschoolers had been vaccinated.[3]

The problems with vaccination came to a head in the fall of 1901 as smallpox swept the country. To complicate the lives of the public health advocates and further undercut the rationale of vaccination, two separate, entirely different strains of smallpox were now circulating. In some cities and towns, a deadly strain, *Variola major*, claimed the lives of 3 in every 10 unvaccinated people exposed to it. But sometimes only a few towns down the road, or even on the other side of town, a far more mild form, which would become known as *Variola minor*, might produce only one death for hundreds or even thousands of cases.[4] *V. minor*, scarcely more troublesome than chickenpox, was well known in Africa and Brazil, but had only recently arrived in the United States.

That fall, in every major city and town, teams of vaccinators went from house to house and block to block vaccinating children and adults. For the six- and seven-year-old boys and girls entering the first grade, this was usually their first vaccination. Older children were often revaccinated, or vaccinated for the first time if they had managed to escape the lancet so far. And then came a ghastly epidemic. Over a period of months, scores of children, many of them in and around Philadelphia, fell ill following administration of vaccine that was apparently contaminated with tetanus bacilli. The tragedy would rattle the vaccine world and help lay the foundation for the first piece of the country's food and drug regulation system.

The terrible irony of this crisis was that it coincided with the gradual fruition of the great bacteriological research of the late nineteenth century. At laboratories in Berlin, Paris, London, New York, and other cities, breakthroughs had led to antibacterial treatments with real promise. Beginning in 1894, antibodies harvested from the blood of horses that had been injected with diphtheria or tetanus germs were being used as treatment and sometimes as preventives. A growing vaccine industry was producing these antitoxins, as well as vaccines to protect against a dozen other microorganisms. But the quality of these products was not uniform. While breakthroughs in the laboratory enhanced sanitation and raised the profile and influence of public health, scientific advances were slow to translate into the "magic bullets" that Paul Ehrlich, the German-Jewish genius of immunology, dreamed would control all pathogens while leaving their hosts unscathed. Vaccine makers made great claims for their new products, but in practice few were effective and safe—and that included the supposedly sterile new smallpox vaccine.

City and state officials, leading doctors and drug executives bickered and blamed each other for the deaths and maimings caused by vaccines. In July 1902, Congress passed the U.S. Biologics Control Act, which for the first time allowed the Hygiene Laboratory, a division of the U.S. Public Health and Marine Hospital Service, to regulate vaccines and sera sold across state lines.[5] Four years later, the publication of Upton Sinclair's muckraking classic The Jungle led to a nationwide uproar over the unsanitary conditions of slaughterhouses. The Pure Food and Drug Act of 1906 passed amid great fanfare and stern declarations from the nation's Progressive president, Theodore Roosevelt. Yet the Biologics act, the first law to regulate drugs in America, passed without hearings and with little publicity.[6] The nation's shock over bad vaccines preceded its horror over rotten meat, but there was apparently less concern about the vaccine entering millions of children's bodies than there was about the sausage they ate.

There were, to be sure, some exceptions to this national indifference. As early as 1899, at an annual meeting of the American Medical Association,

Surgeon General George Sternberg had recommended careful bacteriological examinations of vaccine. But there was no legal basis for compelling such inquiries. Before 1902, reported John F. Anderson of the U.S. Public Health Service, "anyone could make a product, label it vaccine virus and place it on the market."[7] The risks of vaccination were not fully respected until 1986, when the National Vaccine Injury Compensation Program established the right of vaccine-injured children to file for government payments. Eighty-five years earlier, as Progressive-era government stepped in to demand that parents educate their children and protect their health, it also gingerly began assuring the safety of the product it was forcing upon them. Regulation had modest beginnings. The red-brick Hygienic Laboratory stood on a hilltop over the Potomac near what is now the Kennedy Center. Initially, the handful of government scientists did little more than make an annual visit to each vaccine farm, and periodically examine the product on sale in pharmacies. Vaccination was a very messy business, nowhere more so than in Philadelphia, center of the nascent pharmaceutical industry and soon-to-be headquarters of the antivaccination movement.

Filthydelfia

In 1901, Philadelphia was run by one of the most corrupt political machines in a country whose cities were rife with corruption.[8] John Ashbridge's Republican Party bought and sold votes, robbed the till, and traded the spoils from all public institutions, while overseeing a fast-growing city with a gruesome mortality rate. To muckraking journalist Lincoln Steffens it was the most shameful of cities. Thousands of deaths from typhoid, tuberculosis, and other infections gave Philadelphia a medieval atmosphere. For three decades reformers had pointed out the risks of supplying the city's water directly from the Schuylkill River, which was contaminated with every imaginable filth. But the funds for a filtration plant, not completed until 1909, were rich fodder for the GOP trough.

The city's smallpox eradication program offered fewer opportunities for graft but was every bit as ineffective.[9] In 1865 the city was divided into 10 vaccination districts, each with a public collector and a vaccinator, paid 25 and 50 cents, respectively, for every successful immunization. The inspectors frequently charged for vaccinations they didn't perform and failed to check whether those they did perform had taken. In 1871 the city nominally vaccinated more than 50,000 people, but there were still 1,879 smallpox deaths, and another 2,285 deaths in 1872. A decade later, another epidemic claimed 1,758 lives, but it wasn't until 1895 that the Pennsylvania state legislature required the exclusion of unvaccinated pupils.

In New York, Chicago, Boston, Minneapolis, and Providence, European-educated doctors were building healthier cities by applying science and common sense. Men like William M. Welch of Johns Hopkins University, William Prudden of New York, and the Flexner brothers, Abraham and Simon, were overhauling American medicine from scratch, integrating the laboratory into clinical practice and medical education. In Manhattan, William H. Park and Herman M. Biggs, who had whipped cholera in 1892, took on diphtheria and were starting to attack typhoid and tuberculosis by enforcing the sale of pasteurized milk. These were rough-and-tumble times for public health, which was struggling to control the chaos of rapidly expanding cities where thousands of unregulated vendors sold contaminated products to people who lacked the educational or financial wherewithal to question or challenge them. Tough measures were required to clean up the streets, enforce clean food and drink, and get people to vaccinate.

Not every city was up to the task. Philadelphia wasn't. As the city expanded from 30,000 in 1830 to 674,000 in 1870 and 1.3 million in 1900, its "most disconcerting aspect," according to one account, was not the intolerable sanitary disorder but rather "the lack of support and moral indignation of the physicians, engineers, educators, clergy and businessmen" who might have provided leadership to improve conditions.[10] The leaders of the smallpox eradication campaign were an unimpressive bunch. They included William H. Welch, an elderly physician who had run the Municipal Hospital for more than 30 years. He was no relation to the William M. Welch of Hopkins and his hospital was notorious. Located at 22nd and Lehigh in the Strawberry Mansion neighborhood of North Philadelphia, it was a poorly ventilated wooden building with primitive medicine and no control over its patients' comings and goings. Thousands had entered its ghoulish infectious disease wards, and thousands had come out in boxes, while others left it only to spread infection elsewhere. Welch's partner in the promotion of vaccination was Benjamin Lee, the dour secretary of the state health board whose defining career experience had been as an orthopedic surgeon during the Civil War.

Philadelphia was the hub of the nascent pharmaceutical industry, with three of the country's largest vaccine farms—the progenitors of pharmaceutical giants Wyeth, Merck, and Sanofi-Pasteur—located within 40 miles of the city. But the conditions under which vaccines were made, distributed, and administered were part of the "intolerable sanitary disorder" of the time. No less an authority than Walter Reed decried the vaccine farms' unreliable certificates of purity. Reed—who discovered the cause of yellow fever, reformed Army medicine, and then died of appendicitis at age 51—examined needles from six of the biggest vaccine companies in 1895 and found colonies of pathogenic bacteria in all of them.[11] This was news few in the profession wanted to hear, especially with an epidemic lurking. Smallpox would sicken more than 5,000 people and claim 900 lives in Philadelphia by the time it faded out in 1904. It first arrived, however, as the disconcertingly mild form, *Variola minor*.

V. minor—also known by a collection of vivid nicknames including eruptive grip, bean pox, yaws, yarrows, and the bumps—was no trifling matter. It caused fever and an eruption of lesions on the face, though fewer than regular smallpox did. When it first appeared in the southern United States in 1896, contemporaries doubted whether the disease was even smallpox. People called it "Cuban itch" or "Manila scab" under the false impression that U.S. troops had brought it back from imperial missions to Cuba or the Philippines. But the Spanish-American War hadn't begun until 1898; in all probability the beachhead for the disease was Pensacola, Florida, where it may have arrived aboard a ship from South Africa or Brazil that docked in 1895. The new disease was easy to catch but not regularly reported. Flashing through the population, it confounded medical men who had seen thousands of cases of smallpox in their careers.

In Ohio, the new smallpox arrived in 1898 with a theatrical traveling group—Uncle Tom's Cabin Show—on its annual tour of the western counties. An adult and two children in the cast had contracted what physicians at first pronounced chickenpox. The kids weren't too sick and the one playing Topsy continued to take her part on the stage each night and to mingle freely with the crowds at a county fair. The disease quickly spread to six towns. In Wapakoneta, the first cases cropped up in May, but the outbreak wasn't declared until October. "If this disease we are having here is smallpox," an old physician said, "the whole subject of smallpox as given in our textbooks must be rewritten."[12]

Dr. Welch, who had probably seen as much smallpox as any doctor in the country, was taken aback. The first men he treated, black laborers from Virginia, called the ailment "elephant itch," because one of them was a circus roustabout who had slept in the hay with elephants in Norfolk. Welch had never seen nor heard of a smallpox epidemic as mild as this one. "After the appearance of eruption, the patients would frequently go about their work as usual," he wrote. "Not infrequently they were employed gathering cotton and preparing it for market while the eruption was developing or the scabs falling off, and it is believed the infection was spread to distant localities by this article of commerce. It was a novel sight for me to see smallpox patients, Negroes, unvaccinated, at about the eighth or tenth day of the eruptions, engaging in a game of baseball." Often, the patients weren't ill enough to be confined "but on the contrary mingled freely with the public by visiting dispensaries, riding on trolley and steam cars, walking and driving on the streets and the like." Within a year he had seen 128 patients and not a single death, although only 17 of the patients had been vaccinated. In 1899 the count was 990 cases, with 14 deaths. During the city's 1871 epidemic of *V. major*, two-thirds of the unvaccinated smallpox cases had died.

The Philadelphia doctors were confused by the disease and worried that it was a Trojan horse. They told reporters that if this was a mild tropical smallpox, it was sure to revert once it had lingered in northern climes for a while. In fact, *Variola minor*, like *V. major*, bred true. Because the disease was mild enough to occur without being identified as smallpox, *V. minor* patients were seldom quarantined and spread the disease easily, making *V. minor* a better-adapted organism than its predecessor. *V. minor* quickly became the dominant form of smallpox.[13] Statistically, the change it wrought looked like this: In the United States in 1895 there were 2,826 cases of smallpox registered, with 589 deaths, or 21 percent—a typical rate in an indifferently vaccinated population.[14] From 1906 to 1912, approximately 22,000 cases were reported each year, with an average of 230 deaths, a mortality rate of about 1 percent. Most of the deaths occurred in localized outbreaks of *V. major*. In *V. minor*–only areas, deaths averaged 1 in 300. In North Carolina in 1911 authorities made note of 3,294 cases of smallpox, and not a single death. High rates of smallpox in America were a source of shame, but the less severe form of the disease— which protected against future infection by *V. major*—may have been just as reliable an immunizing agent as vaccine, though no one recognized this fact at the time. Eventually, severe smallpox strains were isolated and stamped out, while *V. minor* established itself as the dominant strain of the disease in the United States. (It remained so until the last case of smallpox was reported in the country in 1949.)[15]

Still, the old, deadly smallpox made an occasional appearance, usually slipping in by ship or over the Mexican border. Welch and Lee may have been wrong about the nature of *V. minor* in 1898, but they were correct in one sense: its appearance had given the public a false sense of security. In 1900, there were only 27 cases of smallpox in Philadelphia, and no deaths; by the end of the year the city appeared to be smallpox-free. In January 1901 a new case turned up in West Philadelphia, then two more, and the disease quickly began to spread. This, however, was not the mild form of smallpox. A virulent imported virus had crept in through a New York port, an investigation later showed, and traveled surreptitiously along the New Jersey rail lines through Newark, Trenton, and Camden.[16] By the fall it was bubbling through Philadelphia. The first week of October saw 40 new cases of smallpox and 6 deaths in the city. On Wednesday, October 9, the newspapers ran the story of the Coyle family of 3752 Randolph Street, in which the father had been a brakeman on the Philadelphia and Reading Railways. Smallpox had killed the father, mother, two-year-old son, and infant daughter, leaving brothers John, 5, and George, 4, in an orphanage. The following week brought 60 new cases, and panic. Mrs. John Wynn fled through the streets with her two daughters to avoid being quarantined at Welch's hospital, only to be captured and sent there with her girls.

Smallpox tragedies were nothing special in those sickly times. The

same week that Mrs. Wynn made her dash for freedom, 49 Philadelphians
died of tuberculosis, 10 children died of diphtheria and 3 of scarlet fever,
and 4 babies died of whooping cough. Unclean milk and unfiltered raw
sewage claimed a large toll, with 9 typhoid deaths and 14 by diarrhea.
With so much death everywhere, bogus cures were rampant, and vaccination may have seemed like just one more. The *Philadelphia News
American* was filled with ads for patent medicines, such as Dr. Slocum's
Spiro Power and F.W. Kinsman's Swamp Root—"good for kidney and
liver troubles." Mystical healers like Professor Thomas F. Adkin, president of the "Institute of Physicians and Surgeons," might pay for an
entire page to stake a claim to a "secret new treatment" that "conquers all
diseases." The state board of health was almost as clueless. Secretary Lee
grimly waggled his finger "I told you so." He had been pleading for more
funds he said, but everyone had shrugged off his sage predictions and
now they were paying for it. Smallpox, he claimed, "is essentially a cold
country malady; the germs do not live and preserve activity in high temperature."[17] But once this "West Indian smallpox" arrived on American
soil, it had slowly mutated and its mortality steadily grew. "Now," he
stated, "we are within the grasp of a serious epidemic."

Lee wanted a general vaccination requirement, but even the vaccination of schoolchildren was under challenge. By wide margins, the
Pennsylvania state legislature in 1907 and 1909 passed bills to overturn
or limit compulsory vaccination. It was only the governor's veto, under
strong pressure from doctors and the pharmaceutical firms, that kept the
law on the books. Only a small percentage of the population volunteered
for vaccination, even when it was free. Opposition was equally strong in
the cities and the towns, the mining camps and the farms. In Greenville,
Pennsylvania, in the fall of 1899, the local health board demanded that all
children be vaccinated, but school officials allowed exceptions for children with medical excuses. About 260 pupils—a third of the school
district—produced such certificates.[18] Fear and loathing of vaccination
made health inspectors persona non grata and recruitment a constant
challenge. "To be a health officer," said Dr. Morris Cawley, an inspector
in Allentown, "throws an odium on a man."[19]

Opposition to vaccination was strong in 1901 and was about to get much
stronger. Philadelphia was in the midst of a deadly smallpox epidemic, but

smallpox was a familiar disease. Many parents would soon be more worried about catching a deadly disease from vaccination. Tetanus could strike down and kill a child within a few hours after horrible suffering.

Never Again Shall I Vaccinate Any of My Children

On November 11, the *North American* newspaper ran the first report of vaccine-related tetanus in the Philadelphia area. The 16-year-old patient, William Brower, was being treated with tetanus antitoxin, which was manufactured in horses by the Philadelphia branch of the Pasteur Institute. The doctor overseeing Brower's case, Dr. William I. Kelchner, reported that the boy had been vaccinated against smallpox on October 22. On November 1 he fell ill and by November 8 he was running a fever of 105, with faint respiration and a convulsive arching of his spine. Despite injections of antitoxin, the prognosis wasn't good. " 'Never, never again shall I have one of my children vaccinated," Mrs. Brower lamented. "Willie was vaccinated last spring by the same doctors when there was a smallpox scare in Camden. It didn't take then, so now that there's so much smallpox around again I sent him to have it done over. This time, it's taken." "In fact," the newspaper report added tactlessly, "it's nearly taken him."[20]

The doctors doubted vaccination caused William's misfortune, but their reasoning was based more on affinity than logic. "Never for a minute did I believe vaccination was responsible," Kelchner was quoted as saying. "I believe in vaccination as one of the greatest preventives ever discovered for a deadly disease. I vaccinate people every day and expect to continue to do so." Indeed, the doctor who performed the vaccination went so far as to blame William for his illness: "I saw the Brower boy last Sunday. He had a bad arm but no worse than many others. . . . There are a hundred ways that tetanus microbe might have been introduced. No boy ever lived who doesn't say 'See my vaccination?' and uncover the wounded arm a dozen times a day. And boys are not careful always to keep a sore place absolutely clean. Vaccination doesn't produce tetanus, that I know. And tetanus, when it has been produced, is incurable—that I believe." Attended by a nurse, the boy lay in bed on the second story of the house. Downstairs, his mother and father paced and tried to keep their younger children from making too much noise.

Nine-year-old Edward Dougherty of Camden died of tetanus two weeks after vaccination. The boy's grandmother sadly described the course of his illness: for a time he had seemed to recover and even asked his mother to take him down to the Labor Day parade. After his outing the boy "ate a hearty supper and was in excellent spirits all the evening. He had only been in bed a short time when he called to his mother, who went to see him. She saw that a remarkable change had come over him. He was suffering great pain." That night, Edward died. Joseph Goldie, 11, of Bristol, died in the Hahnemann Hospital despite tetanus antiserum injections, some of them dripped into a hole drilled in his skull.[21] By November 14 there were at least six tetanus deaths in the Philadelphia vicinity, with two other children in critical condition. The doctors claimed that 11-year-old Thomas Hazleton had gotten lockjaw as a result of wearing a woolen shirt over his vaccination wound. His parents, however, denied such a garment was used and were pondering legal action against the firm that produced the vaccine. Six months earlier, the Hazeltons' other child, Laura, a bright little girl of 6, had "died suddenly of brain fever caused by jumping rope."

While doctors insisted that vaccination could not be to blame for the deaths, Dr. H. H. Davis, president of the Camden Board of Health, asked all physicians in the city to sterilize instruments carefully before vaccinating, lest lockjaw follow the use of unclean instruments. His city was awash in fear. Only 5,000 of Camden's 8,000 schoolchildren had been vaccinated by December. Children were terrified, and parents were keeping them home from school.

Across the Delaware in Philadelphia, about 4,000 parents of unvaccinated children filed a petition opposing a compulsory vaccination order. The doctors, meanwhile, continued to deny, against all evidence, any connection between vaccination and the tetanus deaths. Lee said the reports of tetanus were "probably newspaper exaggeration. The idea that the very means on which we rely to combat or prevent disease may in itself prove a cause of death is too terrible to be entertained."[22] Since it was too terrible to be entertained, Lee refused to entertain it.

On November 29, the Camden Board of Health announced that the vaccines it tested were free of tetanus bacteria. Due to a "long period of dry weather with high winds" there was tetanus in the air, and thus the deaths "resulted from neglect on the part of the patients." With no dissent, the board upheld the return to compulsory vaccination for schoolchildren. So did Philadelphia. On December 3, the city's health board sent out 500,000 letters to parents ordering them to vaccinate their children. Forty doctors were hired to sweep the area between Broad Street and the Delaware River, vaccinating every citizen "willing to be vaccinated."[23]

In a December 15 newspaper interview, Lee warned that 127 Philadelphians had died of smallpox, with 300 hospitalized in the Municipal. "Any man or woman who does not submit to vaccination is no better than a criminal," added Welch, "because that person is endangering the lives of the community." In its only nod to public concern about tetanus, the Pennsylvania Board of Health authorized Robert L. Pitfield, assistant bacteriologist for Philadelphia, to conduct an investigation. Pitfield simply tested samples of vaccines on sale at local drug stores early in 1902 by injecting them into guinea pigs, which survived. His report gave no indication that he had sought out the lots of vaccine potentially contaminated with tetanus.[24] Lee concluded that Pitfield's examinations "showed a marked improvement in all the samples over those examined by the bacteriologist of the board some years since. In not a single instance was a germ found which could communicate any disease other than vaccinia."

The epidemic of severe smallpox that winter spread along the eastern seaboard from Boston to Philadelphia, with New Orleans and other cities also afflicted. The Pennsylvania Board of Health sent the Philadelphia Free Library the names of new smallpox cases each day.[25] If they were library borrowers, any books they bought back were to be burned. All students at the University of Pennsylvania had to present certificates of vaccination before leaving on Christmas vacation. In Camden, the failure to vaccinate gave smallpox a Christmastime boost. Sixty people were crammed into a makeshift isolation hospital. In the pox-ridden African American neighborhood of Snow Hill, residents looked starvation in the face because well-to-do whites, in a panic, had fired their servants and workmen.[26] Boston, where *V. major* struck in May, approached the epidemic with the same systematic thoroughness it applied to outbreaks in the eighteenth century. By December 1901, when 504 cases and 72 deaths had been recorded, more than 400,000 Bostonians—fully two-thirds of the city—had been vaccinated.[27] The zealous Board of Health began a house-to-house campaign to vaccinate anyone not "successfully vaccinated since January 1, 1897." Refusers were subject to a $5 fine or a 15-day jail sentence. One refuser, an antivaccine minister named Henning Jacobsen, sued the state of Massachusetts. His case would end in the Supreme Court with a powerful victory for public health authorities and the principle of compulsory vaccination.[28]

Dr. Samuel Holmes Durgin, a Harvard Medical School professor and chairman of the Boston Board of Health, sought to silence the antivaccinists once and for all by inviting 60-year-old Immanuel Pfeiffer, an antivaccine homeopath who insisted that smallpox was not contagious, to tour the pesthouse at Gallop's Island in the Boston harbor. Pfeiffer visited the hospital on January 23, 1902, escorted by the physician in charge, Dr. Paul Carson, who encouraged Pfeifer to smell a patient's breath. The doctors clearly wanted the antivaccinists to try their own medicine, and they succeeded. A month later, Pfeiffer was observed at home in Bedford, seriously ill and covered with sores.

The Cleveland Experiment

One good-sized American city seemed to have vanquished *Variola major* by Christmas 1901, although its streets had been alive with smallpox only six months before. Much to the vexation of East Coast public health authorities, the growing Midwestern metropolis of Cleveland, with a population of 380,000, had apparently done it without vaccination.

Smallpox had entered Cleveland quietly in 1898, with 48 cases that year and 993 in 1900. By the spring of 1901, Detroit and Buffalo were threatening a blockade.[29] Vaccination was unpopular in heavily German cities like Cleveland, where immigrants favored homeopathy and tended to associate government-mandated vaccination with Prussian oppression.[30] In Cleveland, the reticence was particularly strong in 1901 but, then, the vaccine in use was particularly nasty. Cleveland Mayor Tom Johnson, a well-known reform Democrat, hired a new chief public health officer, Martin Friedrich, on Monday, July 21, and Friedrich quickly won the mayor's approval to use disinfection and isolation to fight the epidemic, which had sickened 1,200 people. At the time, there were 17 cases of smallpox in the city. The next Monday, July 29, Friedrich and a team of 40 medical students began their campaign of disinfection, which they continued until November 9. Friedrich, a handsome, thickly accented immigrant from Bavaria, had studied medicine at Western Reserve Medical College and returned to the city after postgraduate study in Europe. By late August, Cleveland had only a handful of new smallpox cases, all of them imported, and by Christmas the city was smallpox-free.

Some would question, later, whether the epidemic hadn't simply followed its natural course. But in a widely distributed speech to the Cleveland Medical Association in December 1901, Friedrich credited his own unorthodox approach. As an experienced vaccinator, Friedrich said, he had noticed that the practice was becoming increasingly ineffectual. Either the vaccine was improperly made or it was spoiled in delivery or the doctor didn't debrade the skin enough to allow the lymph to enter the body. "Last June, I was out in Newburg" (a village south of Cleveland), he recounted, "I met a crowd of children calling to each other: 'Are you vaccinated? Are you vaccinated?' I knew then that the vaccinators were in the schoolhouse and I slackened my pace to hear the comments. Pretty soon I knew what they were up to. The groceryman on the corner had told some of them they must wash their arms to prevent them from getting sore." As soon as they were vaccinated, the kids rushed to the water pump to wash off the vaccine.

Manufacturers responded to complaints about ineffective vaccine by sending out ivory needles that were guaranteed to take, Friedrich said. "I tried one of them last year and it took all right, but, Oh what an arm!" Of the hundreds of children he vaccinated prior to taking over as chief officer, Friedrich said, 27 percent developed severe infections. "Some arms swelled clear down to the wrist joint, with pieces of flesh as big as a silver dollar and twice as thick dropping right out, leaving an ugly, suppurating wound which to heal took from six weeks to three months," he said. Four children died of tetanus. Friedrich's assessment of these deaths cut through the establishment's cant on the subject. While no one had demonstrated the presence of tetanus baccilli in vaccine, "no one can doubt that there is some connection between tetanus and vaccination. The vaccination seems to prepare the soil for the tetanus baccilli by causing suppuration." In any case, with Clevelanders growing suspicious of the health department—hiding cases, jumping quarantine, and shutting their door to inspectors—vaccination had become counterproductive. "So," he concluded, "I dropped it."

Instead, he sent a crew of 40 medical students out to disinfect entire districts where smallpox had been reported, using a gas-driven fumigator that rendered a house uninhabitable for several days. They sprayed all the houses in neighborhoods where smallpox had been—"every room, nook and corner of a house, paying special attention to the winter clothing which had been stored away, presumably full of germs." He put barbed wire and two guards around the smallpox hospital, to make sure people stayed put until they were cured. He shot dogs and cats living in neighborhoods with smallpox cases. On one poignant occasion, he snatched away a dog that a sick child was hiding under the covers. "Almost six months have elapsed since the source has been exterminated from our midst," Friedrich told the doctors. "The death-blow was dealt by formaldehyde."[32]

In the discussion that followed Friedrich's speech, doctors were clearly uncomfortable with its drift. A Dr. J. H. Belt probably summed up the feelings of many when he said that Friedrich's antivaccine edict was "furnishing aid and comfort to the enemy." In a war for the public health this was a serious charge. Friedrich responded that he knew vaccine could eliminate smallpox, "provided we get reliable virus. . . . But I cannot and will not uphold it when done with virus full of pathogenic germs." Although three-quarters of Cleveland's physicians were said to have opposed Friedrich's decision to drop vaccination,[33] his description of the ills of bad vaccine must have been familiar to many of them:

> A man would have to have a heart of stone at the sight of the misery it produces. Visit a happy family with our Pandora gift and make your appearance at the same house two weeks later and you will be horror-stricken. Instead of a smile they will receive you with a curse. The father has been thrown out of employment on account of a sore arm, every child is crying with pain, shrieking as soon as they see you come, the mother frantic with fear that next week the family is going to starve, that some child

may lose an arm or even its life and you stand there and witness the tears and cries and pains and misery of which you have been the cause. The man who can stand all that is no man.[34]

Friedrich's approach to smallpox was not entirely novel; the city of Leicester, England, had become the darling of the antivaccination movement in 1873 when its public health director stopped compulsory vaccination and instituted a brisk sanitation campaign. But in both Leicester and Cleveland, it was tough isolation practices, rather than disinfection, that were effective in defeating smallpox, and the disease would make a ferocious comeback in Cleveland the following spring. Friedrich, once satisfied that the city's vaccine supply was clean, returned to mass vaccination. In fact, formaldehyde gas is of little use against pathogens such as smallpox. Although the disease could occasionally spread through articles such as blankets, it was far more contagious when transmitted directly by a patient's cough. The disinfection of a room and its objects was "burning incense to the false Gods of pre-scientific sanitation," as the Yale public health expert W.-E.A. Winslow put it a few years later.[35] But the antivaccinists trumpeted Friedrich's early success in their propaganda—and ignored his subsequent return to vaccination.

Dissatisfied Parents Get Together

In comparison to Britain, organized antivaccinism arrived late in the United States. In 1879, William Tebb, a leading British anti, arrived in New York to form the first U.S. society, but floundered because homeopaths, struggling for their piece of the medical pie, were reluctant to get involved.[36] "For a 'regular school' physician to declare against vaccination" was to "place himself under a professional boycott," an early opponent of vaccines wrote. In 1882, Henry Bergh, founder of the Society for the Prevention of Cruelty to Animals, published a strongly antivaccine article in the popular *North American Review* in which he denounced Pasteur for "poisoning the flocks and herds of France after the fashion of his predecessor, the notorious Jenner of England, who nearly a century ago commenced inoculating his countrymen with a noxious mucus taken from diseased animals."[37] The vaccine league called on Bergh and had him immediately elected president, but, perhaps feeling it was politically risky to link the two movements, he resigned without explanation a week later.

Despite Bergh's cold feet, the two movements retained strong ties; animal rights arguments were a strong weapon in the arsenal of the anti-vaccinists. Jessica Henderson, the leading Massachusetts opponent of vaccination, was also president of her state's Anti-Vivisection Society. When the movement was reforged at a 1908 conference as the National Anti-Vaccination League, delegates to the founding conference in Philadelphia included Henderson and Josephine Redding, the editor of a column on animal protection at *Vogue* magazine.[38] It was easiest to attack animal experiments in a period of slow medical progress, when the ultimate achievements of laboratory research were generally too far off to be glimpsed. "The vivisectionist, for the sake of doing something that may or may not be useful, does something that certainly is horrible," wrote G. K. Chesterton, "Whether torturing an animal is or is not an immoral thing it is, at least, a dreadful thing."[39]

Later in the 1880s, leadership of the Anti-Vaccination Society passed to Montague Leverson, a Brooklyn "eclectic" physician who claimed to have seen syphilis transmitted by vaccination. Although Leverson was aggressively antivaccination, he brought few members into the group. The 1885 certificate of incorporation lists 15 trustees, most of them alternative practitioners or vendors of patent medicines.[40] In June 1895, the group sent letters to boards of health proposing a controlled experiment in which 5,000 children would be vaccinated and 5,000 unvaccinated, with the health of the two groups checked every year. Letters mailed at the same time urged boards of education to cease compulsory vaccination "until the result of the experiment . . . in your city or state has been ascertained." No one took the group up on its suggestion—which in recent years has been repeated by the vaccine skeptic Barbara Loe Fisher.[41]

Leverson presented model compensation legislation in which a sore arm for more than two weeks, or "the occurrence of disease in any vaccinated person within a period of ten days, for which no other cause can be proved, shall be conclusive evidence that such disease was caused" by vaccination. In such cases, physician and vaccine maker would be liable for injuries, damages, and attorney fees. But the movement attracted few adherents, and those it did attract squabbled, as Leverson noted in his letter of resignation in August 1898. People were failing to grasp that smallpox was "the mildest and safest" of diseases when treated with homeopathy and "hot air baths." Leverson was succeeded by I. H. Piehn, a small-town Iowa banker whose six-year-old daughter had died after vaccination. Piehn damned the procedure passionately and incessantly in conversations with the governor of Iowa, judges, lawyers, merchants, and virtually anyone else he met.

One of his recruits was Reuben Swinbourne Clymer, founder of the esoteric Rosicrucian Fraternity in Bucks County, north of Philadelphia. Clymer, a chiropractor, believed that vaccination caused syphilis, tuberculosis, and leprosy, and made your teeth fall out. He was troubled above all by the threat vaccination posed to a man's vitality, to his blood: "no matter how hygienically a man may live, or how pure his blood, if compelled to submit to vaccination, his blood is thereby rendered impure and his health quite often irretrievably ruined."[42]

Natural disease, Clymer argued, had its place. "Smallpox is not a mysterious visitation to be dealt with or dodged by medical artifices, but a crisis of impurity in the blood induced by foul conditions of life, which cannot better be disposed of than in the course of nature by eruptive fever."[43] Clymer continued to publish mystical health and philosophy books until his death in 1966.[44] A self-proclaimed prophet, and quite a wealthy man, Clymer preferred the quiet penning of books and shepherding of his occult flock to the boisterous rabble-rousing of the antivaccine movement.

In August 1898, in the midst of the Spanish-American War, Piehn lowered the Anti-Vaccination Society's dues from $1 to 25 cents for a life membership, but that failed to rouse the masses despite the fact that "the whole army is being poisoned," and "the murder of schoolchildren once again approaches."[45] The movement's followers were scattering, but the elements of a new, more professionally organized group were starting to take shape. On September 8, 1898, eight-year-old Victoria Field, the daughter of an English-born antivaccinationist named Charles J. Field, was barred from Keystone Public School in central Philadelphia because she was not vaccinated. Field enlisted the services of Oscar Beasley, a well-to-do lawyer and former city councilman, who filed suit. Though it was eventually dismissed by the state supreme court, Beasley himself become convinced of the fallacy of vaccination and joined the fray. He was joined by Porter F. Cope, youngest son of a leading Philadelphia financier, who said he had become suspicious of vaccination after the birth of his first son in 1901, when he dug up a copy of Jenner's book from an antiquariat.[46]

As the 1901–4 smallpox epidemic raged, Cope, Beasley, Field, and others entered into furious combat against compulsory vaccination. They lobbied and protested at City Hall in Philadelphia, at the state capitol in Harrisburg, and wherever parents and school boards fought mandatory vaccination. Cope, a litigious, genealogy-obsessed amateur journalist and art collector,[47] had a gadfly personality. Despite his family wealth, as an antivaccinist he seems to have been cast adrift by his parents, and the movement was chronically short on cash in the early years of the century. But it did not lack for excellent political contacts. Even as smallpox spread around Philadelphia and the state, many politicians joined in opposing compulsive vaccination as unjustified state coercion. Safety concerns obviously fed the cause.

In Philadelphia, 1,159 cases of smallpox were reported in 1901, with 156 deaths, a 13.4 percent fatality rate that reflected the mixture of *V. minor* and traditional smallpox. In 1902, *V. major* struck in Philadelphia and neighboring Delaware County, and an outbreak spreading from Ohio killed 61 people in Pittsburgh. The rest of the state reported plentiful smallpox, but of the mild kind: 59 deaths out of 3,388 additional cases.[48] A typical epidemic hit Auburn, a factory town northwest of Philadelphia. A medical inspector, A. H. Halberstadt, visited the pesthouse that winter and found its shabby inhabitants happy but bored in a tumbledown tenement heated by wood fires:

The nurse met me at the gate and after learning my business, asked if it was her family I wanted to see. On my reply she walked to the door, gave a signal that was followed by a rush of a happy crowd ranging in age from 3 to 50, some with scabs, others marked, and many without evidence of ever

having had the disease. Where the 29 inmates, with nurse, maid and
orderly, were packed, I could not imagine, but the joy and novelty of hav-
ing a visitor made them curious and jubilant and I assure you it was a
scene I shall never forget.[49]

In his 1901 annual report, Lee bitterly attacked the legislature for
failing to authorize emergency funds to fight smallpox, and the people for failing to vaccinate. He took satisfaction that smallpox had
scared the legislature out of repealing the compulsory vaccination
law: "Those mistaken, although well meaning and occasionally intelligent Meddlesome-Matties, the anti-vaccinationists, as usual, sent many
distorted statistics and long petitions to the legislature in the hope of
obtaining the repeal of the [1895] law." Happily for Lee, two children
from Colorado introduced smallpox to a village outside Harrisburg,
and before long the disease appeared in the capital, "creating considerable anxiety among the legislators. The members of the assembly having charge of the matter had the good sense to appreciate that the time
was inopportune for attempting to force the measure, and the bill . . .
never saw the light."[50]

State health inspectors found that it was usually the "native" Americans
who put up the most obstacles to vaccination. The Board of Health frequently received appeals from townspeople warning that smallpox had
appeared among the "ignorant foreigners" in the next town down. But
their investigations "invariably found that it was not the filthy foreigners,
but the spick and span, enlightened Americans among whom the infection
spread. The reason was not far to seek. The foreigners had nearly all been
vaccinated at home in infancy, and most of them were revaccinated under
the laws of the United States when they embarked for this country. This
effectually disposes of the argument that ordinary hygienic conditions and
personal cleanliness are sufficient to prevent the spread of smallpox."[51]

Yet even the state board's officers acknowledged that the vaccine was
faulty, and in its faultiness threatened to invigorate the antivaccine movement. In a July 16, 1903, report, the director for communicable disease,
Dr. George Groff noted that an outbreak near Scranton had been so mild
many doctors had diagnosed it as chickenpox, with no deaths after more

than 200 cases had been diagnosed.[52] The vaccine, on the other hand, was harsh. "In an experience of nearly 30 years and including a large experience in the West Indies, I have never seen such universally bad results following vaccination," Groff wrote. "It would seem that practically all the arms are infected, some of them fearfully so. It is constantly remarked by the people that the vaccination is worse than the prevailing smallpox. . . . Vaccination has received here a very hard blow, for if the people are to be sick for days and even weeks . . . it cannot be expected that this valuable means of preventing a fearful and loathsome disease will be employed as it ought to be."

At that year's meeting of Pennsylvania public health officials, many groused about the publicity generated by Friedrich's work in Cleveland. "He takes to himself very great credit for stamping out smallpox by means of formalin," Welch said. "But from his figures, it is plainly evident that . . . the disease had run itself out. Epidemics do not last forever. They come to an end in Ohio as well as in other states."[53] Welch was right. In May 1902, virulent smallpox returned to Cleveland, and the following month the city's doctors decided to resume vaccination, with vaccine guaranteed by Friedrich's new bacteriology laboratory.[54] About 300,000 Clevelanders submitted to vaccination, apparently mollified by their reformist leaders' vaccine clean-up efforts.

In Philadelphia, however, city officials had taken Friedrich's work to heart. In 1902, the city health bureau's division of disinfection added 36 auxiliary disinfectors and purchased 225 formaldehyde generators with hopes of imitating Cleveland's success.[55] Police officers collected requests for disinfection, and fumigators proceeded to spray the interiors of 5,541 houses—houses in which there had been no smallpox or other contagious disease! Fruitless as this exercise was—except, perhaps, as a fund for graft—"in all cases the utmost satisfaction was expressed by those who availed selves of the service," wrote the doctor in charge of it.

The smallpox epidemic of 1901–4 was tremendously frustrating to public health officials whose self-estimation had been pumped up by sanitarian reform. In 1901, the Pennsylvania legislature had passed bills to regulate plumbing, improve sewers, penalize the adulteration of milk or fruit juices, prevent the marriage of first cousins, prohibit child labor in bakeries, regulate food additives, and provide for safe disposal of sick livestock. Yet just as public health's leading lights threw themselves into the battle against diphtheria and typhoid, hoary old smallpox came tapping on the door like a ghost they were sure they had put to rest. The 3,392 people treated in 1902 at the Municipal Hospital was the highest number on record. In his annual report, Welch noted that he had been forced to vaccinate scarlet fever patients on admission, due to their proximity to smallpox beds in the overcrowded hospital. Even for the rough-and-ready Welch this was a distasteful procedure, "scarcely fair to these persons while suffering from an acute disease."[56] That year the city suffered 292 smallpox deaths, 2,845 deaths by tuberculosis, and 2,976 by "inflammation of the lungs," in addition to 435 by diphtheria, 588 by typhoid, and 143 by scarlet fever. Nearly 6,000 corpses were tumbled into illegal graves scattered among the thickly populated wards of the inner city.[57]

In 1903, a nominally reformist mayor, John Weaver, was elected to replace the notorious Ashbridge. The city health bureau, which had been run by a police colonel, was reorganized under Dr. Edward Martin, who shook things up a bit. During the previous year the U.S. Public Health and Marine Hospital Service had begun its regulation of vaccine. Reform was in the air, and Martin seemed determined to bring Philadelphia along. He demanded weekly reports from his employees and tightened paperwork. He also launched immediate investigations of the city's disgracefully high rates of typhoid fever and smallpox. Martin took a much-publicized walkabout inspection of the Schuylkill River and canal leading from Reading and found they were the dumping ground for every kind of waste.

Seeking to ascertain the efficacy of vaccine, Martin noted that while the city's vaccination corps had performed 207,024 operations in the last six months of the year, the disease continued to spread. He conducted experimental trials of all the "main brands" of vaccine and found that "in some products there were only five successful vaccinations out of 50 operations."[58] Welch's Municipal Hospital was "a focus for the dissemination of the disease," Martin added, because infected patients often left the hospital and hopped on a streetcar to the nearest flophouse without anyone stopping them. Martin named a new director of the hospital and demoted Welch to diagnostician. The notorious hospital was shuttered in 1909 after the Philadelphia Athletics bought the land to build Shibe Park.

Tetanus Investigation

In his year-end report in 1902, Lee had wrapped up the state's "investigation" of the tetanus deaths with a one-paged summary. But while the establishment washed its hands of the disaster, a young, European-educated bacteriologist and physician, Joseph McFarland, was conducting his own private investigation. McFarland was a scientist of some renown, having produced the first commercial antidiphtheria serum in the United States for H. K. Mulford in 1894. Around 1900 he left Mulford to teach bacteriology at the Medico-Chirurgical College in Philadelphia, while consulting with Parke-Davis, a Mulford competitor. McFarland helped Parke-Davis produce antitetanus serum, and it was while attempting to save victims with his serum that he became intimately aware of the problem with smallpox vaccine.

The most damning piece of evidence, for McFarland, came from Philadelphia Hospital. A smallpox outbreak at the hospital led doctors there to begin vaccinating all 4,500 patients. They ran short of vaccine, however, and a lot of fresh vaccine was used to vaccinate the mental ward. Within a few weeks, five of the unfortunate men on the ward had died of tetanus, while 11 others had fallen sick and were saved only by McFarland's administration of "enormous doses of antitoxin." This episode induced McFarland to examine 84 other recent cases of postvaccine tetanus.[59] Out of 40 cases in which it was clear which type of vaccine was used, McFarland reported, 30 of the vaccines were from the same manufacturer—which he called "E"—whose vaccine had caused the hospital tetanus outbreak. This product accounted for the deaths in Cleveland, Camden, and Atlantic City, and several in Philadelphia. Contemporaries presumed that vaccine virus stored in glycerine—either on ivory needles or points, or stored in tiny tubes—was the safest form of vaccine. But McFarland found this wasn't so. Whether sold on glycerinized points and tubes, or dry, the manufacturer's vaccine caused tetanus. The pattern, he said, "leads me to conclude that the tetanus bacilli may be contained in the virus and distributed with it."

The outbreak, McFarland said, "seemed a matter of the greatest importance." The deaths in Camden alone, he said, "have been made the subject of so many editorials and comments . . . that they will no doubt form the starting point of numerous future attacks against vaccination." McFarland did not accept the received medical wisdom that the tetanus was caused by secondary infections or weather conditions. As for the argument that "ignorant and filthy children" were to blame, he found that "extraordinary care" of the vaccine wound had been exercised in most cases, and indeed one of the victims was an adult sister of a physician in Cleveland—"refined and cultured people."

Eventually, McFarland and other investigators determined that the addition of glycerine to vaccine, begun in the 1890, had lured manufacturers and their customers into a false sense of security. The intense demand for vaccine during the 1901–4 epidemic had inspired vaccine producers to take shortcuts. Rushing their product to market, they hadn't allowed the fresh bovine vaccine to sit long enough in glycerine, which in any case was but a "feeble" bactericide.[60] "After the manufacturers heard that glycerine killed bacteria," one vaccine maker said, "they sold all the dirty stuff they could gather up, depending on the glycerine to save them."[61] The overly "green" vaccine they sold had contained tetanus spores,[62] which the Hygienic Laboratory eventually was able to detect in the linen shields that vaccine makers sold to protect the sore while it scarred over.[63] Tetanus typically requires only about a week to incubate, while most of the tetanus deaths followed vaccination by three weeks. McFarland hypothesized that the bacteria in the vaccine had remained dormant "until after the development of the vaccine lesion"—generally the eighth or ninth day following vaccination—which provided damaged tissue in which the tetanus could easily grow.

Despite his resourceful detective work, McFarland refrained from revealing the identity of "E" in public. His papers at the College of Physicians in Philadelphia confirm what was assumed by many at the time: "E" was H. K. Mulford, his former firm. Vaccines were a cutthroat business, and McFarland's inquiry stirred up vicious recriminations, particularly between Mulford and H. M. Alexander of Marietta, Pennsylvania, as he probably knew it would. Alexander, a doctor from a well-off farming family, was apparently the first commercial vaccine maker in Pennsylvania;

his establishment took top honors at the 1893 World's Fair in Chicago. Mulford, an itinerant pharmaceuticals peddler, had started his competing farm after frustrated attempts to buy into Alexander's operation.[64]

In their corresondence with McFarland, each company accused the other of culpability for the disaster, and each expressed shock there hadn't been more tetanus in the past, "given how little care is generally used in the process." Each urged McFarland to keep quiet—with the "good of the cause of Vaccine at heart."[65] Mulford in particular responded to accusations with the blunt intimidation of a tough PR firm. McFarland had been "unwise to make statements" that linked vaccination to tetanus, "in view of the fact that the entire country is threatened with an epidemic of smallpox." He must "overcome the tendency on the part of any physician or customer to believe that the vaccine is in any away responsible for the recent cases of tetanus . . . Every drop of our vaccine is tested and is free from germs of all kinds. The highest medical authorities are unanimous in the statement that when tetanus occurs after vaccination it is always due to ignorance or neglect." This brazen letter, with a copy of the Camden Board of Health report enclosed, was apparently sent out to all physicians and pharmacists who purchased the company's products.[66]

The boards of health and pharmaceutical companies were not the only ones to engage in the cover-up. Private physicians, even some who treated patients sickened from tetanus, did their part to hush it up. A Cincinnati physician whose wealthy patient died of tetanus after vaccination responded to an inquiry from Parke-Davis (which was apparently trying to besmirch its rival Mulford) by urging it to "pay no attention to the case. Agitation can only injure the cause of vaccination . . . our [news]papers are considered and sensible enough to suppress any reference to vaccination. Drop the matter. It is dead here." A lawyer examining the case decried a "Masonic spirit" among Cincinnati physicians who suppressed information on the case.[67]

The tetanus deaths were just part of a depressing picture of vaccine impurity. An ad for Alexander's claimed that a study "prompted by the unusually large number of dangerous results reported from vaccination during the present smallpox epidemic in Chicago" had shown that only Alexander's vaccine lacked "pus bacteria" and "saprophytic bacteria."[68] In its propaganda, meanwhile, Parke-Davis boasted that "we are the only propagators who (1) test their vaccine physiologically to determine its activity, and (2) examine it bacteriologically to insure its absolute freedom from any trace of harmful micro-organisms." McFarland's probe showed that science, when applied by a shrewd mind, was up to the task of determining the condition of vaccine. But for the most part, the public health administrators of the day lacked the resources, nerve, or philosophy to take on the doctors and drug manufacturers. The Hygienic Laboratory beefed up its regulation in 1906, demanding that each vial of vaccine be inspected for tetanus and other bacteria, but it had no way to enforce the requirement.[69]

There were a few exceptional individuals who recognized the contradictions of compulsory vaccination and took steps to deal with it. In Rhode Island, Charles Chapin, a leading public health official urged his state to relax vaccination requirements.[70] "What is feasible in one community may not be feasible in another," he said in 1907. "A compulsory vaccination law is a good thing in Germany," he said, because of that country's efficient bureaucracy and obedient citizens, but "they can do nothing of the kind in England. We can do nothing of the kind in Providence, and it would be folly to attempt it."[71] In Massachusetts, the bacteriologist Theobald Smith argued that protection of the vaccine could only be guaranteed if the state supervised the preparation of lymph, and in 1903 the state laboratory he ran began producing smallpox vaccine.[72] So did labs in Michigan, Texas, Illinois, and other states. But most of the public health world ignored McFarland's investigation, and McFarland himself backed off a few years later, when he became active debating the antivaccinists.[73]

The newly empowered Hygienic Laboratory was not shy about using its limited muscle to get rid of shoddy firms and products. Its first wave of inspections led several companies to immediately quit the business, "being unable to comply with modern requirements of asepsis and laboratory control," as one of the lab's scientists noted. Eight companies were licensed to continue producing vaccine. They included Mulford (which later merged with Sharpe and Dohme, and then Merck), Alexander's (taken over by Wyeth in 1942), and Richard Slee's Pocono Laboratories, in Swiftwater, Pennsylvania, an establishment that would be operated over the next century by Merrill-National, the Salk Institute,

Connaught, and finally Aventis-Pasteur (now Sanofi). The other licensed vaccine makers were Parke-Davis and Frederick Stearns of Michigan, the National Vaccine Establishment in Bethesda, Maryland, Fluid Vaccine in Milwaukee, and Cutter Analytic Laboratories in San Francisco. Others, including the New York Bureau of Health, were required to make modifications before resuming production. Federal scientists found that the bacterial colonies in the vaccine were much reduced by 1904.

Despite the examination of enough vaccinia to vaccinate 2 million people from 1902 to 1915, the Hygienic Laboratory never found tetanus germs in the vaccine itself.[74] But as one vaccine critic put it, "what difference does it make to the victim of this terrible fatal disease whether the lockjaw germ was in the virus itself or on the victim's body?" To absolve vaccination in either case would be tantamount to "free[ing] the assassins of Presidents Garfield and McKinley from the charge of murder because, forsooth, the fatal septic infection which killed these victims was not on the murderous bullets when they were fired into their bodies but entered the wounds afterward through their own carelessness."[75]

Progress in regulating vaccine would be slow, as government and industry alike contended with the impossibility of making sterile products with inadequate technology. But the regulations imposed by the Hygienic Laboratory, which was nominally in the Treasury Department, were tougher, in principle at least, than the Department of Agriculture's restrictions on food producers. The vaccination companies pressed the limits and claimed that regulations justified prices far higher than what government-run institutes in Europe were charging their physicians. Skeptical federal regulators proposed producing their own vaccines and diphtheria sera, but pharmaceutical companies effectively fought a law that would have created a government vaccine facility.[76] To forestall investigations of collusion and price gouging, they distributed vaccine and antitoxin at lower cost to state public health agencies for the poor. This material was often of shoddy quality or sold past its expiration date.[77]

Getting Tough with Resisters

In 1904, the Pennsylvania Board of Health issued its own vaccine regulations, committing itself to annual visits to all farms and issuing detailed guidelines.[78] These measures had little effect on the antivaccination movement. Even if vaccine was getting safer, opponents objected to the hamfisted coercion that the authorities used to enforce vaccination, especially among communities of color and immigrants. Forced vaccination was common. At an 1899 meeting of the American Medical Association, Dr. Ravold, the St. Louis bacteriologist, bragged about those he had performed. After an African American man walked into a city clinic with smallpox, the Health Department "vaccinated the whole male negro population of the city, and as many women as could be captured." They raided "barrel houses . . . low, filthy saloons [where black laborers] sleep on the floor," and forcedly vaccinated scores at a time. "It was only in this high-handed manner that the Health Department finally stamped out the disease among the negroes," said Ravold.[79]

U.S. and Texas health officials showed similar arrogance toward Mexican citizens attempting to enter the United States in the early decades of the century. Border officials were instructed to vaccinate anyone who didn't show a recent vaccination scar or evidence of smallpox. Since smallpox scars were not always visible, the general rule was vaccinate first, ask questions later. The policy infuriated Mexicans because it implied they weren't modern or clean, when in fact they were better vaccinated than Americans. In 1932, members of the Mexican Olympic Team were vaccinated against their will in El Paso on their way to the summer games in Los Angeles. When the indignant Mexicans threatened to impose the same quarantine on Americans, U.S. consular officials were shocked at the prospect of "respectable American ladies" being physically handled by Mexican soldiers.[80]

Forced mass vaccination of immigrants was common in Philadelphia. In 1903, a witness saw doctors and policemen in a patrol wagon brace hundreds of Polish, Italian, and African American laborers at work on the water filtration system. Hired to help the city build its defenses against typhoid and other diseases of the gut, they suddenly found themselves forcibly joining a smallpox eradication campaign. A group of Italian men, enjoying their lunch in the middle of a meadow, jumped up and bolted when they saw a doctor with his satchel approaching. One man was vaccinated after being yanked out of a 40-inch water pipe by his heels.[81]

There was a eugenicist element to the public health response. The

Public Health Service, in the early years of the twentieth century, was responsible both for protecting the general health and for inspecting foreigners for infectious and hereditary diseases. Its work fit well with the beliefs of many American eugenicists that hygiene and sanitation had eased the rigors of natural selection and allowed more weakened specimens to prevail. Groups such as the Race Betterment Foundation mixed genetic concerns with preventive health measures, and eugenicists who advocated thinning out the unfit often won the support of public health officials. For example, when from 1915–18 Chicago surgeon Harry Haiselden permitted at least six infants with conditions such as spina bifida and hydrocephalus to die after diagnosing them as defectives, he won the support of New York bacteriologist William H. Park, FDA founder Harvey Wiley, and other prominent men. Heredity and contagion were linked in public health lingo. Both germs and germ plasm "enabled disease to propagate and grow," wrote the historian Martin Pernick. "[B]lood, the age-old heredity metaphor, became identified as a vehicle for infection as well." When Congress imposed ethnic restrictions on immigration in 1924, it drew upon earlier restrictions adopted to fight infectious disease. And when the Supreme Court ruled in *Buck v. Bell*, the infamous case allowing sterilization of the "unfit," it drew heavily upon *Jacobson v. Massachusetts*, its 1905 decision backing compulsory vaccination.

Henning Jacobson had challenged a vaccination ordinance passed by the Cambridge Board of Health in 1902, as well as the state's compulsory vaccination law. A minister of the Swedish Evangelical Lutheran Church, Jacobson claimed his family had been harmed by previous vaccinations and refused the procedure and the $5 fine. He argued that the town and state laws conflicted with "the inherent right of every free man to care for his own body and health in such a way as to him seems best." But the U.S. Supreme Court ruled that while the state could not require vaccination in order to protect an individual, it could do so to protect the public from an epidemic disease. The Court declared at the same time, however, that vaccination could not be physically forced upon anyone. In *Buck v. Bell*, the case that prompted Justice Oliver Wendell Holmes's notorious comment that "three generations of imbeciles is enough," Holmes cited *Jacobson* in concluding that "the principle that sustains compulsory vaccination is broad enough to cover cutting the fallopian." In sterilization as in vaccination, he found, preventing disease was better than coping with its consequences, and the collective well-being of society outweighed individual choice. Furthermore, the Court found that state power could compel compliance when persuasion was inadequate. Both public health and eugenics, in this instance, set their sights on "final solutions" for eradicating infectious and hereditary diseases.[82]

The smarter antivaccinationists picked up on Progressive-era public health's intolerant streak. "A bull in a china shop is a gentle, constructive creature compared with a lot of prim and more or less pious folks when they start in to clean up society and the world," wrote the activist Lora Little. "Mr. Sudden Reformer sees something he does not like in some of his fellow citizens. Very likely it is a reprehensible thing. Plenty of evils exist in the lives and habits of all classes. This would be a thing of which Mr. Sudden Reformer is not himself guilty, therefore he hates it with a mighty loathing. Dwelling on it, he works himself into a frenzy. He would suppress, eradicate, exterminate and stamp out that evil instantly."[83]

Artists and intellectuals were prone to challenge vaccination as unintended consequences continued to dog the procedure. In 1915, Marcella Gruelle, daughter of the New York City illustrator Johnny Gruelle, became paralyzed soon after a vaccination was administered at school without her parents' permission, and later died. Gruelle believed fervently that the vaccination had killed his daughter, although the medical record blamed a heart defect. He created a special cloth rag doll for her during her illness, a floppy doll with hair fashioned from red yarn. Gruelle called it Raggedy Ann. The doll, with its limp limbs, became a symbol of vaccine-damaged children, and Marcella was the heroine of the Raggedy Ann stories that Gruelle went on to illustrate.[84] The editor of *Life* magazine, John Mitchell, was so strongly antivaccine that he published a prayer in the magazine to the effect "that our children may in future be born immune from all diseases of the kinds for which toxins and serums are injected in their blood—most especially, dear lord, smallpox, for the supposed prevention of which the ancient, useless, dangerous and filthy rite of vaccination is performed."[85]

Even workers at the vaccine plants seem to have been skeptical. In late December 1915, three employees of H. K. Mulford Co. called on

Porter Cope at his Philadelphia office, asking if he could help them keep their jobs if they refused to be revaccinated. The three "stated that intense hostility to vaccination prevailed" among the company's 1,800 employees. "The firm's paymaster had declared he would forfeit his position rather than submit. His arm is not yet healed since he was last vaccinated."[86]

A Shot in the Arm for the Antivaccine Movement

Despite the unpopularity of vaccination, the movement was not well-funded until 1906, when it found a deep-pocketed benefactor. John Pitcairn, a Scottish-born oil and steel millionaire, was a self-made man who wore a thick white mustache and a long white beard that made him look a bit like a dyspeptic Colonel Sanders. Pitcairn had emigrated to Pittsburgh as a boy in 1846 and worked his way up in the railroads. In his crowning moment, he was entrusted with the train that carried Lincoln from Harrisburg to Philadelphia on his way to the 1861 inaugural. Pitcairn invested first in oil and eventually built an empire under his Pittsburgh Plate Glass Co., which made paint and glass supplies and employed 7,000 people with a capital of $22.7 million by 1915. A parsimonious junior robber baron without the ambitious social vision of a Rockefeller or Carnegie, Pitcairn focused most of his resources on Bryn Athyn, the Swedenborgian church and community he founded northeast of Philadelphia. While vaguely antivaccine beliefs were as easy to come by as mother's milk in this era, Pitcairn's antivaccinist fervor was particular to his family and origins. His oldest surviving child, Raymond, apparently suffered blood poisoning after vaccination as an infant in 1885.[87]

The Swedenborgians, a mystical early-nineteenth-century denomination, believed that the contamination of the body left a blot on the soul as well, and it was an easy step for church thinkers to link vaccination to sin.[88] Vaccination, offering a quick fix against smallpox, was the medical equivalent of spiritual quackery, said one minister, the Rev. J. F. Potts: "Come here, says the spiritual quack, and I will make you safe. I will safeguard you against evil. In a moment, with a word, just one prick beneath your moral skin, and the great hocus-pocus is accomplished. But none of this spiritual juggling for us! We have no faith in it! We have no faith in any salvation but that which is the result of living according to heavenly laws and divine order." Vaccine matter, like smallpox boils, was "infernal excrement," Potts said, and vaccinating a child was "introducing the foul corruptions of hell into innocent life."[89]

Many early homeopaths converted to Swedenborgianism, and vice versa; the first homeopathic medical college in America was established in Allentown.[90] Pitcairn carried around a wallet-sized kit of extracts so he could concoct homeopathic treatments for various ills. He was skeptical of vaccination's disease-fighting power, but what really bothered him was the notion that it introduced impurities into the blood. Though Pitcairn's beliefs were in some ways specific to his church, they were also characteristic of the vitalism that influenced a sector of the intelligentsia in the nineteenth century—a belief in the indivisibility of mind and body that countered the mechanical view of life in a rapidly industrializing era. Pitcairn broadcast his views in a variety of forums and was eventually invited, with Cope, to join a five-member commission that in 1913 reviewed the safety and efficacy of vaccination for the state of Pennsylvania. "Even supposing the possibility that by vaccination many times repeated one may eventually become immune—that is, provided he does not die from the process," he once wrote, "who would envy him the condition of his blood? He is 'immune' by virtue of blood poisoning."[91]

Pitcairn was drawn to the fight by an incident in his community. In January 1906, a vaccinated teacher at the Bryn Athyn Academy returned from Canada after the holidays, ill with what appears to have been a case of *Variola minor*. The resident physician informed the state Board of Health, which demanded quarantine of all who declined vaccination. Charles S. Smith, a nonvaccinated Bryn Athynite whose business was in Philadelphia, was threatened with prosecution if he continued to take the streetcar into town. Eventually, 115 community members submitted to vaccination, but 171 others refused. In Pitcairn's view the episode exemplified all that was wrong with the compulsory vaccination system.[92] The outbreak had begun in a vaccinated woman, yet it spread to only three other people in a largely unvaccinated population. None of these illnesses was life-threatening, but the vaccine itself caused swollen arms and sickness. What's more, Philadelphia's public health authorities had used intimidation and blackmail to force religious people to accept a practice that controverted their beliefs. In doing so, they energized a formidable enemy. Porter Cope, who was always on the qui vive for potential recruits and cause celebres, wrote to Smith after reading about his threatened prosecution in the February 18 edition of the *Public Ledger*. A meeting with Pitcairn sealed the latter's participation. Pitcairn donated nearly $100,000 to the cause before his death in 1916, upon which he left an additional $10,000. His sons Raymond and Harold, the latter a well-known aeronautics engineer, helped finance the movement well into the 1940s through the New York-based Citizens Medical Reference Bureau.

By the time John Pitcairn got involved, most of the smallpox in the United States was *V. minor*. Of the 15,223 cases recorded in 1906, only 90 were fatal. Philadelphia reported only eight smallpox cases and not a single death, in a year in which 10,000 fell ill, and more than 1,000 died, of typhoid fever—and a measles epidemic killed 344. Smallpox began to seem like less of a menace than the means of preventing it. In Philadephia, Pitcairn sponsored a public meeting on May 16, 1906, that was attended by about 300 people. Cope presented a historical slide show account of vaccination's horrors. Pitcairn had invited people of opposing viewpoints to speak, and the person who rose to challenge Cope was an old schoolmate—Dr. McFarland of the Medico-Chirurgical College. McFarland greeted him, but proceeded to "cast insinuations of ignorance and unfitness to invade the medical field." He then "launched into a denial that tetanus frequently follows vaccination, though tetanus had not been touched upon by Mr. Cope."[93] Not even McFarland, who had done more than the antis to investigate the hazards of vaccination, was willing to cede them any ground.

In September, with Pitcairn's financial backing, Cope, Field, and Beasley created the Anti-Vaccination League of Pennsylvania, just in time for an uprising against vaccination in Erie, where parents withdrew 4,000 children from public schools rather than have them undergo the procedure. Seeing the antivaccinators as a potential voting bloc, the Republican gubernatorial candidate, Edwin Stuart, met with Cope and Pitcairn at a Pittsburgh hotel that month and promised to give their cause a fair hearing if he was elected.[94] With Cope and Beasley as his agents, Pitcairn directed the organization of a new, national antivaccine society. In June, Cope sent him a 10-paged declaration of principles based on those of the American Anti-Slavery Society and an action plan calling for brisk organization of supporters and lobbying. An enemies list included Dr. William L. Elgin, director of H. K. Mulford Company's laboratory and a Democratic candidate for the state legislature.

Stuart was elected but turned out to be an enormous disappointment to Cope and Pitcairn. They successfully lobbied for a bill against compulsory vaccination in 1907, but after it passed 133–9 in the House and 27–11 in the Senate, Stuart vetoed the bill. He vetoed a similar bill two years later, presumably because of the influence of the pharmaceutical industry.

Having lost the battle in their state, the Philadelphia antivaccinists dedicated themselves to the creation of a national league, with a founding conference at Griffith Hall in Philadelphia in October 1908 that drew about 50 well-known antivaccinists. They included antivivisectionists and journalists, state legislators, homeopaths, naturopaths, parents of vaccine-damaged children, and attorneys battling vaccine manufacturers and state laws. William Lloyd Garrison, son of the famous antislavery leader, attended, along with the Rev. Jacobson. Bernarr Macfadden, the Battle Creek, Michigan-based editor of *Physical Culture*, a pioneering alternative health magazine that featured articles on spas, high colonics, weight-lifting, and the nostrums sold by its sponsors, turned up for the conference. (MacFadden had changed the spelling of his name to give it a more masculine, leonine sound—Bernarrr!) So did J. W. Hodge, whose father, John Hodge, had made a fortune selling Merchant's Gargling Oil, a mixture of ammonia, soap, tincture of iodine, benzene, and crude petroleum. The younger Hodge turned Niagara Falls into a vaccine-free bastion of the opposition.

It was the sort of occasion where Rudolph Straube, a homeopath, could stand up and invite Pennsylvania's state health commissioner to "occupy the same bed with me, with a smallpox patient lying between us. I will stake my normal health against your vaccine scars in defying smallpox." After considering the insertion of the word "compulsory" into the group's title, the conferees settled on the Anti-Vaccination League of America. Its main objective, they said, was "to promote universal acceptance of the principle that health is nature's greatest safeguard against disease and that therefore no State has the right to demand of anyone the impairment of his or her health." They also agreed, by means of education and self-help, to "take steps to abolish oppressive medical laws and counteract the growing tendency to enlarge the scope of state medicine at the expense of the freedom of the individual."

The historian Paul Starr has argued that alternative medical systems are generally rehashes of mainstream opinions from bygone eras. Many of the antivaccine ideas about smallpox could certainly be placed in this category. Generally inhospitable to germ theory, the critics subscribed to a variety of earlier beliefs. Smallpox, MacFadden stated, with arguments that could have been drawn from the seventeenth-century physician Sydenham, was "easily curable" and only caught by "those who clothe heavily, bathe infrequently, eat very heartily and exercise rarely." Vaccine *caused* smallpox because it lessened one's "vital strength and power to resist internal inflammatory diseases." Hodge rehashed the arguments of Alfred Russel Wallace: vaccination didn't work; it was old-fashioned and unsanitary. Vaccine's origins were unknown, it weakened the immune system. The opponents of vaccination emphasized individual responsibility and control of one's own body. Healthy people didn't get smallpox, they said—a "state of perfect health . . . resists and repels the assaults of all morbific influences." This was a circular argument, since falling sick was always proof of one's imperfect state of health.

The antivaccinists were fueled by a spirit of rebelliousness, a sense of being right and outside the law. One antivaccinist at the conference stated that the "medical craft of today has usurped the place occupied by the church fathers of former times and generations. The doctors of today are asking as drastic powers as were ever possessed by a Cotton Mather or a Torquemada, and the courts now as then are as willing to turn us over to the tender mercies of these tyrants."[95]

The most charismatic spokesperson for the new movement was Lora C. Little. A feisty opponent of the medical establishment, Little was a sort of granola-belt Mother Jones who promoted whole foods and naturopathy and denounced white sugar and white male medical practitioners before it was fashionable to do so.[96] In many ways she was the prototype of antivaccine and alternative health activists today. Born somewhere in the Midwest, she suffered from poor health—and poverty, based on her appeals for cash from Pitcairn—much of her life. She traveled constantly, agitating against state-run health programs; one scholar traced her to Berkeley, California, in the 1920s, where her trail ran cold. Like Benjamin Rush, Little rejected "specific diagnoses for specific illnesses since the same general principle underlies the cure of all forms of disease."[97]

The tragedy that spurred Little's fight was the death of her seven-month-old son Kenneth in 1896, when Little was 39. Her son died of "measles and diphtheria," but she attributed his ill health to a smallpox vaccination. Little set herself apart with a thoroughly Luddite philosophy in which vaccination was only one of the many health ills brought about by mechanization and state control. Hers was a classic paranoid critique; she anticipated the creation of a state public health agency that would "have charge of vaccination, antitoxin, anti–bubonic plague serum and all the other inoculations relieving school authorities and individuals of all responsibility in the matter." What was needed, Little wrote Pitcairn prior to the 1908 founding conference, was "a National Health Defense League. Otherwise we win upon vaccination only to find ourselves bound hand and foot and subject to other medical rites as foul and superstitious as Jenner's." In 1911, after arriving in Portland, Little posted an ad for her services in the local newspaper that read like a flyer at your local health food store today: "Be your own doctor," it said. "Run your own machine. You can do it better than another, being inside it."[98]

Little was an indefatigable activist. In 1918, she was arrested in North Dakota under the Espionage Act for trying to convince soldiers to refuse vaccination. She led successful campaigns against mandatory vaccination in North Dakota, as well as Minnesota and Chicago, and a statewide movement in 1916 to abolish compulsory vaccination in Oregon that won 99,000 votes—falling just 374 votes short of a majority. She also wrote and published pamphlets such as *The Baby and the Medical Machine*, in which she described the fate of a neighbor who, during a visit to New York, was separated from her two-year-old boy when the toddler fell ill with measles. The boy was put in quarantine with one nurse and 25 other children, and died. Such brutal, heartbreaking practices were common at the time, and not specifically tied to smallpox vaccination, but for Little they were of a piece. In cities like New York, she wrote, "the private physician no longer exists. Every doctor there has become a cog in the medical machine. And once the machine gets its grip on you, you cannot escape, you are drawn in and ground through the mill."

Arguments like these—and the growing prevalence of *V. minor*—were having success in blocking compulsory vaccination. A year after the founding of the Anti-Vaccination League, Charles Higgins, a wealthy Brooklyn activist, gave a breakout on the status of vaccination law around the United States. About two-thirds of the states had no compulsory vaccination laws. West Virginia and Utah—where the Mormon church was skeptical of vaccination—had laws prohibiting compulsion. In most states with compulsory laws, Higgins reported, appeals had been brought to the highest state courts to invalidate the laws, and, in the case of *Jacobson*, to the U.S. Supreme Court. But in most cases "the higher court sustained the state laws—not on the merits of vaccination, but on the mere right of the states to make the laws."[99] Still, in April 1908, the highest court in Illinois declared it illegal to exclude unvaccinated children from school, and governments in California, Minnesota, Wisconsin, and Indiana abolished compulsory vaccination. In New York, compulsory vaccination was the law but enforcement relied on local health boards, which were reluctant to ruffle the feathers of school boards that generally hated to have to vaccinate. A 1915 survey found that few health officers had the courage to force vaccination or quarantine when the public opposed it. In Kentucky, fewer than half of the people had ever been vaccinated, with 10 percent vaccination rates in some counties.[100]

To boost the antivaccination cause, Pitcairn arranged for the *Ladies' Home Journal* to run essays in favor and against vaccination in 1910. Pitcairn himself prepared one of the essays and the other was written by Dr. Jay Frank Schamberg, an infectious disease specialist who had succeeded Welch at Philadelphia's Municipal Hospital. The two men, who would serve on the state vaccination commission together, appeared to have a cordial relationship, and their viewpoints were moderate by the standards of the time. But Schamberg, chosen by the AMA's Council on Defense of Research to defend vaccination, was far more convincing— because unlike most doctors, he did not claim that vaccination was always effective. Conceding that the procedure was not devoid of risk, Schamberg agreed with Pitcairn that compulsory vaccination laws were largely counterproductive. "People have a natural antipathy to coercive measures," he said. But he stressed that evidence made no dent on his foes—a statement that would still hold nearly a century later. "The opponents of vaccination first emphasized their aversion to coercion, then sought to prove that vaccination did not protect against smallpox, and finally, that it was injurious," Schamberg wrote. "The opponents of vaccine . . . are on a constant search for ammunition. They accept as true every alleged accident after vaccination as resulting therefrom, but assiduously shut their eyes to the evidence offered of the efficacy of vaccination."[101]

Even as Pitcairn debated vaccination with Schamberg in the pages of the national press, he was faced with a more pointed and less theoretical vaccination question inside his own business empire. In the fall of 1910, *Variola major* broke out in Michigan near two sugar beet factories owned by Pitcairn. As the state health department tightened vaccination requirements, Pitcairn corresponded on the matter with Charles W. Brown, who ran his Owosso Sugar Co. On November 21, Pitcairn urged Brown not to take the initiative in vaccinating the employees, but did not forbid it. "Vaccination would probably cause considerable illness and interfere with the operation of our plants," he wrote. Brown replied that while he agreed in principle, the local board of health had threatened to shut the plants at the first sign of smallpox, which had already killed 40 people. Cope, who assumed Pitcairn's correspondence at this point, wrote that the epidemic was probably caused rather than prevented by smallpox vaccine; Michigan law did not require Owosso to vaccinate its workers until smallpox appeared at the factory. Brown replied that while "personally I do not believe in vaccination," the company had "several thousand dollars'" worth of sugar beets that were likely to rot on the ground if the company's operations were halted. "As a question of business policy it is in my opinion decidedly unwise for us to antagonize the health authorities." While a legal challenge to quarantine might triumph eventually, Brown wrote, "I believe that the beets would rot faster than the courts would act."[102] Smallpox killed at least 86 people out of 373 Michiganders who contracted it that season. Luckily for Owosso, Pitcairn, and Charles Brown, the epidemic never reached the plant.

Over the next few years, Cope's group busily gathered evidence of the vaccine's harm to present to the Pennsylvania State Vaccination Commission. The activists compiled a 183-paged dossier of vaccine injuries and deaths gathered from interviews in poor neighborhoods of Philadelphia and its suburbs, and it provides a vivid picture of working-class life. There are reports of midnight raids by doctors and police who burst through doors in search of virgin arms; of amputated legs and dead babies; of seven-year-olds vaccinated for school and then struck down by lockjaw; of corrupt and incompetent vaccinators and medical inspectors; of desperate workingmen seeking to dodge the lancet so they wouldn't have to forgo work on account of a sore arm.[103] One report tells of Dr. A. A. Cairns of the municipal vaccination squad, who raided a tenement house in the Laboratory Hill section and vaccinated two of John McFarland's children. McFarland (no relation to Joseph) himself fled, saying, "You won't put your rotten stuff in me." For a few days McFarland hid out with a friend, and a storekeeper brought him food. "Where's Jack?" the vaccinators asked his wife. "He's out," she replied. "Jack's a socialist, isn't he?" a doctor asked. "Yes he is." The doctor replied, "We'll get him and make a good socialist out of him." One workman told of waking up in his boardinghouse to discover that he had been forcibly vaccinated—for the third time in three months.

In their 1913 report, the commissioners came out in favor of continued vaccination by a 3–2 margin, with Cope and Pitcairn in dissent. While the antivaccine movement continued to capture a great deal of public sympathy, by the late teens the public health establishment was beginning to gain more respect. World War I helped to link patriotism and public health in the public mind and inspired greater fear, if not respect, for government mandates. And while in many localities few efforts were made to vaccinate, the *Jacobson* ruling enabled state authorities to respond to epidemics with hard-nosed tactics.

"Radical Measures" at the Falls

Vaccination was infrequent in northwestern New York, whose Chautauquas and spas were havens for utopian lifestyles and unorthodox medical practitioners. Leading antivaccinist J. W. Hodge claimed that his opposition to vaccination began during an epidemic that swept through the area in 1882, when he was practicing medicine in Lockport. After vaccinating about 3,200 people, Hodge said, he was "confronted with sundry cases of smallpox which had 'broken out' . . . in duly 'protected' subjects upon whom I had recently operated." Furthermore, parents claimed that the vaccinations had actually caused smallpox. "I have never vaccinated another victim since that memorable event."[104] Hodge moved to Niagara Falls and his propaganda soon won over the town, including its board of education, the city attorney, and two newspapers. "During the last 20 years, as far as one can learn, practically no vaccination has been performed in Niagara Falls," wrote Linsly R. Williams, the deputy state health commissioner, in his account of the 1914 epidemic in the bustling town of 32,000.

The first smallpox case appeared in April 1912, but it was not until December 14, 1913, that the town's health officer, Ed Gillick, asked the state Board of Health for help, after authorities in Buffalo and Niagara Falls, Ontario, threatened to blockade the town. On January 3, 1914, the board sent a letter to "leading citizens" and manufacturers urging them to persuade employees to be vaccinated. It didn't work—in one factory where smallpox occurred, the management offered free vaccination or a two-week layoff without pay, yet remarkably, only 7 of 97 employees accepted vaccination. While Hodge's claims about the hazards of vaccination were surely exaggerated, the Niagara Falls outbreak was clearly *V. minor*. Philip Wagner, a frail 55-year-old, was the only fatality reported out of 550 cases. The stress on sanitation and healthy living that was Hodge's credo had to have hit home in a population where other risks, unchecked by the authorities, were so routine. Diphtheria, measles, childhood diarrheas, and tuberculosis were common in Niagara Falls, and for the working class, occupational hazards dwarfed the threat of smallpox. During one five-week period at the heart of the epidemic, a 15-year-old boy died after a machine at an iron foundry threw white-hot metal into his left eye, two men were run over by trains, and a worker at a knife factory plunged into a vat of scalding water.

From Niagara Falls, smallpox spread to six other counties, setting off epidemics in towns such as Tonawanda, Villenova, and East Holland, where it spread at a Holy Rollers convention. Waves of compulsory vaccination orders and antivaccination agitation swept through the area, with thousands of parents withdrawing their children from schools. Gillick, who repeatedly assured the public that the epidemic was over even as it continued to grow, finally let down his hair on January 23, telling reporters that it was the antivaccinationists who caused the epidemic. Not more than 10 percent of the town had been vaccinated, Gillick said—and even the town notables had been a tough sell.[105]

A week earlier, Hermann Biggs, New York City's chief microbe fighter for the past 22 years, had been named state health commissioner. He wasted little time lowering the boom on Niagara Falls, ordering "radical measures" in a January 28 letter to health officials at the Falls—"people of other portions of the state," he thundered, "must be protected against the follies of any local community at whatever cost." He demanded strict quarantine, closing public facilities to the unvaccinated, and the printing of notices titled "Smallpox at Niagara Falls." Unless they quickly complied with his demands, Biggs warned, the signs would be posted on all trains of the New York Central Railroad.

The city felt Biggs was bluffing, but evidently Biggs did not bluff. A week later he sent a new letter urging quarantine of all cases, by force if necessary. Anyone exposed to smallpox who refused vaccination was to be quarantined 16 days from the date of the last exposure. On February 18, the mayor and a delegation of Falls businessmen arrived in Albany to complain about the bad publicity surrounding the smallpox epidemic. Briggs brushed them off. "The policy I have always pursued and intend to pursue is complete publicity in everything done in the department," he said. "If you have smallpox, admit it and let others know it. Through concealment you delude yourself and lay others open to dangers from which they cannot protect themselves." When the delegation pleaded that quarantine would be bad for business, Biggs said that Niagara Falls's business was no concern of his. My job, he told them, is to control smallpox.[106]

The chastened business community returned to Niagara Falls and began to press city officials to vaccinate.[107] Biggs's demands provoked outrage among workers in the city's paper, aluminum, and other industries. With Hodge seated on stage, about 500 working men rallied at a hall to protest the decision, claiming that mandatory vaccination of workers was unfair, "particularly on account of the fact that many of the employees daily handle poisonous substances to such an extent that vaccination will be dangerous to their lives."[108] The antivaccinators' case was undercut by the fact that the mayor, William Laughlin, a Hodge supporter, was in political trouble as a result of a state inquiry into bribery by builders and brothel keepers. The last case of smallpox was released from quarantine on April 28. Hodge's postmortem was bitter. He gave the names of three children who he said died of vaccination during the epidemic.

Smallpox, mostly the mild sort, continued to rage in America, but people generally waited as long as possible to have their children vaccinated. A survey of 9,000 American families conducted from 1928 to 1931 found that 26 percent of five-year-olds had been vaccinated. The number jumped to 59 percent by age 7, by which time many schools required it.[109] In cities, where vaccination was a more heavily enforced annoyance, the percentages were 37 and 75 percent, respectively. The last major epidemic, in 1921, brought 102,787 cases and 563 deaths, as imported *Variola major* joined the mix.[110] There would be more smallpox opposition, with some cities and states prohibiting compulsory vaccination, well into the 1920s.[111] In 1926, an armed mob led by a city councilman and a retired Army lieutenant drove health officers out of Georgetown, Delaware, in a successful attempt to prevent vaccination of the townspeople.[112] Vaccination riots were not at all uncommon.

But Biggs's victory in Niagara Falls gave a taste of the world to come. In the 1930s, successful campaigns against diseases like diabetes, tetanus, diphtheria, and tuberculosis gave scientific medicine a legitimacy it had never previously enjoyed. Although vaccination rates remained relatively low, improvements in vaccine technology slowly increased confidence in the procedure.[113] *Variola minor* probably immunized much of the nonvaccinated population. As heavily vaccinated American troops defeated their enemies overseas, and public health authorities successfully tied their campaigns to the nation's victories and progress, sympathy for opponents waned. "The United States is in especial danger of contagion just now, and yet this is the time chosen by the antivaccinationists to exploit their literally pestilent heresies and to utter their mendacious warnings against dangers that do not exist!" *The New York Times* opined during the Niagara Falls outbreak.[114] "What they say about evil effects following vaccination has an element of truth so small as to be negligible." This was premature opinion, but within a few decades it would become the conventional wisdom.

PART TWO

GOLDEN AGE

CHAPTER 4

WAR IS GOOD FOR BABIES

As more and more thousands of men were slaughtered every day, the belligerent nations, on whatever side, began to see that new human lives, which could grow up to replace brutally extinguished adult lives, were extremely valuable national assets.

—Sara Josephine Baker[1]

I am not so sure I am going to Heaven so I am counting on the vaccine.

—Maj. Gen. Leon A. Fox

For three weeks in April 1947 the streets of New York were jammed with people waiting in long lines outside police stations, doctors' offices, and public health clinics. More than 6 million New Yorkers were immunized against smallpox during those three weeks, the largest mass vaccination in the history of the city. This was the smallpox scare to beat all others in a city whose last death by smallpox had occurred 35 years earlier. No longer a familiar threat to Manhattan, the disease took on the guise of a grandfatherly legend, a bogeyman that could still put a chill into the springtime air. But when the authorities ordered vaccination, the public submitted.

Eugene LaBar, a rug importer from Maine who had lived six years in Mexico, had fallen ill on a long bus ride from Mexico City and arrived in Midtown on March 1 with fever, a rash, and a terrible headache. His illness was misdiagnosed as a drug reaction, Kaposi's syndrome, and finally as bronchial pneumonia before he died on March 10 at the Willard Parker Hospital for Infectious Diseases. A janitor on the ward came down with the rash a few days later, and a young New York University infectious disease specialist named Saul Krugman was brought down to have a look. Krugman, nephew of Albert Sabin—then a well-known Rockefeller Institute virologist—immediately diagnosed smallpox.[2] A microscopic analysis confirmed that LaBar had died of the disease, although his arms showed two vaccine scars, the most recent less than a year old.

The smallpox that LaBar brought to the city also killed Carmen Acosta, the 26-year-old, pregnant wife of janitor Ismael Acosta. Nine other people, including a nun and a baby, became sick. By the time Dr. Joseph Smadel of the U.S. Army Medical School Laboratory in Washington had made the definitive diagnosis, all of LaBar's contacts had been isolated and vaccinated, and there would be no spread of the disease beyond the first 12 patients. It had been stopped cold in the third generation—a few contacts of a few contacts of LaBar. For all its fabled contagiousness, LaBar's smallpox did not affect any of the other riders with whom he had chatted and shared meals as the bus trundled through Monterrey and Laredo, Mexico; Dallas, St. Louis, Cincinnati, and Pittsburgh, taking a week to reach New York.

Although this smallpox outbreak was rather minor, its impact on the minds of the officials and the people of New York was not. Israel Weinstein, the city health commissioner, initially announced that 300,000 people would be vaccinated, but at the urging of Tom Rivers, a famous virologist, director of the Rockefeller Hospital, and member of the city's Board of Health, the health department announced April 4 that it intended to vaccinate the entire city. By then the chain of infection had been cut. But within hours of Mayor O'Dwyer's news conference, lines began snaking through the streets of all five boroughs as vaccine, rushed to town by the army, the navy, and every pharmaceutical company that made it, was distributed throughout the city. By radio, flyers, and newspapers, the authorities spread the word: "Be safe, Be sure, Be vaccinated." The 175,000-member Civilian Defense Volunteer Organization, which had been used to spy on suspect aliens and to supply provisions to war widows during the war, sent its members door to door encouraging residents to be vaccinated. The people of New York were obliging in a most unified way. In all the newspapers of the day, from the high-brow *Times* and *Herald* to the *News American* and the tabloid *Daily News*, not an inch of criticism could be found—not even any discussion of the campaign's potential risks. The skeptics kept their mouths shut, with a few exceptions.[3] The campaign was entirely voluntary.

Such a response would only have been possible, perhaps, in a country that was still militarized. Deadly smallpox was unknown or a distant memory to most citizens. The city's last smallpox death had occurred in 1912, the last case in 1939. Yet New Yorkers enlisted in the battle against smallpox with all the unquestioning self-sacrifice they had summoned up to fight Hitler and Hirohito. For a time, vigilant citizens saw smallpox popping up everywhere. Along Broadway the night of April 16, a rumor spread that someone had been stricken in a West 57th Street cafeteria. The report, which turned out to be false, emptied the Rio Cabana and other Midtown clubs. Among the night crawlers who showed up at Polyclinic Hospital for vaccinations at 4:30 A.M. were Gloria King and her 15 chorus girls. The next day, at the height of the scare, 500,000 people were vaccinated. Everyone was joining "the Order of the Itching Arm," as the *Daily News* called it.

The vaccine used to snuff out the LaBar outbreak was a more sanitary product than that administered half a century earlier. But it was still scraped from the underside of a cow and preserved in jars of glycerine. While it apparently caused no tetanus or serious bacterial infections, the vaccination drive did result in numerous casualties. By the time the campaign ended, at least six people had died of brain inflammation or generalized vaccinia and perhaps 100 others had suffered serious injuries, including encephalitis and full-body rashes; some of the injuries were lasting.[4] In hindsight, vaccinating 6 million people was overkill, using a hammer to squash a fly. Vaccination and quarantine of a few hundred would have stopped the outbreak. This was the approach Britain used whenever smallpox appeared on its shores, and it worked well enough.

Yet this "collateral damage" story was simply not big news in 1947. No one complained. "The strange thing is that the people of the city of New York paid little attention to these deaths," recalled Rivers, who shared a measure of responsibility for the outcome. "No one even batted an eye. The people of New York accepted the fact that they had to be vaccinated against smallpox no matter what the price."[5]

Indeed, the smallpox outbreak was taken as a sign of the small world Americans lived in, were threatened by, and were compelled to protect themselves against. Some postwar observers felt that the decline of diseases like smallpox in America created the peril of infection from abroad. Disease was only a plane ride away, noted Brig. Gen. James Steven Simmons, who had run the army's preventive medicine program during the war and was now dean of the Harvard School of Public Health. "A wave of hysteria swept through the cities of the North Atlantic Coast" that could have been prevented "by an effective national vaccination program in the United States."[6]

Not law but "moral suasion,"and fear, had convinced the men and women of New York City to submit to the same discipline in force during the wartime years of blackouts, meatless days, and white-butter rations. On April 12, in the middle of the smallpox scare, 2 million New Yorkers packed the sidewalks of upper 5th Avenue to cheer an enormous Army Day parade. Although it had seen fit to vaccinate the entire city, the health department did not postpone this patriotic event, whatever the risks of smallpox contagion. Something more than the city's health was being promoted and protected here. Neither parade attendance nor vaccination were duties that could be shirked.

The public's patriotism was bolstered by confidence in leadership that had led them to victory with superior technology—not only the A-bomb, the B-52 bomber, and the M-1 rifle, but medical technologies as well. DDT saved the lives of millions threatened with typhus and malaria in postwar Europe and Asia. Penicillin had cured our boys and our babies by the thousands. Vaccination had put a sanitary shield around our men, protecting them from the scourges of previous wars: typhoid fever, tetanus, smallpox, cholera, typhus, and plague. Yes, the shots hurt and even caused illness sometimes, but the soldier survived. Returning from the war he wanted his children to have the same protection.

The behavior of New Yorkers in 1947 marked a turning point in the history of public health in America, and its significance did not go unnoticed. The *New York Herald-Tribune*, in a front-page April 20 story, remarked that the public had willingly surrendered to a procedure it had previously shirked and attributed the turnaround to the vigorous immunization of 11 million servicemen, which showed that shots were not to be feared. A queue for vaccination "became no more than a chow line to them," an officer was quoted as saying. Vaccination had been spottily enforced in years past, but that was about to change.

Vaccination and Modern Medicine

The new spirit of compliance was partly the result of militarization, but it also signified a heightened public trust in medicine and its genuine advances. More confidence in medicine meant more confidence in vaccination, which improved the market for vaccines, and the more they were used, the better they worked. In an era before mass media, the bad reputation of vaccines had been hard to shake, and coercion had been necessary to create a vaccine market. But coercion of the type that public health officials in turn-of-the-century Philadelphia relied upon could only go so far. School officials didn't like playing policemen, especially when they wanted to keep children in school. Unpopular demands could backfire when they were exerted upon a community held together by trust rather than hierarchical relationships. The one institution in which a "necessary" coercion could be carried out with few second thoughts was the military.

Immunization of soldiers had two obvious strengths. First, soldiers confined in small barracks were an ideal breeding ground for infectious diseases. As the vast influenza epidemic in the closing days of World War I had shown, armies were both uniquely susceptible to epidemics and especially good at spreading them from one civilian population to another. Immunization of soldiers was a necessity for any nation that wished to protect its fighting force—and its population. The second strength was that soldiers were under orders and free of anxiously hovering mothers and fathers. Once in the hospital they were useless to the army, so they were lined up for their shots and they kept their mouths shut, even if the shots were practically experimental.[7]

As it happened, the two great wars of the twentieth century occurred when immunization technology was just beginning to achieve success.

Beginning in 1911, when typhoid fever vaccination became mandatory for American troops, the coerced vaccination of millions built a strong case for the efficacy of certain vaccines. The military boot kickstarted vaccination into a national habit. At the same time, infectious disease death rates were beginning to plummet. Medicine and public health were gaining in prestige. As the mortality and morbidity of germs declined through the first half of the twentieth century, faith in medicine grew, and so did faith in vaccination. By 1945, many of the deadly infectious diseases that haunted the tenements and farms of turn-of-the-century America—diphtheria, scarlet fever, whooping cough, and smallpox— were no longer commonplace tragedies of childhood. Although vaccination was only partially responsible for this bright turn of events, it enjoyed the same status as other weapons in the arsenal of scientific medicine—antibiotics, isolation, sanitation, and better disease treatment in general.

Pasteur's Triumph

Most of the early groundbreaking advances in microbiology took place in Europe, and it is impossible to discuss vaccines without acknowledging Pasteur's role. Pasteur believed that Jenner's vaccine was smallpox that had been weakened—*attenue*, Pasteur said—by repeated passage through a cow. In the 1880s, after astonishing the world with his flashy public demonstration of the germ theory, Pasteur set out to conquer new worlds by making vaccines. He did this employing a combination of serendipity, a handful of blunt scientific principles, savvy public relations, and a lot of hard work. Pasteur lived at a time in which foolproof science was a great deal harder to conduct than it is today. Like Edison, his contemporary, and like Albert Sabin, a successor, Pasteur was a tireless worker who by dint of thought and experiment knew more than anyone else about his subject. Thus armed, he was not above the verbal bulldozing of opponents whose legitimate questioning interfered with his theories. His demolishing of the notion of spontaneous generation is a case in point. As every high school biology student learns, Pasteur devised a series of ingeniously curved glass tubes to show that when germs were prevented from entering a sterile broth, it would not give rise to more growth.

Pasteur's opponent, Felix Pouchet, had shown otherwise—in all likelihood, Pouchet's "sterile" broth was not sterile at all, but contained spores of an organism like anthrax or tetanus that had remained dormant despite heating. But Pasteur's ridicule of Pouchet, and his influence in Parisian scientific circles, helped bury those inconvenient results until they could be understood in the proper context.[8]

Pasteur's scientific reputation was already well established when he devised the first post-Jennerian vaccine—against chicken cholera, an affliction of the French poultry industry. This was an accidental vaccine. With Pasteur on vacation at his family home in Arbois in the summer of 1879, his lab assistants neglected to follow his instructions for culturing a bottle of the bacteria, and the organisms spontaneously mutated. The concoction now sickened chickens instead of killing them, and afterward they were immune.[9] Pasteur then tried to attenuate anthrax to create a vaccine for sheep. Again he succeeded, although he secretly used potassium bichromate, rather than air, to weaken the germs. In his 1882 demonstration at Pouilly le Fort, 24 of 25 vaccinated sheep survived injection with the virulent bacteria, while the 25 unvaccinated controls died.

The crowning experiment of Pasteur's life, the experiment that convinced the world that these mad scientists were onto something worthwhile, was the injection of rabies vaccine into Joseph Meister, an Alsatian schoolboy, in 1885. The boy had been savaged by a rabid wolf. His frantic parents brought him to Pasteur, who had been attempting to attenuate the rabies virus by passaging it through dozens of rabbits—infecting the creatures, cutting their brains into strips, drying them in a jar, grinding them up to make vaccine. Each day Meister was administered a ground rabbit brain that was fresher and theoretically more virulent, in a gradual sensitization process. While there was no way of knowing whether Meister had truly been infected with rabies—only about 1 in 10 rabid animal bites transmits the disease—the boy lived and his rabies injections were the shots heard "round the world." From Russia to Portugal to Japan and points in between, Pasteur was hailed as the savior of children.

America's expanding newspaper industry was just waking up to the revolution in bacteriology, and the rabies shot gave it a powerful new hook. On December 2, 1885, four kids from Newark were bitten by a stray dog. With funds partly raised by newspaper campaigns, the boys were shipped to Paris to get the new treatment. The press at home breathlessly followed their progress, and afterward the four survivors toured county fairs as living proof of the saving power of Pasteur's science.[10] Certain wiseacres hinted at the possibility that neither dog nor patient was actually rabid. "Now is the time to get bitten by a rabid dog and take a trip to Paris!" read the caption in the December 23 issue of *Puck*, under a cartoon depicting a line of well-heeled ladies and gents boarding a ship for France. The Society for the Protection of Children and Animals claimed that the Newark dog had no symptoms of rabies and that hundreds of dogs were being needlessly slaughtered.[11] But American horse sense, in this case, could not compete with the sense that something new had been achieved, a genuine hope for humankind. By November 1886, 2,500 people had been subjected to Pasteur's treatment in Paris alone.

Rabies, caused by a virus with an affinity for nerve cells, was a rare, inevitably fatal disease in which the patient ricocheted between interludes of eerie calm and of madness, during which he or she might become sexually aroused, foam at the mouth, or bite. Such disfiguring diseases have always frightened the public more than humdrum killers like tuberculosis, malaria, and whooping cough. The rabies vaccine itself may have been a serum rather than a vaccine—that is to say that antibodies in the rabbit brain, rather than attenuated virus itself, were probably the agents that produced immunity. In most countries of the West, rabies would be controlled not by vaccine but by the trapping of rabid dogs and wild animals. There were doctors who doubted the safety and value of Pasteur's rabies vaccine. But Pasteur's status as a scientist and a nationalist defender of the intellectual honor of *la France* against his *boche* competitor, Robert Koch of Berlin, made him almost untouchable.[12] His vaccine, as historian Gerald Geison notes, became a rallying point in the struggle for the cultural authority of scientific medicine. All criticisms of Pasteur (or, by extension, of vaccination and animal experimentation) were lumped together as obscurantism. "One can be an anarchist, a Communist, or a nihilist" in France, wrote critic August Lutaud, "but not an anti-Pasteurian."[13] With the elevation of Pasteur to national hero, public health had been enlisted as a rifle-bearer of nationalism. This same elevation would occur in America. Critics of vaccination would discover they were also attacking mom and apple pie.

Using Science to Fight Diphtheria

The first great conquest of immunization in this new era was diphtheria, which claimed the lives of thousands of little children each year in turn-of-the-century America. The battle against diphtheria employed bacteriology, quarantine, and active tinkering with the immune system and its by-products. In 1884, Friedrich Loefller, a scientist in Koch's laboratory, cultured diphtheria and grew it in guinea pigs. His Berlin colleagues Karl Fraenkel, Emil von Behring, and Shibasaburo Kitasato—and Emile Roux and Alexandre Yersin in Paris—followed with a series of experiments proving that a diphtheria toxin, when injected into guinea pigs, and later, dogs and horses, produced a substance in their blood that could be used to treat diphtheria in another creature. In early September 1894, Roux created an enormous stir at the Eighth International Congress of Hygiene and Demography in Budapest when he announced that his anti-serum had cut diphtheria mortality from 56 to 24 percent in sick children at Paris's Necker Hospital. Doctors in the conference hall flung their hats in the air at the results.[14] Science had proven it was possible to diminish the effects of a killer disease.

Biggs, chief of New York's bacteriological lab and one of the 3,000 or so Americans who had studied bacteria in Europe during the 1880s, was visiting friends there at the time of Roux's speech. Sensing its importance, he wired home for William Park, whom he had hired in the previous year, to start making the stuff immediately. The first horse antitoxin factory in America, established in the veterinary college on East 72nd Street, provided its first doses in January 1895. Although Park, an aspiring ear, nose, and throat doctor from a wealthy family, was not a great scientist, he was a clever public health engineer and a persistent lobbyist.[15] Working out of two basement rooms in a tenement building at 42 Bleecker Street, he first designed an effective diphtheria swab test kit and organized a network of depots where doctors could drop off the filled kits. Within a year, he had collected specimens from 5,611 cases. Thus began a campaign that cut the diphtheria death rate in New York City from 150 per 100,000 inhabitants in 1894 to 2.8 per 100,000 in 1928.[16]

Antitoxin was no miracle drug. It was frequently given too late to effect the course of a diphtheria case, and for a while it was given intraspinally, which probably did more harm than good. What Park and his colleagues found most striking was the degree to which it inhibited the spread of the disease in siblings of the severely ill. As they sampled the throats of thousands of New York youngsters, his lab workers discovered that the city was full of healthy diphtheria carriers who could spread the disease. This was controversial. Prior to the Park's work, many critical cases of strep throat and other upper respiratory illnesses had been misdiagnosed as diphtheria.[17] But defining the disease by the presence of Loeffler's bacteria improved the look of statistics when it came to talking about how effective antitoxin was at preventing death. Patients who, while carrying diphtheria germs, showed no symptoms of the disease, of course tended to remain well after antitoxin at a higher rate than those whose disease was already well advanced. Many prominent doctors, angry at having their diagnostic skills usurped by the use of a bacterial culture, were skeptical of antitoxin's value. Diphtheria's toxicity rose and fell, they said—pointing out that in the year Roux announced his dramatic decline, a hospital in Britain reported only a 29 percent fatality rate, without the use of antitoxin.[18] But as diphtheria mortality gradually declined, few could deny that Park's diagnostic aggression and antitoxin therapy were having a positive effect. By 1905, the diphtheria mortality rate had declined to 10 percent in New York City. Park was using more purified, potent antitoxin, and doctors and the public health system were able to respond faster to cases of the disease.

War on diphtheria opened a new door in science by introducing the massive, almost industrial use of animals to test and produce biological products. It was diphtheria investigations that gave rise to the term "guinea pig" to describe an experimental subject. Thousands upon thousands of the adorable little Andean creatures were slaughtered in the great European bacteriological labs after Loeffler discovered in 1884 that, unlike mice and rats, guinea pigs were highly susceptible to the germ.[19] George Bernard Shaw, an animal lover and a skeptic of the Pasteurian worldview, as we've seen, described the extension of such experiments to human subjects as "the folly which sees in the child nothing more than the vivisector sees in a guinea pig: something to experiment on with a view to rearranging the world."[20] Thus "guinea pig" entered the English language, reflecting a new social risk brought on by medical progress.

"Guinea pig" was not the only neologism to emerge from this Promethean era. As scientists such as McFarland in Philadelphia and Milton Rosenau of the new Hygienic Laboratory were getting a handle on the bacterial contamination of smallpox vaccine at the beginning of the twentieth century, young scientists in Vienna became aware of another disease process that could be set off by immunization. The syndrome occurred in children following the injection of diphtheria antitoxin. To make the substance, a horse was injected with increasing dosages of diphtheria toxin until its body had produced enough antibodies to be harvested. The horse was then bled, the red blood cells separated out, and the clear yellow serum, which contained the antibodies, was heated to kill bacteria. Of course, the serum contained more than just diphtheria antibodies. It contained antibodies to other germs the horse had been exposed to, and billions of other proteins that the human immune system would recognize as foreign. The injection of these substances into the blood of thousands of people created a new arm of the growing field of immunology—the study of allergy.

When practitioners injected crudely separated horse serum into a small child, it often produced an allergic response—and the child might respond with a full-out anaphylactic reaction to a second injection. Neither "allergy" nor "anaphylaxis" were terms that existed before the twentieth century. Clemens von Pirquet, a scientist working at the Universitaets Kindkerklinik in Vienna, first invoked allergy in his 1906 publication *Klinische Studien über Vakzination und vakzinale Allergie*, to refer to responses observed in some children vaccinated against smallpox. There were two different responses to antigenic stimulation, von Pirquet noted: immunity, and "altered reactivity"—*allergie*, in German. The term anaphylaxis, or "against therapy"—was coined in 1902 by two French scientists, Charles Richet and Paul Portier, who found that a tiny amount of jellyfish toxin produced an inflammation in a dog dramatic enough to kill it. Upon the second injection of antitoxin, some children responded with an almost instantaneous rash and skyrocketing fever. Von Pirquet and his younger colleague, Bela Schick, called it *Serumkrankheit*, serum sickness. They theorized that the first injection had led to the synthesis of antibodies, whose rapid response to horse proteins after the second shot led to a systemic reaction. This completed the basic picture of allergic responses.

The two scientists recognized that a paradox was at work. In general, the body's quick response to a foreign antigen was a healthy thing—it fought off an infection. But when responding to an inert protein of a type and quantity it rarely encountered in nature (other than insect stings and snake bites), such a reaction was unhealthy.[21] In the largest contemporary study of diphtheria antitoxin's side effects, Dr. E. W. Goodall of the Northwest Fever Hospital in Hampstead, England, reported that 40 percent of the serum recipients he observed suffered reactions ranging from rash to fever. More than 4 percent endured serious reactions like joint pains or vomiting.[22] The secondary reactions were always worse than the first, Goodall noted.

The next step in the application of science to fighting diphtheria came in Massachusetts, where in 1909, 40-year-old Theobald Smith[23] demonstrated that he could protect guinea pigs with a mixture of diphtheria antitoxin and toxin. The purified diphtheria toxin induced the body to mount its own immune response, while the antitoxin buffered the body's response to the toxin with passive immunity. This advance was coupled, to great advantage, with Bela Schick's discovery that the immune status of a given individual could be discovered by injecting a droplet of diphtheria toxin under the skin and waiting for a reaction, which signalled the presence of antibody. Schick had found that while 93 percent of newborns had maternal antibodies to diphtheria, immunity waned among 63 percent of two- to five-year-olds. By 1916, Park had used 12,000 Schick tests, and in 1918 the Health Department began attempting to administer them to all New York children, vaccinating with toxin-antitoxin all those in whom tests indicated a lack of antibody.[24]

The bacteriological revolution epitomized by Park and Briggs's New York laboratory "immeasurably strengthened the position of the medical profession and firmly ensconced physicians in charge of public health," the medical historian John Duffy has written.[25] With its victories against diphtheria and typhoid fever—banished through filtered water and pasteurized milk in most of the big cities of the country—the legitimacy of state medicine was greatly enhanced, and its involvement in children's welfare widened.[26] "Eventually," wrote Dr. Hermann Baruch in the New York *Medical Record*, "it will be possible to breed a race of human being progressively immune to all the acute infectious diseases. The longevity of the race would then be vastly increased."[27] Reformers suggested that the death rate of children was "a touchstone of community welfare; a test of civilization" and made the reduction of childhood disease a shining goal of Progressive-era politics.[28] In 1908, New York created the Division of Child Hygiene under Sara Josephine Baker, and the city opened its first 15 safe-milk stations in 1911.[29]

The 1921 Sheppard-Towner Act, a piece of vintage Progressive reform, provided matching federal funds for states that funded health services for mothers and children. Herbert Hoover's 1930 White House conference on child welfare helped cement this intervention in what had formerly been considered family matters, as did the 1935 Social Security Act, which provided for an expanded national public health program. Private foundations were also increasingly involved in public health. The Rockefeller Institute for Medical Research had been established in 1901 and quickly acquired a reputation as the premier scientific institution of its type in the world. Nobel prizes rained down on its scientists for discoveries on the nature of bacteria and viruses, cell biology, biochemistry, and anatomy. The Rockefeller Foundation and its International Health Board were created in 1913 to translate basic research into advances against infectious diseases that plagued America and the world.[30]

Success brought publicity. In 1925, sled dogs managed to transport lifesaving toxin-antitoxin from Anchorage to Nome, where an outbreak of diphtheria had sickened children. Leonhard Seppala, the Norwegian driver of the run since immortalized in the annual Iditarod race, toured cities in the lower 48 the following year with 40 Siberian Huskies, including Togo, his big lead dog (Balto, who led the final stage to Nome, gained most of the celebrity, including several film portrayals and a statue in Central Park). Seppala and the dogs mushed through Seattle, Kansas City, Dayton, Detroit, and Providence before reaching New York City, where the explorer Roald Amundsen presented Togo with a medal at Madison Square Garden.[31]

Saving children became a celebrity cause, and big philanthropies now wanted to get on board. In 1927, the Milbank Fund bankrolled a $1 million project to provide healthcare to children in poor East Side neighborhoods and funded a survey that showed only 9 of 235 families were aware of toxin-antitoxin. The results spurred state intervention. That year, more than 82,000 city children were immunized with toxin-antitoxin shots. Two years later, the Metropolitan Life Insurance Co. funded a citywide diphtheria immunization blitz. Flyers came in monthly utility bills; placards went up on store windows, and articles on the Diphtheria Prevention Commission's activities appeared in 36 newspapers, including the foreign-language press.[32] Children wore "Yes, I've been Schicked" buttons, an early "branding" of a public health intervention. More than 500,000 New York City kids were immunized from 1929 to 1931, when the city switched to toxoid, a true diphtheria vaccine. By 1936, fewer than 1 in 100,000 New Yorkers contracted the disease. In the crucial period of the antidiphtheria campaign's success, immunization was entirely voluntary. It was only after it had been whipped that compulsory vaccination against it began.

To be sure, the fight against diphtheria and other diseases did not advance evenly. William H. Park was a unique character who kept his laboratory going in the thick of Tammany Hall corruption with a mixture of prestige, connections, and his own wealth. Most of his research funds came from the New York University Medical School, where he taught bacteriology. Wearing two hats, he was able to recruit good researchers to the Department of Health by offering them dual posts as instructors at NYU. Without a Park to wrest funding and respect for them, the public health departments of most other cities did not progress as fast. The doctors in Philadelphia's public health leadership were distinct mediocrities; William H. Welch, for example, was "virulently" opposed to setting up a bacteriological laboratory, apparently fearing it would diminish his influence. Joseph McFarland, the most microbiologically advanced Philadelphian of the era, said he was "far too busy to be attracted" to work in public health.[33]

The publicity generated by the antidiphtheria campaign raised expectations in a way that encouraged unscrupulous salesmen, as well. Capitalizing on the public's belief that an era of magic bullets had arrived, the pharmaceutical industry cranked out dozens of vaccines, sera, and other immunizing substances as soon as scientists in Europe, Japan, and the United States gave names to the microorganisms they were supposed to combat. If you were to look inside the Sharpe and Dohme display cabinet from the late 1920s,[34] for example, you'd see: smallpox vaccine, and diphtheria and tetanus antitoxins; as well as pneumococcal, streptococcal, and meningococcal sera; and more than a score of "bacterins," made from killed bacteria. But microorganisms were not as similar to one another as these early vaccine-makers hoped and pretended, and few of these new products worked. Pneumonia proved particularly resistant to serum treatment. Sera "saved a portion of the victims of types I and II pneumonia—if you got them early," wrote Paul de Kruif. "But it turned out there were distinct types of pneumonia microbes almost ad infinitum, and while doctors were waiting around for labs to tell them what type was infecting their patients, death might not wait. Overall, the pneumonia death rate was not going down."[35] Serum therapy was also widely and unsuccessfully used against polio throughout the 1920s. "Hell, they had been using diphtheria antitoxin against diphtheria . . . so why should they believe it wouldn't work against virus?" recalled Tom Rivers. "Trying to sell a new idea in medicine is like trying to elect a completely unknown man to be president of the United States. It takes a lot of doing."[36]

Yet while victory through vaccines was far from common, infectious diseases were declining, and the reputation of public health grew as a variety of measures brought illness to heel. Filtered water and pasteurized milk beat back the typhoid bacterium, which lives only in substances contaminated by human feces.[37] Sometimes a mesh of related factors accounted for a disease's diminished mortality. A 1937 article in the *Journal of the American Medical Association* (*JAMA*) found that measles in the pre-antibiotic 1930s was only about one-thirtieth as deadly to children in Massachusetts as it had been in 1920.[38] The measles of the 1930s was just as contagious and endemic as ever; but it may have become less virulent, or it may have been that children's constitutions were heartier because of better nutrition and declining rates of tuberculosis and diphtheria. A lower birth rate in the mid-1930s meant fewer very young and vulnerable children caught measles from their siblings; research many decades later would show that measles was deadliest when a child was exposed to a large quantity of virus—the "inoculum effect." With the arrival of penicillin in 1942, measles fatality rates declined again as pneumonia complications decreased.[39]

The early twentieth century witnessed the gradual elevation of the status of medicine in America and the defeat of the pioneer ideal that anyone can be his or her own expert about anything. To be sure, the spoonerisms of a Lora Little—"Be your own doctor. It's your body, you know it better than anyone else"—still held plenty of appeal. France and Germany, with their strong state bureaucracies descended from monarchies, could exercise authority that in America had to be built through trust or enforced by law. But even in 1911, perceptive observers could see the rudiments of trust in medicine. "The central problem in democracy is to give prestige and authority to the elite and to their prescriptions," George Edgar Vincent, president of the University of Minnesota and later the Rockefeller Foundation, said in a speech that year.[40] But while "the tendency to identify opinionated obstinacy with personal liberty" was a character flaw in American democracy, "slowly but inevitably we are selecting a medical elite, experts to whom we entrust a constantly growing body of scientific knowledge and elaborated technic."

Between 1900 and 1930, life expectancy increased from 47 to 60 years. By the end of this period, opinion polls showed that Americans had more respect for doctors than any other professionals—this was something new! Doctors like Park and Biggs became celebrities, while others like Walter Reed became martyrs.[41] The deaths of yellow fever researchers such as Jesse Lazear and Hideyo Noguchi of the Rockefeller Institute (Jimmy Stewart played a gallant researcher–guinea pig in the 1934 Broadway production of *Yellow Jack*) created an exalted, heroic image of the scientist as brave explorer (as did the gallant Huskies of Nome). By 1941, 17 "health soldiers" from the Public Health Service had died of infections contracted in the line of duty—yellow fever, typhus, polio, tularemia.

In the face of sacrifice and progress, Little's homespun verities were no longer culturally decisive. "American faith in democratic simplicity and common sense," wrote Paul Starr, "yielded to a celebration of science and efficiency. Doctors were able to claim a new measure of cultural authority over a public impressed by the achievements of the laboratory and ready to accept that it was complex beyond their majesty."[42] When that authority was not granted, it had to be imposed, which perhaps explains why the United States had and has more compulsory vaccination than any other country. For it was also the American way to impose solutions on an unruly public, as Biggs did in Niagara Falls in 1914. The state, not the individual, "would define the common goal and see to its fulfillment," Biggs said.[43]

The two great wars of the first half-century would present an opportunity for authority and scientific medicine to marry; with the soldier's health tied intimately to his ability to defend the country, the imprimatur of military doctors carried absolute weight. "If typhoid should become as prevalent among our troops as it was during the Spanish-American war . . . every person in this room could expect to be an Axis slave by New Year's," said Gen. Simmons, who did much to promote vaccination, during World War II.[44] Sometimes Simmons inserted smallpox, rather than typhoid, into the same speech. After the war, Simmons would expand his comments to embrace the peacetime nation. Without health, he would say, "America can not fulfill her destiny as the future guardian of civilization. A physically weak nation, like a sick man or woman, cannot hope to function successfully or to remain a leader in this barbaric world." Good health was not just the right of a citizen; it was his or her duty.

The Prevention of Barbarism—and Typhoid

America's first major wartime experience with vaccination had come during the Revolutionary War, when George Washington took command of scattered units of volunteers who were vulnerable to smallpox. Washington resisted variolation, but at Quebec in 1775, and again at the Battle of Boston, well-variolated British troops were able to drive back American forces staggered by smallpox. After 1777, all troops entering the colonial army were variolated. Had Washington started variolating earlier, Canada might well be American.[45]

During the Civil War, the federal army was vaccinated mostly with "crusts," vaccinal scabs that companies shipped to health officers with certificates giving the time, date, and name of the provider—ostensibly healthy vaccinated children. The army required each state to vaccinate the regiments it raised, but few complied. Because vaccination had lagged in the prewar years, the urban regiments, with many foreign-born soldiers, were relatively well vaccinated at the start of the war, while Western frontiersmen were pox-naïve. Whatever its drawbacks, vaccine apparently had ameliorative properties during the war; on the Union side, unvaccinated whites and blacks clearly fared worse than the vaccinated soldiers.[46] The poor vaccination of the Confederate troops became a major liability after smallpox jumped to the army of Northern Virginia during the Battle of Antietam, causing 2,500 cases and 1,000 deaths. In the absence of good lymph supplies, Southern doctors mailed each other scabs from vaccinated slaves; soldiers tried to procure vaccinia on their own, but this often ended badly. Widespread vaccination with "spurious matter"—rumored to have been procured by an off-duty officer from a syphilitic Atlanta prostitute—rendered as many as 5,000 of General Lee's men unfit for duty at the battle of Chancellorsville.[47] Bad vaccinations led entire communities and armies to shun the procedure.[48]

Smallpox was of relatively little import to either side in comparison to the pervasive gastrointestinal disorders, the respiratory invasions of packed sleeping tents, the malarial mosquitos, and gangrene-infested operating wards. Surgery was much improved in time for military engagement with Spain, but diseases of crowding were still serious threats. While fewer than 250 American troops died fighting in Cuba and the Philippines, a fifth of the 107,000 troops stationed in U.S. camps during the war fell ill with typhoid fever, and 2,192 died of it.

As it assumed a growing colonial role after the war with Spain, the United States began aggressively incorporating vaccination into occupation policies. In Puerto Rico, where a smallpox epidemic had hit 3,000 people in 1899, the army quickly set up vaccine farms and within six months had inoculated 800,000 Puerto Ricans against smallpox. Mandatory vaccination was ordered for the canal workers in Panama and for millions of Filipinos starting in mid-1899. The military governor of Cuba, Gen. Leonard Wood, established the mandatory vaccination of children before their first birthdays,[49] a legacy that continued well into the Castro era.[50] An army-commissioned report following the Spanish-American War established that the unsanitary conditions of the camps, the failure even to dig proper latrines, had made for the easy spread of typhoid.[51] The report led to the incorporation of basic medical training for officer candidates at West Point. Its authors—Walter Reed, Victor Vaughan, and Nicholas Shakespeare—also recommended that soldiers be vaccinated against typhoid, just in case the hygienic reforms failed. As they finished their report, a vaccine was just coming on line.

This first typhoid vaccine, made by growing the bacteria in beef broth and inactivating it with heat and phenol, was produced by Richard Pfeiffer of Germany and by Almroth Wright, an eccentric Briton whose brilliance and theoretical daring made him one of the famed scientists of his day. Born in 1861, Wright was a remarkable polymath who had degrees in law, modern literature, and medicine by age 23. He could quote you any of 250,000 lines of poetry over a glass of sherry. A scowling hulk with a bulbous nose and thick, disputatious lips, Wright was also misogynistic, arrogant, and impolitic, an energetic demolisher of the arguments of foe and friend alike, and his friends included George Bernard Shaw, no slouch in argumentation. When Wright began his work in the mid-1880s, the bacteriological world was divided between the humoral school, which argued that immunity resulted from particles floating in the bloodstream—"humors," in the seventeenth-century sense—and the cellular school, under whose model the white blood cells chomped up invading germs. Wright was the first to marry the two theories of immunity, although his failure to master the complexity of the immune system helped earn Wright his inevitable nickname, "Sir Almost Right." But he made other contributions, including typhoid vaccine.

In 1898, during the Boer War, Wright, then a professor of pathology at the Royal Army College, became involved in testing the vaccine on British soldiers. He made no effort to cultivate relations with the brass, a big mistake that helped establish another key principle of scientific work—the value of connections in high places. While the vaccinated soldiers got less typhoid than the uninoculated majority, Wright's haphazard selection of volunteers and poor record-keeping cast the results in doubt.[52] A series of commissions found Wright's vaccine had provided some protection in both South Africa and colonial India, but the army refused to adopt it. Karl Pearson, the father of biometrics and eugenics, attacked Wright's data, and Wright's vaccine caused nasty side effects. When randomized trials were held decades later, more than a fifth of those who received the Wright vaccine couldn't work for a day or more.[53] In an October 1899 account from his wartime passage to South Africa, Winston Churchill, then a young journalist, described the vaccine's unpleasantness: "The doctors lecture in the saloon. One injection of serum protects; a second secures the subject against attack. Wonderful statistics are quoted in support of the experiment. Nearly everyone is convinced. The operations take place forthwith, and the next day sees haggard forms crawling about the deck in extreme discomfort and high fever. The day after, however, all have recovered and rise gloriously immune. Others, like myself, remembering that we still stand only on the threshhold of pathology, remain unconvinced."[54]

By 1914, the United States had gone some distance in cleaning up its cities. But during wars, soldiers and their germs, aggravated by illness, wounds, indifferent nutrition and stress, were thrown together in the same trench. "An army, on going out on active service, goes from the sanitary conditions of civilization straight back to those of barbarism," Wright stated in a September 28, 1914, letter to *The Times* of London. That, in the view of American military planners, made Wright's typhoid vaccine worthwhile. Reed had returned to Cuba to fight yellow fever after concluding the typhoid report, but Vaughan and Maj. Frederick Russell worked to perfect Wright's vaccine at the Army Medical Museum in Washington. While Russell argued that the highly reactive vaccine should be voluntary, in 1911 commanders ordered typhoid vaccination for a division of troops serving in Texas near a Mexican Army outpost, and in 1912 it became mandatory for all recruits.[55] This strategy was apparently successful. During WWI, the U.S. force of 4.1 million men under arms suffered only 1,500 cases of typhoid and 227 deaths (although more than 35,000 troops were made ill enough from the vaccine to report to sick call).[56] The initially unvaccinated French suffered 12,000 deaths from typhoid fever in the first 16 months of the war.[57]

National mobilizations, in the United States and in Britain, had already made vaccination a patriotic issue. In Britain, where the 1907 conscientious objector law eliminated compulsory vaccination, antivaccinationists organized resistance to typhoid vaccination in the military at the start of the war. But the government responded effectively with a line of argument both scientific and moralistic. Canadian physician William Osler, the world's most respected doctor and a visiting scholar at Oxford, was invited to speak to officers and men at a training camp at Churn in 1914. "Formerly an army marched on its belly," Osler told the men in an address entitled *Bacilli and Bullets;* "now it marches on its brain." Citing statistics that showed the efficacy of typhoid vaccination and the perils of the disease in unvaccinated troops, Osler urged the men to do the right thing and accept the shots. "With a million men in the field, our efficiency will be increased one-third if we can prevent [typhoid fever]," he concluded. "It can be prevented, it must be prevented, but meanwhile the decision is in your hands, and I know it will be in favor of your King and Country."[58]

Such remarks had their desired effect, and as Britain's resolve against Germany grew, so did its willingness to be vaccinated. Ninety-seven percent of the troops were vaccinated, and the Anti-Vaccination League bowed out of the fight. In the United States, where the opponents of vaccination were less organized to begin with, a few feeble efforts were made to turn patriotic arguments on their head.[59]

As the antivaccinists were quick to point out, the biggest decrease in typhoid fever in both military and civilian life had come from better hygiene. There were physicians, notably Ernst Friedberger in Germany, who argued that the vaccine had done little or no good, offering as evidence the fact that other intestinal disturbances had declined even faster than typhoid fever during this period (the Prussian military censored Friedberger's papers).[60] Comparisons of typhoid rates within the vaccinated military and unvaccinated rural population, however, suggested the vaccine played a modest positive role. Among enlisted Americans from 1901 to 1908, the rate of typhoid fever was 5 or 6 per 1,000. The rate fell below 1 per 1,000 in 1911 and continued to decline, even at the height of the war.[61]

All in all, the disastrous typhoid experience of the Spanish War had provoked a healthy response from the army, and vaccination, however limited its real achievement, basked in the afterglow. Whereas "in 1898 a civilian enlisting in the army was quite sure to have typhoid fever," Vaughan noted, "in 1917 the civilian who wished to escape the typhoid could find no safer place than the army."[62] The medical triumph of the war dried up official sympathy for the antivaccine crowd, a shift that was vividly on display during congressional hearings to discuss the Economic

Security Act of 1935, a keystone of the New Deal that included $8 million for the Public Health Service. H. B. Anderson, representing the Citizens Medical Reference Bureau, testified against the bill, arguing that bolstering the public health agency would encourage states to expand compulsory vaccination. His message was not well received. Representative Roy O. Woodruff of Michigan, who had suffered typhoid as a soldier during the Spanish-American War, praised the typhoid vaccine and mentioned that diphtheria antitoxin had "reduced mortality of that disease from 73 to 2 percent. . . . How any sensible man, how anybody having those figures before him can be opposed to vaccination and the prevention of disease, is more than I can understand . . . if your ideas prevailed in this country, we would still have smallpox, typhoid and other epidemics which have now almost entirely disappeared."

Yes, World War I was a great success from a medical perspective—until, of course, the great flu pandemic swept through the camps, at which point the public health champions all tasted ash in their mouths. There were even some who wondered whether the new vaccines had somehow made the troops more susceptible to flu.[63] From September 1918 to the following summer, 43,000 American troops died in the great pandemic, which claimed an additional half-million lives in the United States. The navy reported that 40 percent of its members were flu-stricken; the army, 36 percent. "If ever again I attempt to talk about the triumphs of preventive medicine," one doctor wrote in 1919, "my words will choke me."[64] But if the pandemic was a great tragedy, it did have one positive side effect. It helped convince everyone concerned with public health—whether civilian or military—of the importance of the development and use of vaccines.

Greater confidence in medicine did not immediately lead to an expansion in civilian vaccination requirements. The Bacille Calmette-Guerin vaccine, or BCG, a live bovine tuberculosis bacteria that had been modified by French scientists to provide immunity to tuberculosis, never caught on in the United States, although it became a routine vaccination in most of the rest of the world. Between 1929 and 1930 a batch of BCG made with the wrong strain of bacteria killed 72 infants in the city of Luebeck, Germany. That disaster, the inconsistent evidence of BCG's lasting effectiveness, and perhaps a touch of national chauvinism about the vaccine's French origins, undercut support for it in the United States.[65] Nor was typhoid vaccine put into widespread civilian use.[66] The severe adverse events and the need for a three-shot series every three years made it impractical, especially for infants, who in any case got a less serious form of the disease.[67] Clean water and clean milk were surer and safer, and eliminated other germs. Nationwide, the typhoid death rate declined 99 percent from 1906 to 1936, with little vaccination. By 1942, the medical consensus was for vaccination only of people living in flooded or typhoid-endemic areas with unfiltered water, for travelers to such places, for the chronically ill or those living in asylums and elderly homes. 7.5 percent of white Americans had been vaccinated against typhoid by 1930, the largest percentage being World War I veterans.[68]

Vaccines for World War II

As it got ready for the second great war of the century, the military began enlisting the best minds in American medicine to assist in the preparation of its troops, including protecting them with every vaccine available. In December 1940, one of the rising stars in the newly organized medical corps, Lt. Col. James Simmons, drafted a proposal for a new approach to infectious disease. The military's great fear, of course, the bullet it was determined to dodge, was the flu. While the brass pressed for a vaccine, scientists working on influenza were beginning to understand the vexing immunological maze created by the constantly mutating virus. There would be no single vaccine that was effective against all manifestations of flu, just as there had been no vaccine capable of preventing pneumonia. But the cash for research was there, and as is often the case when an unwavering bureaucratic directive issues from the top, clever and industrious underlings made sure the flow of cash, aimed at the flu, went as far as possible to irrigate other productive fields of inquiry. The military campaign against flu would be transformed into a general attack on viruses and would lead to the conquest of polio and other viral diseases after the war.

In January 1941, the surgeon general announced the creation of the Board for the Investigation and Control of Influenza and Other Epidemic Diseases, which became known as the Armed Forces Epidemiological

Board. The board's commissions, created to battle different types of disease, conducted research that was crucial to the future of vaccines—not only for the army, but for the baby boom children of the United States. The commissions were supposed to spot epidemics as they were germinating, but in practice they conducted pioneering work in many areas, providing new vaccines and setting a pattern of military patronage for civilian medical research.

Simmons, who ran the whole operation, was a lean, high-strung and intemperate boss, a "preventive medicine evangelist," in the words of Stanhope Bayne-Jones, his deputy in charge of vaccination programs. Simmons traveled 70,000 miles to inspect the troops and research facilities during 1943–44, and was an army representative on Vannevar Bush's Office of Scientific Research and Development, which oversaw the Manhattan Project and many others.[69] Simmons enticed leading physicians and scientists from the Rockefeller Institute, Hopkins, Yale, Harvard, Vanderbilt, and Pennsylvania to his unit. The schools paid their salaries and donated lab space for research; the army paid a $20 annual consultant's fee, and in the process it established a blueprint for a military-academic technical complex. After the war Simmons became dean of the Harvard School of Public Health (and a fervent advocate of socialized medicine), and his equation of prevention and the imperial imperative reached new heights.[70]

The preventive effort, of course, had a sound logical basis. Despite the typhoid vaccine and other sanitary advances, about half of the 112,855 army soldiers who died during World War I were casualties of disease (mostly flu, in this case) rather than battle wounds. If the army could strengthen the immunity of its men it would be that much more powerful. With the support of the military, the Rockefeller's International Health Division set up a flu listening post in California.[71] But Rockefeller scientists such as Alphonse Duchez, Thomas Francis, and Frank Horsfall, who had been working on a vaccine for 20 years, found that the more they learned about flu, the less confident they were of their success. There were many strains of flu and getting sick, or vaccinated, with one strain did not protect against others. The demand for the vaccine was intense; Rockefeller files from the period are filled with appeals from state and city health departments, mental hospitals and prisons, universities, corporations, and even desperate individuals for something to prevent flu. In 1940, the Rockefeller Foundation shipped 500,000 doses of flu vaccine to Britain, and subsequently provided enough vaccine to immunize several hundred thousand Americans, with mixed results.[72]

Vaccines v. Hell: Tetanus and Typhus

While neither the wartime flu vaccine nor the vaccine against pneumoccal bacteria achieved a great deal—they didn't have to, since no pandemic strains emerged during the war—several other effective vaccines were developed. One high spot was the neutered tetanus toxin known as tetanus toxoid, which was similar to the diphtheria vaccine that American cities had begun using in 1931. The British, Canadian, and French armies adopted tetanus vaccination, and in June 1941, Simmons pushed through compulsory tetanus vaccination of all U.S. troops. The results of the campaign were sparkling: only 12 army personnel suffered tetanus between December 7, 1941, and January 1, 1945—and only one of these resulted from a battle wound. Seventy cases had been reported during WWI and that figure probably reflected underreporting. Spies reported 80 cases of tetanus among unimmunized German troops during the Normandy invasion.[73] Vaccines were also developed for plague, meningitis, and Japanese encephalitis B virus. Antivaccine activists had no luck challenging the program,[74] though the military's smallpox vaccine was at times problematic. Some 5,260 vaccinated troops were hospitalized in 1943, with a rate of 1 severe reaction per 300 vaccinations in August of that year.[75]

Typhus was another of Simmons's big targets. The disease, which killed more than 3 million Europeans in the aftermath of World War I, was the classic disease of war and its disorderly aftermath. Refugees and men at the front seldom changed their clothes, which made them welcoming hosts for lice. Typhus-carrying body lice feasted on human blood at its characteristic temperature. When they had infected their host with typhus and the person developed a fever, or died, the lice would flee the skin and swarm to the surface of the clothing, hopping off to find another body heated precisely to 98.6 F. The disease was endemic to impoverished areas of Mexico and other Latin American countries, where it usually caused a mild infection in children.[76] The possibility of military casualties from typhus, which provokes a state of utter listlessness, created a new interest in a vaccine, and the great powers of the war all put scientists to work on it.

The difficulties in culturing the organism led to weird and revolting production methods. Typhus is a member of the genus *Rickettsia*, small bacteria that, like viruses, need to penetrate living cells in order to grow. The organisms that cause Rocky Mountain Spotted Fever and Lyme disease, carried by ticks, are also *Rickettsia*. To most scientists working on typhus, the only method for culturing the germ seemed to be to grow it in lice. For this one needed large numbers of the unpleasant little vermin—but how to culture the typhus germs in them? Each nation came up with its own approach. The Russians skinned and scalded human corpses and stretched the skins over pools of typhus-contaminated blood, creating farms of lice that sucked blood through the skin. The Germans, operating a laboratory in occupied Warsaw, enlisted concentration camp inmates to work on enormous assembly lines injecting individual lice rectally with typhus cultures. The tiny animals would later be dissected, their stomachs ground up to make the vaccine.[77] In North Africa, the Pasteur Institute attempted to make a live vaccine from flea excrement. Hans Zinsser, the famous Rockefeller scientist who wrote *Rats, Lice and History*, tried to grow the vaccine in an agar culture, while others used the ground-up lungs of mice into whose noses typhus germs had been blown.[78] Postwar investigations would show that the Nazis and the Japanese had experimented with typhus vaccines on prisoners of war, creating "control groups" of prisoners intentionally infected with the disease. Neither of the Axis powers managed to create an effective vaccine, although at a Nazi typhus laboratory at Bergen-Belsen, inmates reportedly gave low-efficacy lots to the SS, while vaccinating their fellow prisoners with the good ones.[79]

A Rockefeller-trained Vanderbilt University scientist, Dr. Ernest William Goodpasture, had developed a more promising culture medium, one that did not require the use of lice. Working with Alice and Eugene Woodruff, Goodpasture in 1931 discovered a way to insert virus or *Rickettsia* into a chicken egg with a sterile needle, then close the hole before bacteria entered. This would lead to the first successful typhus vaccine and, eventually, to measles, flu, and many other crucial vaccines. At the

Public Health Service's Rocky Mountain Laboratories in Hamilton, Montana, Herald Cox incubated typhus in eggs and then killed the purified product with formalin and phenol. His work required vast henhouses, since each yolk sac grew only enough typhus to vaccinate two people, but such were the requisitions of wartime.[80] By 1942 American fighters bound for North Africa were routinely vaccinated against typhus, and eventually most of the GIs sent into either theater got the shot. In the epidemic conditions of North Africa in 1942, in which 60,000 Algerians died of the disease, only 11 of the half-million vaccinated American soldiers got typhus, and none died.

The success of this vaccine was particularly remarkable considering how little it had been tested; like several other wartime vaccines, it would be considered experimental by today's standards. Rockefeller Foundation scientists Jack Snyder and Fred Soper had begun a trial of the vaccine in Spain in 1940, but it was aborted after the two men got typhus and field clinicians were thrown out of the country.[81] In 1943, the U.S.A. Typhus Commission was created to protect troops from the disease. But since the vaccine had effectively rendered the commission's efforts moot, it decided to join the Rockefeller Foundation in fighting typhus among civilians. This effort would be tremendously successful, possibly saving millions of lives and planting a firm anchor of gratitude for the American force occupying postwar Europe.

The typhus vaccine, like many other vaccines at this cusp of the antibiotic age, would have a short life as a major prophylactic, because vaccination was not the principal tool against typhus. Instead, the army attacked the disease vector, the louse, using a most controversial chemical—dichlorodiphenyltrichloroethane, or DDT. The powder had been developed by J. R. Geigy, the Swiss chemical firm, but experiments at a U.S. Department of Agriculture laboratory in Florida showed its potential. Shooting DDT powder out of dusting guns into the sleeves and undergarments of millions of Europeans, the sanitarians mopped up after the military with their own campaign through Europe. They began in the caves of Naples in December 1943, where typhus had broken out among hundreds of townsfolk seeking protection from Allied bombing raids, and later brought the delousing powder to Dachau and Bergen-Belsen, where thousands under Nazi control had died of typhus.

The victory over typhus at Naples was featured in the *Saturday Evening Post*.[82] While only about 27,000 Neapolitans were vaccinated during the campaign, more than 1 million were sprayed with DDT powder. The army had attempted to vaccinate doctors, nurses, priests, police, and others who came in daily contact with lousy Neapolitans. But many of these people, including military personnel, refused. "You can't vaccinate any great civilian population as you can the Army," the *Post* reporter on the scene noted. "In the first place there is always antivaccination propaganda." People preferred to be sprayed with clouds of DDT. Still, the military rarely missed a chance to talk up the vaccine. "The second battle of Naples established for all time the value of our typhus vaccine," said Gen. Leon Fox, the head of the Typhus Commission, in a talk to American troops in May 1944. Not a single American had contracted typhus there and "don't forget the Allied soldiers had plenty of very intimate contacts with some of those good-looking Neapolitans. . . . Without vaccine, at my age it is a four to one bet that I would have a front row seat in Hell," Fox concluded. "I am not so sure I am going to heaven, so I am counting on the vaccine."[83]

After typhus, DDT would be used to wipe out the *Anopheles* mosquitoes plaguing the hotter, warmer areas of Europe, thereby creating vast new malaria-free areas. "To the wonder drugs of war medicine we must now add DDT," *The New York Times* opined on June 4, 1944. "DDT seems almost too good to be true . . . if the tales of DDT miracles are borne out, there ought to be no excuse for any insect borne disease." DDT was one of the most powerful weapons of the war, and the army doctors who organized its use probably saved more lives than anyone since Edward Jenner. It would not be until 1962, with the publication of Rachel Carson's *Silent Spring*, that the world would see any downside at all to DDT. Then it would choose against using a substance that killed powerful enemies but left residues that seeped into the food chain, weakening the eggshells of baby eagles and other charismatic birds of prey.

Yellow Fever, Yellow Peril

As so often was the case in the era of industrial chemicals, we would learn about risks of a particular technology only after we had thrown ourselves headlong into the enjoyment of its benefits. Yet the more technologies

Was there irony in the fact that Cotton Mather, whose name is historically linked with the Salem witch trials and Puritan orthodoxy, would introduce the first important lifesaving medical invention to the American colonies? Yes and no. The politics of variolation—as well as its medical value—were very much part of Mather's complex field of operation. (Courtesy of the National Library of Medicine)

Edward Jenner vaccinates James Phipps, on May 14, 1796. Jenner scraped pus from a cowbelly into an incision on the boy's arm, and six weeks later variolated him with small-pox virus. Phipps did not contract smallpox and within a decade, Jenner, whose first vaccination is depicted in this painting by Georges-Gaston Melingue, was a household name around the world. (Courtesy of the National Library of Medicine)

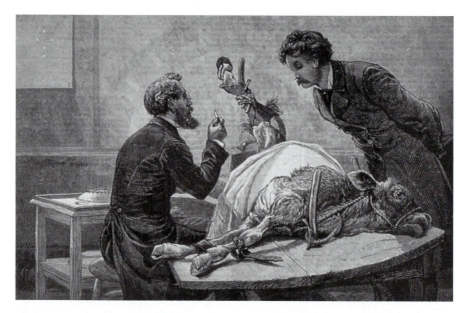

This magazine drawing demonstrates the method for cultivating and harvesting vaccine. Here, a cow strapped to a gurney is being infected with vaccinia by having the substance scraped into its belly. Later the mature virus will be scraped off, cleaned, filtered, and used to protect humans against smallpox. The production technique did not vary much through the 1970s. (Courtesy of the National Library of Medicine)

A newspaper drawing depicts the widespread public disturbances that accompanied vaccination and quarantine efforts in Milwaukee's Polish and German neighborhoods during the epidemic of 1894. Public health officials in the city were all Anglo-Americans, and the cultural differences contributed to mistrust—an interesting contrast with the public harmony that greeted mass vaccination in New York in 1947, as historian Judith Leavitt has pointed out. (Courtesy of the Library of Congress)

Apprehensions about smallpox vaccine were widespread in the United States in the early decades of the twentieth century, as this 1912 cartoon in *Harper's* magazine suggests.

William M. Welch was in charge of smallpox vaccination and treatment at Philadelphia's Municipal Hospital for more than 30 years. He was as dogmatic about vaccination as he was shocked by the mild form of smallpox that arrived in the city in 1898. (Courtesy of the College of Physicians of Philadelphia)

Joseph McFarland in his laboratory with his assistant. McFarland, one of the leading young American bacteriologists at the turn of the twentieth century, gained the enmity of vaccine makers for conducting a private investigation that linked their product to dozens of tetanus deaths in children in 1901. But McFarland was reluctant to give vaccine opponents ammunition for their cause. (Courtesy of the College of Physicians of Philadelphia)

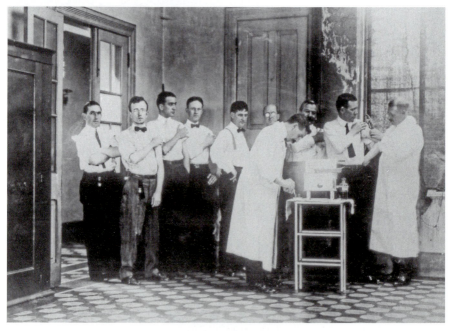

Men lining up for first typhoid shot, U.S. Army Medical School, 1909. Although the vaccine was developed in Germany and the United Kingdom, the U.S. Army, wary of another typhoid epidemic like the one that had caused most U.S. troop casualties during the Spanish-American War, vaccinated all of its troops before World War I against typhoid. (Courtesy of the National Library of Medicine)

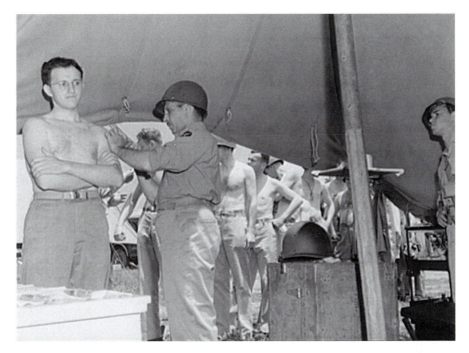

Soldier being vaccinated in a tent during World War II. (Courtesy of the Library of Congress)

Gen. James S. Simmons in Germany. Chief of the Army's Preventive Medicine Service, Simmons laid the framework for post–World War II vaccination campaigns by championing widespread immunization of soldiers and creating the Army Epidemiological Board, which oversaw research into viruses. He also pushed the flawed yellow fever vaccine, and lobbied after the war, unsuccessfully, for a national health system. (Courtesy of the National Library of Medicine)

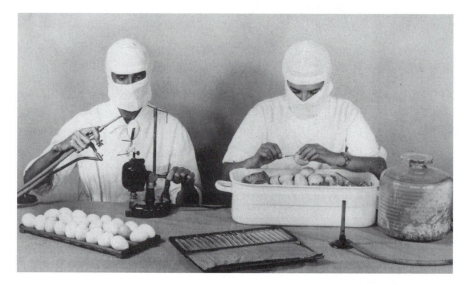

Rockefeller Foundation scientists in Brazil struggled to perfect the yellow fever in the late 1930s and early 1940s as they administered millions of doses to fight the disease, which was endemic to Brazil. Wilbur Sawyer, who led the yellow fever vaccine production, disregarded the lesson on how to avoid contamination of the vaccine with viral hepatitis, however, and the U.S. Army paid the price. Here, the technician at left is using an acetylene torch to open eggs while maintaining sterility. The worker at right is removing the vaccine virus–infected embryos and placing them in a jar. They were then minced and filtered to make vaccine. (Courtesy of National Library of Medicine)

Six million New Yorkers were vaccinated within a few weeks after smallpox appeared in their city in 1947. Contemporaries were struck at the willingness of Americans to accept vaccination in this postwar period—even though the vaccine killed and maimed far more people than did smallpox itself. (Courtesy of Wide World Photos)

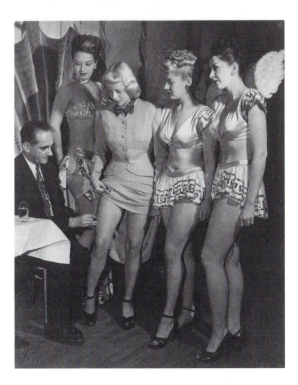

Three showgirls at Billy Rose's Diamond Horseshoe nightclub show some leg while being vaccinated against smallpox in New York City in 1947 (Courtesy of Wide World Photos)

The great humanitarian and the great journalist—both of whom had ample FBI files—in a still taken during Edward R. Murrow's *See It Now* in 1955. Murrow asked Jonas Salk who owned the patent to the vaccine, to which Salk famously replied, "Well, the people, I would say. There is no patent. Could you patent the sun?" He was being fairly honest on this point. (Courtesy of the March of Dimes)

we relied upon, the greater the harvest of unintended consequences. This was also the twentieth-century experience with vaccination—supremely positive, in general, with a few asterisks. Just as DDT would provide a paradigmatic lesson about the buildup of trace chemicals in the food chain, vaccination has repeatedly, accidentally spread diseases through the human population. When viewed from a limited angle, the history of vaccines was a history of all the things that could go wrong when you injected foreign substances into humans. Vaccines systematically exposed large populations to allergic reactions to horse proteins, gels, and chicken albumen. They also exposed people to bacteria, if the concoction was contaminated or if needles were reused or not carefully sterilized. Needle-borne viral diseases were another product of a fatal delay between two technologies, especially because the detection of tiny, ever-evolving viruses lagged behind medicine's ability to vaccinate and transfuse. Although it is difficult to pin down the origins of the global epidemics of hepatitis B and C, the sharing of blood products through therapeutic interventions, including vaccination, was a major factor.

In World War II, for the first time, war was massive and impersonal and fully technological. Bombers left bases in the middle of the Pacific and leveled cities thousands of miles away. Rockets departed from outcroppings on the North Sea and landed in people's living rooms in Coventry. Noxious gases and trains and assembly line death factories produced millions of victims. Like the ordnance, germs traveled far beyond the usual neighbor-to-neighbor transmission route, even beyond the usual wartime vectors of trench and brothel and mess tent. For the first time in the history of warfare, the brass feared and planned for biowarfare, and created offensive germ weapons of its own. In the process, the U.S. military would commit a tragic error, effectively waging biowar against itself.

The yellow fever vaccine, ordered in 1941 for every man and woman in the armed forces, was a vector that entered millions of bloodstreams within a few months' time. As a result of a critical production error, barely two months after Pearl Harbor, American soldiers were turning yellow, laid flat on their backs in sick bay. The yellow fever vaccine, created by some of the best and brightest medical minds in America, was contaminated with what would turn out to be hepatitis B virus. It was a disaster that might have been avoided, but was perhaps the inevitable result of a combination of arrogance, uncertainty, poor communication skills, and fear. It taught a lesson about all these things that the human race never tires of relearning.

Yellow fever was no longer an American disease by the early twentieth century, but the fight against yellow fever had been an American fight, led for decades by the Rockefeller Foundation's International Health Board, later called the International Health Division. From its creation in 1913 until its dissolution in 1951, the agency invested more money and manpower than any other in combating the unattended diseases of the poor—yellow fever but also hookworm, tuberculosis, malaria, and typhus. The health board's yellow fever campaign created many heroes and martyrs. It made gigantic inroads against the disease and immeasurably furthered the reputation of scientific medicine in America and the world.

Yellow fever was a fearsome disease that struck suddenly, causing high fevers, yellow skin, hemorrhages, and the vomiting of black coagulated blood. In colonial times it was endemic to a few parts of the United States, since the *Aedes aegypti* mosquitoes that carry it do not thrive in cold climates. Epidemics would occur from time to time in the cities of the north, brought by mosquitoes that arrived in port at summertime. In 1793, a yellow fever epidemic killed an estimated 2,000 Philadelphians. The first effective campaign against yellow fever began during the Spanish-American War, when Walter Reed and his associates used a series of careful experiments to prove that it was spread by mosquitoes. The mosquito wasn't an obvious vector because few mosquito species transmitted the disease, and it was communicated from *A. aegypti* to humans only during the period 10 to 14 days *following* the insect's bite of an infected patient. The bite of a mosquito that had just feasted on the patient you were visiting in the hospital would not make you ill. Visit the victim's house a week later to pay respects to his widow, and a lingering mosquito might be lethal.

The discovery of this secret cost the lives of Jesse Lazear and Clara Maas, volunteers in Reed's experiments. Reed himself had been called back to Washington while his comrades volunteered to be bitten, though his death by appendicitis in 1902 linked him in the public mind to their sacrifice. ("One realizes," a physician wrote of these experiments years later, "how lucky we are that problems like this were solved before ethical standards changed.")[84] The deaths in Cuba were only the beginning of the suffering that this tremendously contagious virus would inflict on those who sought to conquer it. Five well-known scientists would die tracking yellow fever, and at least 27 scientists and laboratory workers would contract the disease before one of them, Wilbur A. Sawyer of the International Health Board, developed a plausible vaccine for lab workers.

In the early twentieth century U.S. Army Maj. William Crawford Gorgas led the yellow fever eradication campaign in Cuba and Panama, enabling the colonization of Cuba by America and the construction of the Panama Canal. In 1918, the International Health Board hired Gorgas to plan the global eradication of yellow fever, using a combination of mosquito eradication and vaccine development. The Rockefeller yellow fever campaign was probably the largest focused attack on a disease in history at the time. Working with the virologists from its sister organization, the Rockefeller Institute for Medical Research, the health board created a network of operations in South America and Africa that nearly whipped the urban form of the disease, although it eventually became clear that yellow fever could survive in jungle animal hosts indefinitely.

The Rockefeller men—there were few Rockefeller women—were bold, courageous, idealistic, and sometimes arrogant. They worked themselves to exhaustion for the benefit of the masses (and the foreign governments and corporations interested in opening up disease-free lands for investment), attacking their mission with all the Baptist zeal that was the Rockefeller philanthropy's stock-in-trade. The creation of a yellow fever vaccine was a collaborative effort, but no one was more zealous than Sawyer, an Oakland, California-raised Harvard MD, who had been fighting yellow fever since 1926 and had also led the International Health Board since 1935. For more than a year after Sawyer established a yellow fever laboratory in New York City, he and two other scientists did all the work—feeding the monkeys, cleaning and sterilizing the cages, performing necropsies on the dead animals. They didn't hire a lab assistant because yellow fever had killed three of Sawyer's colleagues in Africa, and he didn't want to risk another life. The laboratory was on the top floor of the Rockefeller Institute, with windows looking out over the East River. To enter the lab you went through a series of vestibules. Once inside you were ordered to keep your hands in your pockets, to avoid contamination. No one was allowed to touch the doorknob leading to the outer vestibule, which had to be opened from the outdoors. The monkeys suffered bleeding gums. The laboratory workers suffered being spat at and bitten by the monkeys. Sometimes a monkey would escape; one got into an elevator, fled onto York Avenue, and was picked up in the lobby of the nearby Flower Hospital.[85] The lab men stank of Lysol and bleach so badly that people moved away from them on the subway. Thomas Norton, the wizard of a lab technician who would later join Hilary Koprowski in slugging back the first oral polio vaccine, was one of Sawyer's accomplices. Six of the men became ill with yellow fever.

"Those of us that came down with it but recovered expected long assignments to yellow fever work," said Sawyer in 1949, "for persons who had acquired immunity as the result of an attack were the only ones who could do the research with impunity."[86] He made a virtue of this unpleasantness by concocting a protective treatment that consisted of yellow fever virus grown in mouse brains combined with convalescent sera collected from labmates' and his own blood. The sera contained antibodies that dulled the attack of the virus without eliminating its immunizing power. It was a tremendous boon to yellow fever researchers, who could now be protected from the disease. But Sawyer's vaccine required more than a liter of serum for each patient, which made it impractical for general use. Sawyer tried to replace the human sera with animal sera, but this caused allergic reactions, or rendered the vaccine too virulent, if the animal antibodies were quickly passed out of the system.[87]

No virus is entirely stable, but yellow fever is exceptionally unstable. The first challenge in making a definitive vaccine was to grow the virus. Rockefeller virologist Johannes Bauer brought the Asibi strain of yellow fever to New York from Africa in 1928, and he and his colleagues set about trying to find a good experimental animal. Max Theiler, Wray Lloyd, and Hugh Smith succeeded in growing the virus in mouse brains and eventually created a weakened strain, which they designated 17D, by growing it in ground-up chicken embryos from which the head and spinal cord had been removed. This worked, apparently, because it selected out the organisms that could grow in a medium with less nervous tissue, and hence diminished the virus' tendency to cause encephalitis. Theiler's D series vaccine continued to cause brain inflammations, however, until somewhere after the 89th subculture and before the 114th, after which it was found to no longer grow in monkey brains. What caused the change was unknown. "For all anyone knew," one scientist said, "it could be attributed to a tugboat whistle blowing in the East River at just the precise moment." This was a seminal moment in the history of vaccination, for it established a scientific method of using different tissue media to force the evolution of microorganisms to an attenuated state.[88]

Although the yellow fever vaccine to this day occasionally causes neurological damage, the 114th subculture was safe enough for the Rockefeller to provide a stable vaccine without an accompanying shot of human serum. The International Health Division began mass vaccinations in Brazil, and the vaccine was provided free of charge to London's Burroughs Wellcome Institute, the Pasteur Institute in Paris, and the yellow fever lab run jointly in Colombia by the Rockefeller Foundation and that country's government. Not all human serum was removed from the vaccine, however. Tiny amounts were used to buffer the vaccine virus before it was squirted into the egg for cultivation of vaccine. Paradoxically, the addition of normal human serum—that is, from people who had never been exposed to yellow fever—delayed the inactivation of the virus. There were, however, apprehensions even then about using human serum in a vaccine. Theiler's father, the South African bacteriologist Arnold Theiler, had conducted experiments two decades earlier showing it was possible to create jaundice in horses by injecting them with sera from other horses. Sure enough, soon after the Rockefeller began distributing its vaccine the reports of jaundice began rolling in from the field. In 1937, G. M. Findlay and A. O. MacCalum of London published the first report of 52 cases among the 2,200 people they had vaccinated over a four-year period, with the vaccinees sickening two to seven months after inoculation with the Rockefeller vaccine. By 1939 the Brits had seen 96 cases.

Yellow fever was endemic to Brazil, and by 1941 the Rockefeller vaccine had been provided to 1.3 million Brazilians. Jaundice began to appear there as well, with nearly a thousand cases and approximately 30 deaths by early 1941. It was a devilish problem. At the time no distinction had been made between hepatitis A, which is usually contracted by eating tainted food and is generally transitory, and hepatitis B, or serum hepatitis, which is usually transmitted via sex and blood products, can take weeks or even years to produce symptoms, and is potentially infectious all the while. Nor was anyone aware that hepatitis B could stealthily produce cirrhosis of the liver and liver cancer. Still, most scientists working on yellow fever were worried enough about jaundice to seek an alternative to human serum in the vaccine. "It is difficult to guarantee that no vaccines produced in the future" will cause liver disease, the Brazilian Rockefeller men wrote in a study of the problem. "It was felt that a reasonable assurance might be had . . . if the use of human serum . . . could be avoided."[89]

Fred Soper, who ran the Rockefeller operation in Latin America from his base in Brazil, believed that most yellow fever disease control should be done through mosquito eradication, and the jaundice problems deepened his skepticism of the vaccine. But there was obviously a role for the vaccine, so Soper ordered a new seedlot of the 17D virus from the New York laboratories, and the Brazilians resumed manufacture of it at the end of 1940. This time, the Brazilian team left human serum out of the culture, using raw egg-white juice instead as a buffering agent. By early 1942, 200,000 injections had not caused a single case of jaundice. John Bugher, the Rockefeller's man in Colombia, found another way of making the vaccine safe. He got his serum from prisoners and stored it for two months while their health status was observed. (Later, scientists would realize that long-term storage of blood products at room temperature was an ideal way to inactivate hepatitis virus.)[90]

YEARS AFTER HIS DEATH, Sawyer's colleagues blamed his failure to deal with the jaundice contaminant on his inflexibility and moral self-righteousness. "His defect was that he was always right," said Soper. Sawyer decided that the problem with the British and Brazilian vaccine lay with vaccine production in those countries—he was always skeptical of field operations—rather than the use of donor blood.[91] Based on the scientific evidence at hand this was not an unreasonable supposition, though it was wrong. The laboratory in New York was cooking the serum at 56 degrees for up to an hour, long enough to kill any known virus at the time.

What's more, the army was pressing Sawyer for vaccine, and based on his own experience he was less worried about hepatitis than the possibility that the absence of human serum would render the vaccine unstable, resulting in a failure to immunize—or reversion to virulence. He made a decision, with war approaching, and stuck with it. "He took the scientific attitude," a colleague noted later, "that a chain of causation must not be accepted until proved, whereas a health official always has to assume if the epidemiological picture is not clear, that every potential factor may be involved unless it can be absolutely ruled out."[92] In retrospect, colleagues agreed, it was wrong to have Sawyer simultaneously leading the health division and the yellow fever laboratory. While his judgement was as sound as anyone else's, he shouldn't have had the last word.

As the Rockefeller scientists struggled to perfect their vaccine, the military was preparing for war. To do its part, the preventive medicine division of the army, headed by Simmons, was procuring new vaccines and making sure the troops were vaccinated with existing ones. In October 1940, the tropical medicine committee of the National Research Council recommended that troops being sent to Latin America, the Caribbean, and yellow fever–stricken areas of Africa get vaccinated. A few months later, the surgeon general made this official policy. The perceived threat of bioterrorism is one reason our troops were vaccinated against yellow fever, a disease that had played little role in past conflicts, and one that was unknown in the major theaters of the war. During a posting in Panama in 1934, Simmons became obsessed with the idea that the Canal Zone was vulnerable to the intentional reintroduction of yellow fever–carrying mosquitos. George Merck, the pharmaceutical executive and a fellow traveler in the bioterror scare, cited Simmons's concerns in a 1944 report prepared for the secretary of war.[93]

Simmons's thinking had not gone unchallenged. Maj. Leon Fox, an officer who served on Simmons's staff, had in 1937 described germ warfare as mostly fantasy, "a scarehead we are being served by the pseudo-scientists who contribute to the flaming pages of the Sunday tabloids."[94] Interestingly, Fox made an exception for anthrax, whose hardened spore cases made it "the perfect military pathogen." Fears naturally centered on Nazi Germany, given its advanced bacteriological laboratories. Indeed, a notorious Nazi named Eugen Haagen had worked with Max Theiler on yellow fever at the Rockefeller Institute before returning home, happily, to Hitler's Germany in 1933. (In 1952, Haagen would be condemned as a war criminal in France for experimenting on concentration camp inmates. Sentenced to 20 years hard labor, he was freed two years later and returned to his research at the University of Heidelberg.)

Alarms were raised about the other major Axis power in February 1939, when a Japanese scientist, Ryoichi Naito, approached a Rockefeller lab employee seeking to acquire yellow fever virus, claiming it was for Japanese vaccine development. Another Japanese bacteriologist approached Sawyer directly, and both requests were reported directly to Washington. In a mysterious episode later that month, an eastern European–born technician in the yellow fever laboratory was approached by a man with an American accent and invited to a rendezvous. Thinking it was some-one who knew the fate of his relatives in Europe, the technician met with the agent, who got into his car during a thunderstorm outside the foundation's gates. The agent offered the technician $3,000 for the yellow fever virus; after being rebuffed, he yanked the car keys out of the ignition, got out, and fled in his own late-model Buick.[95]

These episodes added weight to Simmons's conviction that the Axis powers intended to attack with yellow fever, and he slowly gathered enough anecdotes to make a convincing case with his initially skeptical superiors in the military.[96] Some of his yarn was strung together from what turned out to be a fabricated report by an unreliable refugee in Switzerland—an Ahmed Chalabi of the times, perhaps—who claimed the Germans were preparing biological weapons. The shaky rationale behind this rush to vaccinate will sound familiar to public health officials work-ing in the post-9/11 world. Bioterrorism had been broached and was now impossible to ignore. There was no way to quantify the threat, and seem-ingly no way to put it in perspective or assess the risk. Describing the bioterrorist threat as he understood it during the war years, Bayne-Jones's comment has a refreshingly contemporary sound to it: "All very interesting, and as alarming as you want to make it, and as foolish as you want to make it."[97]

Bioterrorism was not the only justification for the vaccine. The Brits wanted lots of it because a yellow fever outbreak in Sudan, near a British base, had killed 1,700 people in the summer of 1940.[98] But with Simmons riding herd, the surgeon general expanded the vaccination recommenda-tion to all military personnel stationed in tropical regions, stressing the "possibility of the introduction of this disease into areas where it did not exist . . . by the willful introduction by enemy action." Since no soldier knew where he was headed, all would be vaccinated after signing up.[99]

"Sterile Blood from Healthy White Adults"

In January 1941 the Rockefeller Foundation provided its first lot of vaccine to the army, and vaccinations started the following month. In the meantime, Johannes Bauer had been communicating regularly with Thomas Turner, a former Rockefeller scientist who taught pathology at Johns Hopkins University. Turner had promised to provide human serum for the vaccine at $30 per 500-cc bottle. He would eventually provide hundreds of liters of serum from more than 900 Baltimore donors, most of them associated with the Hopkins medical center. Initially, Turner asked the donors whether they had ever had hepatitis, but waived that question after being told by Bauer that the serum was being sterilized. "I assume all you will need," he wrote in one communication, "is sterile blood serum from healthy white adults."[100] The blood Turner provided was, to be sure, the best available in Baltimore in eugenic terms. It had been tested for syphilis and bacterial contamination. But no effort was made to avoid pooling the donations and, as a result, scores of individuals' serum was sometimes included in the same lot of vaccine, and the serum of single donors sometimes ended up in hundreds of thousands of shots.

In December 1941, while on a visit home, Soper warned Simmons, recently promoted to general, about the problems with jaundice and encephalitis in the yellow fever vaccination campaign in Brazil. Despite these reservations it was "assumed that all troops will be vaccinated against yellow fever," Soper wrote in his journal on December 22. Four days later the Surgeon General's Office ordered all 11 million American servicemembers vaccinated. The Rockefeller Foundation had promised to deliver the vaccine, free, as a patriotic service.[101]

By now the Brazilian experience was common knowledge on the fourth floor of the Rockefeller Institute. Theiler believed Sawyer was playing with fire and told him so. But Sawyer outranked Theiler and was not to be moved.[102]

In December, Sawyer got a final report from Soper and Fox on the jaundice investigation in Brazil and sent back a lonely message of thanks: "The very extensive application of vaccination without the occurrence of jaundice both in Africa and in the United States [up to then] suggests that the precautions which have been taken to prevent new contamination may be adequate. I certainly hope so as the vaccine issued this year from this laboratory alone has exceeded 3 million doses . . . we should feel very much safer if we knew the nature of the agent or agents responsible for postvaccinal jaundice and we are grateful to you and Dr. Fox for continuing to follow every possible lead."[103]

And so, the vaccinations intensified, and requests for Baltimore serum mounted. The laboratory men on the fourth floor of the Rockefeller Institute worked day and night, seven days a week. By April 1942 the health division had supplied 4.6 million doses of the vaccine to the War Department and another 1.8 million to the British government.[104] For a few months there was no trouble and then, suddenly, there was big trouble. The jaundice outbreaks in Brazil, it turned out, were a "full dress rehearsal for what was coming in New York."[105]

A "Dark Page" for Vaccines

The reports started trickling in from California in late February 1942. At bases around the state where troops were mobilizing for war in Asia, large numbers were reporting in at sick call, all with the same classic symptoms of jaundice—yellow eyes, yellow skin, exhaustion, nausea without vomiting, urine the color of tea, and putty-colored feces. At first, some feared a reversion of the virus, that yellow fever itself had broken out. Mosquito netting was laid over military hospital beds. The army notified Karl Friedrich Meyer, a tough, Swiss-born virologist who headed the George Williams Hooper Institute in San Francisco and saw himself as the West Coast competition for the Rockefeller boys. Meyer had devised a sterilization method that saved the California canning industry from botulism, discovered two encephalitis viruses, published 800 papers, and was a one-man fire brigade for public health emergencies.[106]

He got a call from Simmons's office asking him to investigate an outbreak of what seemed to be leptospirosis, a bacterial liver infection, at the Victorville Army Air Field in the Mojave Desert. Simmons said nothing about the yellow fever vaccine. Meyer described what happened next: "I went down, took plenty of blood specimens, stood by for two days during sick call and saw the cases rolling in. Typical jaundice." Meyer had passed through London a few years earlier and watched officers being vaccinated against yellow fever, with Findlay giving them gamma globulin shots afterwards to prevent hepatitis. Suddenly, it clicked. "Lying on the cot one evening with the whole business going through my head, the London experience came back. I said, 'By George, Sure! This crowd is all getting yellow fever vaccine.'"[107]

Next morning, Meyer asked for the records and saw that the outfit had been immunized in New York and New Jersey before shipping out to California. The incubation time "was just beautiful, between sixty to ninety days"—precisely what one would expect with a viral liver infection. Meyer headed back to San Francisco and found Findlay's articles on serum-associated jaundice. Meanwhile, the calls kept coming in from other base doctors. Initially, all the cases were linked to 2 of the 141 lots of vaccine used up to that point—lots 335 and 336. Within a few months, 50,000 hospitalizations due to jaundice would be reported subsequent to yellow fever vaccination; about 100 people died. Scientists eventually arrived at an estimate of 300,000 soldiers injected with contaminated vaccine. Around the country and the world, a muffled panic crept over the newly mobilized American forces. With American forces in the Pacific just recovering from Pearl Harbor, more than 2,400 men were now hospitalized with hepatitis in Australia, another 2,400 in Hawaii. Lt. Gen. Joseph G. Stilwell led his famous retreat from Burma while struggling with a bout of vaccine-linked hepatitis and would die in 1946 of liver cancer. "It was more shocking to the commander of the Army, I think, than any bombardment," Bayne-Jones noted.[108] "At the moment that was going on so many people were sick, and there were so many critical things in the offing and going forward . . . The Battle of Midway was won by air pilots some of whom were suffering from jaundice." In North Africa, U.S. tank commanders laid low with hepatitis B fought to a draw with Rommel's tigers, who were sick with hepatitis A.[109]

In afflicted American units, 10 percent or more of the troops were hospitalized or bedridden. In London, a health officer reported that soldiers and senior officers were falling like flies. "The personnel situation here is terrible at best . . . I think I ought to tell you that newly arrived units from the United States have disseminated information to all ranks as to the probable cause of this epidemic. British health agencies are tremendously concerned over the possibility of this epidemic spreading among British troops and the British civil population." When soldiers were pushed back into service too quickly, they relapsed. Some officers recommended the affected troops get a high-protein, high-carbohydrate diet. A base commander asked Meyer whether it was safe to send slightly ill pilots on training flights. "Are they going to crash?" Meyer took the question seriously. "I said, 'the answer may be yes, so better go easy on this thing.'"

This was the only major epidemic to sweep the U.S. military during the war.[111] It is difficult to assess its effect on the war effort, but in all likelihood it caused a small delay in the liberation of Europe. Things might have been worse. It is said that Winston Churchill was to have been vaccinated from a contaminated lot in 1942 but took ill with a cold. Considering the condition of Churchill's liver, it might have been a shot that changed history.[112] On March 20, 1942, Meyer sent a note to the army epidemiology board saying he believed the jaundice resulted from yellow fever vaccine. He went over the head of Simmons, who had trouble facing the facts. Bauer and Sawyer arrived in San Francisco the next day and stayed at the St. Francis Hotel. They met with Meyer late into the night, then set off on a tour of stricken bases.

It was a ghastly road trip through rainy California, a journey through a human calamity produced by a tiny group of benefactors of humanity. Here was a nation preparing for war, with men being shipped, flown, and trucked around the country and the world to fight two cruel regimes, and suddenly tens of thousands were ill because of a vaccine made by the most illustrious scientific institution in the world. To Meyer, the cause of the epidemic was obvious in all its gloomy dimensions. But Sawyer was in denial. The scientists spent two weeks on the road, driving in an army car from San Francisco to Moffatt Field near Palo Alto, then out to the Central Valley, and finally as far south as Riverside, Fort Ord, and San Diego before boarding a train back to Berkeley, where the Rockefeller Institute had an office in the state public health department. In Stockton on March 23, they found 48 cases. "One gained the impression as if all were repeating exactly the same words, so exactly similar were their answers," Bauer noted.[113] At the Mojave Anti-Aircraft Range, a godforsaken camp 30 miles outside of Barstow where the men slept in tents in the desert, 10 percent of the regiment was sick, and the commander, a major, was frantic. Even the infirmary was shorthanded because the surgeon had come down with jaundice. Everywhere they went the scientists found hundreds of new cases and collected mountains of paperwork. When Sawyer phoned Washington every day he received news of new outbreaks—in Missouri, at the Jefferson Barracks, at Fort Belvoir outside Washington, DC, at West Point, at Fort Polk, Louisiana. All, in a sense, Wilbur Sawyer's handiwork.

"During the whole time there wasn't a word said about the vaccine," Meyer said. He found Sawyer's silence eerie. "I had known him since 1913 and he was always very frank and open with me." This time, though, "we were together for four weeks daily, and although he knew of the observations . . . in Brazil, he said nothing."[114] The weight of the world rested on Sawyer's shoulders. At one point the group found an unvaccinated secretary with hepatitis, and the news made Sawyer almost giddy. He was sure the jaundice, which he repeatedly characterized as "mild," was from a virus brought on base by civilians. "Gradually the evidence accumulates," he hopefully wrote to his assistant.[115] If Sawyer had acted quickly to stop the vaccinations, tens of thousands of infections and many deaths might have been avoided.

Simmons, who had bought into the program every bit as vigorously if not more so, believed as late as mid-April that vaccination "should be continued by all means."[116] But after a round of high-level meetings with his superiors and top scientists at the Rockefeller Institute, the U.S. Public Health Service, and Johns Hopkins, Simmons was forced to retreat. In an April 15, 1942, memorandum to the army surgeon general he recommended suspending production of the Rockefeller vaccine. The record suggests that the existing vaccine continued to be used for up to two months. Dr. E. Harold Hinman, a public health service physician based in San Salvador, reported coming down with jaundice three months after being vaccinated on April 20 at the Marine Hospital in Baltimore. Hinman's son, Alan, would later become a leading CDC vaccination official.[117]

In June, Goodner and Theiler threatened to resign from the Rockefeller Institute if more vaccine was made with human serum.[118] While debating what to do with the Rockefeller vaccine, the military turned to Dr. Mason V. Hargett, an unassuming Kentuckian who was using Theiler's strain of the virus to produce a serum-free vaccine at the Public Health Service's new Rocky Mountain Laboratories in Montana.[119] The site had been chosen for safety reasons—there were no mosquitoes in the Montana Rockies that could carry yellow fever, should the virus somehow escape the lab. While Sawyer had used human serum to stabilize the vaccine before it was inoculated into the eggs, Hargett found that if he freeze-dried it under a strong vacuum—lyophilizing was the term of art—and kept it stored in dry ice, the vaccine was stable and dependable.[120] Noting the trouble the Rockefeller people had with jaundice—Hargett had visited their operations in Brazil in 1939—he had decided in 1940 to eliminate serum. "As soon as the hepatitis epidemic happened, the Rockefeller switched to vaccine without the serum," Hargett recalled in a 1985 interview. "I don't know why they didn't do it before."

In the postmortem that Simmons, Meyer, Sawyer, and others wrote up on the hepatitis outbreak, they found that most of the cases stemmed from five lots of vaccine. Seven of the Baltimore donors had a history of hepatitis; scrutiny focused upon two of them. John Franklin, a 26-year-old Hopkins internist, had been hospitalized with jaundice in December 1941, scarcely a month before being bled for the program. His serum was associated with at least one heavily jaundice-producing lot. Another donor, J. H. Morrison, had donated a total of 250 milliliters of serum—a coffee-mug full—in three sessions beginning on January 30, 1942. His serum was used, incredibly, in eight separate lots of vaccine, including all of those associated with high jaundice rates. Although some of the hepatitis may have been passed along by other donors, Morrison's serum may have infected 300,000 people[121]—an amazing indication of the replicative power of the hepatitis B virus.

The supreme irony of it was this: Morrison was director of the Baltimore Red Cross blood donation program.[122]

The investigators reasoned that efforts to kill the virus had been inadequate. The serum was supposedly heated for an hour at 56 degrees centigrade, but a review of the technique showed that it was only *bathed* in 56 degree water that long. Since the serum had been refrigerated before being plunged in the sterilizing bath, it probably took half an hour to reach 56 degrees.[123] Later investigations showed that it could take more than an hour at 56 degrees to kill hepatitis B virus.

News of the epidemic was suppressed until July because of its "potential value to the enemy" and because the disconcerting news "might adversely effect the morale of troops being vaccinated at staging areas and ports of embarkation."[124] Secretary of War Henry Stimson discussed it at a July 24 news conference, but press coverage was muted and criticism of the program was rebuked as unpatriotic. The army's circular on the outbreak said "[T]he disease is not contagious and hence does not constitute a danger to the public health."[125] When the *Chicago Tribune* opined that the army had been "guilty of a grievous error of judgment" in failing to test the vaccine adequately, *JAMA*'s editor, the powerful Morris Fishbein, lashed out at the paper.[126] The 20,585 cases and 62 deaths reported up to then were "a small price to pay," Fishbein wrote, considering how deadly yellow fever could be.

The jaundice epidemic had little impact on the careers of the men responsible for it. Simmons became dean of the Harvard School of Public Health after the war. His assistant Bayne-Jones wrote the official history of the affair for the army, returned to Yale, and worked on biological weapons. Sawyer retired from the Rockefeller Foundation in 1944 at age 65 and became the health director of the U.N. Relief and Reconstruction Agency. In 1951, when Max Theiler won the Nobel Prize in Medicine for his contribution to the yellow fever vaccine and viral cultures, Sawyer was shocked, believing that he had done as much as Theiler to create the vaccine. Soper agreed that since Sawyer "shouldered the blame" for the disaster, he should have been given credit as well for creating the vaccine.[127] Sawyer died exactly a month after Theiler won the prize. "It killed my husband," said his wife, Marjorie.[128]

In a postscript to the affair, a group of doctors at the National Institutes of Health and the Veterans Administration surveyed the health of the men vaccinated with the contaminated vaccine in the 1980s and found to their relief that only 0.5 percent continued to be carriers of the virus. The death rate among the survivors of the initial infection appeared to be no higher than that of other veterans, although the risk of liver cancer was slightly elevated.[129] To what extent these men infected others has never been studied. Transfusion hepatitis was an enormous problem during the war, and in the immediate postwar period, 25 to 30 cases of hepatitis were reported among 130,000 American troops each week in Germany. "American forces in Europe represent a group in which more cases of this disease [hepatitis B] are occurring than will be found anywhere else in the world," wrote Ross Gauld, who had been involved in the yellow fever vaccine investigation.[130] Thousands of people died each year of hepatitis contracted from plasma, surgery, or infected needles.[131] In 1970, a *Lancet* article reported there were 120,000 cases of post-transfusion hepatitis per year in the United States.[132]

As for the yellow fever vaccine, Hargett's laboratory cranked out 4 million serum-free doses by the end of 1942 and continued to make the vaccine until 1957, when the government turned it over to the National Drug Co.[133] None of the 11 million Americans vaccinated against yellow fever during the war got yellow fever. Then again, none was challenged with yellow fever. Simmons's fear of biologically trained killer mosquitoes was not realized. It was all somehow in vain.

If the jaundice epidemic had a positive outcome it was in spurring research into hepatitis that would eventually lead to vaccines against both forms of the disease. Scientists from the National Institutes of Health were dispatched in 1942 to conduct experiments on the yellow fever vaccine. At the Lynchburg State Colony in Virginia, doctors injected 190 mentally retarded people with varying levels of yellow fever vaccine and other substances, sickening at least 30 of them. Other experiments were conducted among conscientious objectors, orphans, and inmates of federal prisons. The aim was to determine the agent in the vaccine that caused infection, the dosages required for infection, and how the virus could be eradicated from blood products. These experiments, while key to finding a way to provide clean transfusions, had high costs. In a September 1952 speech following the jaundice deaths of three prisoners, Pope Pius XII chided scientists for stretching moral limits in their experimentation for medical research. Around the same time, the American Medical Association began puzzling out its own standards on experimental ethics, a question that had been examined in the Nuremberg trials.

The American public got few details of the yellow fever fiasco. Even full disclosure probably wouldn't have made much of an impact on public opinion, as we saw in New York in April 1947. Overall, preventive medicine had worked wonders during World War II. During the Spanish-American War, 13 men died of sickness for each battle death; in World War I the ratio was 1:1. In the European theater during World War II, 1 man had died of sickness for every 85 who died fighting Nazis. This kind of success was unheard of.[134] The medical elite entered peacetime with a powerful sense of mission. "While the primary purpose of the army's health program is the conservation of manpower during war, it will ultimately exert a profound influence on the health of the civil population," Simmons predicted. "After the war millions of young men will return to civilian life who are physically stronger, who are immunized against various diseases and have received training in the fundamentals of hygiene." More than 100,000 officers had served in the Army Medical Department and were returning to the states "with new knowledge and new interests. All these factors will strengthen and stimulate the further development of civilian public health."[135]

Vaccination would be the key weapon deployed against viruses after the war. In the new cold war, public health could be pitched as something like a military necessity—a matter of national security. Doctors who had done heroic things during the war were insistent public health taskmasters in its aftermath. The means were there but organization and laws, compliance and authority were lacking. Great things had been accomplished, and vaccinology was readying its attack on the next great foe: polio.

CHAPTER 5

—

THE GREAT AMERICAN
FIGHT AGAINST POLIO

*The dead are soon forgotten, but man mutilated, man paralyzed, is a source
of guilt and shame to everyone who regards him. He cries out for action,
in a way that the dead can never do.*

—JOHN ROWAN WILSON

ON APRIL 26, 2004, the stormy air was clotted with cherry tree petals
and their fragrance as 70 people gathered in the cafeteria of Franklin
Sherman Elementary School, a sheet cake–shaped red brick building in
McLean, Virginia. The school's principal, Marty Smith, and about three
dozen 12-year-olds were packed into the front rows. Toward the back
were a dozen or so baby boomers in varying states of fadedness and several trim, white-haired men who looked like doctors from central casting.
At the podium, a raised board at one end of the linoleum-tiled floor,
stood Anna Eleanor Roosevelt, Franklin Delano Roosevelt's winsome
blue-eyed, blonde-haired granddaughter. She was flanked on one side by
a yellow iron-lung machine, on the other by an incubator of the kind that
keeps tiny babies alive in neonatal wards.

Polio disappeared from America long ago, and the March of Dimes,
the fabled charity that organized its conquest, has moved on to funding
research into preventing premature births. But on this day, the March of
Dimes was celebrating the Salk vaccine field trial, which had begun at the

Franklin Sherman school 50 years earlier with the vaccination of a classroom of second-graders. Richard Mulvaney, a family practitioner from down the street, administered those first few shots, and he and several children vaccinated that day came forward to share their recollections. The public attendance, as represented by the news media, was sparse for this history lesson. There were a few local TV journalists, a Metro section reporter from *The Washington Post*, and a few colleagues whose stories would run deep inside their papers the next day, if they ran at all. There were no polio patients at the celebration. Polio was gone and forgotten, although there were as many as 440,000 American polio patients still struggling with their disabilities.

Listening to Anna Roosevelt and the graying graduates of Sherman Elementary describe the experiment they had joined was like unearthing old, dusty campaign buttons from a box in the attic. Their ideals had come down pretty much unscathed from the Great Depression, and this was a chance to take them out, polish them, and wear them on the lapel. They had trusted their government because it stood for the community of the nation. Nothing else could. That was the idea preserved here at Franklin Sherman. It seemed like a radical idea, although it had been easy enough to sell to the public in 1955. A great wartime leader, crippled himself, and millions of fellow Americans had contributed to the fight against polio, which culminated in a successful vaccine that quickly wiped out that fearful disease. The conquest of polio had been a great moment in American history, and the pinnacle of vaccination's success.

THE CHILDREN who participated in the Salk polio vaccine field trial in 1954 were called "pioneers" rather than trial subjects. And if, viewed through a contemporary lens, this was a violation of informed consent, the name was more than mere euphemism. Parents volunteered their children for the trials because they desperately wanted protection from polio, for themselves but also for the community as a whole. "The country needed us," said Anna Roosevelt.

What the government, and the health professionals who represented it, had done for the public was perfectly evident in 1955. Better nutrition and housing, improved sanitation, vaccines, antibiotics, steroids and other drugs had radically altered a child's chances of survival to old age. In 1920, life expectancy at birth had been 54, in 1955, it was 70. The California Board of Health reported 42 cases and two deaths from diphtheria in 1955, compared to 110,000 cases and 700 deaths in 1924. Because of vaccination and antibiotics, whooping cough had killed only 89 people in the previous five years across the United States. Tuberculosis cases had fallen from 113 to 9 per 100,000 from 1920 to 1955. The army had driven malaria out of the South, and its malaria control agency, based in Atlanta, was transformed after World War II into the Communicable Disease Center, later the Centers for Disease Control, the world's premier public health agency. With money pouring into his endeavors, writes historian John Duffy, "the white-coated medical researcher came to symbolize all that was good and noble in the brand new world of science."[1]

The epitome of this transformation was Jonas Salk, whose vaccine made it all happen and who always wore his white lab coat when the reporters came to visit. When Salk was born on October 28, 1914, his family lived in a tenement house at the corner of 106th and Madison in East Harlem, the Manhattan equivalent of a little log cabin. His father, Daniel Salk, was a designer of women's neckware and blouses. His mother, Dora, raised her three boys to be ambitious, hardworking professionals. The Salks' neighborhood has seen many generations of immigrants, and their tenement was torn down long ago, replaced by a 1,200-unit housing complex named for George Washington Carver. No plaque marks the birthplace of America's first scientific celebrity, nor does any trace of him remain. Salk eventually removed to the Xanadu of La Jolla, California, and he died there in 1995. He left only the grateful memories of a generation that lived in fear of polio, and the tradition of guaranteed vaccination for all Americans, rich and poor.

The Salk polio vaccine was a feat of technological advance and social engineering that involved hundreds of scientists, technicians, and other workers. But unlike the Manhattan Project, built in secrecy, or the moon shot, carried out by a monastic military-industrial technocracy, the vaccine was created thanks to the largest civilian mobilization in the history of the world. The National Foundation for Infantile Paralysis, which managed the attack on polio and funded most of Salk's work, relied on the contributions of 100 million Americans—a combination of science, optimism, and communal effort. At the heart of this outpouring was the polio-crippled man who led the nation through the Great Depression and the Second World War, who had created the polio foundation while searching for help for his own tragic predicament. Roosevelt oversaw the gigantic military-industrial state whose war powers and resources gave us the edge over the Nazis and Japanese, but also protected the American boys, and later their children, from many of the deadly communicable diseases that had plagued armies throughout the history of man and war. Polio vaccine, presented by the equable, hyper-rational Dr. Salk, was the human face of technowar.

Roosevelt's Other War

The polio vaccine's lineage begins with Roosevelt, whose image has been stamped on all the dimes that have marched out of the U.S. Mint since 1946. Linking the polio campaign with the Great Depression—*Brother, can you spare a dime?*—it is a resilient advertisement for philanthropy. The nation brought its technical resources to bear against the disease, but the spirit and ideal of collective battle originated in Roosevelt's own humbling experience.

At Campobello, his patrician family's resort off the coast of Maine, Roosevelt awoke from a nap on August 10, 1921, with a sensation of numbness in his legs. The next day his legs began to weaken, and by the third day he could no longer get out of bed. Two weeks later, with Roosevelt feverish and paralyzed over most of his body, doctors diagnosed polio. He was 39, quite old for a polio case. At the time three- to five-year-olds were the most common patients, and the virus rarely struck anyone over 9—indeed, some researchers have recently speculated that Roosevelt suffered from Guillain-Barre Syndrome, an autoimmune condition, rather than polio.[2] In any case, the illness fatally changed FDR's fortunes. He went from being a rising star of the Democratic Party, who in 1920 had been Ohio Governor James Cox's running mate in a failed presidential campaign against Warren Harding—to being an incurable cripple.

In 1924, Roosevelt discovered Warm Springs, the center that became one of the symbols of his healing mission. Roosevelt found no cure at the mineral baths of Warm Springs, which was located in the pine woods an hour south of Atlanta, but swimming and soaking in its baths and playing cards with the other disabled men and women gave him both physical strength and emotional solace. A local blacksmith worked out hand controls that allowed Roosevelt to drive around the red-dirt roads in his convertible, and he would often stop to talk to people—black sharecroppers, white farmers and shopkeepers, the druggist who came out to the curb to bring him a soda.[3] These chats broadened his contact with everyday working men and women. The bracing humor and pugnaciousness of his fellow patients, a group of unfairly fettered underdogs, reinforced Roosevelt's vision of a nation that used brains, determination, and solidarity to overcome obstacles.

To be sure, Roosevelt tried everything—swimming, hanging from a parachute harness, hot towels, physical therapy—in an unsuccessful effort to walk again. That was a sobering, politicizing fact. "He'd had a golden life. Nothing had ever tarnished young Franklin," says Hugh Gregory Gallagher, a biographer of FDR's who has polio himself. "One of his sons said once, 'Let's face it, my father was a playboy until he got polio.' Then he got polio and he was determined to defeat it, and he worked 7 years and he never defeated it! That was a great lesson. It made him far more sympathetic to other people's problems."

In 1926, Roosevelt bought Warm Springs, and like most of his business ventures it lost money. The arrival of polio patients drove out the high-end consumers of spa culture. After being elected governor of New York in 1928, the aristocratic Roosevelt fobbed the business off on his new law partner, Basil O'Connor, a feisty and dapper Massachusetts Irishman with a monocle. It was a fob that changed history. "If it had not been for the financial burden Roosevelt assumed when he bought the Warm Springs property," one writer has noted, "Basil O'Connor would never have started building the great fund-raising machine of the National Foundation for Infantile Paralysis."[4] Roosevelt had never been much of a businessman, but Warm Springs was at the center of his heart—he would go there to die eventually. O'Connor, who'd started his professional life as a Taunton newspaper delivery boy but made his first million by age 35 as a Wall Street lawyer, took Roosevelt's vision and built it into a smooth fund-raising machine.

In 1932 the nation chose Roosevelt as president—"a crippled man," as Gallagher put it, "to fix a crippled economy with a crippled government."[5] Roosevelt hid the extent of his disability, concealing the fact that he could not walk unaided and the daily suffering and discomfort his crippled leg and abdomen muscles caused him. But though he never spoke of his own case Roosevelt often remarked on the suffering of the "polios," as these patients called themselves, and used the pulpit of his own affliction to proselytize on behalf of care for others. He applied the example of polio to teach Americans, unified in wartime, what made their cause a just one. "The generous participation of the American people in this fight," he said during his last March of Dimes radio address, in 1944, as GIs battled the Nazis at Anzio, "is a sign of the healthy condition of our Nation. It is democracy in action. The unity of our people in helping those who are disabled, in protecting the welfare of our young, in preserving the eternal principle of kindliness—all this is evidence of our fundamental strength . . . How different it is in the lands of our enemies! In Germany and Japan, an individual's usefulness is measured solely by the direct contribution that he can make to the war machine."[6]

With the United States in Depression and federal funds in short supply, O'Connor organized a series of dances on Roosevelt's birthday to raise money for the cause. They employed the Democratic Party apparatus to put on the Birthday Balls, with thousands of postmasters—each a federal appointee—in charge of them. The first Birthday Ball took place on January 30, 1934. "Dance so that others might walk," was the motto that inspired hundreds of thousands of people to join a dance-a-thon at 6,000 halls across the country. "It was Roosevelt's conception that the March of Dimes be a national organization but also a neighborhood organization," says Gallagher. "They had movie stars by the gazillion. Presidential Balls. But Roosevelt had this extraordinary ability to make Americans feel they were all in this together." The first event raised over $1 million, more than O'Connor knew what to do with, and he and Roosevelt created an 11-member commission to dole out grants. The commission members included Navy Secretary James Forrestal, Alice

Longworth—Teddy Roosevelt's daughter—and Jeremiah Millbank, a health philanthropist. The only member who knew anything about research was Paul de Kruif, the manic former Rockefeller Institute bacteriologist whose 1926 bestseller, *Microbe Hunters*, had inspired a generation of bright young men and women, including Albert Sabin, to enter medical science.

De Kruif suggested that Birthday Ball money be put to work funding research into a vaccine. "Why do you use all that dough to dip cripples in warm water?" he asked Arthur Carpenter, manager of Warm Springs. "Why don't you ask the president to devote part of that big dough to research on polio prevention?" De Kruif, who tended to get swept up in the hype of the latest scientific findings, had a delightful time doling out cash to scientists for the president. He used his position on the commission to trade grants for tidbits of scientific gossip that he would later turn into breathless prose for *Reader's Digest, Country Gentleman*, and other mass-market publications.[7]

In the late 1930s, the Commission for the Birthday Balls was renamed the National Foundation for Infantile Paralysis, and fund-raising efforts intensified. More than 2.6 million dimes marched into the White House at singer Eddie Cantor's urging to honor FDR for his birthday on January 30, 1938. Cantor, who'd started his career as a Tammany Hall shill, coined the "March of Dimes" as a clever take-off on the "March of Time" newsreels shown in movie theaters. The National Foundation used the media, public relations, and a corps of volunteers in novel ways, and its overhead was large. O'Connor received no salary but traveled in high style, staying at the Waldorf-Astoria when he was in New York. The legions of public relations hacks he hired enjoyed liberal expense accounts for putting their huckster skills to a noble cause. Hollywood, which had thrown its support behind Roosevelt in 1932, joined in the fund-raising effort with celebrity balls and publicity films such as *The Crippler*, in which Nancy Davis, the actress who would marry Ronald Reagan, played a young National Foundation volunteer fighting polio. Louis Armstrong blew his horn to help polio research, and Jimmy Stewart, Mickey Rooney, and Judy Garland made radio appearances. James Cagney and Humphrey Bogart played "tough guy" dimes in one on-air show. Hollywood's relationship to polio wasn't condescending—it had its own victims, including Helen Hayes's daughter, Mary, who died of polio at 19.

Polio — a Mysterious Crippler

It was entirely appropriate for polio to become the focus of postwar medicine in America, because polio was the quintessential late-twentieth-century disease. This frightening crisis of the nervous system, an assault on the motor neurons that struck suddenly in mansions and slums alike, was partly a curse of modern sanitation. Polioviruses had probably infected human beings for millennia; glyphs depicting crippled pharaohs are etched on Egyptian tombstones. But no one seems to have paid much attention to the disease before the nineteenth century. The first outbreak in the United States was reported in Vermont in 1894. Prior to this episode, poliovirus was rarely a widespread problem because it was so omnipresent that babies were routinely exposed to it. Maternal antibodies protected them, and if they became infected as infants they were apparently shielded from polio's worst effects by an immune response that remains mysterious. Babies generally suffered nothing more than a bad case of diarrhea with headache. Polio epidemics in American cities coincided with the arrival of clean filtered water, which had the unintended consequence of safeguarding babies from routine exposure to the virus. Over time, as communities were protected from dirty water, polio infected older and older victims.

The suffering of polio patients was vast. The virus destroyed their motor nerve cells but left the patient perfectly able to feel pain, to live in his or her own exquisite hell. "It was an out of body experience. Everything hurt," says Roosevelt's biographer, Gallagher, who was stricken with polio while a student at Haverford College outside Philadelphia in May 1952. "I found myself floating above my body. They couldn't give me any painkiller because it would kill my breathing. I sat on the outside of this cave where everything was so inviting, looking down at my suffering body. I waited for my parents, they were standing out in the rain and I said, 'I really want to thank you for being such wonderful parents.' And my mother said, 'Oh promise me you'll be here when we come back.' And I said, 'Oh all right, I'll try.'"[8]

The disease made its dramatic entry into American consciousness with an epidemic in the summer of 1906, not long after the Rockefeller Institute of Medical Research opened its doors. Karl Landsteiner, a Rockefeller scientist and Nobel laureate, showed in 1908 that polio was a "filtrable virus," the term given, in an age before there was any clear sense of what a virus was, to infectious particles too small to be bacteria. Simon Flexner, who founded the great institute, staked out polio as his disease, but his research bore little fruit. In 1916, the virus paralyzed 27,000 people and claimed 6,000 lives in 26 states, including 2,400 New Yorkers. By now, public health officials understood the difference between germs and dirt, but viruses were too small to be seen under a microscope—there was considerable debate about whether a virus was alive at all—and the polio crisis threw them into confusion. Some saw no option other than to return to their old practices. Like tuberculosis and typhoid, and even smallpox before it, polio was declared a disease of "filth," a spawn of wretchedness to be battled with attacks on dirt and poverty that had characterized the sanitarian campaigns of the previous decades. The fact that poor black neighborhoods suffered less polio than affluent white ones was accomodated by a racialist assumption that blacks were somehow genetically immune.[9]

Urban health commissioners fought polio with the tactics they had used against cholera half a century before: mass fumigation (which New York had abandoned in 1915), flushing the streets with lime, forcing landlords to clear their stoops of garbage—none of which did anything to halt polio's spread. The health commissioner of Pennsylvania posted guards at the state border to keep out New Yorkers, particularly immigrants.[10] It was round-up-the-usual-suspects time. Like the man who searches for his keys under the streetlight, instead of in the dark alley where he dropped them, public health, in its futility, stayed busy.

Throughout the interwar period, the depression, and World War II, polio plagued the cities. Its ravages were rarely on the scale of earlier epidemics, but its unpredictable spread was even more terrifying. Struggling with the medical mystery, the public tried to explain polio as the result of "unnatural" and "newfangled" aspects of modern life.[11] Some theorized that flies on bananas, a popular new food for children, were spreading the germ. Others became convinced that polio grew in the new store-bought wheat flour and bread, or was generated by electrical fields or auto exhaust. In a letter to the Rockefeller Institute, one concerned citizen contended that the digging required for New York's new skyscrapers was stirring up germs—a fresh gloss upon Sydenham's "epidemic constitution." Or perhaps, others theorized, pasteurization had removed something wholesome from milk that had protected kids from polio.[12]

This final suggestion was the closest to the truth. In a way it was not filth but its absence that caused the polio epidemics. Most of the endemic diseases of New York City—whooping cough, measles, scarlet fever, tuberculosis, typhoid, and infant diarrhea—were declining so fast, thanks to better public health and living conditions, that the city's overall death rate was lower in 1916 than in 1915, despite the polio epidemic. Just as typhoid's spread had preceded modern methods of filtration, polio epidemics occurred *after* filtration methods had arrived. The epidemics worsened as the century went on. Women of childbearing age had not been exposed to polio and their placentas and milk did not contain protective antibodies to share with their infants. When polio did appear, it attacked a vulnerable field. Inside the family it spread through the bathtub, the viral particles shed by one bather clinging to the porcelain or towels used by the next. Only about 1 in 200 people infected with polio suffered paralysis to any degree; most simply had a gastrointestinal upset or fever for a few days, or no symptoms at all. Thus polio spread unseen like a stealth invader.

Polio was not the greatest killer of postwar America. It was not even the greatest microbial threat. The worst polio epidemic year in America, 1952, saw 57,000 cases, with 3,145 deaths and roughly 20,000 cases of permanent paralysis that could be mild to severe. That same year, by comparison, 24,000 people died of tuberculosis and 46,000 died of pneumonia and flu. In single years earlier in the century, diphtheria, whooping cough, scarlet fever, and even measles had each caused more death and disability. But in 1952 there were ways to deal with flu, tuberculosis, pneumonia, and the other infectious diseases—vaccines, antibiotics, and quarantine measures honed by decades of experience. If you followed medical advice and made a concerted effort you might avoid the worst effects of these diseases. Polio, on the other hand, struck randomly among the rich and poor, from childhood into young middle age. It was an "un-American" disease that showed no respect for those living clean, wholesome, hardworking lives. Its mysterious arrival every few summers undercut the nation's sense of technological advancement. Polio cases multiplied in the 1940s and 1950s, and the National Foundation took every opportunity to publicize these tragedies, keeping the disease in the public's mind.

There were treatments for polio, too, but they could seem as awful as the disease itself. The iron lung, invented in 1929, saved countless lives but seemed like a living coffin, and only intensified public fears of polio. The patient lay immobilized, flat on her back—a head attached to a tin box. A pump connected to the machine mechanically breathed for the patient, changing the pressure inside it with the rhythm of breath. Some patients lived a few months in an iron lung, then emerged to a wheelchair. Others lived a year or two in an iron lung then died; some stayed in the iron lung for 15 years. The machine terrified generations of children and their parents. "Do you want to spend the rest of your life in an iron lung?" became last-resort ammunition used by parents to get their children to wash their hands, brush their teeth, go to bed on time. Polio cripples filled the nation with a vague loathing. Other epidemics, however horrible, took their victims away. Most polio patients survived, "limping through life as a constant reminder to their fellows of the terrible visitation which occurred," the British doctor John R. Wilson wrote. "The dead are soon forgotten, but man mutilated, man paralyzed, is a source of guilt and shame to everyone who regards him. He cries out for action, in a way that the dead can never do."[13]

Life in the iron lung was difficult and lonely, especially for children, who were separated from their parents, though perhaps no less for young adult patients separated from their children. Nurses had to periodically remove patients from the machine to wash them, give them an enema, feed them, or to suction mucus from their throats. Sleep was elusive; no painkillers were dispensed for fear of suppressing the feeble respiratory impulse. "In the night," Kathryn Black wrote in a harrowing reconstruction of her mother's experience with polio, "patients lay listening to the liquid noises of other inmates, the squish of rubber-soled nurses retreating in the corridor, and the rattle of orderlies pushing IV stands through an empty hall. When sleep eventually came, it stayed two, perhaps three hours . . . the mind with no peaceful resting place tortured itself with its own desires. . . . No one could tell Mother when her pain would end; no one could control it or infuse it with meaning, or even offer an explanation for what, beyond the word polio, was causing her pain. . . . how haunting it must have been for Mother to be surrounded by the small, sober faces of suffering children, crying in the night for their own mothers."[14]

Hospitals typically segregated polio patients and limited their visiting hours, compounding the trauma for all involved. Parents were expected to bring in their children and leave them, with daily visits of only an hour or two. In the first two weeks of infection, no visits were allowed at all. After delivering his young daughter Joanne to the hospital, New Yorker Joseph Boettjer and his wife drove around the grounds to an area where they thought she'd been taken. "As we sat quietly for a moment in the back of the building we heard a child cry out several times, 'I want my Mommy.' We were almost certain it was Joanne. What a heart-wrenching scene! We would have been far better off not to have gone there."[15] Newspapers from the era depicted parents standing on ladders outside hospital wards, tapping on the windows to draw the attention of their children trapped inside.[16]

Polio left the patient's thinking and willpower unaffected, and many patients struggled to lead normal lives. "Polios," as they called themselves, created the disability movement by demanding that society accommodate their desire for independent living. Long before the movement became political, many polios matter-of-factly went about their lives, just as Roosevelt had done. Dr. Carl Weihl had just opened a pediatrics practice in 1949 when polio suddenly struck his middle-class Jewish neighborhood in Cincinnati. Weihl and at least four others on his block came down with the disease. At 29, he entered an iron lung and spent a year at Warm Springs before graduating to a wheelchair. Weihl's legs were permanently crippled and the only limb with full motor coordination was his right arm, though he could pull with his left. Thus handicapped, Weihl recommenced his career, directing the Babies Milk Fund, a clinic for poor children in the Over-the-Rhine neighborhood. A tinker modified Weihl's car so he could control the brakes and accelerator with his hands. "He drove all over the place in that car. He drove me to look at colleges," recalled Weihl's son, Albert, a physician at Yale University Hospital. As a clinician, Weihl tested several vaccines, including the oral polio vaccine created by Sabin, and a pioneering pertussis vaccine that Lilly introduced in the early 1960s. "He would always say, 'Well, I don't want to become disabled.' That was the standing remark with my father," Albert Weihl recalled. "'I don't want to become disabled.'"[17]

Error and Trial in the Polio Vaccine

The research that Paul de Kruif funded was not successful, though that was scarcely his fault. Researchers at the time were working with the wrong virus and the wrong theory of infection. Flexner, who had launched de Kruif's writing career by firing him from the Rockefeller, had managed to grow a strain of poliovirus in monkey brains; Peter Olitsky and Albert Sabin, two Rockefeller scientists, set out to grow Flexner's MV strain in rabbits in 1935 but found it would grow only in nervous tissue. These experiments seemed to back Flexner's hypothesis that polio spread along the olfactory nerve, not through the bloodstream. Many of the Rockefeller's polio resources went to testing the efficacy of picric acid and zinc sulfate, which were squirted into thousands of children's nostrils to prevent poliovirus from propagating. They had no effect other than to ruin many of the children's sense of smell. It turned out that the Rockefeller scientists were using the wrong strain of poliovirus. The monkeys Flexner had at his disposal could only be infected with the MV strain. If Sabin and Olitsky had worked with a strain of polio that grew in other tissues, like intestines, "chances are we would have had a breakthrough of major proportions in making a vaccine," Tom Rivers, then director of the Rockefeller Hospital, recounted. "As it turned out, we had to wait 14 years."[18]

De Kruif, who was happy to tweak Flexner, decided to put his money into scientists who were ready to ignore the Rockefeller's findings and plunge ahead with efforts to make vaccine. William H. Park, Flexner's scientific rival and bête noire, had thrown his weight behind Maurice Brodie, a young Canadian researcher at New York University with a promising polio vaccine. De Kruif decided that Brodie was going to conquer polio and arranged for Park's lab to receive $65,000, the Birthday Ball Commission's largest grant of 1934, to produce a vaccine made from a solution of polio-infected monkey brains inactivated by formaldehyde. In a separate effort, John Kolmer, a Philadelphia dermatologist famous for his failed attempt to save Calvin Coolidge's son from a bacterial infection, created his own bathtub vaccine out of a live virus that he had "tamed" by exposure to a castor bean extract. Both efforts were heralded in the press.[19] Both failed spectacularly. The two vaccines were tested on about 20,000 children, mostly in California, North Carolina, and Philadelphia. Kolmer's vaccine killed nine children, while Brodie's caused allergic encephalitis. At a November 1935 meeting of the American Public Health Association in St. Louis, Rivers and James P. Leake of the U.S. Public Health Service delivered the bad news: Leake is said to have called Kolmer a "murderer," although his remarks were cleaned up for the official transcript.[20] "All hell broke loose, and it seemed as if everybody was trying to talk at the same time," Rivers recalled. "Finally, Kolmer got up and said, 'Gentleman, this is one time I wish the floor would open up and swallow me.'" Kolmer continued to teach at Temple University until 1957; Park faded into a senile fog but died in 1939 with his reputation intact. Brodie, relegated to a Detroit hospital laboratory, died that same year at 36. His obituary blamed a heart attack, but many colleagues believed he had killed himself.[21]

The failure of these early vaccines was the nadir of the Birthday Ball Commission. By 1936, it was on the verge of throwing in the towel, but O'Connor didn't surrender easily.[22] To save the commission, he expanded its scientific advisory board to include Rivers and several other virologists and polio experts. Rivers figures in all histories of the polio vaccine because he was an outspoken, outsized, old-school physician from Georgia whose scientific judgment was highly respected. His memoir is a classic in the literature of infectious disease.[23] Rivers's first job was to deny Park another grant, an assignment he took on with great relish. "They needed me because I was a roughneck," Rivers said. The new advisory board turned to the basics. Of the funding priorities it drew up in 1938, "production of a good vaccine" was eleventh on the 11-point list. Earlier objectives included understanding how the virus was transmitted, where it grew, and whether it might be susceptible to antiviral treatment. Gradually, these goals were achieved, except for the treatment.

The Kolmer and Brodie vaccines were the polio war's bridge too far—

an attempt to link the crude pragmatism of the Pasteurian age with an era of more-studied vaccine manufacture that was beginning to characterize the production of viral vaccines. Most of the great names associated with the polio vaccine had worked with the Armed Forces Epidemiological Board during the war. Salk, Sabin, and Thomas Francis investigated influenza; Theiler, John Fox, and others focused on yellow fever. John Enders worked on mumps; Joseph Stokes studied measles; Rivers and Joseph Smadel, typhus. Joseph Melnick, David Bodian, Dorothy Horstmann, and John Paul deepened the foundations of polio research. Along the way enormous risks were taken, huge errors were made, and lives were lost. But all in all, like the storming of Normandy, it was an enormously successful campaign. Assigned to protect the troops from viral pandemics, these young scientists gradually developed more sure-footed ways to make vaccines.

The key to progress on viral vaccines was to find the proper medium and culture in which to alter and grow the virus. Bacterial vaccines were relatively simple by comparison. You found the proper medium, brewed up a nice batch of bacteria, and inactivated the organisms with formalin or phenol or some other chemical agent. There you had it—a whole inactivated vaccine, such as whooping cough or typhoid, or inactivated components such as diphtheria and tetanus toxoids. But viruses didn't grow in broth. It was not until the 1920s that Rivers and other Rockefeller scientists established that viruses grew only inside cells, sometimes very specific types of cells. Well before this dependent relationship was fully understood, pragmatic vaccinologists found ways to grow viruses and even to make vaccines from them—smallpox, after all, was a viral vaccine. But the first trick in making new viral vaccines was to figure out where to grow them—to kill a rabbit, as one author put it, you had to first catch the rabbit.[24]

Experience upon rueful experience had piled up procedures and principles for making vaccines while avoiding disaster. The makers of the yellow fever and Japanese B encephalitis vaccines had learned from smallpox vaccine debacles to make sure their vaccines were antiseptic; they also learned that growing vaccines in mice brains was a sure-fire way of provoking allergic encephalitis—the vaccine contained myelin basic protein from the animal's brain, which could cause the human immune system to begin attacking its own brain tissue. Adventures with diphtheria anti-serum had taught medicine the pitfalls of anaphylaxis and serum sickness. The Rockefeller's jaundice disaster instructed the wise vaccine maker to avoid human blood components, although the hepatitis virus itself was imperceptible until well into the 1970s. Rivers constantly worried about the risk of jaundice and allergic encephalitis. His policing of the field, on boards and commissions and from the Rockefeller bully pulpit, left a deep imprint upon the vaccine establishment.

Until 1986, when Merck's hepatitis B vaccine became the first vaccine derived from modern biotechnology, all viral vaccines of common usage were either killed or modified versions of the original virus—a choice one writer likened to "strangling a parrot or teaching it to talk."[25] When designing a killed vaccine, scientists might look for the most potent strain of the virus found in nature—Salk did this—then develop a way of inactivating it so that it would stimulate a good immune response without causing infection. If, on the other hand, the intention was to make a live-virus vaccine, the job was to get a weak version of the virus, but one that wasn't so weak as to be ineffective—this was what Sabin did. And how, exactly, was one to do it? Through trial and error. You took a stab at figuring out which animal flesh would serve as the best place to grow a good strain of virus, and then you grew it there. If it grew well, you weakened it by growing it in a series of other cell types, until it seemed weak enough to try out on a child. Theiler and his colleagues had established the principle. It was "rational empiricism," said Maurice Hilleman of the Merck Institute for Therapeutic Vaccines.[26] He also called it "guts and guesses," although Hilleman's science was always more rigorous than his off-hand manner suggested.

Trial and error is evolution by another name. Whether through studied manipulation of DNA to create transgenic species, or the stumbling steps of eighteenth-century variolators like the Suttons and Aspinwall, the viral vaccines in our arsenal have all been the products of artificial evolution conducted with more or less conscious intent. Whatever Jenner's vaccine began as, arm-to-arm and arm-to-cow-to-arm passages of vaccinia caused it to evolve into a virus that scarcely resembled anything in nature. Pasteur, honoring Jenner, tamed rabies, intentionally, by growing it in a series of rabbit brains. When Pasteur made his vaccine for rabies he was not aware the microorganism was a virus. By the postwar period, scientists were aware that they were harnessing evolution, but their technique—like natural evolution—still bowed and scraped before Serendip. Science relied upon the chance exertion of mutations, if not foghorns blowing in the East River, then something similarly dubious. The idea was to grow the virus in a tissue that was unfamiliar enough to create evolutionary pressures on the virus, which might, by chance, result in something humanly useful. Passaging a virus through a particular tissue—which meant growing the virus in a tissue, then harvesting it, diluting the harvest and implanting it in the same type of tissue, and repeating the process sometimes scores of times—might kill the virus or increase its virulence or diminish it. There was no way to know without experiment, although hypotheses would of course guide the scientists' hands.

Eggs had been used to grow typhus and yellow fever virus, but polio didn't grow well in eggs, at least not in commercial quantities. It did grow in monkey brains, but these were attached to monkeys, and you couldn't grow millions of doses of a vaccine in living monkeys, particularly since there would be no way to test the vaccine's consistency until the monkeys were killed. Another possibility was mice, despite the dangers. Under military contract, Lederle, Parke-Davis, and Squibb grew millions of doses of a Japanese B encephalitis vaccine in mouse brains for the troops stationed in Asia before and after the war. Working for Squibb during the mid-1940s, Hilleman was given 30 days to produce a viable Japanese B vaccine from scratch. He went to the company library, pulled out a 1943 Sabin article from the *Journal of the American Medical Association*,[27] and got to work. A few days later the plant manager "took out a 3-by-5 card and with a broken piece of pencil scribbled around it in circles and said, 'OK, we'll bid $3 a dose.' " The next question was where to put the lab. Squibb decided to use a barn. They carted out the horse manure and put in a concrete floor, where a group of women in shower caps sat harvesting the vaccine from mice. "I knew how to take these mice, pinch them the right way, pull the heads around, cut the scalp open, bingo!" Hilleman recalled. "The head would pop open, you take a scissors, cut out the brain and put them in a Waring blender. Fred Waring developed them for making cocktails. The virus would seep out of the blender, people would get infected with it. It was really something. Awful stuff."[28]

The military vaccinated hundreds of thousands of soldiers with this vaccine. The scientists who prepared it weren't shedding any tears for the GIs—they knew all too well what it was like. Beginning in 1948, when Hilleman started a decade of work under Joseph Smadel of Walter Reed's division of communicable diseases, "we got shot after shot of yellow fever, western equine, eastern equine. We just got the most crude crap!" Hilleman recalled.

As it turned out 1947 was not a bad polio year, but in 1948 more than 20,000 cases were reported. This was the year of the Berlin airlift, the Communist takeovers in Poland, Czechoslovakia, and Hungary. Cold war brinkmanship and nuclear test mushroom clouds were the daily fare of the newspapers. The fact that FDR had had polio "merely symbolized a more general perception that polio was a peculiarly American malady," wrote medical historian Alan Brandt.[29] Adding to the sense of a national campaign against the disease, the Military Air Transport Service beginning in 1950 transported hundreds of patients to special treatment hospitals and took equipment such as iron lungs into epidemic areas.[30]

There was new hope, that year, on the preventive front. Thanks to the work of physician-scientists such as William Hammond of the University of Pittsburgh and Joseph Stokes at the Children's Hospital of Philadelphia, immune gamma globulin (IGG), a refined human serum containing billions of polio antibodies, was used for the first time to protect contacts or anticipated contacts of polio patients. Hammond produced enough IGG to protect 55,000 children, and O'Connor in 1951 spent $14.5 million to corner the American market for the antibodies, which he turned over to the Office of Defense Mobilization for free distribution to public health officers in communities threatened by epidemic.[31] The problem was that polio had often spread before the symptoms and need for immune gamma globulin were recognized. If ever there had been a disease for which vaccination was the answer, polio was it.

The experiments that would enable the creation of a vaccine were the work of many men and women. In studies conducted in Egypt and Malta, John Paul demonstrated the centrality of the gastrointestinal tract to transmission of polio infection. He returned to Yale after the war with evidence that the disease was endemic in the world's poor countries, a finding that was key to understanding its epidemiology. Meanwhile, Bodian, at

Johns Hopkins, and the Australian scientist McFarlane Burnett, determined that there were at least three immunologically distinct forms of
polio.[32] Bodian and Dorothy Horstmann of Yale, working independently,
showed that antibodies against polio could protect against paralysis, which
proved that polio was carried to the nerves via the bloodstream.

The breakthrough came in the Harvard laboratory of John Enders, a
Connecticut yankee and heir to the Aetna insurance fortune. Enders had
trained pilots in World War I. After his wife died in the flu pandemic of
1918, he drifted into graduate English literature studies. One night he
went along with a roommate to look at the biology lab of the charismatic
Hans Zinsser and became hooked. Unlike most of the up-and-comers of
his day, Enders was a gentlemanly, eccentric fellow. Arriving at his lab at
10, he would work until 6 or 7, donning one of his lucky hats when it was
time to check an important experiment. He gave away his polio and
measles samples to anyone who could use them—and parsimoniously
returned unused grant money. Enders typically sent the younger men
and women in his lab to present important papers at conferences; he did
not crave the limelight and feared flying as a result of his wartime experiences.[33] When in 1954 the Nobel committee called to announce that he
was being awarded for his work, Enders insisted that two junior assistants, Thomas Weller and Frederick Robbins, receive a share of the
prize. He was a well-seasoned 57, and "impressed his colleagues because
he doesn't try to impress anyone," a contemporary noted.[34] Sam Katz,
who helped create the current measles vaccine as a junior member of
Enders's lab in 1954, calls his master a "person of the Old School. He
didn't believe you built a huge lab to show how many people you could
attract. You had a small enough group that you had intimate contact with
each person. He'd stop at your desk each day and say, 'What's new?' and
you were driven by that because you wanted to have something to present him." The annual office Christmas party was at Enders's house. He'd
put on a velvet smoking jacket and play the piano while the guests sang
carols. And he encouraged collegiality with parlor games and poetry
reading.[35]

Enders's key scientific accomplishment, which would open the way
for the mass production of a polio vaccine, occurred in 1948. In the
1920s, scientists Hugh and Mary Maitlin had started growing viruses in

tissue culture—plates or test tubes full of living cells kept alive with a variety of chemical foods—but they were usually overgrown with bacteria before long. Since their day, penicillin and streptomycin had been introduced, and Enders used both drugs to eliminate bacterial contamination from the cultures, which allowed him to keep viruses growing for up to 40 days. One day, the 32-year-old Weller was seeding chickenpox virus in a culture of diced fetal arm tissue in flat-bottomed Ehrlenmeyer flasks. Weller had four leftover flasks and Enders casually suggested he try to grow poliovirus in them. The stuff took, which Weller was able to prove by diluting and regrowing the culture several times, still managing to paralyze mice with it each time. Eventually, the lab was even able to grow polio on pieces of sterilized intestinal wall. The discovery showed the way forward for virus growers and "incited a restless activity in the virus laboratories the world over," as Swedish virologist Sven Gard remarked at the Nobel ceremony. None was more restless than Jonas Salk.[36]

Salk and the Four-Letter Vaccine

In 1949, the year the Enders lab reported its results, the National Foundation had just contracted with Salk to complete a vital but tedious step on the road to a polio vaccine. His laboratory and four others were testing 100 different strains of polio to ascertain whether they fit into three immunologically separate types. If the work proved the thesis of three types, scientists would know which components to put in an effective vaccine. Salk would find shortcuts to that conclusion.

Salk trained under Thomas Francis, the world's premier influenza investigator, and was hired as a full professor by the University of Pittsburgh in 1947. He was 33. While not a genius student, Salk from a young age stood out for his calm and steady appetite for work. He attended the Townsend Harris High School and the City College of New York, entered NYU medical college at 19, married a well-off Smith girl the day after receiving his MD, and believed to a pious extent in his own ability to achieve noble things. He spoke, with wholesome orotundity, of the things that "men of goodwill" could achieve in a "non-partisan" environment. In the positivist world Salk inhabited, science was a temple, and he was one of the not-so-humble priests. On other occasions he declared himself a "foot soldier" in the war on disease.

Working under Francis at the University of Michigan, Salk had produced a killed flu vaccine given to millions of U.S. troops. Though it achieved mixed results because of the multiple, shifting strains of flu, the work established the feasibility of killed virus vaccines at a time when most polio experts favored a live vaccine. The National Foundation, which dominated polio research—it spent more than $2 million in 1953, while the federal government spent just $75,000[37]—expected its researchers to type the polioviruses using live monkeys. A monkey would be infected with a poliovirus that was known to belong to one of the three types—say a type 1 strain. Then it was exposed to 1 of the 100 samples the labs were testing. If the sample failed to infect the monkey and was neutralized by type 1 antibodies, then it could be categorized as type 1 strain. And so on—a cumbersome, time-consuming process.

The foundation's attention and money meant Salk could build a professional staff, which soon included Byron Bennet, a hard-drinking Texan ex-army major who had been Smadel's chief lab technician. Julius Youngner, a young microbiologist who had worked on the Manhattan Project studying the toxicity of uranium, was hired as senior research assistant. Bacteriologist James Lewis and zoologist Elsie Ward rounded out Salk's inner circle with secretary Lorraine Friedman, a "tall, equable Pittsburgh girl" who would remain with Salk his whole career.[38] The virus-typing work was drudgery, and Salk rebelled. The stately pace of scientists like Enders was not for him. If Enders returned parts of his grant money, Salk burned it all up and demanded more. Before long his science was on an almost industrial scale.

Salk was a driven, cerebral scientist, more soft-spoken than outsized characters such as Sabin and Rivers, yet instilled with a supreme confidence in his own ideas and judgment. He suffered when his work was under attack, but clung to his goals tenaciously, keeping his head down as he carefully picked his way through the profane, sharp-elbowed polio research crowd. His gentle manner masked ambition that was white-hot, yet almost impersonal. As a junior scientist at Michigan he had pestered Francis interminably to be listed as the first coauthor of their flu papers. His letters to National Foundation officials are full of elaborate, almost Talmudic arguments for more money and material. Yet at Pittsburgh, according to Youngner, who presents a somewhat dark picture of his master, Salk neglected to credit his subordinates, though it was Youngner who devised many of the important technical steps for the vaccine, including the use of monkey kidneys to grow the virus and dyes to distinguish infected cells. "I was Mr. Inside," says Youngner. "Jonas was Mr. Outside." Salk, he maintains, put on the white coat for photographers. The bench science was done by others.[39]

Salk's image captivated a public ready to worship science, particularly when the bespectacled, prematurely balding man was so evidently devoted to whipping the era's nightmare disease. Without seeming too patronizing, Salk could refer to his vaccine as a recipe "very much like the one a housewife uses when she wants to prepare a new dessert, say a cake. She starts with an idea and certain ingredients, and then experiments, a little more of this, a little less of that, and keeps changing things until finally she has a good recipe. In the process she will have deduced certain universal laws." At times his reassurances veered into all-American hubris, as when he said that his vaccine was "safe, and you can't get safer than safe."

Tiring of the virus-typing methods recommended by the graybeards of the National Foundation's Immunization Committee, Salk found shortcuts, including the use of Enders's tissue culture, that allowed him to type faster and more efficiently. But Enders had used his network of medical school colleagues to secure human fetal tissue to grow viruses and Salk wanted to make *gallons* of virus—an unrealistic quantity given the amount of fetal tissue required. "If you think abortion is a problem now, imagine it in the early 1950s," Youngner said. "It was a crime in Pittsburgh." By the end of 1950, Salk felt that he knew enough about the varieties and vagaries of polio to start working on a vaccine. Within two years he had convinced Rivers, Smadel, Weaver, and O'Connor that he had a vaccine. He would fail to win over many other scientists, including Nobel laureates Wendell Stanley, Weller and Enders, Stokes, Paul and Gard, as well as competing vaccine makers Albert Sabin, Hilary Koprowski, and Herald Cox. They tended to view Salk as more of a cake baker than a scientist.

The skepticism partly grew out of Salk's close relationship with his funder, O'Connor, whom the great brains of Rockefeller, Harvard, Hopkins, and Yale scorned as an influential upstart. While O'Connor lavished support on many worthy scientists (including more than $1 million to Salk's archrival, Sabin), by 1952 Salk was the focus of the foundation's attention. He adopted O'Connor as a kind of second father, and O'Connor took on Salk as the son he never had. Their friendship began on the *Queen Mary*, sailing home from Copenhagen in 1951, where Salk charmed O'Connor's adult daughter, Bettyan Culver, who was recovering from a paralytic bout of "daddy's disease."[40] The next day in the ship swimming pool, Salk and O'Connor had a wide-ranging discussion of ethics and science, and O'Connor discovered that Salk was no mere young technician, but a mensch who cared about democracy, social justice, and the advance of humankind. (The FBI had noticed Salk's idealism as well, and nearly put the word out to blackball him at universities.)[41]

To their critics, O'Connor was a ventriloquist and Salk his talking dummy. O'Connor's promotional methods seem low-key by the standards of some contemporary disease advocacy groups. But it was true that the National Foundation would morph from an organization that funded a variety of polio research and treatment to one whose entire structure, with hundreds of thousands of volunteers, was devoted to testing the Salk vaccine. Salk did not deny that he and O'Connor were using each other, but he saw no problem with that. "From the standpoints of science and society, O'Connor is an invaluable facility, a rare item of human equipment," Salk told a biographer. "In him are combined self-interest and social interest in ideal proportions. He can't satisfy the first without fulfilling the other. I honor his qualities but it is irrelevant to speak of 'indebtedness.' We do not do favors for each other like a pair of ward-heelers. We try to get work done in biology." As for O'Connor, his foundation had given millions to scientists he couldn't stand because their work was important and "not a penny" to scientists who were friends, because their work wasn't qualified. "I don't know how Jonas has been able to tolerate the abuse and obloquy he's suffered in his profession," O'Connor said. "He shows the world how to eliminate paralytic polio and you'd think he had halitosis or had committed a felony."[42]

Salk's Competitors

Salk is the polio vanquisher who became a household name, but each of the four contenders in the vaccine race was a remarkable figure. Of the three oral polio vaccine makers, Cox was the best-established vaccinologist. He had discovered Q fever and developed shots against Rocky Mountain Spotted Fever, hog fever, and typhus. But Cox, a homespun Hoosier who suffered bouts of depression, lacked the personality to jostle for position with Sabin and Koprowski, who were polished, urbane, ambitious, and ruthless. Koprowski was a concert pianist and physician, an intellectual who wrote short stories and composed avant-garde music and kept fierce Beauceron dogs that frequently bit his guests and children—peculiar pets for a man who dedicated much of his career to making rabies vaccines. After being chased out of his native Poland by the Nazis in 1939 Koprowski had gone to Brazil where he was hired by the Rockefeller Foundation to work on yellow fever under John Fox and later learned about viral vaccines from Theiler. But the Rockefeller didn't want to hire another Jewish émigré, so in 1944 Koprowski went to Lederle Pharmaceuticals in Pearl River, New York, to work under Cox. He was only 28 at the time, but soon developed his own research agenda, and left Lederle on unfriendly terms in 1957.

Sabin was the most widely admired scientist of the four, and probably the most difficult man. He had emigrated from Bialystock in eastern Poland (or western Ukraine, depending on who was in charge) with his family in 1921 when he was 15 and studied dentistry because a rich uncle agreed to support such studies. After tiring of teeth and running out of money, he convinced William Park to take him on at the New York University bacteriology department.[43] He published his first paper on the 1931 polio epidemic, and in 1934 he went to work with Olitsky at the Rockefeller Institute.[44] In 1939, Sabin left for Cincinnati, which offered him twice the money and an associate professorship in pediatrics. During World War II he made vaccines against Japanese B and dengue fever. Sabin was a "spare, tough, sharp-eyed man with snowy hair and a vulpine countenance, a small moustache and a resemblance on occasions to Groucho Marx," wrote John Rowan Wilson. "Like Groucho he is intelligent, serious and extremely businesslike."[45] He was also aggressive, selfish, and irascible. His first wife committed suicide by putting her head in a plastic bag on Sabin's sixtieth birthday. During one of Sabin's shouting fights with the couple's daughters, the family dog—defending the girls, apparently—lunged for Sabin's crotch and after being shaken off, chewed through Sabin's Achilles heel.[46]

As a scientist, Sabin was a lone wolf—he hired assistants for a few years and worked them until they quit—but the graduates of his lab were an illustrious bunch. Joel Warren and Anton Schwarz went on to make measles vaccines; Robert Chanock discovered the respiratory syncytial and Norwalk viruses and developed many vaccines, including a West Nile virus vaccine currently in clinical trials. Ben Sweet helped discover the SV-40 virus in Hilleman's laboratory, while Ed Buescher isolated rubella. "He can decimate you, he can wilt you with a few words," recalled one virologist.[47] The polio committee meetings were a savage business, nothing gentle about it. Someone would give a presentation, Rivers would jump up and criticize it, then Sabin would attack it, then somebody else would chime in and "boy you were looking for a hole to crawl into."[48]

Whatever his difficulties as a human being, however, Sabin's brilliance, and his contribution to humanity, are uncontested. Gifted with a fine, clear writing style and an amazing memory, he would startle colleagues by responding to the presentation of a paper with the recitation, by memory, of lab notes from an experiment he had conducted 20 years earlier. According to one account, Sabin personally inoculated 20,000 research monkeys during his work on polio and also did the clinical evaluations of their health.[49] His concept for the oral polio vaccine was one that withstood the test of time. He championed the oral, live vaccine because it flooded the gut with virus and induced the prompt secretion of antibodies. Sabin wanted to flood the world with his virus. The World Health Organization came to share that goal.

The National Foundation's immunization committee met every several months to keep abreast of progress. But the talk of vaccines was quite muted and theoretical until a March 1951 meeting in Hershey, Pennsylvania—a town O'Connor selected because it was a good midway point for the scientists and because he liked it—when Koprowski dropped a bombshell. As the scientists dozed in a postluncheon fog, Koprowski presented the results of the first trial of a polio vaccine since the Kolmer/Brodie fiasco. He and lab technician Norton had tested an attenuated poliovirus grown in ground-up mouse brains. They drank the gray-white sludge themselves, then fed it to 20 children and two adults in a home for the mentally disabled in Sonoma, California. The other grantees were shocked out of their somnolence. "What's this, monkeys?" Francis asked Salk. "No, children!" Salk whispered. Sabin, who at the time still believed that polio did not produce a blood-borne infection, demanded of Koprowski, "Why have you done it, why?"[50]

Koprowski's daring act was the opening volley in a vaccine war that would last for a decade. Whose vaccine was the most effective? The safest? Who was the wisest scientist, the most ethical? Despite the triumph of polio vaccine, none of the vaccine makers escaped the wars with reputation entirely intact. At the 1951 meeting, Rivers expressed his moral opposition to experiments on disabled children and the stigma stuck, for Koprowski's experiment was riddled with ethical and scientific problems. "An adult can do what he wants but the same does not hold true for a mentally defective child," Rivers later said. "Many of these children did not have mommas or poppas, or if they did their mommas didn't give a damn about them."[51] Rivers also felt—and contemporary scientists would certainly agree—that the brains of mentally ill children were different from those of normal kids, which made them inappropriate test subjects for a disease that affected the nervous system. Koprowski's first paper on his vaccine described the children as "volunteers," leading *The Lancet* to comment sarcastically, "one of the reasons for the richness of the English language is that the meaning of some words is constantly changing. Such a word is 'volunteer.' We might yet read in a scientific journal that an experiment was carried out with 20 volunteer mice and that 20 mice volunteered as controls."[52] This was to be the first of many controversies that Koprowski would weather in a career lasting more than 60 years.[53]

Of more immediate relevance to Salk were two other papers presented at the Hershey conference. Isabel Morgan Mountain and Howard Howe of Johns Hopkins had inoculated monkeys and chimpanzees, respectively, with poliovirus inactivated by exposure to formaldehyde—the same principal Salk would use. Neither of the scientists suggested it was time to start vaccinating children, but the proof in principle for a killed-virus vaccine was there. Salk liked to say that Enders had "thrown the football and I caught it and ran downfield." Although the Harvard boys—Enders had no desire to make a vaccine himself—felt Salk was out of bounds, he decided to start running for longer yardage. He was more stealthy than Koprowski but moving just as rapidly, and with significant resources from the National Foundation.

Salk's lab chief, Youngner, had figured out how to use an enzyme, trypsin, to chemically chop up kidney tissues and make polio cultures on them. Whereas 17,500 Asian monkeys had been imported simply to type the poliovirus, a single monkey's kidney might now grow 1,000 test tubes full of virus. Before long, Salk and Youngner had harvested a liter of polio. The move into monkey kidneys as a source for testing viruses and a medium for growing them was a fateful breakthrough. By the mid-1950s, more than 100,000 rhesus monkeys were being netted in the wilds of the Philippines and India, thrust into cages, and shipped to virology laboratories each year. Along the way they shared their infections and so arrived at their destination screaming, biting, and often dying of various diseases. They hurled their waste products at the scientists and technicians who took care of them and ultimately sacrificed them. In 1931, a Canadian scientist in Park's laboratory named William Brebner was bitten on the hand by a monkey and died of paralysis. Sabin cultured virus from his friend's brain and called it "monkey virus B" in Brebner's honor.[54] After reading Sabin's paper in 1949, Salk asked for a $10,000 insurance policy for each lab employee who worked with monkeys. It was refused.[55] Robert Hull, a Lilly scientist, was starting to catalogue the monkey viruses his and other firms were finding. By 1960 he had counted 40 of them.[56]

At a December 1951 meeting of the immunization committee, Salk laid out his idea for a vaccine. On April 18, 1952, Rivers predicted that a vaccine would be ready within a year or two, but warned that it would have to be done carefully. "We've got something big here," he said, "and unless we handle it right it will be terrible."[57] Two months later, on June 12, 1952, William F. Kirkpatrick, the 16-year-old son of an industrial executive, became the first child experimentally vaccinated with Salk's vaccine at the D.T. Watson Home for Crippled Children. Rivers, though leery of such experiments, seems to have waived his reservations—largely because the D.T. Watson children already had antibodies to polio, apparently because of exposure as infants, and were therefore at little risk. But Salk summed up the feelings of this jittery group of high-strung scientists very well when he said, "when you inoculate children with a polio vaccine, you don't sleep well for two or three months."[58]

In January 1953, the Immunization Conference was back in Hershey for another explosive meeting.[59] "Hold onto your hats, boys," Rivers told the assembled group, "we're going to toss a bomb at you." The bomb was Salk's report, to be published in the *Journal of the American Medical Association* (*JAMA*) on March 28, on the successful vaccination of 161 people with various polio preparations. Now, at Hershey II, the polio vaccine battle lines were thickly raked out in the sand. Rivers, Smadel, and a few other graybeards lined up with Salk—the young Jew, in Rivers's blunt formulation—in support of the inactivated vaccine. Most of the working virologists were with "the smart Jew," Sabin. Although Sabin up to then had gotten about as much research funding as Salk, he was outraged at the foundation's endorsement of Salk's approach. Sabin thoroughly trashed the *JAMA* paper, as did Enders. Both men said that monkey kidney tissue could stimulate autoimmune-mediated organ damage. They would lodge many other criticisms of the Salk vaccine later in the game.

The attacks on Salk were a mixture of professional disparagement and serious fear of repeating the Kolmer/Brodie fiasco. Rivers was not the only virologist anxious to find a practical answer to a scientific and medical problem that had dominated his career.[60] But the path was thorny. It was time, he and Basil O'Connor agreed, to call in a higher committee. On February 26, they convened a roomful of notables at the Waldorf-Astoria Hotel to discuss how to proceed with the vaccine. Salk was the only polio vaccine worker invited to the meeting. Suddenly on the spot, he went humble, telling the gathering, "I don't know that we even have a vaccine yet . . . we have a preparation which has induced antibody formation in human subjects." River responded, "I think you have a vaccine, Jonas."

As Rivers asked for the group's blessing to move forward with thousands of experimental vaccinations, Salk realized the vaccine was being taken out of his hands. "I felt like someone driving a team of wild horses and being whipped at the same time," he said.[61] In the end, Salk did a pretty remarkable job. But his reputation suffered among peers, and O'Connor only hastened the process by nudging him into the limelight. On March 26, Salk appeared on a CBS program titled, "The Scientist Speaks for Himself." For half an hour he led the interviewer on a tour of his laboratory, explaining his work in simple language. From that point on, Salk became a celebrity, and the vaccine he was working on became a four-letter word: the Salk vaccine. Salk would say he had been naive about the intense attention that would follow his work, but another scientist said, "naive my foot. . . . Jonas went on the air that night to take a bow and become a public hero, and that's what he became."[62] Salk's scientific colleagues were shocked. Scientists were supposed to speak through journal articles and at conferences. To address the public directly was to break a code of honor.

In April 1953, Salk inoculated his wife Donna and their sons, Peter, 9, Darrell, 6, and Jonathan, 3. National Foundation leaders Harry Weaver and Hart van Riper brought their kids to Pittsburgh for injections. In May, Salk resumed experiments on local schoolchildren whose blood had no antibody. He spent many hours figuring out the perfect temperature and mixture of formalin in water to kill his virus while leaving it capable of producing antibodies. Eventually, Salk derived what he called a "margin of safety" in the inactivation of polio. Salk found that when a million viruses had been bathed for six days in a solution of 1 part formaldehyde to 4,000 parts water, only a single infectious particle would be left. He left the solution another three days to further ensure that no live polio survived in it. Graphically, the relationship of time, temperature, and viral presence could be mapped as a straight line—in chemical terms, as a first-order kinetic reaction—with live virus gradually disappearing as it was exposed to formaldehyde. But some, such as Sven Gard, found that the line curved at the bottom—inactivation slowed as time went on, and it would take at least 12 weeks to get the result that Salk claimed occurred in a week. Gard and Enders believed there would always be live virus in Salk's vaccine.[63]

Rivers, O'Connor, and Weaver, the foundation's scientific director, went about plotting the largest field trial in the history of medicine. To make it happen, they sidelined the immunization committee, full of thoroughbred scientists jockeying for power, and in May 1953 created the Vaccine Advisory Committee. Its members were older, more sedate, and favorable to polio vaccination. "We formed the committee to break a logjam," said Weaver. Carryovers from the immunization committee included Smadel, who had trained under Rivers and would later train Hilleman. He was a bold, results-oriented scientist who tended to push ahead once he'd decided upon a course. Other committee members included David Price, who as a representative of the Eisenhower administration had a strong interest in making a polio vaccine available, and Norman Topping, an NIH microbiologist known to oppose the live polio vaccine. Another prominent member was Thomas Turner of Johns Hopkins, of yellow fever vaccine fame. These men would shepherd the Salk vaccine to its creation and stand up for it in its hour of need.

By the time of the polio trial, the virus research world had begun to grasp the enormous complexity of the systems—human and viral—that they were dealing with. It was a realm, as Hilleman put it, in which the Heisenberg uncertainty principle was always a factor. The more science advanced, the more it could do, yet at the same time the more potential for failures and pitfalls became evident. The temperaments of cautious and ambitious men are present in equal measure at the highest levels of science—what might be called the "doers" and the "doubters." Society needs both types. The public wants science to produce goods, so it encourages the leadership of can-do, hypomanic individuals. The doubters are there, in the flesh and in institutional structures, to keep the hypomaniacs from getting too far ahead of themselves. They are also there to say, "I told you so."

In the polio years, the risk of doing nothing seemed greater than the threat of doing something wrong. The creation of the Vaccine Advisory Committee was an opportunity for the doers. What else could it be? Seven million Americans had not emptied their piggybanks and spent their evenings at National Foundation meetings, after all, so that a bunch of scientists could serenely examine antibody titers for years on end. Polio still claimed thousands of victims each year, and the National Foundation on Infantile Paralysis was clearly biased toward the doers, for its mission was one-sided: to cure or prevent polio.

The Home Stretch

Although the Pope and the American Medical Association were starting to mutter about experimental ethics in 1955, it was still a great time to be a researcher. The West Coast microbiologist Karl Meyer would recall with nostalgia the ease with which he picked up experimental subjects from the prisons of California. "At San Quentin, almost everyone at least during wartime and even during the Korean period, was willing to do something in the interests of the country," he said. "When I needed 100 volunteers there were sometimes as many as 300, so I picked the ones who I thought would be best." The prisoners "didn't squawk too much . . . therefore, I tested some preparations which were very, very toxic, VERY toxic." The plague vaccine he injected into 12 prisoners in 1952 gave "every one of them a temperature of 104 or 105, feeling perfectly scandalously miserable, having reactions that lasted for four or five days." But "it cost me nothing because the boys over there were very eager to make a contribution to the welfare of the country. They thought they could redeem themselves."A decade later Meyer found San Quentin prisoners less eager to volunteer. "They just don't want to get sick again," he said. "They're a pampered group."[64]

The "consent form" that parents of nearly 2 million children signed to enter their children in the Salk trial were emblematic of the public's trust in scientific authority. Parents were invited to make their children "polio pioneers" in a "field trial." The phrase "experimental vaccine" occurred nowhere in the one-paragraph sheet. After the trial was over, some county health departments "regretted to inform" the participants that they had received placebo, but reassured them they could come in for a real shot at any time. Such language would never meet the informed consent standards demanded of any experimental drug or vaccine introduced today.

Whether they considered Salk a demigod or a demagogue, 90 percent of Americans knew about the trials in May 1954, more than could give Dwight D. Eisenhower's complete name. In an age in which Madison Avenue sowed nonsense handsomely among the rubes, Salk was a natural advertisement for his own product, a real-life Dr. Kildare. Reporters called him Jonas and he struck them as frank, friendly, and kind. John Troan, a science writer for the Pittsburgh Press, called on Salk early and found him ostensibly reluctant about publicity for his vaccine. Salk telegrammed him after the story was printed to say that he had "done a splendid job." Scientists didn't generally treat newspaper reporters with that kind of respect (nor did journalists, generally, deserve it). "He could sell me the Brooklyn Bridge" one reporter said in 1955.[65] Reporters didn't seem to feel Salk was courting hubris by promising that "we can state flatly that no human being has been, or ever will, in any field trials, be inoculated with any material that has the remotest suspicion attached to it."

The public wanted polio vaccine to be real. It was a grim time, the height of the cold war, with the United States poised to go to war with China over Taiwan, H-bomb tests in the Pacific every few months, and civil defense on high alert. Polio was a real threat. With ambulances bringing up to 17 patients to the Municipal Hospital in Pittsburgh every day in summer, the staff pleaded with Salk to hurry his work. Women gave birth in iron lungs. High school athletes cried for guns to kill themselves. Physicians and nurses worked themselves to exhaustion and beyond. "To leave the place you had to pass a certain number of rooms, and you'd hear a child crying for someone to read his mail to him or for a drink of water or why can't she move, and you couldn't be cruel enough just to pass by," a nurse recalled. "It was an atmosphere of grief, terror and helpless rage."[66]

The fact that paralytic polio did not stike entire communities was part of what made it so difficult to develop a vaccine. The efficacy of a measles vaccine could be proven in a trial of a few hundred children, because measles spread like wildfire in a virgin population, causing an obvious rash and high fever in virtually anyone susceptible who'd been in the same room with a patient. Polio spread less efficiently, mainly from unwashed hands, and only occasionally caused paralysis. You'd need hundreds of thousands of children to conduct a convincing trial, a task for which the National Foundation proved uniquely capable. At the height of the trial, the foundation's trial involved 20,000 physicians and public health officers, 40,000 nurses, 14,000 school principals, 50,000 teachers, and 220,000 volunteers. The 3,000 foundation chapters had more than a quarter of a million full-time volunteers processing 140 million items of information from IBM cards.

But the money, power, and vast volunteer force of the foundation were worthless without their connection to legitimate science, and the polio scientists, for the most part, were as wary of mistakes as the public was optimistic of the fruits the scientists could produce. These men and women were deeply concerned about the problems of the vaccine: contamination of the medium, reversion to virulence, the ever-mutating nature of viruses of all kinds. Weller was sure that Salk's monkey kidney cells would not detect all the live virus in his vaccine. They were only one-tenth as sensitive as human cells, Weller found.[67] He sent his evidence to the National Foundation but never heard back. Sabin and Henry Kempe, then a pediatrician and virologist at the University of California in San Francisco, both ran letter-writing campaigns to stop the trials but ran into the brick wall named Basil O'Connor.[68]

To guarantee that whatever the trial found would be respected, if grudgingly, by other scientists, the National Foundation convinced Salk's mentor, Tommy Francis, to cut short a leave in Europe and run the trial. The well-respected Francis agreed on the condition that the trial include a large double-blind component in which half the subjects would receive a salt-water placebo, and neither the subjects nor the scientists would know who was who. Up to then, this level of objectivity had been sought on a large scale only in a British trial of streptomycin.[69] Salk opposed the double-blind wing—he felt all children should get his vaccine—but was overruled.

In November 1953, O'Connor announced at a news conference that the trial would begin in a few months' time. It was only then that the government got involved, and federal officials were appalled by the protocols for the trial. "The whole NIH was well below standard at the time," the agency's chief deputy, James Shannon, would say later, and NIH officials were intimidated by the foundation and its minions. But the public health service "should not have handled the foundation any easier or less rigidly than they handled any pharmaceutical company."[70] Shannon was right in principle, but who could stand in the way of this trial? Certainly not the laissez-faire Eisenhower administration. Shannon succeeded in getting thimerosal added to the vaccine as an antiseptic. Salk opposed the use of thimerosal, which ended up inhibiting the vaccine's potency—while possibly neutralizing live polio virus inadvertently contained in the vaccine.[71] The NIH director wanted each batch of vaccine tested on 350 monkeys, which would have slowed the trial by months; the National Foundation agreed instead that 11 batches in a row had to test free of live virus before any of the batches would be released. The government's insistence on tougher standards would prove ironic later, when it was responsible for overseeing commercial production of the vaccine and permitted vaccine makers to use much looser standards.

Less than a month before the Salk trial began, it hit a final speed bump. On April 1, 1954, the Ottawa County, Michigan, medical society, whose constituency included Paul de Kruif, announced it would not take part in the trial. De Kruif, "mastermind of polio science's stone age," as a cynic dubbed him, had sleuthed out the fact that live virus had been detected in some of the commercially produced vaccines.[72] He fed the story to the radio personality Walter Winchell. That night, after his customary introductory call, "Good evening, Mr. and Mrs. America, and all the Ships at Sea!" Winchell bellowed into the microphone that the new polio vaccine "may be a killer!" In response to Winchell, thousands of Michigan and Minnesota kids were withdrawn from the trial. In Buffalo, New York, where the health department had expected 50,000 participants, only 30,000 rolled up their sleeves.[73] The Medical Society of Washington, DC, where the opening shot was to have been fired, so to speak, also withdrew its support. It was for that reason that the March of Dimes had to hold its big photo opportunity across the Potomac in McLean, Virginia, a new tree-shaded suburb where farm kids rode to school on horses through the mud of housing developments being built for CIA men and their families.

At 9:35 A.M. on April 26, Randy Kerr, a smiling little boy in a checked shortsleeved shirt, became the first polio pioneer. "The nation unsheathed its newest weapon against the vaunted crippler, polio," *The Washington Post* reported. "And the first battleground was the body of a 6-year-old boy." The scene at Franklin Sherman Elementary was bedlam. Kids screamed and fainted, television cables lay thick as a nest of anacondas across the cafeteria floor. Parents and teachers stood around, a little nervous but proud. Gail Adams Batt was one of those vaccinated that day. Earlier in the year, the little boy sitting behind her in class had stopped coming to school—over the weekend, he had contracted polio. No one had visited the boy or written get-well notes. Batt, who was 8 at the time, thought she had been picked to get the vaccine because she sat next to the boy, whose desk was left empty the rest of the year. She remembers telling herself, "Don't cry, don't cry." Then a boy in the line in front of her saw the hypodermic needle and "whomp, he fainted and fell to the floor. And then all I could think of was, 'don't faint, don't faint.' " After the shots her class got to eat chocolate sundaes at the drug store. As for Winchell, "my mother thought he was a reactionary! She probably had me vaccinated to spite him."

The polio trial required intricate levels of organization. In Buffalo, for example, authorities vaccinated children at 50 schools each day across a county of 50 square miles. They had mountains of forms, conference rooms full of cotton balls and syringes, thousands of boxes each containing six vials—three filled with polio vaccine type 1, 2, or 3 and three filled with saline solution. The vials each had numbers on them, and no one in the field knew the code. About 100 nurses and 1,000 volunteers helped with the trial in Buffalo alone, and the process became the standard for other vaccine trials, recalled Warren Winkelstein, a Buffalo public health official who later taught at the University of California.

O'Connor was faced with a logistical quandary: if the vaccine worked, everyone was going to want it as soon as the trial ended, which meant that manufacturers would have to produce it during the trial. But none of the companies wanted to be stuck holding millions of doses of unusable vaccine if the trial failed. So O'Connor took a calculated gamble: he spent $9 million to pre-order 27 million doses of vaccine. In hindsight, it might have been unnecessary, since the companies were less concerned about efficacy than demand. When asked at a meeting before the trial how good the vaccine would have to be for them to market it, company presidents estimated that 15 to 25 percent effectiveness would do. A congressional inquiry in 1957 found evidence the companies had fixed prices for the vaccine and sold it to the Public Health Service at rates twice what the National Foundation was paying.[74]

As the summer of 1954 came and went, it seemed to the reporters avidly covering the story that polio rates were lower than past years, but the polio epidemic curve had lots of peaks and valleys, so that didn't mean anything. The winter was spent in a flurry of tabulating, and then on April 12, 1955—the tenth anniversary of FDR's death, although Basil O'Connor would swear it was coincidental—the world's eyes were all focused upon Ann Arbor, Michigan, where Francis was to announce the results. This would be a day of jubilation, a great day for the American people who had made the vaccine possible. Yet it was also a day of tremendous foibles and frustration, a jumble of scientific evidence, hoopla, and unanswered questions. On the East Coast it was unseasonably hot, and Ike threw out the first ball for the Washington Senators on Opening Day. Everything that made America shine with promise—her energy, her enthusiasm, her concern for children—was right up there on stage with the waste, the hastiness, the disorder, and fast-buck operators.

While 500 of the world's leading medical scientists waited in Rackham Hall in Ann Arbor, Michigan, for Francis to read his paper—an event carried by short-circuit television to theaters around the country (Eli Lilly paid for this extravagance)—messengers carrying a press release that undercut all of Francis's caution got off an elevator upstairs and were immediately mobbed by 150 reporters who surged out of the pressroom, clutching for the release. The frightened messengers pitched packets to the "hungry dogs at a garbage pail," as Jack Geiger of International News Service described his colleagues, while hundreds of scientists heard Francis carefully read out his report. Many of the scientists complained of being "stage props" or "window dressing." Some got drunk on the train back home. "The vaccine works!" trumpeted the University of Michigan news release. And indeed, the vaccine had proven very effective in preventing polio, although the results were somewhat muddled by the complicated structure of the trial, which had two arms—one with double-blinded placebos, the other with "observed controls," third-graders who got no vaccine while their second-grade fellows did. The vaccine was as much as 90 percent effective in preventing paralytic polio (the vaccine strains preserved with thimerosal were less efficacious). Fifty-two children in the study population had died or become paralyzed badly enough to end up in an iron lung. Only two of those children, both survivors, had received the real Salk vaccine.

All over the country, people tuned in the news on their radios at 10 A.M. EST. Church bells pealed, fire and air raid sirens wailed joyfully. In Buffalo, at the Center Theater, 600 doctors and others rose for a moment of silent prayer following the Lilly broadcast. "This is the greatest day in my medical career, and one of the great days in the history of medicine," said Dr. William J. Orr, chairman of the county's polio advisory committee. And in Philadelphia and around the country, parents flooded the health department with calls demanding vaccine.

Immediately after the announcement, the National Foundation passed the baton of polio vaccine to the federal government. Workman, director of the NIH's Laboratory of Biologics Control, convened a 15-member licensing committee that very afternoon in an Ann Arbor hotel room. The secretary of the new Department of Health, Education and Welfare, Oveta Culp Hobby, allotted exactly one hour for the licensing decision. The committee pleaded for and got another $1\frac{1}{2}$ hours—to consider 2,000 pages of data. The vaccine was to be shipped to pharmacies as soon as the group signed off on the licenses. The delay annoyed Hobby, a tough Texas politico ("Tell them to hurry," one staffer was overheard saying, "Mrs. Hobby's makeup is running").

While the National Foundation had provided manufacturers a 55-paged protocol on the specifications of the Salk vaccine, the federal government's "minimum requirements" were only five pages long. During the vaccine trial, the government, Salk, and the manufacturers had each tested the vaccine, requiring that unless 11 consecutive batches were free of live virus, the whole lot was to be thrown out. But the federal licensing committee required only "a method which [was] consistently effective and reliable in inactivating a series of lots."[75] In short, the licensing committee, in a brief afternoon get-together, approved six separate vaccines, all different from the one that had just been tested. Three of the vaccine makers—Cutter, Wyeth, and Pitman-Moore—hadn't even produced vaccine for the trial. This was a recipe for the disaster that Weller, Kempe, Sabin, and others had been predicting.

Still, it was a day of triumph for Francis and O'Connor, for Rivers and Smadel. Salk, on the other hand, managed to alienate his peers further by claiming that the vaccine could have been "100 percent" effective if thimerosal hadn't been added—which sounded like sour grapes. Too, he neglected to thank his lab associates, an unconscionable error that Youngner and the others never forgave. "Salk should have kept his mouth shut," Rivers said, "But he just had to get into the picture."[76] As Salk became a media darling, scientific colleagues began to withdraw from him. That night, the journalist Edward R. Murrow interviewed Francis and Salk on *See It Now*. He asked who held the patent on the vaccine. "Well, the people, I would say," Salk responded. "There is no patent. Could you patent the sun?"

That timeless quote turned out to be a bit of Salkian self-righteousness. Salk and the university had, in fact, discussed patenting the vaccine. But since Enders's laboratory had pioneered the key discovery, and Bodian and

Horstman tested the inactivated vaccine on apes, the lawyers determined there was nothing to patent.[77]

Salk became the first scientific superstar. Grateful Americans mailed him millions of dollars. A 208-foot telegram arrived from Winnipeg with more than 7,000 signatures. The town of Amarillo, Texas, sent him a new car, which was sold to buy vaccine for the children of Amarillo. He turned down farm tillers, cars, endorsements for baby products, and five Hollywood offers for his life story. The city of New York created eight, $35,000 Jonas Salk medical school scholarships and asked to hold a ticker-tape parade in his honor. Salk refused, but his parents were pressured into marching in a Loyalty Day Parade organized by McCarthyites. Salk told everyone he met, "I just want to be left alone, I just want to get back to my laboratory." But first, Eisenhower had him to the White House. He was supposed to say "Thank you, Mr. President," but instead read a five-paragraph statement praising the people who had worked on the vaccine—trying to make up for his lapse at Ann Arbor. Salk's son, second-grader Darrell, asked Eisenhower what he did besides play golf.

Salk won the Criss Prize, the Lasker Award, the first congressional medal of Distinguished Civilian Service. But he never won a Nobel, and he was never elected to the National Academy of Sciences, unlike most of his less-famous polio peers. And although he wrote several thick philo-sophical treatises, and spent the gifts to finance an institute on a cliff above the Pacific Ocean where cutting-edge research on brain science was done by men like Nobel laureate Francis Crick, Salk himself never again achieved anything of significance as a scientist.

The Cutter—and Wyeth—Incident

Youngner flew to San Francisco for a scientific conference the day that Oveta Culp Hobby licensed the polio vaccine. He received a call at his hotel there from Ralph Houlihan, an official at Cutter Laboratories across the bay in Berkeley. Cutter had been producing the vaccine, on spec, for several months while awaiting the trial results. Things were not going well, Houlihan told Youngner. "When live virus is supposed to be gone, we still find it for days and days." Youngner crossed the Bay Bridge to visit

Houlihan. He was shocked at the state of the production facility. In a crude concrete slab room, live virus was stored together with killed vaccine. The company's notebooks were sloppy; its production plans murky. "[T]hey didn't look like they knew what the hell they were doing," Youngner said years later.[78] Returning to Pittsburgh, Youngner told Salk what he had seen and asked whether he should write to the National Foundation, or the NIH, about the problems at Cutter. Salk promised to write the letter but never did, according to Youngner, who never forgave himself. "I blamed Jonas, but I should only blame myself. I didn't follow up on my request." Years later Youngner reproached Salk, but got no response. "He was passive—said nothing, as if he'd blanked it out."

On April 25, three days after Darrell Salk asked the president about his golf game, bad news started flowing East from Chicago, Idaho, and California. For public health officials, memories of how and where they heard the news are still vivid. They talk about it the way people speak about the hour of Kennedy's assassination. The Salk vaccine, particularly lots manufactured by Cutter Laboratories, was crippling children.

On April 25, the Chicago Department of Health called Workman, of the NIH, to report that a physician's infant had become paralyzed after vaccination. The next day, Edwin Lennette of the California Department of Health called in five Californians paralyzed after receiving the same lot of Cutter vaccine. After meeting with several officials the night of the April 26, NIH officials urged Surgeon General Leonard A. Scheele to withdraw the Cutter vaccine, which he did. Then reports started trickling in from Idaho. In Buffalo, Warren Winkelstein heard the news first from another public health official whose stockbroker had called with an urgent tip: dump Cutter.[79] The first victim to die was Dorothy Crowley. The eight-year-old girl was inoculated April 21 in Clearwater County, Idaho. A few days later she'd begun to feel ill, then her arm had gone numb, a numbness that spread to the rest of her body until she could no longer breathe. The *Seattle Post-Intelligencer* called it "the miracle of modern medicine that failed for a little girl."[80] Eventually, the Cutter vaccine was linked to 164 cases of severe paralysis and 10 deaths.[81]

Epidemiological studies of the populations vaccinated with Cutter's vaccine suggest that it probably infected at least 220,000 people and made 70,000 of them more than a little ill, some with transient muscle weakness.[82] Many of the vaccine-associated polio cases, ironically, occurred in the families of doctors who had used their connections to get vaccine through private commercial channels, rather than waiting for the National Foundation's rationed distribution network. Pamela Erlichman, 7, whose father Dr. I. Fulton Erlichman was a physician in lower Bucks County, was the first vaccinated child in the Philadelphia area to die of polio. She fell ill 11 days after her May 2 Wyeth shot. When Pamela died on May 28, her father pledged his continued allegiance to Salk, saying he had also vaccinated Pamela's brother. In lieu of flowers, Erlichman asked that money be donated to the National Foundation.[83]

This watershed event in medical history, a disaster that could have wrecked the polio eradication campaign and all vaccination programs, in fact inspired major reforms in the public health service, bolstering new structures that would serve the country well over the next five decades. The CDC set up active disease surveillance units; investigations by the CDC's new Epidemic Intelligence Service established the credibility of that group, which would become the crown jewel of the agency. Seeing the need for better vaccine safety regulation, the government replaced the Laboratory of Biologics Control with a much larger, somewhat more powerful Division of Biologics Standards, and expanded the scientific staff from 10 to 110. What's more, lawsuits resulting from the disaster would permanently alter liability law.

It would come to be known as the *Cutter incident*. But Cutter was not the only company to produce Salk vaccine contaminated with live virus. Although Alex Langmuir, the chief CDC investigator, established a precedent during the episode by the public release of most of his findings, one important result was suppressed.[84] By May 18, five people vaccinated with Wyeth's vaccine had developed polio (six other Wyeth cases were eventually reported). Each time, paralysis began in the vaccinated limb, strongly suggesting a causal relationship. The Wyeth problem, though evident, "wasn't statistically significant. It wasn't clearcut like Cutter was," says Neal Nathanson, a coauthor of the report. "My guess is—and nobody told me this—there was a lot of pressure not to publish it because the conclusion was ambiguous and the sense was, the powers that be in Washington, the surgeon general and so forth, had said, all you are going to do is alarm the public." Keeping Wyeth out of the public eye kept the narrative of the disaster more straightforward. "The idea of a manufacturer producing a defective product is hardly a new idea," said Nathanson. "The manufacturer stopped making this product, and that was it. People still buy cars despite the occasional recall of a model." On the other hand, the idea that the government had recklessly approved a flawed product could have had a major impact on public confidence in polio vaccine and vaccination overall. "Everybody was talking about that," Nathanson recalled. "If the conclusion had been that the process was flawed, the whole pendulum of public opinion could have swung in a different direction."

That may be why the government did not publicize another troubling fact. During meetings at the NIH, it became clear that all the companies were having problems producing a safe vaccine. The monkey test—injecting vaccine into the monkey's spinal cord and later examining the monkey, postmortem, for signs of nerve damage—was not as good at detecting live viruses as tissue culture, just as Weller had insisted. And manufacturers weren't testing large enough samples of vaccine in tissue culture to find live virus particles. When it conducted its own experiments, the NIH found the percentage of lots that still contained live virus ranged from 4 to 33 percent at different manufacturers. Even when the three separate strains of vaccine tested negative for live virus, they would sometimes test positive after being mixed together as the final vaccine.[85]

Nathanson and his boss, Langmuir, were officers of the new Epidemic Intelligence Service, which Langmuir, a former Hopkins professor, had been slowly building at the CDC since 1949. That year the CDC had invited him down to Atlanta to create a new epidemiologic investigative service, although the service didn't really take off until two years later, when hundreds of American and U.N. troops in Korea came down with a mysterious hemorrhagic fever. Fears of a Red Chinese germ attack created a sense of urgency in the federal government, though the hemorrhagic fever turned out to have natural causes. From the start, Epidemic Intelligence Service (EIS) members were part of the Commissioned Corps, a militarized layer of the Public Health Service that originated in the service's birth, in 1798, as a Marine quarantine agency charged with preventing infectious diseases from entering U.S. ports.[86] EIS officers were committed, tirelessly curious scientists who would combat disease outbreaks around the country and the world over the next five decades.

Nathanson would go on to direct the infectious disease department at Hopkins and the federal AIDS research program before "retiring" to head the laboratory on neurovirulence at the University of Pennsylvania. A trim, quick-witted midwesterner, Nathanson gives an articulate, hard-boiled account of the events, like a Dashiell Hammett of medical detectives. Although Cutter broke no rules, Nathanson reserves special contempt for the company.[87] "Whoever was involved in quality control at Cutter was either sloppy or dishonest depending on what you want to believe," he says. "There were some major warning signs that they chose not only to ignore but to cover up. Some of their lots were not passing safety tests. Anybody who was ethical or commonsensical would have contacted the regulatory folks at that point, which they didn't. In fact they washed those down the drain and never sent out the records. So it was partly incompetence and, let's say, absence of behaving in an ethical fashion." Cutter tossed out 9 of the 27 batches it produced after finding live virus in them. By today's standards, quality control would require that the remaining 18 batches be destroyed on the chance that they, too, contained undetected live virus. But Cutter wasn't obligated to report bad batches to the government as long as it didn't sell them.

Why did the commercial vaccines have live poliovirus in them, when under Salk's protocol all the virus was killed? The explanation starts with a simple mechanical problem. Between growing the virus and inactivating it with formaldehyde, Salk filtered out the cellular debris. This was a crucial step, because live virus could "hide" from the formaldehyde in clumps of cell tissue, avoiding inactivation. In his laboratory Salk had employed asbestos Seitz filters, which have even, regular pores. Liquid drains slowly through the Seitz filters, though, so some of the manufacturers, including Cutter and Wyeth, switched to sintered glass—filters made by fusing hot glass, a process that left tiny holes in the tubes. Such filters were not uniform and in some, it turned out, clumps of debris snuck through. Cutter also erred by allowing the filtered material to sit in large jars before it was deactivated, because debris clumped on the bottom of the jars.[88] Finally, Cutter did not conduct enough tests to find live virus that it had failed to inactivate in manufacturing. Salk's inactivation curve was terrific in theory. But in a large batch of virus, the particles inactivated at different rates because of the heterogeneity of the solution. "You could cook it 'til hell freezes over," the NIH's Shannon said, "and you would not inactivate it."[89]

The federal government ultimately was to blame for failing to regulate the production of the vaccine. But Eisenhower had pledged to keep the government's long nose out of business, and the NIH lacked the resources or inclination for close oversight. From the beginning, the polio vaccine had been O'Connor's responsibility. The National Foundation had its own Vaccine Advisory Committee, and it, practically alone, financed the labs, the experimental animals, and the rest of the apparatus needed to produce a safe and efficacious vaccine. After the NIH took over, its tiny staff mostly relied on company promises. Before the completion of the Salk trial, a young West Virginia scientist named Bernice Eddy had detected live virus in the Cutter vaccine. But her superiors at NIH declined to intervene. Shannon blamed National Foundation officials—wiseacres christened O'Connor's group the *Infantile Foundation for National Paralysis*—for keeping the government in the dark. Many scientists felt that Rivers, Smadel, and Salk had compromised the integrity of the endeavor by ignoring those who questioned the speedy delivery of the vaccine. Weller complained, "pressures exerted at very high levels on public officials and on scientists . . . have molded opinions. I cannot remember at any time being aware of a situation in which there has been such a close approximation to an 'official party line' in a matter of scientific import." Of course much of the pressure came from the public.[90]

After Cutter, part of the public recoiled from the Salk vaccine. In Philadelphia, where 43,000 first- and second-graders had been lined up for free vaccination in the schools once the remaining vaccine had been cleared by the government, parents withdrew 17,000 children. In some areas, such as District 3 in South Philadelphia, as many as 68 percent of parents pulled their consent forms. In Camden, across the Delaware, the board of education delayed all vaccinations until the fall.[91] The U.S. Public Health service responded in a number of small and large ways. It changed the minimum requirements for production and testing of the Salk vaccine, and created a technical committee to help improve it. It vastly expanded the NIH regulatory and research program, and established a Polio

Surveillance Unit at the Communicable Disease Center that would track cases of vaccine-related and wild polio.[92] From this start the CDC vastly expanded its surveillance of infectious disease in the United States.

At a seemingly endless round of meetings in Bethesda and Washington—Hull, the Lilly scientist, recalled 17 flights to Washington from Indianapolis in one month—Salk kept insisting that his methods guaranteed a live-virus-free vaccine. At one point, the normally circumspect Enders leaned across the table and said, "it is quack medicine to pretend that this is a killed vaccine when you know it has live virus in it, every batch has live virus in it."[93] The government meanwhile kept changing its tack. Surgeon General Scheele initially said that parents of vaccinated children "have no cause for alarm" even as several new cases of crippling rolled in each day. On May 7, Scheele told the press that polio vaccinations would be suspended, but the next day reversed himself. A week later, after a "recheck" of the product, the Parke-Davis and Lilly vaccines were back on the market. On June 22–23 a panel of experts was called before Representative Percy Priest's Committee on Interstate and Foreign Commerce. It included Salk, Sabin, Paul, Enders, Francis, Rivers, Shannon, Smadel, and seven others. The committee voted 8–3 (Enders, Sabin, Hammonds) to keep using the Salk vaccine. Paul and Salk abstained, as did two PhDs, since they weren't doctors.

Despite the Cutter incident, Eisenhower proudly touted the Salk vaccine, offering "full details of its manufacture" to "every country that welcomed the knowledge, including the Soviet Union." The global reception was mixed. After retesting their batches, the Swedes canceled a campaign to vaccinate 200,000 children. The Danes immediately started using a vaccine that included a less-virulent type 1 strain. Britain dragged its feet. Graham S. Wilson, director of the Public Health Laboratory Service (and the author of *The Hazards of Vaccination*, an admonitory 1967 book that no American public health leader could ever have gotten away with) felt the Salk vaccine could never be made safe and said Britain would wait for a good oral polio vaccine.[94] The Russians, meanwhile, were testing Sabin's vaccine.

Among the public, fear intermingled with frustration at the inability to get vaccinated. Soon the two snowballed into a general condemnation of the administration. On May 15, Democratic National Committee

Chairman Paul Butler attacked the government for failing to lead the polio vaccine effort. "The federal government inspects meat in the slaughterhouse more carefully than it has inspected the polio vaccine offered by the drug companies to the parents of the nation for inoculating their boys and girls," added Senator Wayne Morse of Oregon. Oveta Culp Hobby resigned in July, as did Scheele, and Workman was shifted to another NIH position. "If April was the month of triumph, and May the month of conscience, June was the month of sackcloth and ashes," John Wilson wrote. "The Americans are an extravagantly volatile nation . . . they beat their breasts as readily as they beat their drum and with the same exuberant lack of self-consciousness."[95]

As for the distribution problem, the Eisenhower administration twisted in the political winds. At first, Culp declared herself unwilling to lead a nationwide drive to eradicate polio by buying up and distributing polio vaccine. That would have been socialist medicine—which Eisenhower's predecessor, Harry Truman, had unsuccessfully championed—and "once that precedent was established, the next vaccine discovery would go the same way," an Health, Education, and Welfare official said. "We could not permit that. Our administration had a different philosophy." The American Medical Association's (AMA's) House of Delegates issued a grumpy statement that disapproved of the purchase and distribution of the Salk vaccine by "any agency of the federal government except for those unable to procure it for themselves." However, Eisenhower, guarding his flanks, also promised "there will never be a child in the United States denied this protection for want of ability to pay." And when the Senate protested that it would not endorse any legislation requiring a means test of prospective vaccines, Eisenhower agreed. His bill established a fund of $30 million for the states. "The day that the Senate refused a means test for free vaccination," says William Foege, the former CDC director, "was an enormously important day for vaccination in this country. It was the first day that public health was accepted as a right for all."[96]

It was a halting promise, fulfilled in fits and starts. By 1957, the National Foundation was having trouble getting people to vaccinate their children—whether out of poverty, fear, or the conviction that a better vaccine, made by Sabin, would be available soon. In the fall of 1957,

37 million Americans under 40 still had not been vaccinated, although vaccine was gathering dust on warehouse shelves. Lilly produced only 800,000 doses of vaccine, compared to 14 million the year before. There were 5,787 cases of polio in 1957, mostly concentrated in poor and black neighborhoods. With less polio circulating, black infants tended not to get natural immunity. Yet because they were unvaccinated, they were more vulnerable when outbreaks did occur. In 1959 in Kansas City, for example, the attack rate in blacks was 32 times higher than in whites. Only five years earlier, the rate had been higher among whites than among blacks. The solution was to bring vaccine to the people, but this wasn't being done. Polio continued to spread, Joseph Melnick wrote in a letter to Congress, "because of the failure by physicians and public health officials to administer Salk vaccine."[97]

At a hearing in late 1960, as Sabin's oral vaccine was gaining ground, Salk spoke to the problem that Democratic officials would begin confronting as they reversed Eisenhower's legacy: the successful vaccination of the public was now a problem of public policy, not science. "Any new vaccine," he said, "will not solve the problem of the unvaccinated. . . . there is probably little difference between the two preparations so far as effectiveness is concerned. The most important point is that either one be administered to a sufficient segment of the susceptible population."[98] The Cutter incident had a small but definite impact on the public's sentiment toward the vaccine. While only 2 percent of those interviewed in a 1957 survey felt the vaccine didn't work, 10 percent weren't entirely confident of its safety.[99] Half felt the Salk vaccine was more important than smallpox vaccination, which 91 percent had undergone; two-thirds found polio vaccine more important than the shot against whooping cough, which by then had been injected into roughly half. Eight percent of those interviewed said someone in their immediate family had had polio.

Sabin Steals the Show

By now the country seemed inexorably drawn to the oral polio vaccine as the winning alternative to the Salk. Sabin and Koprowski had convinced federal officials that it would be safer, easier to administer, and offer more lasting immunity. Though only one of the three claims would prove entirely true, oral polio vaccine was cheaper to make. By 1957, the oral polio vaccine makers, while pausing now and again to heap aspersions on Salk, were mainly fighting amongst themselves. Sabin, who spoke a little Russian, had formed an alliance with virologists in the Soviet hierarchy. Before long the Russians were testing his vaccine on tens of millions and marching from conference to conference with successful results—and not so much as a headache by way of side effects. Sabin's feint behind the Iron Curtain was politically risky, but it made scientific sense. Widespread Salk vaccinations had made it difficult to find a vaccine-naïve population in the United States. The Khrushchev thaw was on, and after a quick FBI check, Sabin was cleared to go to Russia.[100]

Herald Cox, of Lederle, was at a disadvantage. Lederle's medications were regularly pirated in Russia and it would have no truck with Communists. Cox had prescient concerns that monkey tissues would be contaminated with viruses, but the chick embryos he used to culture his virus were not the ideal host. As the three scientists trotted the globe from conference to conference, Cox fell behind. After some of Cox's vaccinated test subjects sickened with polio after trials in Dade County and West Berlin in 1959, Lederle canceled its research program and made arrangements to manufacture Sabin's vaccine.[101]

Koprowski, who had angered Lederle by taking its strains when he left the company, charged ahead with his own trials. His first efforts flopped, though, when a prospective partner, George W. Dick in Belfast, recovered viruses from the stool of patients fed two of the Koprowski strains and found that they paralyzed monkeys. Shaken by the episode, Dick became an outspoken foe of the Koprowski vaccine and a voice of caution about vaccines in general.[102] Koprowski scoffed at such worries. Any live polio vaccine had the potential to revert to virulence, he said, but a positive test for neurovirulence was not infallible evidence of danger. In 1958 he began large trials in the Congo.[103] "The natives were called by drum beats," Koprowski recalled, and his assistants sprayed a salty vaccine suspension into each mouth. In six weeks, Koprowski and his associates had vaccinated 244,000 people. But no antibody titers were performed before or after the vaccinations, and since most Congolese had been exposed to polio as infants, the efficacy of the vaccine could not be ascertained.

Koprowski's Congo trial, later followed by trials in Poland, Switzerland,

and Croatia, would be portrayed in a lurid account by the British journalist Edward Hooper as the source of the global AIDS pandemic. In his 1999 book *The River*, Hooper argued that Koprowski had used kidney tissue from chimpanzees—the animal that proved to be the source of the virus that at some point mutated to become HIV in humans—to culture his vaccine. Hooper argued the case on the basis of the pattern of early AIDS cases in Africa, which seemed to appear in areas where Koprowski's trial was conducted. Although Hooper had no solid evidence that Koprowski had used chimp kidneys, the rest of his hypothesis was well enough within the realm of possibility to generate the interest of AIDS experts around the world. It appeared to have been disproven a few years later when tests of the Koprowski vaccine turned up neither HIV nor chimp DNA.

But the Congo trials did fail in a more prosaic way: they did not convince the world's public health officials that Koprowski had the best vaccine. Koprowski's dismissive attitude toward the establishment was part of the explanation. He had thumbed his nose at a 1957 World Health Organization (WHO) statement on standards for oral polio vaccine trials. He was not part of the "old boys" network. Health authorities in the rest of the world chose the Sabin vaccine under the belief it was safer, although Koprowski never accepted this verdict. "There was no statistically relevant difference in virulence between Sabin's vaccine and mine," he told me in 2004. The drumbeat for Sabin's vaccine was growing in the United States. After a small spike in polio cases in 1959, Surgeon General Leroy E. Burney appointed an advisory committee to study quick licensing of oral vaccine. As Sabin happily pointed out, the rapid vaccination of the Soviet bloc seemed to be a terrible reproach to the U.S. system. There was talk of a "polio gap" between Russia and the United States, much like the "missile gap." In 1960 the AMA urged that all American children get the oral vaccine, though it wasn't yet licensed. Basil O'Connor was horrified. A full-paged ad in the July 4 issue of *The New York Times*, paid for by the local March of Dimes chapter, urged readers, "Declare your independence! Don't be confused! Get your polio shots now! You the American people paid for the Salk vaccine through your support of the March of Dimes."

In a July 1, 1959, summary of a WHO conference, Maurice Hilleman summed up the politics of polio vaccine.[104] Despite doubts about the safety data in their trials, the Russians were ready to produce 50 million doses a month and might "make and give away the vaccine to the under-developed nations as a propaganda move." Hilleman was confident an oral vaccine would be marketed in the United States. "[L]ive virus vaccine is coming and someone will prepare it. He who does it successfully will likely have a feather in his cap if not dollars in his pocket."

In March 1961, a year in which the 1,312 polio cases in the United States marked a reduction of 97 percent from 1952, Kennedy authorized a gift of Salk vaccine to the people of Cuba to stop an outbreak there. Sabin wired the White House that the Salk vaccine would not stop the epidemic and noted that the Russian wooers of Fidel had offered his vaccine to the Cubans. Said Sabin: "I think that if we do not ourselves make such material available to other nations, that the Soviet Union probably will. When I was there in Moscow in May 1960 they had been producing at the rate of about 20 million doses a week."[105] Kennedy asked Congress to authorize a $1 million purchase of oral polio vaccine to be held in reserve for epidemics. It was a batty idea—there were no more epidemics in the United States—and Sabin said so. It would make more sense, he said, for the United States to copy Cuba's example and pay for a mass campaign to "do away with polio altogether." Salk, for his part, claimed that the nation could already have done away with polio had it paid for mass Salk vaccination. He was absolutely right.[106] Sabin was by then basking in his own success. He had fed his vaccine to 110,000 children in Cincinnati in 1960, with only a single case of polio reported that summer (the city's health commissioner attributed the children's health to earlier Salk vaccination).[107] Asked at a congressional hearing if the Salk vaccine could complement his own formulation, Sabin replied haughtily: "as soon as we can get mass production to give us the automobiles, I think that horses and buggies have a place for sport."[108]

Salk was by then building his scientific paradise, the Salk Institute for Biological Sciences, at La Jolla, California. His goal was to extend the principle of the killed vaccine "to the point where we may have 100 different viruses included in the same inoculation."[109] It was a beautiful, impossible dream. Congress had other things on its mind. If the Russians had the oral vaccine, we had to have it. So while 15 million doses of Salk vaccine languished in warehouses in February 1961, the Public Health Service voted to license Sabin's vaccine.[110] It quickly became dominant, with Salk's shots sidelined in most of the world, with the exception of the Netherlands.

Before long, though, it was apparent that Koprowski's warning about the tendency of live polio vaccine to revert to virulence had been all too true. By 1964, the Sabin vaccine was giving paralytic polio to about a dozen Americans—vaccinated children and their caretakers—every year.[111] To his dying day, Sabin denied that his vaccine ever caused polio, although he was possibly the only virologist who believed that. D. A. Henderson, who went over a set of vaccine-associated polio cases with Sabin in the mid-1960s, recalled how tenaciously Sabin disputed each case. Because of his lifelong work with polio and other viruses, Sabin could come up with explanations that went beyond other scientists' capacity to respond. Years later, former Soviet scientists confessed to Henderson that their perfect record of safety with the oral vaccine had been a mirage. They, too, had seen polio cases caused by Sabin's vaccine and were disposed to report them. Sabin had talked them out of it.[112]

Monkeywrench

Just as polio became epidemic as a result of hygienic measures created to deal with other diseases, "the measures designed to protect the world from polio may, in their turn, for all we know, lead to some other quite unexpected consequence which may be to man's disadvantage", a contemporary of the vaccine trials wrote.[113] In the tumultuous history of the polio vaccine, filled with outsized scientists and publicists and stuffed with the yearnings of an entire nation, there have been many concerns about safety. The theory that the global AIDS pandemic arose through vaccination is nearly in the category of an urban legend. But another secret hitchhiker was discovered in polio vaccine in 1960, when it was too late to do anything for millions of children who had been vaccinated with the product. In time it would be called SV-40, the fortieth simian virus recorded by Lilly's Robert Hull. Unlike other threats arising from the polio vaccine, the seriousness of SV-40, which demonstrably caused cancer in laboratory animals, could not be ignored. For years, the virus lurked in the kidney cells of thousands of rhesus monkeys imported from India, without causing disease in those animals. It took the work of two scientists—Hilleman at Merck and Bernice

Eddy at the Division of Biologics Standards in Bethesda—to find the virus and show that it was potential trouble.

Eddy was a country doctor's daughter with a deep work ethic, a sharp mind, and the high, twangy voice of her native West Virginia. She had graduated from tiny Marietta College in Ohio and come to the NIH after completing her doctorate at the University of Cincinnati in 1936. A dowdy woman in a man's world, a small-college girl among Ivy Leaguers, she would quietly hold her own in the laboratory for two decades until she discovered some unpleasant truths and in the process suffered the classic fate of the whistleblower. In 1954, during the Salk vaccine trial, Eddy discovered live virus in Cutter vaccine. The inconvenient discovery, in Eddy's account, led to her being transferred out of the polio division into the laboratory's cancer section. While working there with Sarah Stewart, a Mexican-born scientist who was also interested in oncogenic viruses, Eddy ground up monkey kidney tissues that were being prepared for vaccine safety testing, injected them into hamsters, and watched as tumors grew. She repeated the experiment several times and then ground up the hamster tumors and injected them into mice, which also got tumors. This was an alarming finding, and she took it to her boss, Joseph Smadel, recently appointed chief of the NIH biologics lab.

Smadel tended to brush aside ethical concerns about vaccine experiments. Having helped oversee the birth of the polio vaccine, he was now in charge of monitoring its safety. He was remarkably unswayed by what Eddy had to say, although he would be very interested in the same result when it was presented a short time later by his former protégé, Hilleman. "I took the hamsters over to Smadel," Eddy said in a 1986 interview, "and you know what he said? 'They're just lumps, they don't amount to anything.' . . . And when I wrote papers, he held them up. He never gave me any reason, he just held them up."[114] Invited by the American Cancer Society to give a talk in New York on another virus that she and Stewart had discovered, Eddy seized the opportunity to present data on the monkey kidney virus. Smadel hadn't authorized her talk, and when he got wind of it—through the gossipy Sabin—he gave Eddy a royal chewing out. "I think if there was any awful thing in the English language to call me, he did," Eddy recalled with a giggle. "I didn't care because I thought it should be known. This vaccine was going into children!"

Eddy was not the only scientist concerned about the number and omnipresence of viruses in monkey tissue cultures. The monkeys harbored a witches' brew of viruses. Sabin had licensed Merck to produce his vaccine, but Hilleman was so troubled by the potential of monkey kidney cell contamination that in 1958 he was on the verge of urging Merck to get out of the business. Then he went to see William Mann, director of Washington's National Zoo and a world expert on simians. Mann's wife was from Montana and Hilleman used this fact as an icebreaker, and not for the last time. Over dinner at Mann's house, which was filled with beautiful African art and artifacts, Mann gave Hilleman a few useful tips. First, he told him to import *Cercopithecus*—African green monkeys, or vervets—which hadn't been used in the vaccine business. Second, he suggested that Hilleman import the monkeys through the Madrid airport, which wasn't used by other shippers and was therefore free of dirty monkeys and their viruses.[115] Some months later, Hilleman picked up the first batch of African greens at LaGuardia and brought them to Pennsylvania. Tests revealed the monkeys to be remarkably free of indigenous viruses. But when Hilleman attempted to transfer polio vaccine virus from other monkey kidneys into the *Cercopithecus*'s cells, he received an alarming surprise. The African greens' cells essentially exploded. The substance wreaking this havoc would turn out to be the same virus that was giving Eddy's hamsters cancer. Because it did not harm rhesus monkeys, the virus had passed, unnoticed, in cells that had been used as the substrate to produce millions of vaccines. A generation of American children had been exposed to the virus.

Hilleman also found SV-40 in Sabin's vaccine and in seed stocks of the killed adenovirus vaccine, used to prevent a form of the common cold, that he had invented a few years earlier. He assumed that the formalin used to kill polio virus in the Salk vaccine would have also destroyed SV-40, but this was wrong. The virus did not kill macaque, rabbit, human amniotic or heart cells, or the immortal human cancer tissue known as HeLa cells. The question was, would it hurt people? Had it hurt people, and if so, how? The unique characteristics of SV-40, a virus that caused cancerous changes to cells extremely slowly, have made its risks slow to emerge and difficult to define. But in 1960, it was giving lab animals cancer, and it was obviously not an agent that belonged in a pediatric vaccine.

Hilleman presented his findings on the first day of the Second International Live Polio Vaccine Conference in Washington in June 1960.[116] It was like releasing wasps at an anaphylactics convention. Many participants, including the Russian virologists, immediately left for home. "The Russians got out of there fast," Hilleman says. "They had vaccinated about 50 million people by then." A manufacturer came up to Hilleman and told him, half-seriously, to stop finding so many viruses. "I said, 'Please, I'm just the messenger here. I'm a f——ing hero!'" Sabin was very upset, telling Hilleman that publicity about the virus was unnecessary and would upset the vaccination program. "I said to him, 'I have a feeling in my bones and I think this virus is different. I think it may have some long-term effects.' He said, 'What?' and I said, 'Well, cancer. You may think I'm nuts, but I have that feeling.' Well by that time we'd put it in hamsters. The joke was going around that we'd win the Olympics because the Russians would be loaded with tumors."

Later, Hilleman gave hamsters cancer by injecting them with Salk vaccine. "That caused great consternation," he said. "I'll never forget the day that Joe Smadel listened to the review of this."[117] Smadel would die in 1963. Eddy, who retired in 1971, would later be vindicated and gain a special recognition from the government. Koprowski came out on the angel's side of the SV-40 debate. As early as 1961, he began urging the U.S. government to require that polio vaccine be prepared in human cells since monkey cells contained "innumerable viruses, the number varying in relation to the amount of work expended to find them."[118] Sabin, on the other hand, never stopped denying the significance of SV-40, just as he continued to deny that his vaccine ever caused polio. Hilleman saw it as wishful thinking. "My God, when somebody hits you in the face with a lemon pie, you ought to know you've been hit in the face with a lemon pie," he said. "This is real." And because Hilleman believed it was real, Merck, alone among the major vaccine producers, shied from the polio vaccine. It had pulled its inactivated polio vaccine out of the Salk trials in 1954 after Betty Lee Hempil, a Merck scientist, found evidence of live virus in the experimental product.[119] In 1959, Hilleman had developed Purivax, a highly concentrated inactivated vaccine, but it was withdrawn in 1961 because of SV-40.

Almost 100 million Americans were vaccinated with the Salk vaccine between 1955 and 1962, when producers had to certify their vaccine free of SV-40. The earliest studies of exposed children showed no higher risk of cancer, but later studies did detect the virus in tumors. While the evidence is still sketchy,[120] some researchers believe SV-40 may have caused mesotheliomas, an aggressive cancer of the chest wall often associated with asbestos poisoning.[121] Polymerase chain reaction, a sensitive molecular detection test, has found SV-40 in many types of cancerous cells. The virus is a common laboratory contaminant and may be contagious, which makes it hard to directly link Salk vaccination with a cancer. Hilleman, who gave a lot of thought to the issue, was agnostic about whether real harm had been done. "The problem is, how do you prove it?" he said.

Polio Redux

The polio matadors became celebrities who jetted around the world to give their advice to its leaders. Salk got divorced, married one of Picasso's ex-lovers, and fashioned himself into a West Coast sage, a Carl Saganesque biologist who emitted a series of philosophical treatises about man's inevitable conquest of himself.[122] Sabin, the more hardheaded of the two, pushed for polio eradication, coming up with a system of national immunization days that was used, first in Cuba, later in the rest of Latin America, to eradicate the virus from the Western Hemisphere in 1991. He also dabbled in saving the world, meeting with Tito in November 1967 in an unsuccessful effort to get the Yugoslav leader to intercede in the Arab/Israeli conflict.[123] Koprowski created a successful rabies vaccine and his Wistar Institute worked on the virology of cancer and multiple sclerosis; in the late 1990s the institute kicked him out but he landed at Thomas Jefferson University, where when last seen, at 93, he was growing SARS, West Nile, and AIDS antigens in tobacco and tomato plants, hoping to create a new generation of edible vaccines. Many vaccinologists were skeptical that such formulations would ever pack enough antigenic punch to make it through the gut into the bloodstream, but Koprowski, supremely confident as ever, shrugged them off.

Public health officials had hoped to eradicate polio in time for the fiftieth anniversary of the Salk vaccine trials, but it was not to be. The numbers of cases were in the hundreds now, although there was a troubling resurgence in Africa in 2002 that eventually spread to the Mideast and even Indonesia. As for the "polios" themselves, they had been pushed far to the margins of national consciousness, although they were visible everywhere if you looked hard enough. Funding for the treatment of polio dried up even in the late 1950s, as the nation moved on to the next issue and tried to forget about the people for whom the vaccine was too late. "Once lionized as heroic examples of human fortitude, the thousands of polio survivors who continued to need medical and financial help were suddenly ignored as embarrassing emblems of their own poor timing," wrote Kathryn Black. "As veterans of other wars would continue to discover, the same civilians who pay for them in the heat of battle don't like to be reminded of the wounded and the dead after the war is over."[124]

The hundreds of thousands of polio survivors still living today pulled themselves out of iron lungs and hospital beds and grew new nerve cells and muscles to control the ones they lost during the acute stage of infection. But many decades after their original symptoms, about one-third of all polio patients begin to suffer recurrent muscle weakness, known as post-polio syndrome. The nerve and muscle connections they have reestablished start to wear out, and the problems of weakness, severe fatigue, and even limb atrophy return. In 1984, Lauro Halstead, an internal medicine specialist at Baylor University in Houston who had contracted polio in 1954, began to suffer pains that were similar to his acute illness. Halstead organized a conference at Warm Springs on the problem that year. But just as there was no cure for polio, there seems to be little to ameliorate the sufferings of polio victims once again falling ill with their conquered disease.

In September 1982, Carl Weihl—the Cincinnati pediatrician who "didn't want to become disabled"—found he was having difficulty breathing. He refused to go on a respirator and died a few days later, at age 62. Like many of the battling polios, he is remembered at least as much for his achievements as for his agony. The generations he left would never have to remember—would in fact, forget—that their parents had lived in mortal terror of the virus that lurked in swimming pools and movie halls, until a young scientist from the Bronx, with the whole nation behind him, chased that bad dream away.

CHAPTER 6

—

BATTLING MEASLES, REMODELING SOCIETY

Measles! said the doctor. Mumps! said the nurse. Nonsense! said the lady with the alligator purse.

—NURSERY RHYME

ON NOVEMBER 7, 1961, the National Institutes of Health in Bethesda hosted the world's leading virologists for a gathering that heralded the swelling optimism of that medical era. The participants in the first International Conference on Measles Immunization were men and women rushing to use the suddenly limitless technologies of virus growing. Ambitious scientists and pediatric humanitarians saw a future of lifesaving vaccines for the world's children. Drug companies saw a significant new profit arena. President John F. Kennedy, who opened the conference with a letter of welcome, envisioned vaccination as a wholesome symbol of the nation's strength, prosperity, and generosity. "Your purpose is to advance the health of all people everywhere," Kennedy said, "and especially to protect children against a formidable and widespread threat. Such meetings and such goals transcend national boundaries and unite mankind in our common striving toward a better life."[1] The conference-goers included leading U.S. scientists from government, academia, and industry—Salk, who had withdrawn from vaccine making to loftier pursuits, was a notable absence—and from India and

Japan, Egypt and England, France, Germany and Brazil, Yugoslavia and the Soviet Union.

In many ways the conference was a tribute to John Enders, the Nobel laureate who had found the key to polio while searching for a measles vaccine. Paralytic polio cases in the United States had declined to 892 in 1961, and Enders and his young associates had in the meantime cultured measles on a new food and trained it to adopt milder manners. Their raw vaccine was being perfected and tested by hundreds of scientists with whom Enders shared his virus. In June 1960, the *New England Journal of Medicine* had published eight articles on vaccines derived from Enders's virus, a bold splash that was warranted, the editors said, by "the effectiveness of this vaccine, which seems assured in spite of the small numbers of patients so far tested."[2]

The Enders virus was derived from a single case of measles in an 11-year-old boy, David Edmonston. Years later, Edmonston would attend the March on Washington and, inspired by Dr. King, became a schoolteacher in Mississippi. Just as the 1960s transformed Dave Edmonston, the measles virus that started out in him helped transform the nation. Derivations of the Edmonston measles vaccine became a cornerstone of the Great Society, a little helper to mothers that would become a guaranteed right to all. Edmonston became the McDonald's of man-made viruses, with more than a billion served worldwide. Children would generally rather eat hamburgers, but the Edmonston was better for them.

In 1963 the first measles vaccines were added to polio, diphtheria, pertussis, and tetanus on the schedule of childhood vaccinations. Mumps and rubella would be added within a few years.[3] The development and application of these vaccines unfolded very much in the transformative way that Gen. Simmons, Vannevar Bush, and others had envisioned during the war. Between 1955 and 1980, American culture underwent a sea of change: Daycare centers and small families became a commonplace; the civil rights movement introduced new expectations of fair treatment and access to society's resources; the influx of women transformed the workplace; and people began taking global travel for granted. The conquest of common, highly contagious diseases played a major role in these changes. Vaccines and antibiotics produced healthier, better-protected children, which helped women working outside the home. As we'll see, vaccines also played a crucial role in the legalization of abortion, the disabled rights movement, and the creation of the welfare state.

In this era of miracle drugs and vaccines, enthusiasm was especially high for a preventive against measles, a disease whose morbidity and mortality was greater than polio, if not as visible. Measles was the world's most contagious common disease and a defining rite of childhood. Unlike smallpox, which spread only when the patient bloomed with a visible rash and was usually bedridden, a child with measles could be shedding germs for four days before the rash appeared until four days after it had disappeared. The virus could survive in the air for up to an hour and a half; you could catch it from someone you'd never even seen.

Several days after exposure a vivid red rash flared up on the child's face, gradually resolving into a blotchy red mask, like a severe sunburn, accompanied by puffy, watery eyes. Children with measles got diarrhea and vomited, and they stayed sick for two weeks. Mortality was estimated at somewhere between 1 in 500 and 1 in 10,000 cases in the United States; in the 1964 epidemic there were 4 million cases and 400 deaths. It was far deadlier in poor countries, where it killed millions every year—less so among well-cared for, well-nourished children living in uncrowded housing. Even so, measles hospitalized an average of 48,000 Americans each year in the early 1960s, with 4,000 cases of encephalitis and 7,000 seizures that left up to 2,000 children brain-damaged or deaf. Unlike the polio-damaged, who retained wits and willpower enough to struggle with their withered limbs, measles encephalitis victims often could not even feed themselves. Many became wards of the state.[4]

The prospect of vanquishing the disease fired the imaginations of American public health crusaders in this era of miracle drugs and the moon race. "Any parent who has seen his small child suffer even for a few days with a persistent fever of 105, hacking cough and delirium, wants to see this disease prevented if it can be done safely," said Alexander Langmuir, the chief CDC epidemiologist, in a speech to the 1961 conference.[5] "In the United States measles is a disease the importance of which is not to be measured by disability or deaths, but by human values and the fact that tools which promise effective control and early eradication are becoming available. To those who ask me 'Why do you wish to eradicate measles?' I reply with the same answer that Hillary used when asked why he wished to climb Mount Everest: 'Because it is there.' To this may be added, 'and it can be done.' "

There was another exigency driving the country toward vaccination against measles. By the early 1960s, pharmaceutical companies were producing high-quality immune gamma globulin by centrifuging and filtering out that portion of the blood containing antibodies to disease. IGG could provide short-term protection against measles. Therefore, a physician who withheld it from a child exposed to measles accepted "a degree of responsibility for complicating encephalitis, which although statistically small, looms high when it occurs in his own patient," Dr. Edward Shaw, a San Francisco pediatrician, told the conference. But IGG was expensive, and measles was common. If immune gamma globulin was to become the standard of care, it was only a matter of time before measles vaccine replaced it. An effective vaccine was clearly the safer, surer, longer-acting, and cheaper alternative. Parents were hesitant about loading their children up with vaccines, Shaw said—but measles would make the cut. "Whatever the effects of uncomplicated measles . . . the ultimate results of central nervous system complications are decidedly worse than the overall effects of paralytic polio," he said.

But the vaccine would have to be safe. Measles and other viral diseases were so rife that up to a few decades earlier, it had been public health policy to urge the infection of children. In the 1930s, Tom Rivers wanted the New York Division of Health to require parents to "see to it that boys and girls were exposed" to mumps and rubella, until the city attorney pointed out a law (a prostitution control measure) that prohibited the intentional spread of disease.[6] In 1935, University of Pennsylvania professor Theodore Ingalls, whose wife had lost a child after contracting rubella during pregnancy, tried to get the city of Philadelphia to organize chickenpox and rubella parties so children could be exposed in a controlled way.[7]

The Cutter incident had had an impact on pediatricians, and a few, such as C. Henry Kempe, by then at the University of Colorado, were anxious about whether enough was being done to ensure the safety of vaccines. Kempe, who hosted the measles conference, was a virologist and pediatrician who would coin the phrase "battered child" and create the field of child abuse prevention. Kempe had been born in Breslau and fled the Nazis in 1939, at age 15. Separated from his parents and siblings, Kempe worked his way through medical school at the University of California in San Francisco during the war and later became a fellow in Smadel's Washington, DC, laboratory. There he assisted in lab work for the 1947 New York smallpox outbreak, and began work on a safer small-pox vaccine and on an immune gamma globulin for vaccination injuries. The casualties of the mass vaccination of 1947 left Kempe with a pro-found appreciation for the risks of vaccination as well as its benefits.

Indeed, Kempe even scheduled a British skeptic of the measles vac-cine, Dr. Graham S. Wilson of the Public Health Laboratory, to speak during the opening session of the conference. No mother in Britain would accept a new jab, Wilson said, unless it was extremely safe and effective. Measles killed fewer British than American children, according to Wilson's data; Brits wouldn't put up with a vaccine with obvious side effects—especially for a disease that his stout-hearted fellow citizens didn't particularly fear. The vaccine being tested in Britain at the time seemed to be a woeful product. Alan Goffe, a virologist at the Wellcome Institute, documented raging fevers in most of the 152 children he injected with a supposedly attenuated vaccine in early 1961. Three of the kids had become delirious, and more than 50 were so sick they needed sedation or antibiotics.

Like most of the scientists meeting in Bethesda, Goffe had created his vaccine from Enders's Edmonston strain, which was captured from Dave Edmonston at the Fay School for boys in a western suburb of Boston. Edmonston's father worked for the federal government in suburban Maryland on secret projects such as a computer that would run the econ-omy in case of a nuclear disaster. Edmonston himself was a searching, depressed 11-year-old who had been packed off to the boarding school to get a dose of stiff-upper-lip. A measles epidemic brought him together with Thomas C. Peebles, a fellow in Enders's laboratory, in January 1954. Hearing that cases had been confirmed at the Fay School, Peebles rushed there and asked for volunteers. Dave, lonely and bored, with a full-blown rash on his face and chest and a fever of 102, was only too happy to let the friendly, intelligent ex-navy pilot take throat swabs, blood, and stool sam-ples. Peebles worked them up in tubes of human kidney cells and after a few days saw clustering that indicated a cytopathic, or cell-killing, effect—an indication that the measles germ had been captured. The first paper was published in May 1954. The men and women of Enders's laboratory continued to put the Edmonston virus through its paces until they had a strain of measles that was strong enough to produce immunity to wild measles virus, yet mild enough not to sicken children, at least not too much. That was the Edmonston strain. While Enders was working on it, the Nobel committee called with the news that he'd been awarded the 1954 Nobel in Medicine and Physiology. Harvard, which hadn't thought much of Enders until then, hastily promoted him to full professor.

It was all very murky, very Promethean, this talent for transforming viruses into vaccines. More than one virologist pointed out that the essentials of the search for an effective viral vaccine were given in Macbeth: "Round about the cauldron go, in the poison'd entrail throw . . . cool it with a baboon's blood, then the charm is firm and good."[8] The development of the Edmonston measles vaccine relied upon Enders's hunches, his stately pace, and gentleman's network of friendships. The virus was passaged through kidney cells from hydrocephalic youngsters whose kidneys were removed as part of their treatment. A neurosurgeon friend provided them to Enders.[9] After a while the neurosurgeon no longer had extra kidneys, but Enders had another friend at the Boston Lying-In Hospital, now Brigham and Women's Hospital. Enders guessed, correctly, that the cells lining the placentas his friend discarded would have great growth potential, and a single placenta produced enough amniotic cells for weeks of laboratory work. Enders's observation was that since children with measles didn't have amniotic membranes, "moving [the virus] away from a normal situation maybe you would select out some kind of variant," recalled Samuel Katz, then a well-spoken New Hampshire lad who would have the wits, extroversion, and good fortune to become a central player in the enterprise.

"It was just the blind leading the blind, but the virus grew well in amniotic cells. Well if it grew well in the amniotic cells of humans how about chicks?" The virus didn't do anything visible to chick eggs, but it grew in them, and when Enders and Katz passaged the virus from chick eggs to chick tissue, it started to make changes. They then injected the virus into monkeys, and the monkeys didn't get rash or fever, but they developed antibodies. "So we said, 'let's see what happens.' We inoculated one another. No one got sick, fevers, sore arms. So, 'Oh, it appears to be safe.' But we'd all had measles so we had no evidence it was immunogenic. The only way we could find that out was to inoculate children." Enders put on an old felt hat for the final, successful tests. At a conference in New York, Smadel rose to his feet and said, "John, you've done it again."

Then the Harvard researchers distributed the virus to clinical investigators around the country to test on children. Enders didn't patent the vaccine. "Anyone who came to the laboratory who was seriously interested and a legitimate person we gave any and everything to—virus, serum, tissue culture," says Katz. "His attitude was, the more people working on the problem the sooner you'd get an answer." The man looking hardest for the answer was Maurice Hilleman, the crafty, strong-willed workaholic who headed the virological research department at Merck Pharmaceuticals, which by the end of the 1960s would be the indisputable leader of the vaccine industry. Hilleman, who died in 2005, is a legend in the world of vaccination and public health, and for good reason: his products undoubtedly saved more lives than those of any other individual of the past half-century. In a field in which brilliant men have worked their entire lives without managing to bring a single vaccine to market, Hilleman developed three dozen. Not all of them, in truth, were Hilleman's from start to finish—he was as willing to beg, borrow, or steal a good virus as the next man. Some failed, were shelved, or fell by the wayside. But Hilleman pushed Merck to make vaccines, and he had the clout, intelligence, and resources to make it happen, while other companies failed.

More than half the shots that American children receive in their first two years of life either started or spent a significant period of their development in Hilleman's lab. Along the way he also made important scientific observations about the behavior of flu and the common cold, protected billions of chickens from Marek's disease, and helped Merck produce synthetic interferon, the anti-inflammatory and antiviral drug. "Most scientists would be thrilled to have achieved only one of the scores and scores of Maurice's accomplishments," says Anthony Faucci, the longtime director of the National Institutes of Allergy and Infectious

Diseases.[10] "One can say without hyperbole that Maurice has changed the world." Hilleman might have won a Nobel for some of his discoveries were it not for the fact that "I work for dirty industry,"[11] as he jokingly told associates. There was no scientist in the vaccine industry who came close to matching Hilleman's record. This is testimony both to his brilliance and to the half-hearted resources most pharmaceutical companies have invested in vaccines.

In a sense, it was Hilleman's world to change. Merck, the company where he worked starting in late 1957, had smart people who grasped the importance of viruses. One of these was Vannevar Bush, the inventor and administrative genius who oversaw the Manhattan Project and other feats that transformed America's wartime technology. Roosevelt had appointed Bush to head the Office for Scientific Research and Development, which organized wartime science by contracting out projects to private and academic scientists. Bush's recognition that civilian scientists could be funded to overcome problems of national importance presaged an important shift. During the immediate postwar period, before the Salk vaccine forced the federal government to strengthen its regulation of vaccines, much of the nation's virology talent was clustered at the Walter Reed Army Institute of Research. By 1960, many of these scientists had transferred to the National Institutes of Health or to academia, or were cherry-picked by a pharmaceutical industry looking for well-connected talent. The company that benefited most from this web of relationships was Merck and Co.

During the war, the company's president, George W. Merck, had headed the innocuous-sounding War Research Service, which investigated the possibilities of defensive and offensive biological warfare. At the height of the biowarfare program, nearly 4,000 soldiers and scientists made anthrax bombs, vaccines, and drugs at Fort Detrick, Maryland, and at labs in Utah, Idaho, and Indiana.[12] The expanding work of virologists at these labs showed the limits of antivirals, and this insight helped nudge Merck to focus on vaccines. (Another company, Pfizer, leased the government biowarfare laboratory in Indiana in 1947 and turned it into a civilian vaccine plant.) Vannevar Bush joined Merck's board of directors in the early 1950s and in 1955 became chairman of the board. Merck and Bush "knew there were these things called viruses that would be important," Hilleman recalled.[13] In June 1955, Bush headed a Merck research committee that met with virologists at the Rockefeller Institute, and he emerged with a "grand feeling of the importance of the field and the worthwhileness of pursuit of an aggressive program."[14]

Merck's reputation rested on its chemistry. Max Tishler, head of the new Merck, Sharpe, and Dohme Institute for Therapeutic Research (created after Merck merged with Sharpe and Dohme in 1953), had synthesized cortisone, B vitamins, and antidiuretics, and developed new processes for making antibiotics.[15] Tishler was given free rein to find a good vaccinologist and start making vaccines. He was quickly drawn to Maurice Hilleman. The young Hilleman had written a virology textbook even before receiving his PhD in microbiology from the University of Chicago, then surprised his colleagues by turning from a shining academic path to sign on with industry in 1943, as chief of virology at E. R. Squibb. He was 24 at the time. Four years later Hilleman went to Walter Reed to work under Joe Smadel and learn more science. At Walter Reed, Hilleman oversaw blood tests during the Salk vaccine trial and created the first effective vaccine against the adenovirus, which caused the common cold and was a nemesis of armies everywhere. And he made one particularly canny prediction. In 1957, while leafing through a *New York Times*, Hilleman's attention was drawn to a photograph of some children in Hong Kong. The photo accompanied an article about a flu outbreak, and Hilleman was convinced by something in the glassy-eyed listlessness of the children that they were suffering from a new pandemic strain of influenza. For raising the alarm about this second flu pandemic of the twentieth century, as well as his contributions to understanding of the principle of antigenic drift and shift, the Pentagon awarded Hilleman the Distinguished Civilian Cross.

According to the company's account, Tishler was at Walter Reed interviewing Hilleman for the job when the phone rang. Hilleman picked it up and proceeded to chew out an army colonel. Hilleman, it was clear, had the chutzpah the company needed to get the jump on its rivals.[16] Tishler decided on the spot to hire him. Though formally retired at age 65, Hilleman still showed up for work at least six days a week well into his 80s.[17] Time had worn down a few of his rough edges, but when I interviewed him in 2004 he was still an ornery, imperious figure with a piercing intelligence and a terrific sense of irony. Hilleman was the antithesis of the studious, detached Salk. He came from a line of tough, profane virologist cowboys, pioneers who had to fight for their ideas and often trampled on those who disagreed with them. His strongly voiced opinions, bluffs, and wisecracks changed the tenor of many a scientific meeting. At times this behavior seemed at odds with scientific principles of judicious objectivity. One scientist recalled being uncertain whether Hilleman was joking when he suggested, during a meeting on a rotavirus vaccine in the 1990s, that the scientist "go to the poorest area of town and try out your stuff on 20 kids there."[18] As it happens, Hilleman managed to make fewer errors than any of his more-cultured competitors. He produced vaccines that worked *safely*.

From the beginning, Hilleman's life was a sort of tall tale, full of Bunyonesque close scrapes and big hauls. Like Enders, his life was profoundly affected by the great flu pandemic. Born in 1919, he survived both his mother and twin sister, who died in childbirth of flu. The eighth of eight motherless children growing up in Miles City, on the freezing and scorching plains of eastern Montana, Hilleman was cared for by an aunt. Life on the farm was all work—seven days a week, with time off Sunday morning to go to church. Hilleman had all the usual childhood illnesses and narrowly escaped being run down by a train while riding his bicycle over a trestle. But the farm was a cauldron for science. There was a blacksmith shop and plenty of animals, plants, and chemicals. After high school, Hilleman had signed up to become a salesman at J.C. Penney, selling bolo ties to cowboys, when a scholarship to Montana State College came through at the last minute. After college he was accepted to graduate school at the University of Chicago, which left him smart but also hungry. Only 128 pounds clung to his six-foot-two frame. "They expected me to live on $28 a month," he says, explaining his jump to the drug industry. "Hell, I had a fiancée waiting for me back home."[19]

By the time the pandemic flu of 1957 arrived in the United States, Hilleman was working for Merck. His first job was to produce 20 million flu vaccines in chicken eggs. Hilleman set about building a team that was up to his standards—which meant working seven days a week. As the onetime caretaker of his family chicken coop, Hilleman had a leg up on his more urbane East Coast competitors. "I am very indebted to chickens," he said—wryly, but a little seriously, too. "Chickens helped me out so many times." As Hilleman told it, he had nearly despaired of viral research by 1960 because of the contamination of monkey tissues. It was then that he met William Mann in Washington and brought over the African green monkeys, which led to the discovery of SV-40 and also became a new substrate for experimentation and attenuation of viruses. Although the vervets were clean compared to the rhesus monkeys from India, they had their own viruses. In 1986, Hilleman joked to medical historian Edward Shorter that in importing the vervets he "didn't realize I was bringing AIDS into the United States." It was a typically freewheeling Hilleman remark, one that he later regretted since it was taken as truth by antivaccine fanatics.[20] (The global AIDS pandemic is caused by variants of the HIV1 virus, which at some point passed to man from chimpanzees, not vervets. HIV2, though similar to a virus found in the vervet, is not the type of HIV circulating in the United States; nor did it jump to humans through vaccines.) But the viruses monkeys harbored certainly posed dilemmas for virologists, especially the ones who produced vaccines for millions of children.

Hilleman was proud of the fact that he used Havertown, an upper-middle-class, highly educated community, as the testing ground for his measles, mumps, and rubella vaccines in the 1960s, as well as for varicella, hepatitis B, and other viruses in later years. "The big problem with measles was, you had to test it in children. We had guts. We used Havertown. Not the poor people, but a high social level and they volunteered." For these trials Hilleman was fortunate to have the help of his old friend Joseph Stokes Jr., the venerable chairman of pediatrics at the University of Pennsylvania Medical School. Stokes, born in 1896, was a tall, smiling man who drove an old Plymouth stationwagon like a bat out of hell. It was said that he could hear more through a stethoscope than any other teacher at Penn. He was also liberal, pro–civil rights, pro-choice, and opposed to the Vietnam War. At home, Stokes and his family addressed each other as "thee" and "thou." "Joe was really a great guy," says Hilleman. "He was a devout Quaker, you know, but when his wife wasn't around I could get him to drink." Stokes had worked with the American Friends Service Committee, whose members volunteered for medical research in lieu of military service during World War II. The trial ethics were his responsibility. "He said, 'We'll do something that has possible benefit, and to the best of our ability we'll do no harm,'" Hilleman recalled. "That's real nice, you know. I used to lay in bed at night and sweat. Because everything was just guess and guts. Always following logic, of course, but it was rational empiricism. If you want to describe it, that's what it was. Checked out, mind you, in human beings—in children." Hilleman once said that he never felt entirely confident of a vaccine until it had been given to 3 million children.[21]

When Hilleman arrived at Merck, scientists were working on a killed measles virus vaccine, but Hilleman was skeptical—a good thing, it would turn out. He took Edmonston virus and tested it on children but most of them got fevers, and a few had seizures. So he ran the virus through a series of additional passages, keeping the general formula used by Enders. How did Hilleman choose what materials to use? "Well, they were available! If I took a virus and it was adapted to chicken fibroblast and I put it in guinea pig cells, mammalian cells, well that's a different environment, there'll be a certain degree of selection, and you weed out the wild virus." But there was no way to test the vaccine except in children.

For three years, Hilleman kept cycling the virus from lab to clinic and clinic to lab. The man on the ground in Havertown was Dr. Robert Weibel, who contacted local doctors and pastors to drum up 600 five- and six-year-old subjects. In December 1960, Merck started its first measles trials. The children were injected with the vaccine in one arm and measles immune gamma globulin in the other. The placebo children got a single shot of killed measles vaccine that Merck knew didn't work. "We told people we wanted to know the relative effects of the two vaccines. Truth is, we knew those that got the killed vaccine were not going to get protection," Weibel recalled recently. "You might say, 'Did you misrepresent that?' Well, I felt badly about any child with measles, but I had to look at the greater good. I was trying to eliminate a disease and I thought it was the right thing to do." In September 1961, Children's Hospital put out a press release saying that the "200-year search for a method of vaccination against measles is now over."[22] That wasn't strictly true. At the November 1961 conference, Merck was confronted with a challenge from Anton Schwarz, a portly German scientist who was working for Pitman-Moore, a division of Dow Chemical. Schwarz had come up with a superior vaccine.

The mention of Schwarz elicits a chuckle of pleasure from the scientists who knew him. A submarine officer during World War II, Schwarz was captured and came straight from a POW camp to Sabin's laboratory in Cincinnati.[23] One of his notable behaviors was making himself a big sandwich every hour. "I asked him why," Katz recalled, "and he said that he'd always been hungry during the war. He made up for it in a most gigantic way." Albert Sabin was hard on his subordinates; sometimes he'd literally pound on them, screaming, if he was disappointed with their work. The junior scientists took to hiding in a broom closet when Sabin was on a rampage. But that wasn't possible for Schwarz. "Once Albert started pounding him and he ran down the hall to the closet but he couldn't fit through the door," a colleague recalled. "He was too big!"[24]

Despite the harassment, Schwarz learned important things from Sabin, including a method of passaging viruses at a lower temperature to render them milder. He relied on this approach to create the second-generation measles shot that is the basis for the shots used around the world today. The Schwarz vaccine, licensed in 1965 as Lirugen, caught on quickly because it caused fewer reactions than other live vaccines. Since it did not require a shot of immune gamma globulin it was clearly a superior product to the cumbersome, two-arm Merck vaccine.[25] By 1966, Lirugen had captured two-thirds of the measles vaccine market in Los Angeles, for example. "Hilleman was working on a more attenuated measles vaccine constantly because of the Schwarz strain," Weibel says. If the company couldn't come up with a shot that didn't require immune gamma globulin "we were done for."

The government licensed Merck's further attenuated vaccine in 1968. It was obvious to others in the field that Hilleman had gotten the idea—if not the virus itself—from Schwarz. Genetic testing later showed the Schwarz and Merck vaccine viruses to be identical. But there was no way to prove you'd plagiarized a vaccine in the pregenomics era, and Hilleman denied stealing the virus to his dying day. "Hilleman gave it a different name. He called it Moraten, which meant More Attenuated Enders—he figured we'd like that," says Katz. "Then he went and

patented the vaccine. Well that really frosted me. I don't think it bothered Doctor Enders very much."

Expanding Vaccination

Vaccination was an increasingly routine part of middle-class life. One
pediatrician at the 1961 conference quoted a colleague's daughter as saying: "Monday is French, Tuesday music, Wednesday dancing, and Thursday, shots." Before too long, vaccination would transcend class boundaries,
becoming an entitlement and a pillar of the Great Society, and a decade
later it would become almost universal. In the late 1950s, poorer
American children were lucky to be vaccinated against smallpox, the only
immunization required for school entrance in most of the country. By the
end of the 1960s, most children would get shots against eight separate
diseases. A decade later protection against these organisms was considered not only the right of every American child, but a parent's legal
responsibility. And the diseases against which Americans were vaccinated
had mostly disappeared.

The same can-do spirit that had guided the National Foundation in
its pursuit of the Salk vaccine would be applied to other new vaccines for
children. Children's health was an endless frontier, and the lone rangers
of the Public Health Service were ready to ride, rope, and brand every
germ in sight. Bolstering the planetarium dreaminess and ambitions, of
course, was the implacable logic of the cold war. It was being fought with
shiploads of grain and pharmaceuticals as well as proxy armies. The polio
story had already shown that vaccines could be something glorious and
powerful and valuable, elixirs sought by kings and princes, commissars
and tinhorn dictators. The Russians took Sabin's vaccine, copied it a mil-
lionfold, and claimed it as their own invention. In 1960, with *Sputnik* and
Fidel traumatizing American technocrats,[26] secret cables hinted darkly at
a Soviet plan to distribute Sabin's vaccine free of charge to "non-aligned"
nations—a sugar cube to sweeten client status.[27] Cuba's militant government happily played along. In February 1961, Health Minister Jose
Machado Ventura announced that Cuban children would be mass vaccinated with the "Soviet" vaccine, which had "proved more effective" than
the American Salk. In March, Kennedy told a press conference that 161

Cuban children in Guantánamo City had been vaccinated by U.S. medical personnel to stop an epidemic. The Cubans hotly denied there had been any polio outbreak and charged Kennedy with "imperialistic objectives." A month later came the Bay of Pigs invasion.

Thus, with the Alliance for Progress carrot and cold war stick, Kennedy made vaccination a landmark of his administration. His interest in the immunization program established a pattern. Each time a Democratic administration took office over the next 32 years, vaccination got a boost. It was an arena in which the two political parties tended to be true to their ideological roots. Under the GOP, steeped in free-market beliefs dictating that medical goods such as vaccines should neither be freely awarded to nor forced upon anyone, vaccination funding tended to be flat or to get the knife. Democrats, on the other hand, were eager to finance a program that improved the health of children in the United States and overseas. It was altruism in the liberal tradition, and good politics. In 1961, Kennedy took the first steps toward creating a national immunization program with the Vaccine Appropriations Act, or P.L. 317, which subsidized polio and measles vaccine purchases by the states and would become the basis for federal vaccine funding for four decades. During a hearing on the bill, a skeptical Arkansas Senator Oren Harris asked, half in jest, "Now you would not want to deprive a child of the wonderful experience of measles, would you?"[28] Yes sir, said Abraham Ribicoff, secretary of Health, Education, and Welfare. Kennedy was framing vaccination as a right, failure to vaccinate as a national disgrace. Organized labor fully supported the program. Among children under 5 in Atlanta, AFL-CIO representative Andrew Biemiller noted at a 1962 hearing, 78 percent of the "upper socioeconomic group" had received three or more Salk shots, while only 30 percent of the "lower sector" were fully protected. "Cases of paralytic polio are concentrated in a city's central core," he said. "This is where the poorer, the less privileged, the minority groups are to be found."[29]

An amendment to P.L. 317, which passed in 1962, spoke darkly of the need for tetanus vaccination to protect people in a nuclear war, in which most injuries would be "penetrating wounds contaminated with dirt." Diphtheria immunization would be of "immeasurable importance" for a population crammed into bomb shelters in time of disaster.[30] The

National Institutes of Health, recognizing that vaccines were a costly, risky business, in 1965 set up the Vaccine Development Board to issue grants for promising vaccines.[31] Ribicoff predicted that polio, diphtheria, pertussis, and tetanus could be eliminated by 1965. A few years later, President Johnson announced the goal of adding five years to the life expectancy of the average American within the next decade, as he signed a bill for $33 million in annual federal grants for immunization.[32] His goals included reducing the infant mortality rate from 25 to 16 per 1,000 births in 1969, in part by eliminating measles.[33] Eradication, though, proved as evanescent as the light at the end of other tunnels.

During the May 1962 hearings for Kennedy's bill, Republican congressmen from southern and western states took note of the coercion implicit in it and sought reassurance that it wouldn't supersede state clauses allowing religious or philosophical exemption. So did three vaccination skeptics—Christian Scientist J. Buroughs Stokes, Frances Adehlardt of the Natural Hygiene Society, and Clinton R. Miller of the National Health Federation—who were invited to testify against the bill. The critics foresaw, rightly, that increased federal funds would mean increased pressure to vaccinate. At the time, only 19 states plus the District of Columbia had mandatory vaccination laws, and only eight—Hawaii, Kansas, Kentucky, Michigan, Missouri, North Carolina, Ohio, and West Virginia—required vaccination against any disease besides smallpox. Miller, a lobbyist for naturopaths and a John Birch Society member best known for fighting the fluoridation of water, gave the most cogent critique. A mass vaccination program, he said, "carries a built-in temptation to oversimplify the problem, to exaggerate the benefits, to minimize or completely ignore the hazards, to discourage or silence scholarly, thoughtful and cautious opposition, to create an urgency where none exists." It also tended to "extend the concepts of police power, of the state in quarantine far beyond its proper limitation."[34]

Miller's alarmism reflected the capacity of vaccination to stir paranoia in those inclined to mistrust mass medicine, but his perception of the circle-the-wagons mentality of the vaccination authorities wasn't entirely off-base. By the early 1970s, problems would surface that showed the vaccination system to be lacking in the necessary safeguards. Smadel's handling of the SV-40 issue was a case in point. Although he had helped fight polio and trained a generation of brilliant scientists, Smadel's handling of the affair would end his career—he died of throat cancer in 1963—on a sorry note. Though quietly recalled in 1963, SV-40-contaminated vaccines continued to lurk for years in the minds of leading scientists. The first major epidemiological study of children who received the contaminated vaccine[35] showed that they were not growing horns. The Public Health Service assumed, or hoped, that the choice of monkey kidneys had produced a close call, not a disaster, because the amount of SV-40 was too small to start an infection leading to cancerous changes. But Nobelist Fred Robbins, who later became president of the Institute of Medicine, told colleagues that he couldn't help fearing the virus "is going to come back and grab us just when we think we've got everything going."[36] Smadel's refusal to accept the potential gravity of SV-40, some would argue, established a tone at the Division of Biologics, which dragged its feet about withdrawing questionable vaccines.

Other problems were unforeseeable, but could have been addressed more quickly when they surfaced. The killed-virus measles vaccines produced by Lilly and Pfizer were a case in point. The vaccines were not nearly as effective at producing immunity as the live-virus vaccines, but they appeared safe, and that was enough for the Division of Biologics to approve them in 1963. Two years later, two Cincinnati pediatricians published the first in a series of reports on a peculiar disease syndrome that appeared during measles epidemics in children previously vaccinated with the killed-virus vaccines.[37] Drs. Louise Rauh and Rosemarie Schmidt, who worked in Carl Weihl's clinic, had participated in a Pfizer trial in early 1961, vaccinating 277 children with three doses of the inactivated vaccine at the Babies Milk Fund Association. When measles swept the city in early 1963, exposing 125 of the vaccinated children to the virus, 54 of them became ill—including 10 with severe pneumonia, persistent high fevers, swelling, and unusual brain-wave patterns. The patients had deep purple rashes, blood clots, and pus-filled blisters. Soon other scientists who'd taken part in the killed-vaccine studies found similar cases, and a new disease was born: atypical measles syndrome.[38] The killed-virus vaccines were taken off the market in 1967. Canada kept using them through 1970 and suffered outbreaks of atypical measles for many years.[39] Nearly 1 million American children were vaccinated with killed-measles vaccine, and as many as 160,000 developed atypical measles, according to one study,[40] although only a single death was reported in the literature.[41] The disease was frequently misdiagnosed as streptococcal pneumonia, Rocky Mountain Spotted Fever, or chickenpox.

Studies indicated that formaldehyde used by Pfizer and Lilly to deactivate the measles virus had denatured the F protein, a spiky protuberance that the virus uses to move from cell to cell. Vaccinated patients failed to develop antibodies against the F protein and were thus vulnerable to infection. At the same time they experienced violent allergic reactions to other components of the wild virus.[42]

Though no one could be blamed for this weird turn of events, it was not a happy chapter in the new era of viral vaccines. What's more, the vaccine establishment was reluctant to accept the bad news. Vincent Fulginiti, who with Kempe presented one of the first studies of the atypical measles syndrome, found some of the reactions alarming. "Katz and Krugman said, 'You're crazy,' " he recalled. "I've been very good friends with Sam Katz, and was friends with Krugman, but they could not accept that a vaccine could do harm, and it came not out of scientific examination but out of the strong belief that you were damaging vaccine use and faith in vaccines to criticize them."[43] The demise of its measles prophylactic knocked Pfizer out of the human vaccine business.[44] Hilleman, meanwhile, did not break his stride. "Who's got time to celebrate? I'm working seven days a week, night and daytime. We had a party for the nurses, that's all," he said. "Back on the farm, we'd have a cup of water to celebrate."

Mumps, Rubella, and Abortion Liberalization

The same month Hilleman's measles vaccine was licensed, his six-year-old daughter Jeryl Lynn wandered into her father's bedroom one night complaining of a sore throat.[45] Hilleman's first wife had just died, and a housekeeper was taking care of Jeryl Lynn. Hilleman himself was preparing to leave on a business trip overseas, but when he recognized that his daughter had mumps, he responded—though not in typical fatherly fashion. Instead of comforting the girl and going back to bed, Hilleman drove to his laboratory at West Point and picked up cotton swabs and a culture medium. Returning home he swabbed his daughter's throat, drove back to Merck, and dropped the mumps culture into a freezer, then got on a plane and flew to Central America for a research meeting. Back at Merck a few days later he began work on the mumps vaccine. After a trial of 1,337 children in Havertown, many of the same ones who'd gotten his measles shot a few years earlier, the vaccine was licensed in 1967.[46] Hilleman named it the Jeryl Lynn strain, in honor of the daughter he'd left at home that night.[47]

Of the three components of the MMR vaccine—measles, mumps, and rubella—mumps was perhaps the least urgent. To be sure, children occasionally suffered deafness or aseptic meningitis as a complication of mumps, and in teenaged and adult males it sometimes caused painful swelling of the testicles. Mainly, children with mumps got swollen glands that made them look like chipmunks, and they stayed home from school for a week. But the third and final vaccine of the trio, rubella, would prove to be the most avidly sought after, although the disease it caused was even milder—in most cases. Rubella, or German measles, caused little more than a minor rash and low fever for a day or two. Sometimes patients didn't even notice they were sick. But to a percentage of the unborn, the disease was mortally serious.

Medicine began to connect the dots to this danger in 1941. About a year after an epidemic of rubella swept wartime Australia, three women sitting in the crowded Sydney office of ophthalmologist Norman McAllister Gregg began to talk to each other. They all had babies with cataracts and other defects, and as they confided to one another about their difficult lives the women realized that each had suffered a fever with rash and joint pain during the first trimester of pregnancy. Gregg, who was capable of putting two and two together, compiled a list of 78 babies with cataracts born to women in Sydney who had been ill with German measles early in their pregnancies.[48] It wasn't long before Gregg's observation was confirmed. If infected during the first trimester of pregnancy, a woman had a better than 2 in 3 chance of giving birth to a child afflicted with congenital rubella syndrome, whose symptoms included deafness, blindness, autism, and mental retardation. Rubella had a special affinity for fetal tissue. When the movie star Gene Tierney gave birth to a severely deformed baby eight months after being kissed during a USO tour by a fan with German measles (the baby was placed in an institution and Tierney was hospitalized with severe depression), Americans got their own sense of rubella's repercussions.

Deafness was the most common complication of congenital rubella, affecting two-thirds of all children born to women infected up to the fifth month of pregnancy. Cataracts and heart disease were common symptoms when the mother was struck earlier on. But while the virus attacked a relatively small number of cells in the eye and heart, it "delivered a shotgun blast to the brain."[49] The fetus might suffer anything from mild autism to profound retardation. In some cases, damage to the frontal cortex caused children to become aggressive and rage-filled in their teenage years.

Developing a vaccine for rubella carried its own set of challenges. Everyone knew that German measles was a viral disease, but unlike polio, measles and other viruses that destroyed cells in characteristic ways—thereby proving that they were growing in a cell culture—German measles left no clear, visible signature of its growth in cell culture. It was not until the early 1960s that the virus was cornered. Once again, the child of a scientist played a role. When his 10-year-old son Robert got a severe case of the disease in 1960, Nobel laureate Thomas Weller decided to culture the virus.[50] The boy had fever of 104 degrees, joint pain, and swollen lymph nodes. Weller centrifuged his son's urine and grew the fluid extracted from it in human amnion cells or human embryonic skin muscle tissue. Nearly a month after inoculating these tissues, he found some evidence of growth.[51] Weller called it the RW strain. Around the same time, at the Walter Reed Army Institute for Research, Paul Parkman and Mal Artenstein, two young doctors performing their military tours of duty, stumbled onto a German measles infection on a back ward while conducting a study of adenovirus at Fort Dix, the largest recruit intake center on the East Coast. German measles was a childhood disease, but a large group of men had apparently missed out on it until they were jammed together for training. Since no one had isolated rubella yet, Parkman and Artenstein decided to try.[52] The Walter Reed doctors detected the growth of their rubella samples using the phenomenon of viral interference, in which one virus emits chemicals that halt the growth of another (these are called interferons, and they are now important drugs). When their success was announced at a fall 1961 meeting, Sabin noted that Weller had also recovered the virus. But before either group could publish, the NIH issued a news release taking credit for discovery of rubella by its scientists John Sever and Gilbert Schiff.

Despite the bruised egos, the isolation of rubella virus inspired clinical virologists around the country to begin testing potential vaccines. The man for the job in New York was Saul Krugman, chairman of pediatrics at New York University. Krugman had already tested many measles vaccines, including Schwarz's, and in 1969 he created the first crude vaccine against hepatitis B. Krugman and his partners did much of their clinical research at Willowbrook, an enormous home for the mentally retarded on Staten Island. Since they were busy with hepatitis and other projects, they hired Louis Z. Cooper, a Yale-educated internist from southern Georgia, to take over the rubella work in the old Bellevue Hospital on East 29th Street. Work with rubella would transform Cooper into an early advocate for disabled children. He later became president of the American Academy of Pediatrics.

Rubella is slightly less contagious than measles, and rubella epidemics tend to be less frequent, occurring about once every five-to-nine years. But better sanitation and less crowded housing altered the postwar epidemiology of rubella in a pattern similar to that of polio. As fewer children contracted rubella in infancy, the vulnerable population of older children and young adults began to accumulate, like dry brush awaiting a spark. In 1964, the fire ignited. Initially, the epidemic was taken lightly. *Time* magazine called it "the best of times for a German measles party. The rules call for lots of kissing games in an unventilated room so that all the little boys and more especially all the little girls get the infection." There was "just one vital precaution," the article said. "No infected child should be allowed anywhere near a woman who is or even may be in the first three months of pregnancy."[53]

Despite such warnings, nearly 50,000 women in vulnerable periods of pregnancy were infected. These women had two options: an illegal abortion or delivery of a child at risk for severe birth defects. A June 4, 1965, cover story in *Life* magazine entitled "The Agony of Mothers about their Unborn" focused on this terrible dilemma, noting that "conscientious doctors of highest integrity" were willing to abort the damaged children "in defiance of community convention and state law." Dolores Stonebreaker, a Catholic mother living in Merced, California, found it hard to explain to her anguished 12-year-old son Billy that it was not his fault she caught rubella from him. Although the fetus she aborted probably would have miscarried anyway, a priest who visited after the operation made it clear she was a sinner, and Mrs. Stonebreaker turned away from him. "No need for discussion?" he asked sarcastically. "Well, it's only murder."

According to one estimate, about 5,000 therapeutic abortions were carried out in 1964 and 1965, when the rubella epidemic reached California. Another 10,000 women miscarried, but an estimated 25,000 babies were born with severe damage.[54] The medical cost alone was in the billions, and by 1980 at least 6,000 of these children, with hearing impairment secondary to congenital rubella, were enrolled in special programs.[55] About 1,000 rubella babies were born in New York City alone. "It was overwhelming," recalled Cooper. I'd get calls from doctors saying, 'I have this baby and I don't know what's the matter. Will you come and look?' I'd be off with my bucket trying to find some little hospital in Bay Ridge, wherever Bay Ridge was. My family said I never was seen without a Styrofoam bucket in my arms." The bucket was filled with dry ice and ready to be refrigerated with the latest specimen of infected tissue that Cooper had collected from some hospital.

Cooper, who was involved in clinical trials for a rubella vaccine, began the definitive study of congenital rubella pathology. It was compelling, shattering work. "Dealing with these families that were devastated with the awfulness of having profoundly injured children, at that point I went from being an internist to being a pediatrician," he recalled. Cooper's wife, Maddie Appel, a social worker, noticed early on that many of the children had odd behaviors. They twirled and obsessively fiddled with objects and refused to look you in the eye. "My wife said to me, 'I've worked with deaf kids. These kids are different. They're like what I read about autistics.' " A study found that 6 percent of the children with congenital rubella syndrome in Cooper's research center were autistic.[56] That came out to about 1,200 children nationwide. This was a major insight into the still-mysterious etiology of autism, establishing that the disease stemmed from something that occurred in brain development in the first trimester—at least in some cases. It also laid the foundation for the first effort to get comprehensive care for severely disabled kids. With

$13,000 in seed money from the March of Dimes, Cooper built a staff of 45 people to provide services to rubella-crippled children. The investigation his unit did "showed me what a tremendous gap there was in our service delivery system for families struck by lightning," he says, and it opened his eyes to their multiple needs. Joining forces with directors of schools for the deaf and blind, and aided by the popularity of *The Miracle Worker*, the play based on the life of Helen Keller, Cooper lobbied Congress to pass the first special education program in 1968, which eventually led to many federal disability programs. "Thank you, Helen Keller," said Cooper.

As he testified before state legislatures on the ravages of the syndrome, Cooper also was thrust into the abortion debate. In fact, the rubella epidemic of 1965 probably did as much as any other single event to make the legalization of abortion plausible. In many states, doctors became outlaws for aborting rubella-damaged fetuses in defiance of laws that banned the procedure except to save the life of the mother. Their persecution created sympathy for women seeking abortion and the doctors who performed it. Paradoxically, abortion might be much more widely accepted in the United States today were it not for the fact that the rubella vaccine took away one of the most obvious justifications for the procedure. It was said that hundreds of thousands of women had abortions yearly in the United States at the time, yet only 18,000 were done by qualified surgeons. Probably 1,000 women died each year as result of "back alley" abortions, though reliable figures do not exist. In early 1965, about 1,200 New York obstetricians signed a petition urging that abortion be legalized when the mother's health was threatened or the child could be born with serious mental or physical defects. "I testified in a number of legislatures," Cooper recalled. "I tried to show the pictures of the kids and tell the stories of some of these families. They were stories people could relate to."[57]

Before *Roe v. Wade* reached the Supreme Court, many state abortion laws already had been liberalized, starting with New York's law, in 1970. In Cooper's view, the congenital rubella cases were one of the forgotten catalysts for the change in abortion law. "It came out of facing these families and what these women had to endure," Cooper said. He recalled one mother who heard she could get an illegal abortion for $150. When the intermediary told her the cost was $250, she couldn't afford it. As a result she had a profoundly retarded, blind and deaf child who placed a burden on the family that it was unable to shoulder; the couple divorced. Said Cooper: "I saw those stories over and over. I had all these children with multiple handicaps who couldn't find anywhere to get services." The devastating cost of the epidemic may also have helped push Congress to approve the Medicaid provision of the Social Security Act of 1965, which made it the right of every American child to receive comprehensive pediatric care, including vaccinations.[58]

With the congenital rubella tragedy as the backdrop, the race was on among vaccine makers. At Merck, Hilleman started out with the expectation that his chickens would provide a good culture medium, but the virus wouldn't grow on their eggs. On a hunch, Hilleman tried duck eggs, figuring that since they developed faster, they would provide enough "meat" for a cell culture while still in a relatively undifferentiated state— similar to the first trimester of pregnancy—that was suitable for viral growth. "Talk about a dumb idea—that was fantasy! But it worked!" Hilleman said. After repeated passages in duck eggs, the virus was attenuated enough to try out in children, and it produced the desired effects. Hilleman was in the process of getting into vaccine production when Mary Lasker paid a visit. Lasker, widow of the advertising millionaire Albert Lasker, was a canny philanthropist who concerned herself with health research funding. She "had her nose into everything in a positive way," as Hilleman put it, and like many health policy leaders she was concerned about the likelihood of another rubella outbreak as early as 1969. Lasker noted that Paul Parkman—by then at the Division of Biologics Standards—was working on a vaccine with Harry Meyer, another leading scientist there. Hilleman wasn't very impressed with the Parkman/Meyer vaccine, but Lasker warned him, " 'You should get together and make one vaccine or else you'll have trouble getting yours licensed.' Well, I listened," recalled Hilleman.

Hilleman, who had worked with Meyer in Smadel's laboratory at Walter Reed, modified the Parkman/Meyer vaccine by passage through duck tissue and by 1969 they had a vaccine. The Merck/NIH arrangement raised a few complaints from scientists like Stan Plotkin, who had a competing vaccine. But most people in the field just shrugged. It was not unusual for vaccine regulators to wear two hats at the NIH. "It was a different time," says Cooper. "There were a limited number of kids to test on, a desire to get a vaccine out, and there were Meyer and Parkman in the government regulation agency. From my own data, which showed a buildup of rubella vulnerability in the community, I was dying to get a vaccine. I didn't give a damn how you got it." A decade later, Hilleman decided to switch to Plotkin's superior vaccine, which Plotkin grew at the Wistar Institute in the cells of an aborted fetus.

There was a larger debate over the strategy for rubella vaccination, one that would surface in years to come over other vaccines. The decision to vaccinate toddlers, rather than young women, was extremely controversial, and with good reason. "When you work on vaccines," says Gilbert Schiff, a member of the CDC's immunization advisory committee in the early 1970s, "you try to test for everything. But there's probably a thousand things you don't know how to test for." One such intangible was the duration of immunity the rubella vaccine provided. Rubella disease provided lifetime immunity. Scientists worried that the vaccine's protection could wear off by the time the girls who got it were of childbearing age. It was a terrifying concern, but with another rubella outbreak in the offing, the decision was made to go ahead and vaccinate kids. The die was cast, and it turned out to be the right thing to do. "But in a sense," Schiff says, "we were lucky."

In the history of vaccines, critics have repeatedly warned that upsetting the epidemiology of a virus or bacteria could have unforeseen, perilous effects. The use of measles vaccine raised concerns that older people would come down with nastier versions of the disease once the vaccine wore off. Later, the chickenpox vaccine would trigger similar misgivings.[59] So far, fortunately, such worries have not been realized in any significant way.[60] But rubella was the first vaccine given to children primarily to protect a group other than children, and it was not an easy sales job. To convince parents to go along with it, New York's Department of Public Health created a campaign based around the idea of a "rubella umbrella"—a vaccine that protected both the babies and their mother. The agency released posters and television ads in which a red and white polka-dotted umbrella sauntered down the street while a narrator explained the need for children to be vaccinated.[61]

The MMR vaccine, which had always been Hilleman's goal, was licensed in 1971 and had an immediate payoff since it fought three diseases with a single shot. Britain, meanwhile, decided to control rubella through the selective vaccination of schoolgirls only. By 1983, there were 25 cases of congenital rubella syndrome in the United Kingdom, and only 7 in the United States with a population four times as large. The American rubella fighters' strategy had proved sound.[62] In the late 1980s, Britain would also adopt the MMR vaccine for toddlers.

As the fight against measles and rubella proceeded, scientists were also at work on vaccines to protect against two other common childhood infections. One of the targets was group A streptococcus, *Streptococcus pyogenes*, which normally causes "strep throat" but also occasionally invades the lungs or other tissues, resulting in scarlet fever, toxic shock, flesh-eating disease, and even neurological damage. The other was respiratory syncytial virus, or RSV, a common cause of wheezing and bronchiolitis in small babies that was fatal to several hundred a year and linked to asthma in thousands more. Both of these vaccines failed disastrously. The group A strep vaccine generated antibodies against the patient's own tissues, including brain and heart muscle. At least five people died during the trials of the vaccine in the mid-1960s, and no effort was made to introduce a new shot against this common afflication for another 40 years. The RSV vaccine, created by Robert Chanock, also failed spectacularly. Two children vaccinated in trials of the vaccine died because vaccination had somehow enhanced subsequent infection.[63]

Overall, however, the 1960s was a decade of fabulous success for vaccines. In 1964, for the first time, not a single case of polio was reported in Philadelphia; only 121 cases were reported in the whole country. By 1968, 20 million American children had been vaccinated against measles, and only 22,000 cases were reported. Measles "have just about had it," *The New York Times* declared, and in this new era of optimism, "freedom from all these diseases will be taken as a matter of course."[64] Vaccines were making major inroads on diphtheria, mumps, rubella, and whooping cough. When President Nixon declared war on cancer in 1970, it looked as though infectious diseases were scarcely an enemy worth the name.

Closing with Cowpox

It was in this context that Henry Kempe did a brave and unexpected thing: he launched a campaign to end routine smallpox vaccination. Unsurprisingly, it did not immediately win him great support among his peers. But that meant less to Kempe than it might have to others. As a teenage refugee from Nazism, Kempe had a special empathy for children and was staunchly independent. In 1952, without the support of the NIH or any other institution, Kempe set out for Madras, India, to establish a smallpox research institute. At the Madras Hospital, Kempe saw more than 10,000 smallpox patients and conducted remarkable studies on the transmissibility of smallpox, the value of different vaccines, and the role of nutrition and vaccination at different ages on the mortality of the disease. While they weren't always statistically impressive, his studies established that smallpox was not nearly as communicable as measles. Madras, with its heat, poverty, and tremendously lethal smallpox, was a far cry from the sterile confines of American medical centers. Crows sometimes flew into the hospital and alighted on patients, ripping off their smallpox scabs. (The crows did not spread smallpox this way—but Kempe had screens put on the windows to keep them away from terrified patients.)[65]

By 1962, when Kempe published his first article on the "battered child," he had begun to question whether smallpox vaccine represented a kind of child abuse for the small percentage of children who reacted badly to it.[66] "What connected the two issues," says Richard Krugman, Saul Krugman's son and Kempe's protégé at Colorado, "was that he was always trying to help pediatricians and others challenge their assumptions and see how an evolving world made their previous assumptions no longer valid." Kempe liked to tell a story from 1948, when he was making hospital rounds at Yale as a resident and presented a child with eczema vaccinatum—a severe allergic reaction to smallpox vaccine—to his mentor, Grover Powers. "I pointed out fairly blithely that occasional serious complications were bound to occur with any procedure, and 'this is a risk we have to take.' " Dr. Powers looking very sad, gazed at the baby and then at Kempe and said, "Who has to take? Who asked him?"

Eventually, Kempe came to the conclusion that the risk of an

American child contracting smallpox from an accidental importation of the disease was much smaller than the known risk of routine vaccination.[67] To reduce the risk, Kempe first attempted to attenuate the New York Board of Health vaccinia strain, passaging it 78 times through chicken tissues. That vaccine, called CVI-78, was administered without serious side effects to more than 1,000 people on an experimental basis, but the Division of Biologics Standards showed no interest in licensing it. Kempe also created an immune serum that was used to treat hundreds of sick children and made the University of Colorado a center for the study and treatment of smallpox vaccine reactions.

At the May 1965 meeting in Philadelphia of the American Pediatrics Society, Kempe prompted a debate on the wisdom of routine smallpox vaccination, which most American children received between the ages of 1 and 4. Since 1949, when the last smallpox case appeared in the United States, Kempe said, an estimated 200 to 300 children had died as a direct result of smallpox vaccination, and 5,000 had been hospitalized with complications including generalized rash, secondary infections, and encephalitis. The standard argument was that smallpox introduced from elsewhere in the world remained enough of a threat to justify these casualties. But most cases of smallpox imported into Europe over the previous two decades had infected adults, mainly physicians and nurses. Given that protection from smallpox vaccine often did not last into adulthood, what was the point of continuing to vaccinate American infants against the disease? "Our vaccination practice is not only obsolete but also, in the light of our present knowledge, really reprehensible," agreed Margaret Smith, a Tulane physician.[68]

Kempe pointed out that the mass vaccination campaign in New York City in 1947 had not influenced the course of the epidemic, and "there is no evidence that it was justified" since once the outbreak had been identified, only close contacts of diagnosed patients became ill. Having seen more smallpox than almost any American physician, Kempe felt justified in challenging the dogma that smallpox was a wildly contagious disease. But Smith was the only doctor at the meeting to second Kempe's idea. Luminaries such as Krugman argued that not all the complications Kempe cited were clearly related to vaccinia and worried that smallpox epidemics might return if vaccination were stopped. The World Health

Organization feared America would set a bad example that would hamper its work of global eradication.[69] Pediatricians recoiled from the idea. Hadn't they told their patients again and again that smallpox vaccine, the oldest of humankind's successful medical interventions, was a perfectly safe weapon against a dread disease? "Hopefully the medical writers present will avoid the headline tomorrow morning that will say, 'Pediatricians against vaccination,' " said Joel Alpert of Children's Hospital in Boston. Kempe's idea was tossed out.

Slowly, though, American doctors recognized that the risks of smallpox vaccine were greater than the risks of smallpox. In 1972, with the approval of the CDC and the Academy of Pediatrics, most states abolished routine smallpox vaccination. "Kempe was a visionary in many ways," says William H. Foege, the CDC director under President Carter, and a leader of the global eradication campaign.[70] "But you can imagine what trouble we all would have been in if we stopped vaccination, had an importation of smallpox and a real problem. Kempe was just a few years ahead of his time." Foege feels a special debt to Kempe; his young son had had eczema and was therefore at risk of a severe vaccine reaction, posing a dilemma for Foege's family as they prepared to journey to Africa to fight smallpox in the 1960s. Foege vaccinated the boy with Kempe's CVI-78 strain, which boosted his immunity and allowed him to subsequently be vaccinated safely with regular smallpox vaccine. Thus immunized, the boy traveled with his father to Nigeria. There, Foege employed "ring vaccination" to eradicate smallpox. With this method, which the British had used for years during smallpox outbreaks, public health officials would surround a case of smallpox by vaccinating and quarantining any possible contacts.[71] Cheaper, safer, and easier than mass vaccination, the ring system would unltimately enable the eradication of smallpox from the world.

Tightening the Grip

While the World Health Organization was rolling toward the eradication of smallpox, holes were starting to appear in the U.S. vaccination system, which, like many aspects of U.S. healthcare, wasn't very systematic at all. Diseases that crashed in the 1960s were making a comeback.[72]

This was particularly true of measles. After dropping from 500,000 to 20,000 cases from 1964 to 1968, the disease began to creep back. Measles was an illness whose incidence tracked with levels of public investment and commitment to an uncanny degree. Investment in vaccines was patchy, and so were measles outbreaks.[73] Democratic administrations increased federal purchases and distribution of vaccines in the 1960s, but in 1969, the Nixon administration switched all federal money from measles to rubella vaccine, leading to a measles resurgence. In 1974 only 40 percent of the 14 million preschool children were protected against vaccine-preventable diseases.[74] Polio vaccination rates were at 60 percent, compared with 84 percent a decade earlier. Although there were only seven cases of polio that year—an all-time low—waning vaccination had led to 32 diphtheria cases on a Navajo reservation.

Vaccination laws were also spotty. In 1969, 24 states, including all of the Great Plains, Mountain, and Pacific Northwest states, still had no compulsory-vaccination requirements. Of the 26 states with some compulsory vaccination, only eight had specific penalties for not vaccinating.[75] But, under pressure from the CDC, this was changing. By May 1977 only Wyoming and Idaho had no compulsory-vaccination law.[76]

Public health inquiries found that apathy rather than active resistance to vaccination was responsible for low rates. African Americans and Hispanics were the least-vaccinated groups. In Los Angeles in 1966, half the measles cases were among African Americans, another 20 percent among Hispanics, although blacks and Hispanics made up only a quarter of the city's population.[77] Similar demographics were noted in St. Louis, Philadelphia, and other cities. Measles, which had spared no one in the past, was now "primarily a disease of the urban ghetto."[78]

The quality of the vaccines was also part of the problem. The killed measles vaccine had been a fiasco, and the early versions of the live vaccine also failed if they were left out of the refrigerator for too long before being administered. In children vaccinated before they were a year old—the usual case through the 1960s—maternal antibodies from placenta and breast milk would neutralize the virus and prevent long-term immunity. By 1972, scientists began to ask whether a second measles shot would be required to eradicate the disease.[79] But the biggest problem with measles was low immunization rates, and as public health officials began to squeeze unvaccinated communities, they got results. After a 1976 outbreak in Alaska, more than 7,000 unvaccinated children were excluded from school. Within a month, only 51 of them hadn't been vaccinated. "Fears that excluded students would not return to school proved to be unfounded," wrote Walter Orenstein, who joined the CDC's vaccination efforts in the 1970s and led them in the 1990s. Since it was hectic and frustrating to enforce laws during epidemics, public health officials began trying to promote tougher and better-enforced laws in all states.[80] They got the support they needed from the new administration of Jimmy Carter.

When Carter came to Washington in 1977 he was accompanied by another former southern governor, Dale Bumpers, who had just been elected senator from Arkansas. In the early 1970s, Arkansans had suffered through measles and whooping cough epidemics, and Governor Bumpers had enrolled churches, the national guard, and school officials in vaccination programs. In Washington, Bumpers's wife Betty would launch a campaign of her own. "Betty and Dale went to the White House for dinner and converted the Carters shortly into the administration," says Foege, who was the CDC's new director. "Next day I hear from [HEW Secretary Joseph A.] Califano and he's been given the job to improve immunization. He came down to Atlanta and said we're going to have a big push on childhood immunization. 1977. And I remember distinctly when he turned to Don Millar, head of the immunization program and said, 'I would like to see us reach 90 percent immunization by school entry' and Millar said 'I'd hate to see that on my job description.' The next day he saw it on his job description."

The Carter administration increased spending on immunization— $88 million in vaccine grants in 1978 and 1979, four times higher than the previous two fiscal years.[81] And Califano mobilized forces outside the CDC to get it done. Eight million slips placed in welfare check envelopes reminded mothers to immunize their children. HEW's Office of Education asked schools to talk to children about immunization, and Califano pressed governors to enforce or strengthen their laws. The professional football, baseball, and basketball commissioners ran public service ads, as did *Star Wars* movie characters.[82] "No shots, no school!" exhorted Darth Vader and Muhammad Ali. "It is unconscionable," Califano said in launching the program on April 6, 1977, "for a nation spending more than $170 billion a year on health care to neglect a proven and inexpensive method of safeguarding its children against preventable suffering and death." Califano set out to establish a "permanent system to provide comprehensive immunization services to the 3 million U.S. children born each year."[83] While his goal of measles eradication by 1982 failed, his larger goal took hold, although many parents still waited to vaccinate their kids until they were school-age—because they could easily get free shots then.[84]

To BOLSTER COVERAGE, Foege set aside 30 minutes each week for his immunization officials to report their progress on stopping measles transmission. Among these officials was Orenstein, who had trained to be a pediatric urologist but was converted to vaccines during the smallpox eradication campaign in India, as were many young doctors. Orenstein and Don Millar ordered the review of 28 million schoolchildren's immunization records.[85] Federal officials did not have the power to change immunization laws, but they offered free vaccines and cash to help states run their programs. The disincentive was shame. "They were good people, and you publish in *MMWR* [the CDC's weekly epidemiological bulletin] the number of cases by state. People don't want to be in that situation," says Foege. "It was an unbelievable amount of work. But it turned out to be cost-beneficial."

On March 31, 1977, after measles was detected in Los Angeles, the county director of health services, Shirley Fanning, announced that unvaccinated kids would be excluded from school by May 2 if they didn't get their shots. Fifty thousand students were vaccinated in the following month, and there were few absentees. The Alaska and Los Angeles experiments showed that strict enforcement was acceptable to most parents. "It wasn't forcing vaccination on people that opposed it. It was making vaccination a priority for people who didn't have it as a priority," said Orenstein. "And the feeling was that if you can do it Los Angeles, you can do it anywhere."

School laws, Orenstein and Alan Hinman wrote, "establish a system for immunization that works year in and year out regardless of political interest, media coverage, changing budget situations and the absence of vaccine-preventable disease outbreaks to spur interest."[86] Parents and physicians trusted the CDC to make the right recommendations, and the use of school nurses and school-flyer exhortations allowed public health departments to get the job done with fewer of their own resources. Shame worked. By the end of the Carter administration, all 50 states had school requirements, and many required immunization of preschoolers. By the end of 1979, an astonishing 90 percent of the nation's schoolchildren were immunized. Measles had declined 78 percent, rubella 43 percent, mumps almost 30 percent. There were no cases of diphtheria or tetanus in children. This was an important story, but not a sexy one. Rosalynn Carter kicked off another measles eradication drive at a Health, Education, and Welfare news conference that year. None of the television networks showed up.[87]

One of the few American parents who decided not to vaccinate his child against measles in those years was Dave Edmonston, the man for whom the virus was named. After teaching high school science for a while, Edmonston became a carpenter and joined a Hindu religious community in northern Virginia. He called himself a "spiritual thinker; a scientist of the heart." Edmonston took no particular pride in the role he had played in bringing the measles vaccine to life. He tended to view it as a karmic event that was incidental to his own life's path. I met Edmonston, a tall, handsome man with a long face and the big, agile hands of a craftsman, on a winter evening at an Italian restaurant a few miles from his home in suburban Maryland. It was his late wife's decision not to vaccinate, he said, but he went along with it. "She had a masters in public health and her attitude was, 'In this country, measles isn't that bad a thing.' She felt it was healthier for his immune system not to be vaccinated." Edmonston's son, born in 1978, never got measles, as it happened—a fortunate thing, in his father's view, since the boy had some psychiatric problems. But in some ways Edmonston and his wife were ahead of their times. A few decades later, thousands of parents in the United States and Britain would blame the MMR shot for their children's mental problems, and refusing a vaccine because it was "unhealthy for the immune system" would become a position with some cultural cachet.

PART THREE

CONTROVERSY

CHAPTER 7

—

DTP AND THE VACCINE
SAFETY MOVEMENT

As in all wars, some soldiers are injured. . . . At present, the draftees injured
in the war on infectious disease are in effect told by the conscripting
authorities, "Thank you for your contribution to the war effort, and best of
success in coping with your disability."

— LEROY B. WALTERS, 1978

ON APRIL 19, 1982, the NBC-TV affiliate in Washington, DC, aired a
harrowing hour-long documentary about the dangers of the whooping
cough vaccine. *DPT: Vaccine Roulette* was not for the faint of heart. Small
children were filmed in the throes of convulsions. A quadriplegic struggled to hold his head up off the floor. Silent stills of babies who had died
appeared on the screen. Brain-damaged teenage twins, the cousins of a
U.S. congressman, grinned agitatedly at the camera. The filmmakers put
together a convincing narrative from interviews with scientists, lawyers,
and policymakers, as well as angry, weary parents, who blamed the damage to their children's brains on the whole-cell pertussis vaccine—the "P"
in the DTP shot that also protected children against diphtheria and
tetanus disease. The filmmakers maintained that the DTP vaccine (DTP
was the correct name, though people often called it DPT) was harming
children while the government, industry, and pediatricians covered up
the evidence or looked the other way. Not only that, the documentary

argued, the vaccine often failed to protect against whooping cough, which in any case was no longer mortal enough to risk the lives and health of so many children. Without explicitly saying so, the film suggested that children might be better off if their parents stopped immunizing them against pertussis with the current vaccine.

Vaccine Roulette was a serious blow to the gleaming reputation of vaccination in the postwar period. Jonas Salk's conquest of polio had enshrined vaccines as a ritual of American life, and the advent of the MMR shot had only added to their luster. A baby cringing at the approach of the pediatrician's gleaming needle was something that most parents viewed with no more alarm than they did a Norman Rockwell painting of childhood antics. Shots were one of those minor unpleasantnesses, like swimming lessons in cold pools and tying loose teeth to door handles, that little children suffered on the road to safe, healthy, modern lives. But *Vaccine Roulette*—which aired at a time when the news media were fewer, and more trusted—broke the picture frame. It delivered a disturbing, disruptive message. Sometimes, it turned out, those childish fears were right. Vaccines, unlike mother's milk and apple pie, could hurt you. Not the companies that made vaccines nor the government bureaucrats who regulated them nor the family doctor who administered them could ensure that they were positively safe. That was an ugly fact. It was something worse than ironic "to take your daughter in to protect her health," as a parent, Jeffrey H. Schwartz, put it, "and have that be the agent that destroys her."

For Schwartz and his wife Donna Middlehurst, *Vaccine Roulette* was a revelation. Schwartz was a liberal environmental lawyer who in 1969 had come to Washington to work at the Health, Education, and Welfare Department after taking a degree at the University of Minnesota. Middlehurst was a securities lawyer. Their daughter, Julie, was born on February 27, 1981, and got her third DTP shot one morning the following July. Schwartz was taking care of her that afternoon at the couple's home in Silver Spring, Maryland. Julie was fussy, and Jeff was holding her when he noticed "a sort of startle." The startle became more of a twitch, and then something more pronounced and regular, and he realized she was having a grand mal seizure.[1] He called 911, and the emergency squad came and took them to Holy Cross Hospital, where her convulsions continued for a total of 40 minutes until they were brought under control with medication. "We asked the doctor about the DTP. And she said, 'No, it's usually fevers that produce these things.'" No one wanted to believe that pertussis vaccine could cause brain damage. After that day, Donna and Jeff could think of little but their daughter's perilous neurological state. After watching *Vaccine Roulette* they looked at each other. "We said, 'Oh my God. Now we know what happened.'"

Twenty-seven-year-old Kathi Williams, who lived in a one-bedroom apartment across the Potomac in Fairfax, Virginia, was also watching the show. A slender, blue-eyed brunette with a calm, poised manner, Kathi took care of her 18-month-old Nathan, while her husband worked in her family's glass business. Nathan had screamed for eight hours after getting his second DTP shot, an event that was near the top of Kathi's mind since it had occurred just four days before the airing of *Vaccine Roulette*. Before then, "he never, ever cried," Williams would say many years later. "He had no reason to. He had me 24-7. I didn't work, and he was this great little happy kid." By evening Nathan had calmed down, but his leg was sore and he limped for two days. Later, he developed allergies and behavioral problems, which Kathi blamed on the shot. Right after the show ended, Kathi's mother phoned and said, "Call the TV station and call your congressman. This is my grandchild they're messing with."

In nearby Alexandria, another mother, Barbara Loe Fisher, watched a rebroadcast of the show the next day. Fisher, 34, who had worked as a publicist, had a four-year-old son, Christian. In 1980, on the afternoon after his third DTP shot, as Fisher has recounted in many interviews and speeches since then, she walked into Chris's room to find her son staring vacantly into space. She picked him up, talked to him, took him to the bathroom, but Chris remained foggy for the better part of the next 24 hours. Fisher would come to believe that the shot had overwhelmed his immune system, although the symptoms she described were typical of the shocklike state that occasionally occurs after DTP and which safety studies indicate rarely has long-term health repercussions. Before the shot he was a bright, even precocious child, she said. Afterward he was sick all the time and "regressed mentally." Like Schwartz, Fisher had not linked the vaccine to her son's condition until she saw the TV program— a year and a half after his vaccination. After watching the documentary she was desperate for facts and wanted to be connected with other parents with similar experiences.

Lea Thompson, the reporter on *Vaccine Roulette*, won an Emmy for the show. When parents called the station to say "thank you" for airing it, they often asked for other parents' contacts, and a phone tree of a dozen or so parents sprouted. After a brief phone conversation, Barbara drove over to Kathi's house, initiating a relationship that would energize a movement, one that would ultimately question the safety and value of practically every vaccine. But the parents who came together over the pertussis vaccine, and decided to call themselves Dissatisfied Parents Together, or DPT, had the simple objective of safer vaccines, better medical education on the risks of vaccination, and parental rights to decide whether to vaccinate their children.

Fisher and Williams were temperamentally quite different, but they were well-suited as partners. Kathi was cool and quietly organized, a good listener, and a comforting presence. Barbara was passionately articulate, a conservative libertarian with large, unblinking blue eyes. She could be fierce, with an unusual ability to bring controlled anger to bear in public. She also had a background in medical writing for nonprofits. "I was looking for studies, I wanted the studies," Fisher recalled 22 years later in her office at the National Vaccine Information Center (DPT was rechristened in 1991) in northern Virginia. "I knocked at her door and Nathan, her little boy, was in diapers. Kathi came to the door. And I was like *in her face*."[2]

"She expected me to have all the answers to her questions and I didn't know anything either," Kathi says with a laugh.

On April 28, nine days after the show first aired, Kathi and Barbara and Jeff and several other parents assembled for the first time. They met, auspiciously enough for their movement, in the office of Representative Dan Mica, a Democratic congressman from Florida whose twin cousins, Anthony and Leo Resciniti, were depicted in the show. Each was said to have become disabled and retarded after undergoing a severe reaction to a DTP shot, one year apart from one another, and Mica probably put Lea Thompson onto the story. Mica's brother John, who would later become a Republican congressman, was working at the time as a staffer in the office of Florida Senator Paula Hawkins. The fledgling citizen's action group had some extraordinarily able and connected members. Dan Mica and Jeff Schwartz knew each other well, both having worked in the office of Paul Rogers, the influential former chairman of the House Commerce health subcommittee. Mica was elected to Rogers's seat after he retired. Schwartz, a staffer for Rogers on the Commerce committee, had worked on the Clean Air Act. Other members of "DPT" were lawyers, accountants, management consultants.

Schwartz wondered: "Can severe reactions to the DTP vaccine be that rare, if two people on one congressman's staff have direct family experiences with them?"[3] Chance or not, this coincidence gave clout to the vaccine program reform movement. When important things happen in the family of a congressman, the congressman tends to make important things happen.

Vaccine Roulette mobilized a parents' movement that would change the laws and expose U.S. vaccine policies to systematic questioning for the first time in nearly a century. As journalism, the documentary had the strengths and drawbacks typical of its genre. It went for the jugular more than the complex truth. Some of the scientists who participated in the show claimed they were questioned on camera repeatedly until their cropped comments could be woven into a supporting case for exaggerated allegations about the vaccine's dangers. The statements of parents who blamed the vaccine for stripping them of their healthy children were taken as fact. To make the legitimate case that the whole-cell pertussis vaccine was unnecessarily dangerous, the film underestimated the hazards of whooping cough itself, although the disease had been a scourge of childhood for decades before vaccine began to control it in the 1940s.

Still, in its impact, *Vaccine Roulette* was a devastating piece of muckraking. And although the film oversold its story, it was a story that had been waiting to be told. It is quite possible that the whole-cell pertussis vaccine has saved the lives of more children than any vaccine currently in use. Yet problems with the vaccine were well known to all who had closely worked with it. It was difficult to make pertussis vaccine of consistent quality, and the vaccine had a somewhat notorious association with side effects. Most children who got the shot had nothing more than a touch of fever or a sore arm, but a small percentage would scream inconsolably for hours. Convulsions and episodes of a shocklike state, in which the child stared into space for hours, were also part of the record, as were, more controversially, deaths and permanent brain damage. As trial lawyers were discovering, serious adverse reactions to the pertussis vaccine had been recorded in the medical literature regularly since 1933. Not only that, but apparently safer vaccines had been discovered and discarded along the way. No one had given much thought—or rather, not enough people had given enough well-funded thought—to these drawbacks because whooping cough was a serious disease and the vaccine seemed to have made it go away. In that sense it was an amazing, largely unsung triumph.

But now it was 1982. The smallpox vaccine was no longer given, and the whole-cell pertussis vaccine was the most dangerous, or at least the most reactive, shot given to American children. There was a problem in sorting out the risks of the shot, however. It was given to tiny babies, at two, four, and six months of age. Infants got DTP shots because infants were most vulnerable to the deadliest whooping cough infections. Yet infancy was also a time of life in which serious conditions appeared in vaccinated as well as unvaccinated children—seizures, rashes, developmental delays, and sudden infant death syndrome. Separating DTP reactions from the other illnesses of infancy would be a difficult task. Over the years, the official message of public health had been simplified, stressing the dangers of the disease and ignoring or even denying the risks of the vaccine. If one took pertussis vaccination as an either/or proposition, the choice was clearly in favor of vaccination to protect the individual, and the people he or she came in contact with, from the germ. But now the germ was almost conquered, and *Vaccine Roulette* suggested that the prevention was worse than the cure.

As Dissatisfied Parents Together was organizing to get answers about DTP and, eventually, a law to protect vaccine-injured children, hundreds of parents began to file lawsuits against the vaccine makers. Parents were starting to refuse to vaccinate their children, and manufacturers were threatening to stop making vaccines. Suddenly, a system that relied upon the collective goodwill of the government, vaccine makers, doctors, and the public was thrown wide open, exposing its fragility. This was sad, but not surprising. Many people in public health had seen the crisis coming for years. The laws and regulations that governed the creation, sale, and use of vaccines were proving inadequate for their use in mass immunization campaigns. And as infectious diseases disappeared, in part thanks to vaccines, the risks of vaccination itself were thrown into relief.

To some extent these risks were intrinsic, a product of the lag between empiricism and understanding. As the pediatrician Vincent Fulginiti wrote in a 1977 commentary on vaccination, "man's attempts to prevent disease have always been a little further ahead of his basic knowledge of the agents and their properties. . . . Technology allows us to put attenuated or killed infectious agents in bottles, but voids in our biologic knowledge often result in unexpected or untoward effects of letting these modern 'genies' out of the bottle and exposing humans to them."[4] Pandora was an undeniable presence in the history of vaccines, and Fulginiti had seen children stung by concoctions sprung from this box, whether they were defective polio and measles vaccines, vaccinia, or pertussis vaccine. The other side was the failure of authorities to systematically look for, or correct, damages caused by vaccines—though of course this was true of all medicines.

Vaccine Safety Comes of Age

The DTP controversy marked the first time that consumers had gotten involved in vaccine policy since the early twentieth century. But it was not the first shakeup of the vaccine program. In 1972, Nader's Raiders had cracked open the Division of Biologics Standards, then located in the National Institutes of Health, and a whistleblower's complaints had helped force the federal government to switch vaccine regulation from the NIH to the Food and Drug Administration. The Division of Biologics had been led for 17 years by Roderick Murray, a mild-mannered Australian who had taken over in the reorganization following the Cutter incident.[5] Murray, who had a staff of 260 and a budget of only $6 million, was reluctant to push for improved products and timorous to the point of absurdity about enforcing efficacy standards for existing vaccines. A five-day hearing held by Abraham Ribicoff, by then a Connecticut senator, showed that flu vaccines sold in the United States in the mid-1960s had been next to worthless. Regulators confessed that under pressure from the White House and Pentagon, which wanted flu vaccine—any flu vaccine—to give to troops, they had permitted the sale of tens of millions of substandard shots. One senior scientist told investigators that the vaccine makers would sell water if they could get away with it.[6] Another glaring error that came to light was the continued licensing of 75 biologic products known to be ineffective and sometimes dangerous.[7] These were brazenly crummy concoctions like the bacterin licensed in 1956 that contained six killed organisms and was sold for the treatment of "upper respiratory infections, bronchitis, infectious asthma, sinusitis and throat infections." This "vaccine" had been associated with systemic allergic reactions, diarrhea, and other side effects, yet remained on the market. Murray claimed he lacked the authority to regulate biologics for effectiveness.

The 1972 probe had begun as a result of a job complaint by J. Anthony Morris, a regulator of viral vaccines. Morris, hired by the division in 1959 and a Smadel protégé, had become convinced that SV-40 was an agent of human cancer. Depending on who was describing him,[8] Morris was either a dogged, fearless investigator of government negligence or an incompetent victim of an idée fixe.[9] He was represented in his grievance procedure by James Turner, a Nader Raider who turned Morris's complaint into a broad critique. The agency had a tendency "to downplay an open public discussion of the possible harmful effects of vaccines," as one scientist noted in testimony before Ribicoff. Partly as a result of the investigation, Ribicoff, who had previously been Kennedy's secretary of Health, Education and Welfare, sought to create an independent Consumer Safety Agency to replace the FDA. Instead, the Nixon administration merged Biologics with FDA under the continued auspices of HEW.

By pushing universal childhood vaccination in the late 1970s, the Carter administration had inevitably exposed failings in vaccine safety. All in all the system had worked well at reducing vaccine-preventable diseases. There hadn't been a major vaccine safety disaster since the Cutter incident of 1955, but there had been problems, including SV-40 and atypical measles. There were few smoking guns but, then, there was no system in place to look for them. Millions of parents were being required to inject their babies with vaccines, but there was scant effort to keep tabs on their side effects. After switching to the FDA, successive reorganizations of the vaccine branch created the Bureau of Biologics (1972–82), the Office of Biologics Research and Review (1983–88) and finally the

Center for Biologics Evaluation and Regulation. But none of these agencies was exactly loaded with government scientists making sure vaccines were safe, especially once they were on the market.

By 1976, the "bacterins" and other quackish formulas had been delicensed, but there were still vaccines on the market whose safety and efficacy were uncertain. Leading their ranks was the whole-cell whooping cough vaccine. By the late 1970s, many scientists in the pertussis field had lost confidence in the methods established 30 years earlier to test the vaccine. The CDC, which was responsible for monitoring disease in the United States, in 1978 created the Monitoring System for Adverse Events Following Immunization, but it was voluntary, small, and did little postmarketing evaluation. Adverse events were volunteered by doctors to the drug companies, which were supposed, but not obligated, to pass them along to the FDA. The agency never recalled any lots of DTP vaccine.

This passive monitoring system was established in response to the swine flu fiasco, a series of disconcerting events that began in February 1976, when an alert New Jersey state epidemiologist detected an unusual strain of influenza virus in the bloodstream of a soldier who died of a respiratory illness at Fort Dix. Influenza is in many ways the bane of the vaccination industry, although the flu vaccine is also the most commonly used shot. Flu vaccine is hard to make well because the most common strain of influenza changes every year. With new influenza antigens popping up, manufacturers are forced to start production each spring on a product for the next fall based on their best guesses about what the coming flu season will hold. Major shifts in two of the virus's key proteins occur every few decades, further complicating the job. In addition, although flu mainly kills the elderly, their immune systems generally respond feebly to the vaccine—a dirty little secret that the government seldom mentions in its annual "get out for the flu shot" campaigns. It continues to hold these campaigns because while flu vaccine is not widely effective, it is the only preventive available. In addition, the government wants to maintain a healthy flu vaccine industry so that it can be prepared for pandemic shift years, when a flu vaccine may be all that lies between a deadly strain and a catastrophic tithing of the human species.

In any case, the Fort Dix soldier had been infected by a flu strain that appeared similar to the deadly flu virus of 1918–19, which was believed to have jumped from pigs to humans. (Not even the name "swine flu" was correct, it turns out. Recent studies have shown that the Great Pandemic flu jumped from birds to humans).[10] The major flu shifts since then—in 1957 and 1968—had been followed by pandemic influenza. Scientists at the CDC predicted that the Fort Dix strain would dominate the flu season of 1976–77, and their initial memo to the secretary of Health and Human Services invoked the 1918 pandemic. That was all it took to set off a panicked political response in the midst of a presidential campaign. Flanked at a news conference by Sabin and Salk, President Ford on March 24, 1976, asked Congress to appropriate $135 million to produce enough vaccine to immunize "every man, woman and child" in the United States against the swine flu. He announced a crash program by the nation's four flu vaccine manufacturers—Merck, Parke-Davis, Lederle, and Connaught—to rush out 200 million doses, roughly ten times the usual number vaccinated against flu. Few in Congress raised questions, with the notable exception of a young Los Angeles Democrat named Henry Waxman, who suggested that the campaign was a "pharmaceutical industry ripoff."[11]

Problems soon developed. First, the insurance industry refused to underwrite the manufacture of the shots, assuming, correctly, that a massive campaign would lead to real and imagined side effects and lawsuits. Vaccine makers immediately halted their production and demanded that Congress assume the risk. Then the vaccine, in preliminary tests, proved to be missing a key antigen. Meanwhile, mid-August arrived without a single additional case of swine flu reported in the United States or anywhere else. Studies showed that the young soldier at Fort Dix was 1 of 500 people infected with the same flu strain, and none of the others had even been seriously ill from this supposedly deadly bug. The government might have cut its losses and scrapped the vaccination campaign at that point had it not been so afraid of admitting a big mistake. Then an entirely unforeseen event—the outbreak of a deadly disease among American Legion veterans attending a convention at a Philadelphia hotel in late July—made mass flu vaccination seem like a good idea. By August 5, swine flu had been ruled out as the source of "Legionnaire's disease"—which was caused by a bacterial infection—but by then the horse was out of the barn. Driven by fear of a mystery germ, Congress rushed to pass the swine flu bill, making the government liable for the program. By year's end 48 million doses had been administered, and all for nothing. Not a single additional case of swine flu was reported in the United States.

The sad denouement to the story began in late November 1976 with the first reports of a paralyzing neurological illness, Guillain-Barre syndrome, possibly triggered by an immune response to egg proteins in the swine flu vaccine. The government established a special branch of the federal Court of Appeals to deal with 4,000 injury claims. Although many of the injuries were unrelated to vaccination, the government would ultimately pay flu shot recipients about $100 million.

The fiasco could not have come at a worse time for American public health. In the aftermath of Watergate, confidence in government and medicine, and government medicine in particular, was at a historic low. Many blacks had avoided the swine flu shot out of suspicion that it was part of a white genocidal conspiracy.[12] The debacle opened public health authorities to the kind of criticism and even ridicule they had never had to countenance before. The zeitgeist was summed up well by *Washington Post* columnist Richard Cohen, who wrote, "An injection is nothing to smile about. Sometimes it hurts, more often than not it smarts and sometimes it is followed by aftereffects—in this case the lingering suspicion that you are now immune to a disease that does not exist." Cohen said swine flu skepticism showed that "something had gone out of American life—our unbridled faith in science. . . . for too long we believed uncritically in science, swallowing whole what we were told. Sometimes science wrapped itself in the flag and was called 'American knowhow' and did things like eradicate yellow fever in Panama. Years later, science was used to sell us a war. . . . the thing about Vietnam, you will recall, was that it was going to be scientific—clean and surgical for us . . . so when it comes to the swine flu vaccine, I feel the cynicism is healthy. Any program conceived by politicians and administered by scientists comes to us doubly plagued."[13]

In the wake of the Guillain-Barre litigation, vaccine makers, who were already leaving the industry in droves, began to balk at making any vaccines at all. A polio vaccine shortage began in September 1976, after Wyeth stopped production and Lederle held up deliveries of the vaccine until the federal government promised to get parents to sign informed consent statements before vaccinating their kids. The NIH's William S. Jordan proposed the construction of a National Production Facility for new vaccines.[14] The idea was seconded by an expert panel convened to consider the issue—and immediately shelved.

The swine flu episode was a classic example of good intentions gone bad, of a failure to calmly assess risks amid changing circumstances. A few good things came of it, though. One was a generational shift at the CDC, away from the older, cold war set toward the younger idealists who had honed their skills as vaccinators in the third world. David Sencer, who though brilliant had a top-down management style, was fired as director of the CDC, to be replaced under President Carter by Bill Foege, a 41-year-old, six-foot-seven former missionary doctor who had battled small-pox in Africa and India. Sencer had consulted many flu experts before making his decision, but was faulted for failing to consult with those who might not favor the program—such as consumer advocates or state health officials who had other priorities. Though a staunch supporter of vaccination, Foege was more on the wavelength of the swine flu skeptics.[15]

The swine flu episode was the climax of a series of troubling developments that were driving manufacturers out of the vaccine business. The Ribicoff inquiry sparked by J. Anthony Morris had required the FDA to establish panels to review all vaccines, and the tightened regulations that were anticipated drove many out of the field, as many executives felt it was becoming too expensive to make vaccines. At the same time, new liability risks had cropped up as a result of a successful lawsuit against the polio vaccine maker Wyeth. Thus, the Guillain-Barre litigation made it clear that some kind of special compensation program was needed for all vaccine injuries, to protect the remaining manufacturers or at least encourage them to stay in the business. The August 1976 law that authorized the swine flu program required the Department of Health, Education, and Welfare to generate a compensation plan for all public health measures within a year. Nothing much became of that plan, but a committee was established to study the problem. Its chairman was Richard Krugman, a pediatrician at the University of Colorado Medical School.

Krugman was a good choice for such an inquiry. His father, Saul Krugman, along with Sam Katz, Sabin, and other pediatricians, had often batted around the idea of a compensation program. In 1973, Saul

Krugman told a magazine interviewer that a no-fault system was needed for vaccine makers and takers to "protect the tiny minority who are harmed by vaccines."[16] The elder Krugman had undoubtedly been sensitized to parental concerns by the ugly dispute over the vaccination experiments he had performed at the Willowbrook school.[17] Between 1956 and 1971, Krugman fed many of his patients hepatitis virus obtained from the stools of other students. In what would become the story that launched him to stardom, Geraldo Rivera of the local ABC affiliate exposed what Krugman was doing. Rivera's programs led New York state officials to investigate Willowbrook, and the student left vilified Krugman. At a 1972 ceremony honoring him with the pediatric society's Howland medal, 3,000 people picketted. Many of his colleagues, as well as his son, felt Krugman was unfairly pilloried, but the controversy offered an opportunity for rumination over the rights of patients and their parents.

In a 1975 editorial in *Pediatrics*, Richard Krugman wrote that the country's immunization practices had put it on a "collision course."[18] While many states had passed compulsory immunization laws for schoolchildren, "what recourse does the one person in tens of thousands suffering an adverse reaction have? . . . We should not wait for a series of expensive legal suits to be filed before acting. We immunize our children to prevent disease. We should do no less than to immunize ourselves with preventive legislation." Yet when it came to establishing such a program, Krugman and his colleagues on the compensation committee wrote later, they were constrained by the lack of good data on serious adverse effects following vaccination. There was no reporting system for these occurrences "and as a consequence no one can estimate the cost of compensation."[19]

Compensation systems existed in Japan, Hungary, West Germany, and other countries. In the United States, the only remedy was a lawsuit. In December 1961, Cutter Laboratories announced that it had settled 44 of 50 polio vaccine lawsuits for a total cost of $3 million. In the most important trial stemming from the Cutter episode, attorney Melvin Belli won $125,000 for Anne Gottsdanker, a paralyzed five-year-old from Santa Barbara, California. It was an unusual verdict. The district court judge in San Francisco had instructed the jury that Cutter's product had an "implied warranty," which meant the company was liable for her injury whether or not it had been negligent in producing the vaccine.[20] A second wave of vaccine lawsuits culminated in 1968. Eric Tinnerholm, whose parents alleged he had been left quadriplegic by the Quadrigen vaccine, which included diphtheria, pertussis, tetanus, and inactivated polio antigens, won $651,783 from Parke-Davis. Quadrigen, put on the market as a four-in-one convenience shot in 1959, was troubled from the start by the company's difficulty finding the correct preservative for it. Thimerosal, used commonly in pertussis vaccine, ruined polio antigens, while formaldehyde, used to inactivate polio, damaged the pertussis antigens. Parke-Davis used a third preservative, benzethonium chloride, which tested safely in the lab. But under real-life conditions of heating and cooling between the factory and pediatrician's office, the pertussis element released toxins, while other antigens in the vaccine were weakened. The company's response was to increase the bacterial concentration of the vaccine, which apparently made it more toxic. Parke-Davis removed Quadrigen from the market in 1968 after a series of lawsuits. It had been administered 3 million times.[21]

From the vaccine industry's perspective, however, the most important liability precedent was set in 1974. In *Reyes v. Wyeth Laboratories*, the parents of Anita Reyes, of Mission, Texas, near the Mexican border, won a $200,000 judgment after the seven-year-old became paralyzed following vaccination with oral polio. By then, Sabin's oral vaccine—the liquid solution that children drank from paper cups or ate in sugar cubes—was the standard of care. The Salk vaccine had been phased out in the 1960s (only to return in the late 1990s). By now it was well established that the Sabin vaccine occasionally reverted to virulence, paralyzing perhaps a dozen children or their caretakers each year. That point wasn't at issue; oral polio vaccine was, in legal parlance, "unavoidably unsafe." The liability issue established by *Reyes v. Wyeth* was far more troublesome to the vaccine industry. The Reyes family argued, successfully, that they had not been fairly *warned* of the risk of oral polio vaccine. Although Wyeth's product insert mentioned the tiny risk, the public health nurse who administered the vaccine had not warned Anita's semiliterate mother. The court found that the nurse was not qualified to interpret the product insert, and thus Wyeth was guilty of "failure to warn"—even though it was established that Anita Reyes was infected with naturally occurring polio virus—and not the vaccine virus at all!

The punitive reasoning of *Reyes v. Wyeth* came as a shock to the vaccine companies. For years, vaccines had been administered in many settings by all kinds of healthcare professionals, dispensed like candy with much celebration and little fuss and certainly no threat of legal action. After *Reyes*, the vaccine companies demanded and obtained a change in the Public Health Service law. Now, the CDC was required to print up information leaflets to be distributed to families when children were vaccinated in the public sector. In private settings, the failure to warn would be the responsibility of individual doctors. It was the very rarity of vaccine-preventable diseases like polio—Anita Reyes was 1 of only 31 paralytic polio cases in 1970—that had suddenly put a new onus on the vaccine establishment. People did not expect to encounter these once-common diseases anymore and were horrified when they did. There was something else, too—a change in the social contract mediated through the legal system. The truth was, the state had required for years that children receive sometimes dangerous vaccines. Vaccinia had been administered for years after influential doctors like Kempe had pointed out that its routine use was unnecessarily risky. Yet there had apparently never been a damage lawsuit over a smallpox vaccine injury.

Why? "We weren't nearly so litigious," responds Alan Hinman, a senior CDC immunization official during the late 1970s and 1980s. "And there might be a question of supply and demand in terms of lawyers." The consumer movement had generated plenty of riled-up lawyers looking for juicy cases and encouraged parents who questioned medical and governmental authority. Quietly suppressed vaccine risks weren't the same as the Pentagon Papers, but they weren't to be overlooked in an antigovernment climate, amid growing interest in alternative medicine. There were fewer safe havens, now, from the litigators. Public health veterans like Hinman, Foege, and Katz, who had nobly fought disease in the darkest corners of the world and America, noticed the change, resented it to a certain degree, but knew they had to adjust to it. "Back in the old days you had a lot of disease around, you made a vaccine and everybody said, 'Whoopee!'" said Paul Parkman. "Nowadays there isn't a lot of disease around and people say, 'I'm not sure I want my kid stuck with that stuff.'"[22] Then, too, the documents that laid out the history of the pertussis vaccine showed a long trail of fairly cold-blooded assessments of its dangers.

Pertussis: Germ, Vaccine, and Litigators

"You're taking on your mother and apple pie when you take on vaccine," says Victor Harding, whose Milwaukee personal injury firm made millions suing Connaught and Lederle over the whole-cell pertussis vaccine. The *Gottsdanker* and *Reyes* decisions, which gave overwhelming benefit of the doubt to the unintended victim, seeded the clouds for a downpour of litigation. But the key change seems to have been cultural rather than purely legal. Vaccines were a victim of their own success—they had wiped out the diseases they were designed to defeat. Therefore, the social good of vaccination was no longer automatically accepted. The social contract of vaccination, once implicit, would now have to be codified. The consumer movement had its guns trained on any industry that didn't consider consumer rights.[23]

To be sure, trial attorneys saw a special target of opportunity in the whole-cell pertussis vaccine. The fact that it was so much easier to work up righteous indignation over an allegedly bad pertussis vaccine than a polio vaccine probably stemmed from the peculiar history of the two vaccines. Polio had been willed out of existence by the voluntarism and marching dimes of millions of Americans. Its enemies—Roosevelt, O'Connor, Salk, and Sabin—were fairly unblemished heroes. There was no Salk or Sabin for pertussis vaccine, which had been developed over decades by a series of imperfect steps. The pertussis vaccine makers were unsung, except among their colleagues. It took trial lawyers like Victor Harding to bring the vaccine developers' story out of storage. And the light he shed was not always rosy.

I interviewed Harding in the shiny penthouse suite his firm occupies overlooking Cathedral Square in downtown Milwaukee. Harding is a silver-haired fellow with bushy eyebrows, and he has the informal, casually profane manner of a big-city pub owner. In trying pertussis cases, Harding was knowledgeable and folksy—to help jurors understand the complex biochemistry, he'd refer to the bacterium as a "peach," while its cell-wall proteins and interior were sometimes "the fuzz," or "the juice." Looking back on the 20 or so cases his firm won—in court or in settlements—Harding put his feet up and rubbed his forehead fretfully.

He didn't want to be blamed for wrecking the nation's vaccination program. "There's no question that the whole-cell vaccine didn't do some good over time. That's not an easy take-on. But it makes it much easier to say, not only is it defective because of this, these sons of bitches have on the shelf a formula that does what we're saying. That's why we spent so much time digging up stuff on the acellular vaccine." The tort attorneys used the discovery process to compile a damning case against the manufacturers by showing that they had known about and even patented potentially safer vaccines while continuing to market their own products. The documents he and other attorneys found were, for juries, convincing indications of wrongdoing, but they were also impressive as historical documents, spelling out a chain of decisions and scientific problems that frustrated the arrival of a better vaccine.

Whooping cough is a small bacterium that infects the upper respiratory tract by embedding itself in mucus. Toxins released by the bacteria cripple white blood cells and cilia, the tiny hairs in the upper respiratory passages, and the cilia, through genetic reprogramming, actually promote the growth and production of toxins by the bacteria.[24] The initial 10 days of whooping cough are like a typical respiratory infection, with some fever and malaise. The second phase is characterized by a paroxysmal cough that can last for up to three months, giving the disease its name in Japanese—"100-day cough." At the height of the coughing stage, the bacteria are mostly dead. It is the toxins that do the damage.

From 1922 to 1931, about 1.7 million cases of whooping cough were reported in the United States, and 73,000 deaths, mostly children.[25] In a typical paroxysm, one author wrote, "the child, who can usually foretell it, will often run for support to the lap of the mother or nurse, or seize a chair with both hands. There now occurs a series of explosive coughs, from ten to 20 in number, coming in such rapid succession that the child cannot get its breath between them; the face becomes a deep red or purple color, sometimes almost black. The veins of the face and scalp stand out prominently, the eyes are suffused and seem almost to start from their sockets; there follows a long drawn inspiration through the narrowed glottis, producing the crowing sound known as the whoop, and then another succession of rapid coughs follows and another whoop." Each cycle lasted about three minutes. After a severe attack, stricken by a feeling of suffocation, the child was too exhausted to stand. Children who survived the disease were often malnourished and suffered convulsions, coma, paralysis, disturbances of sight or hearing, and brain damage. There are numerous accounts of behavioral problems developing in children after whooping cough.[26] But the disease is most dangerous to infants, who do not whoop. Unable to breath, they turn blue and sometimes die.

Whooping cough vaccines began appearing shortly after 1906, when the Belgian scientist Jules Bordet cultured the bacteria for the first time in a petri dish. *Bordetella pertussis* was a tricky organism to grow, and it didn't easily infect mice or other test animals. The organism's signal characteristics—its ability to produce immunity, its toxicity—often changed in successive cultures. Many unsound vaccines were produced, sold, and junked between 1914 and 1931, when pertussis vaccine was removed from the list of *New and Nonofficial Remedies* recommended by the American Medical Association.[27] In the 1930s, Pearl Kendrick and Grace Elderling of the Michigan State Health Department at Grand Rapids set about making the definitive vaccine.[28]

The two women found fertile fields for research in Depression-era Grand Rapids. "As soon as the laboratory closed in the afternoon we set out to find new patients," Elderling recalled. "Many of the families we visited were very poor and their living conditions pitiful. Our watchword became 'round to the back and up the stairs.' We listened to sad stories told by desperate fathers who could find no work. We collected specimens by the light of kerosene lamps, from whooping, vomiting, strangling children. We saw what the disease could do." The women made their vaccine by growing whole pertussis bacteria, then killing them with phenol. In 1939, their first study was published in the *American Journal of Hygiene*. The study included 4,212 children between eight months and five years of age. Four hundred cases of whooping cough were reported during the four-year observation period—52 in the vaccine group and 348 in controls. No one died, and the vaccinated children got less sick. In 1940, Michigan started distributing the vaccine to physicians, and Kendrick began testing combinations of pertussis with tetanus and diphtheria toxoids—what would become the DTP shot. During World War II, Kendrick and Margaret Pittman of the National Institutes of Health created the first method of standardizing pertussis vaccine, the mouse potency test. Kendrick and Pittman found they could sicken mice by injecting pertussis directly into the brain. If effective vaccine was injected first, it protected the mouse.

Pittman was the third of the female triumvirate who developed the pertussis vaccine. As an undergraduate in 1926, she wrote a term paper titled "Whooping Cough: Cause and Mode of Transmission," and in the mid-1990s she was still a force to be reckoned with in the field.[29] ("When she was 92," a colleague noted, "she got a speeding ticket on her way to work. You've got to take someone like that seriously.") With the men in her laboratory away making plague and cholera vaccines for the military during World War II, Pittman took command of this child killer. She and Kendrick also worked together on the opacity test, a way of determining how many bacteria were in the vaccine by shining a light through it. Kendrick subsequently developed the mouse weight gain test, a way of screening out toxic vaccines. If there was too much toxin in the vaccine, mice injected with it didn't gain weight. Another of Pittman's innovations was to establish that thimerosal was the best preservative for whooping cough vaccine, and later for *Haemophilus influenza*e (Hib) bacterial vaccines as well. Pittman and her NIH colleagues had now standardized a shifty product, one whose value to children cannot be underestimated. Whooping cough killed about 3,000 children a year in the United States in the 1940s—more than measles, scarlet fever, and diphtheria combined; it also contributed to chronic lung conditions that killed thousands of adults. Vaccination against whooping cough could now proceed with a great deal of confidence.

But the safety problem persisted. In practice, the weight gain test wasn't a reliable way of testing for toxicity. If the witches' brew used to grow the bacteria was altered in the slightest, it would alter the behavior of the bacteria; mice in different labs reacted differently to the same vaccine.[30] Even the temperature and light conditions in the mouse's cage, or the size of the cage, could change the result. And out in the real world, the "sporadic manner" in which children reacted to the vaccine indicated how difficult it was to screen out a "hot" vaccine lot. "Even when a reaction is observed," Connaught Canada scientist Jack Cameron said at one conference, "it is still likely that thousands if not hundreds of thousands of doses of the same lot have been injected without incident." The government's adoption of the uniform standards developed by Pittman and Kendrick in effect quashed initiatives to develop a better product. Even though scientists suspected the safety of the vaccine, their companies had no incentive to buck the regulatory system established by the Division of Biologic Standards.

Before 1949, Lederle had made an "extract vaccine" by using chemicals and centrifugation to break open bacteria. The cell walls were discarded and the remaining liquid was used as vaccine. This crude "acellular" vaccine was probably freer of toxins than whole-cell vaccine, but it had the disadvantage of containing human blood cells used as culture media, and it couldn't be tested with the Pittman standards. Lederle dumped the vaccine and started making a whole-cell shot.[31]

Vaccine-related encephalopathy was a staple of the world's pediatric literature beginning with the earliest whole-cell pertussis vaccines, and it continued to crop up after the vaccine's standardization. A 1933 study from Copenhagen contained references to two deaths following vaccination. "One half hour after the last injection, contractions in the arms and legs occurred, followed by cyanosis, hiccup, convulsions and death within a few minutes," the researchers reported in the *Journal of the American Medical Association (JAMA)*.[32] Two hours after a second dose at 11 days of age another child "died suddenly with slight cyanosis." An influential case series written up in the first volume of the journal *Pediatrics* in 1948 reported on 15 previously normal children admitted to Boston Children's Hospital from 1939 to 1948 with seizures following vaccination with eight different companies' DTP vaccine.[33] Two years later, only one of the children was neurologically normal. To be sure, nearly 40 children with encephalopathies that resulted from pertussis illness itself were also admitted in this time period. Still, the *Pediatrics* authors recommended modifying the vaccine for greater safety. And the report caused a small uproar. A few months later, pediatrician John A. Toomey of Cleveland reported at an American Medical Association (AMA) meeting that he had received 38 case reports of severe reactions to pertussis vaccine, including five deaths.[34] Doctors at the meeting noted that the vaccine, by virtue of its fever-producing qualities, often provoked seizures in children predisposed to them. The same healthy-appearing children who suffered convulsions at the onset of a strep infection, one researcher stated, were probably just as likely to seize after a DTP shot.

Doubts about the vaccine slowed its update considerably. In Seattle, for example, school authorities did not implement regular pertussis vaccination until 1958, following a whooping cough oubreak.[35] The British, tempered by their smallpox vaccine difficulties, also moved cautiously on the whooping cough vaccine. A major British trial in 1942 found nothing of value in the Kendrick vaccine. Larger trials, begun in 1951 and lasting five years, provided favorable results, but the 1964 edition of a leading British reference book concluded that vaccination wasn't advisable in countries where whooping cough mortality was low.[36] Such sentiments were bolstered by two reports in *The Lancet* by a Swedish infectious disease specialist, who found 36 neurological reactions among 215,000 vaccinated children in the 1950s, with 13 deaths. Only 17 cerebral attacks were recorded among the 725,000 Swedish kids who got the disease during those years, he said.[37]

One of the more telling anecdotes about the risks of whole-cell vaccine was revealed in a paper by Carl Weihl, the polio-stricken Cincinnati pediatrician. Weihl and his associates administered a new "extracted" pertussis vaccine, produced by Eli Lilly, to the children of two physicians who had suffered severe fever and seizures following the second injection of whole-cell vaccine. Doctors shunning a vaccine for their own children was always a bad sign. "When the extracted pertussis antigen was utilized, neither responded with any systemic reaction," Weihl wrote. The advantage of the newer vaccine, "which produces an effective antibody response without a high incidence of reactions, seems obvious."[38]

Complaints about whooping cough vaccine continued, sotto voce, for decades. At a 1975 colloquium on pertussis vaccines at the Bureau of Biologics, Charles Manclark, whom the FDA had hired to develop an acellular vaccine, predicted chillingly that "we may be approaching a time in which more vaccine-related problems than those due to the disease will be experienced." The whole-cell vaccine, he noted, was "one of the more troublesome products to produce and assay." It had one of the highest failure rates among products submitted to the Bureau of Biologics—roughly a fifth of the vaccines submitted to the bureau failed, and it was difficult to know how many slipped through the cracks. Adverse reaction rates were "not accurately reported, but more adverse reactions are probably experienced with the use of pertussis vaccine than with other biologicals."[39] About two-thirds of pertussis vaccine recipients had some negative reaction to it. Because of its toxicity and "the lack of appropriate parameters to detect and control it," an industry scientist added presciently, pertussis vaccine "has to some extent become the next target for the antivaccination lobby."[40]

For all the problems with it, pertussis vaccine was so cheap to produce—less than 5 cents a dose—and the blueprints for building a new one were so vague, that most manufacturers weren't interested in an alternative. With manufacturers unwilling or unable to make a better vaccine, the collision that Krugman had predicted was on its way. Whole-cell pertussis was a vaccine, Manclark concluded in 1975, "for which it is strongly urged that legislation be enacted to provide reasonable federal compensation to the few individuals injured and disabled by meritorious public health programs."[41] These were sagacious words. Eight years later, U.S. researchers had made little progress toward a more refined pertussis vaccine. Sixteen million parents were still being asked to have their children injected with a potentially toxic product every year. And when a group of these parents stood up to demand that the system make amends for damaging their children, the litigation crunch hit.

In 1984, Kevin Toner, a paralyzed child in Burley, Idaho, won a $1.2 million judgment from Lederle on the grounds that the whole-cell pertussis vaccine was a defective product. A brain-damaged Chicago girl named Melanie Tom got $7.5 million from her doctor and Wyeth Laboratories a few years later. In 1987, Victor Harding and his boss, Ted Warshafsky, won a $15 million judgment for a Kansas girl, though the award was reduced on appeal and settled out of court.[42] The jury in that case found that Wyeth was negligent in testing the vaccine and failed to adequately warn of its potential danger. By 1985, 219 pertussis vaccine lawsuits had been filed in U.S. courts, with an average compensation request of $26 million.[43] When the lawsuits began in 1981, the total size of the pertussis vaccine market in the United States was only about $2 million.

The company filing cabinets were clearly full of damning evidence of the whole-cell vaccine's reactivity and the companies' knowledge of how to make safer acellular vaccines. In a May 20, 1957, letter to another Wyeth scientist, Dr. Howard Tint noted that it was obviously possible to make a pertussis vaccine "less reactogenic and just as effective as the product we're currently making."[44] Indeed, such a product was being tested by Eli Lilly, which put the vaccine on the market in 1962. Weihl had given the Lilly vaccine to 1,248 patients over two years. When a pertussis epidemic broke out in Cincinnati in 1958, only 2 of them became ill. Presumably to bolster sales of its own vaccine, Lilly commissioned one of its scientists, C. N. Christensen, to study side effects in recipients of its competitors' whole-cell pertussis vaccine. At the 1963 International Symposium on Pertussis, Christensen reported finding 21 children hospitalized with pertussis reactions from the years 1955 to 1961. Fourteen were severely retarded. "It is obvious," Christensen said, "that severe neurologic reactions have occurred in children after immunization with pertussis vaccines that have passed the toxicity and potency tests."[45]

Merck marketed a whole-cell vaccine in 1962, then dropped it a few years later "for fear of lawsuits," according to one account.[46] Cutter announced in 1965 that it was leaving the whole-cell vaccine market, for economic reasons that probably also explained the reluctance of vaccine-makers to change. In its 1966 annual report, Cutter explained that the FDA had set a shelf-life limit of 18 months for the DTP vaccine. As time wound down on lots in storage, companies would try to unload them on the federal government, which bought in bulk for its immunization programs. Bidding wars for government sales had driven DTP's price down too low to profit from it, Cutter reported.

Lilly's acellular vaccine, sold under the trade name Trisolgen, was considerably more expensive to produce, but pediatricians seemed willing to pay the difference. Lilly owned between 20 and 50 percent of the pertussis vaccine market from 1972 to 1976, according to different accounts.[47] It was popular enough for Lederle, Parke-Davis, Wyeth, and Richardson-Merrill to quietly work on their own acellular vaccines in the late 1960s.[48] Lloyd Colio, a scientist at Richardson-Merrill's Swiftwater, Pennsylvania, plant, said he had "concluded there was something in the [whole-cell] vaccine that was causing those cerebral accidents." But the yield of the acellular vaccine he'd developed, he said, was only about a quarter of the whole-cell vaccine, which made it uncompetitive.[49] Trisolgen was safer and just as efficacious as the whole-cell vaccines, according to John Robbins, the intellectual New Yorker who was in charge of regulating pertussis and other bacterial products at the FDA for a time. But in 1976, Lilly decided to get out of the vaccine business. Its Indiana vaccine plant, built in 1913, was out of date. And according to Robbins, Lilly was responding to an illustrious adviser: "Jonas Salk told them, 'Don't work with these old products. Work with the new immunologics.' Now Jonas Salk was a god. He walked on clouds. Jonas said get rid of the vaccine, and they did." There was nothing the FDA could do about it, Robbins adds. "Do you know what Eli Lilly is? Eli Lilly is a country. And there are no jerks there. Those guys are first-rate chemists and decent human beings. A little FDA official couldn't tell them what to do."

Lederle considered stepping in with its acellular vaccine to fill the void, but didn't. The company might have stood to increase its sales from $150,000 to $500,000, according to a company memo, but that would have required vigorous marketing and testing, since FDA regulators were tightening approvals of new vaccines.[50] The project was put on hold, to be reactivated should an acellular vaccine become "unexpectedly favorable." Finally, Wyeth took over Lilly's vaccine, and after finding a way to boost the product's yield, tried to get it licensed. But the FDA was discouraging.[51] It wanted the company to do new comparative safety and efficacy trials before licensing the reformulated Trisolgen. And it wanted better science behind the new pertussis vaccine. Thus the FDA hoisted itself—as well as kids who were being injured by the old vaccine—by its own petard. To get rid of the reactive vaccine it had licensed for decades, it set the bar for the plausible alternative too high for industry to deign to meet. For the companies, it was cheaper and easier to make the bad old vaccine than to make a better new one.

Within a few years, of course, the cost of making acellular vaccine would be moot, as lawsuits led companies to increase the price of pertussis vaccine 100-fold. At that point, some producers would get out of the pertussis vaccine business altogether. But for those that stayed, production costs were now only a small percentage of the cost. Not coincidentally, Wyeth and Connaught now began searching for useable acellular vaccines. Their eyes turned to Japan, where Yuji and Hiroko Sato, scientists at the National Institute of Health—Japan's version of the CDC—had developed a vaccine. To do so, they had struggled mightily with the devious biochemistry of pertussis.

The Peach and Its Juices

Scientists understood as early as the 1940s that you could remove the pertussis organism's cell wall, which contained poisonous endotoxin, without affecting the capacity of the vaccine to produce immunity to whooping cough. This was essentially what Lilly had done with its "extracted" acellular vaccine. It centrifuged out the cell walls, threw them away, and made vaccine out of the "supernatant"—the leftover juice, to use Victor Harding's vocabulary. There were other "factors" in that "juice"—"factor" being a kind of scientific placeholder to indicate non-specific substances with specific effects. Histamine-sensitizing factor was something in the bacterium that produced allergic responses in the aftermath of a bout of pertussis, or a vaccination; lymphocytosis-promoting factor stimulated another branch of the immune system. Islet-activating protein stimulated the pancreas. By the late 1970s, scientists realized these three factors were all the same thing, a protein they renamed pertussis toxin. It would clearly be a major element of a purified vaccine.

Yuji Sato, a diminutive microbiologist with bushy hair and a fiery temper, was hired in 1967 by the Japanese NIH to isolate that substance. He and his wife, the biochemist Hiroko Sato, spent their careers pursuing it, and left as a legacy the acellular pertussis vaccines used today in many countries of the world. There was to be plenty of rancor along the way among the Satos and their colleagues over who understood what and when, though perhaps no more than could be expected for a scientific enterprise this grueling and contested. In a series of experiments,[52] Yuji Sato and his colleagues isolated elements of the dead pertussis bacteria by centrifuging the supernatant in a sugar solution.[53] Later, the Satos and others determined that the isolated material mostly consisted of two distinct proteins—pertussis toxin and filamentous hemagglutinin antigen, or FHA. The Satos realized that mice vaccinated with both substances were able to clear pertussis bacteria from the lungs without the inflammation that generally followed pertussis inhalation.[54]

Meanwhile, something had happened in Japan that gave powerful new relevance to the Satos's work. In 1970, in an uproar over smallpox vaccination, the Japanese government established a compensation program for vaccine injuries. That drew increasing public attention to adverse events following pertussis vaccination. In December 1974 and January 1975, two Japanese babies died shortly after being vaccinated with the same lot of vaccine. The Japanese health ministry immediately suspended pertussis vaccination. It reinstated the vaccine a few months later, but changed the recommended age for the first dose from three months to two years. And a majority of Japanese parents chose not to vaccinate at all.

Around the same time, Britain was seized by a pertussis crisis, following the publication of a paper that identified 36 epileptic children whose illness was traced to the vaccine.[55] When several well-known British doctors suggested that pertussis vaccine injuries appeared unacceptably high—one of the doctors was George Dick, who had blown the whistle on Koprowski's oral polio vaccine 20 years earlier—vaccination rates plummeted to 31 percent by 1978, and pertussis rates rose.[56] Sweden, where vaccination rates had fallen to 12 percent, withdrew its whooping cough vaccine in 1979. Over the next decade 61 percent of Swedish children got pertussis.[57]

The Japanese government responded to its pertussis crisis by ordering the crash development of a new, less toxic vaccine.[58] "By this time," Yuji Sato recalled in an interview in Tokyo in 2004, "we had already developed a detoxified fraction of pertussis and discovered it had enough antigenicity. So we quickly made it into a vaccine." Sato's boss sent him out to consult with six different vaccine manufacturers. Once the companies were able to copy and scale up the procedure, vaccine production started, and mass vaccination began in early 1981, initially for children over two years old. Vaccination rates had fallen to 42 percent in 1977, and cases of pertussis had gone from a low of 400 to 13,000 in 1979, with 41 deaths.[59] By 1984, 95 percent of children were being vaccinated, and in 1988 Japan started vaccinating infants again. The disease largely disappeared.

In 1978, the Satos traveled to the United States at the invitation of the Bureau of Biologics. The Satos say they were invited to show the American scientists how to make the vaccine; others, such as Parkman, say the Satos learned how to make the vaccine in the United States. Perhaps the teaching went both ways. In any case, the Japanese vaccine was not immediately accepted by the scientists in Bethesda. Manclark, who had his own ideas for making a vaccine, claimed the Satos's product wasn't a purified substance. The Satos felt that Manclark didn't understand which elements of the vaccine were antigenic. "If my English were only better," Yujo Sato said with a sigh, "I could have convinced him." Their disagreement was one of the factors that held up introduction of a new vaccine in the United States.[60]

Another reminder that such a vaccine was needed in the United States came about in March 1979, when the Tennessee State Department of Public Health reported the sudden deaths of four infants within 24 hours of receiving the same dose of Wyeth DTP lot 64201.[61] While it was possible that this clustering of deaths was merely a chance event (later studies showed no link between SIDS and whole-cell pertussis vaccine),[62] the company took precautions to prevent such a cluster from being noticed again. Wyeth decided to ship no more than 2,000 vials at a time to the same location. With no effective adverse event monitoring system in place, "hot lots"—real or coincidental—would be invisible.[63]

And then came *Vaccine Roulette*. The question of how to deal with a bad vaccine would no longer be solely in the hands of public health officials and vaccine manufacturers.

"Simple Justice for Children"

Whether DTP caused all the injuries in the film, and those brought forward later by thousands of parents, remained a matter of debate two decades later. But the film certainly exposed the special horror of vaccine injuries. "[T]he fact that when you do something to a perfectly healthy child to protect them from a future event that might or might not occur, and something bad happens as a result of that, it seems worse than if you let them get the disease and die from it," as Alan Hinman, director of the CDC's immunization program through the 1980s, recalled years later. "This was a vaccine known to be associated with some adverse effects. There were terribly damaged children, children in constant convulsions, profoundly retarded. You can see how totally this has overturned entire families. You bring this into everyone's living room and it is a kind of horror you have never imagined."[64]

Overnight, it seemed, parents angry about vaccine injuries, industry officials in need of liability protection, and pediatricians worried about the vaccine supply were all thrown together in an effort to reform the vaccination program. Within a few years they would create the first vaccine injury compensation system, vastly expand the oversight bureaucracy, and improve the system for monitoring adverse reactions. The Reagan administration opposed the legislation, which ran afoul of its bias against new government programs or taxes. But the logic of the effort, the compelling stories of the parents, and the connections and savvy of the players proved too powerful to stop. Within a few years of its creation, almost everyone agreed that the new program was a good thing.

In addition to Schwartz and the Micas, a third former Rogers staffer was pushing the process forward. Steve Lawton, a handsome Oklahoman who had been Rogers's protégé but now worked at Hogan and Hartson, a prestigious DC law firm, had been contracted by the American Academy of Pediatrics in 1981 to lobby Congress for a vaccine liability law. Now, he and Schwartz immediately set to work to write it. Lawton had been a captain in the 101st Airborne Division during the Vietnam War, but he threw his bronze stars in the garbage after returning home and joined the antiwar movement. Schwartz had been a conscientious objector. They were friends and helped sell each other's team on working together. The president of the American Academy of Pediatrics at the time was Martin Smith, who worked Congress tirelessly to raise support for a compensation system, with the goal of "simple justice for children."

The vaccine compensation issue found traction on Capitol Hill with remarkable speed. On Friday, May 7, 1982, less than three weeks after *Vaccine Roulette* aired, the Committee on Labor and Human Relations' investigations subcommittee, with Hawkins presiding, held its first hearing to discuss the safety and efficacy of the pertussis vaccine. At this and other hearings, it would be the government, not manufacturers, who defended the vaccine. To be sure, public health officials testified, it would be good to have an improved pertussis vaccine—and trials should begin within two years, said Harry Meyer.[65] Parents of damaged children testified, too. Their stories were all heartbreaking, though a few had nothing to do with the DTP vaccine. The mother of a psychotic child, born underweight during a rubella epidemic, blamed DTP. *Vaccine Roulette* had made her "hysterically aware that under the inoculation controversy could be hidden an American Holocaust with thousands of our precious children damaged. . . . are my son and I the survivors of a modern day

Auschwitz?" Marge and Jim Grant of Beaverdam, Wisconsin, were the parents of a 21-year-old quadriplegic, Scott, whose injury they blamed on a reaction to Quadrigen. Mrs. Grant had grown bitter after years of battling the FDA. She belittled the risk/benefit equation of vaccination. "When it happens to your own child, the risks are 100 percent," she said. That phrase, with its profound misunderstanding of science, became the motto of Dissatisfied Parents Together, appearing on the letterhead of its newsletter.

The most heartbreaking cases the senators heard about, the most disconcerting in terms of public responsibility, were the children who had serious but not critical reactions to their first DTP shot, then had a second shot, after which they went into convulsions or shock or even died. These children would not have been revaccinated had their physicians followed the American Academy of Pediatrics' guidelines on contraindications. But since the government paid such reactions little mind, pediatricians didn't make much of them either.

Between hearings, the parents organized and dealt with their own children. Some fell out of the struggle as their marriages broke up or their children's conditions worsened. New parents joined. There were disagreements. "Barbara [Loe Fisher] came basically from a kind of conservative libertarian background. I was sort of a liberal statist," says Schwartz. He had been a voter registration worker in Lousiana in the summer of 1964 and he believed in the importance of aggressive federal mandates. But he did think individual parents should have the right to opt out if they or their doctors thought that a vaccination wasn't in the best interests of a child. Fisher, on the other hand, felt that vaccination was like any other medical procedure—one that people should take only if they wanted to. "Barbara tended to look at it like, 'What is this bullshit?' " recalled Schwartz. "Who's responsible for the health of my child—me and my husband, or the government?' "

Schwartz urged the group to do its homework and avoid making statements that weren't solidly grounded in science. In the course of their research, the parents discovered the untold story of the pertussis vaccine. So while Williams talked to new parents and organized the group's database, Fisher began writing a muckraking book about DTP. She worked with Harris Coulter, a homeopathic doctor and medical historian.

Fisher's thinking about pertussis had been influenced by her own experiences and those of her sister, whose family had all come down with pertussis. The sister's children were bright and happy. Her own son was learning disabled and doing poorly in school. Her instincts told her that immunity produced by the disease was healthier than the vaccine. Later, when thousands of parents starting blaming autism on vaccines, Fisher would say that her son's problems were on the same spectrum of damage.

Coulter's wife, a certified homeopath, started treating Kathi Williams's son, Nathan, and "that seemed to help," his mother said, although he was later diagnosed as having attention deficit disorder. Turning to alternative physicians was part of the paradigm shift for the two women. They didn't just blame the public health and pharmaceutical industry for their children's problems; they blamed their pediatricians for looking away, out of dishonesty or cowardice, from problems with vaccination. Even before their children's putative vaccine reactions, Fisher and Williams had questioned medical authority and advice. They took Lamaze classes, sought out natural remedies, made their own baby food. "We were into doing it ourselves, back to nature," recalled Fisher, "it was part of the idea that maybe we haven't been doing things exactly right. Maybe we need to not have so much medicine in our life, all the drugs, all these synthetic chemicals."

If the medical diagnoses of the Fisher and Williams boys were somewhat open to dispute, there was no question about that of Julie Middlehurst-Schwartz. She had an uncontrolled seizure disorder that her pediatrican said was linked to the DTP shot. The worst of the seizures seemed to reverse the physical and mental progress she had achieved after the last one. It was heartbreaking, especially because Julie was the kind of spirited, optimistic kid that people gravitated toward. "One of her seizures left her with a hemiparesis—the whole right side of her was paralyzed," Schwartz recalled. "For a lot of us that would have been earth-shattering, but she was just this really indomitable spirit who emanated joy and love and happiness and determination. Around that time we went to visit a friend and there was a cable repairman there, and she rushes up to this man, whom we didn't know, and says, 'Up, up!' That was the way she was about the whole world—the world is here to pick me up and for me to love it and it to love me . . . each time she got hit she'd recover and fight back and eventually she learned to walk. She learned some words."

On March 25, 1984, shortly after her third birthday, Julie died during a seizure. Fisher had special words for her in the introduction to *A Shot in the Dark*: "This book is for Julie, and for all the other children whose health has been destroyed or whose lives have been taken from them, and for their parents, who will always love them." It was, of course, an earth-shattering experience, a loss beyond description. Jeff and Donna were in a void, a shrunken world. But Schwartz pulled himself up and continued to work on the vaccine safety bill. "I felt, 'Why me?' but the extension of that was to try to keep it from happening to the next person."

Dan Mica had introduced the vaccine compensation bill, which authorized the creation of a quick, no-fault method of compensating vaccine injuries, giving the benefit of the doubt to claimants. With the assistance of pediatric neurologists, Lawton wrote up a table of injuries compensated under the law, including paralysis after oral polio vaccination, thrombocytopenia following measles vaccination, and encephalopathy, or brain disorder, in the 72 hours after DTP. If a child fell into one of these categories, it created a "presumption of causation," and the government would concede injury and arrange a plan to pay for the child's lifetime needs—or a lump sum payment of $250,000 in the case of death. If the alleged injury fell outside the table, the government could contest the claim and the family would have to prove that the injury was vaccine-induced based on the facts of the case. Among its other measures, the law required that the vaccine lot number be listed on every patient's record and that doctors report major reactions to a public government database administered by the CDC. It also created a National Vaccine Advisory Committee to consider vaccine issues, and another committee to oversee the compensation program, which would be funded by surcharges on the vaccines.[66]

The administration—including top vaccinologists at the CDC and FDA—opposed the bill. The payments were too high, they said, the table of injuries was unscientific, and the compensation of retroactive cases would break the bank. Edward Brandt, the point man on the issue at the Department of Health and Human Services (HHS), said at a May 3, 1984, hearing before Hawkins that "the bill establishes a strong presumption that vaccine is responsible for essentially any adverse condition that happens after immunization unless there is incontrovertible evidence of other causation. This presumption of guilt would undermine public confidence in immunizations." He estimated that the bill would cost $5 billion in its first three years. (The total childhood vaccine market in the United States was about $146 million at the time.)[67] The AMA wanted an expert advisory group, rather than Congress, to define the injuries compensable under the law, with the ability to change them as new scientific evidence appeared. The doctors also wanted to end all litigation in the regular courts. The parents were dead-set against that idea. Recognizing that most parents would be happy to have a hearing in the vaccine court, Representative Waxman, a former trial lawyer, and others such as Senator Stafford, a Vermont Republican, insisted on maintaining parents' right to civil proceedings as a last resort. They felt that the ultimate threat of liability would keep the system honest, since DTP litigation had shown that the legal discovery was more effective than Congress in winkling information out of industry.[68]

The bill's supporters had the government in a tight spot morally speaking, and they drove home the point. Who could challenge the iconography of the vaccine-injured child, a young soldier on the battlefield of public health? After her child's death Donna Middlehurst wrote:

A few months ago on Memorial Day, I was filled with anger and bitterness as I watched the ceremonies honoring the unknown Vietnam soldier. When our nation engages in a military war we count the dead and wounded. We set aside one day a year to remember them and their sacrifices. We build monuments, however belatedly, where people can come and see the names etched in black granite and think about the costs of our wars. But in the war against this particular disease, my child is a casualty and no one cares. . . . It is true that we are winning the war against disease but no one is counting the bodies of the dead and wounded or honoring them in any way. No one is asking, does the cost in this war have to be this great? Is there some other way we could fight it?[69]

"We are not antivaccination generally, nor do we oppose child immunization programs," her husband testified at a September 10, 1984, hearing.[70] "If there are as few serious reactions as the medical community claims, [the compensation court] won't be a big burden, and if there are as many as we believe, then we better know about them. . . . When you take children who are healthy and they seize 4 or 8 hours later, and all possible tests are conducted and all possible explanations ruled out, well you know that saying about it walks like a duck and it quacks like a duck, chances are it is a duck until it is proven to be a cow."

On June 13, 1984, Wyeth announced that it would cease distribution of DTP because of liability risks. By the end of the year, shortages began. In April 1986, Lederle announced that the total sum demanded in DTP lawsuits against the company was 200 times greater than their sales of the vaccine in 1983. As a result, it would no longer produce or distribute DTP. That left Connaught as the sole supplier. A month later, Lederle reversed itself, but tripled the wholesale price of its vaccine.[71] In 1982, DTP typically cost the federal government 11 cents per dose. By June 1983 the cost was $2.33, by March 1984 $2.80, and it would rise eventually to $11. Yet the Reagan administration had asked Congress to cut funds for the vaccine stockpile by $12^1/_2$ percent.[72] Public health officials were in a quandary. Fear that the companies would stop making pertussis vaccine created consternation. No one but the companies could know for sure whether they were bluffing to improve their profits or really in jeopardy.[73]

According to Edward Mortimer, a pertussis expert at Case Western Reserve, in the early 1980s several members of the American Academy of Pediatrics' infectious disease committee agreed to "divide up cases to try to help manufacturers in these lawsuits" as expert witnesses. Mortimer and James Cherry, a UCLA pediatrician, testified in the lion's share of these cases, though both had expressed serious concerns about the DTP vaccine in the past.[74] At an April 1979 symposium, Cherry and Mortimer stated that the vaccine caused 140 to 270 cases of encephalitis every year and clearly needed improvement.[75] Yet in 1990, Cherry titled an editorial in *JAMA*, "Pertussis Vaccine Encephalopathy: It Is Time to Recognize It as the Myth That It Is."[76] When it was pointed out that Cherry had received more than $1.5 million in grants from Lederle and Wyeth, and got $260 an hour to testify for them, the journal required him to write a clarification note, and he was cited in a congressional report about medical conflicts of interest. Harding, whose firm cross-examined Cherry repeatedly, called him "a gold mine, my best witness. . . . it would have turned any jury's stomach. He'd have affidavits saying one thing in one case and testimony saying another in another case. And he was getting millions in unrestricted grants from Lederle and Wyeth."[77]

Most of the cases settled out of court, often at the insistence of insurers. "It's perfectly simple," Mortimer said in 1984. "The kid is pathetic, the manufacturer has a hell of a lot of money. The manufacturer goes in and tries to defend it and no matter what kind of expert testimony you get, the kid comes in and the jury cries and the judge cries and I cry and everybody cries. They all feel terribly bad for the kid."[78] It was easier, and cheaper, to settle. It was starting to become clear that the companies would stop making DTP unless they got relief. Though the Reagan administration didn't want to face it, the free market was not going to take care of this problem. If the vaccine makers had screwed up, by the logic of the market they would have to go out of business. But if they went out of business, the whole country would be at risk from infectious diseases. Someone had to intervene. The invisible hand had arthritis. The Public Health Service was doing its feeble best to get an improved acellular vaccine off the ground, but it wasn't easy with budget cuts and the AIDS crisis to deal with. At a March 1985 hearing before Congressman John Dingell of Michigan, NIH and CDC officials explained that while they had put out a call for proposals on a new improved acellular vaccine, only the Michigan State Public Health Laboratory had responded. The NIH's own scientists, including Robbins, had been working on a vaccine for years, but "it's a little like pregnancy," an HHS official explained. "You would like to deliver that infant 2 months after conception, but you have to wait the whole 9 months."[79] Dingell, who kept thousands of officials on their toes, said it didn't sound like they were trying very hard.

Later that year, a team of public health scientists flew to Japan to see about importing the Sato acellular pertussis vaccine. It was a frustrating trip. The American scientists were able to assemble data showing lower local and febrile reactions to the Sato vaccine, which by then had been administered 20 million times. But serious adverse events were no less common than they were with the whole-cell pertussis vaccine, since the children weren't being vaccinated until the age of 2—long past the age when infants tended to suffer severe medical events after whole-cell pertussis. Moreover, the six vaccines had different ingredients and manufacturing processes, but the Japanese had not broken down clinical data on their effects. "About midway through the visit," said Waxman aide Tim

Westmoreland, who went to Japan on a congressional staff trip, "it dawned on all of us—like the naïve Americans in Graham Greene or Henry James novels—that the manufacturers and government in a cartel-like way had divided up whose vaccine was sold and where, and then the next year they rotated, and two years later they rotated again. . . . There was some suggestion some vaccines were better, but they didn't want to know."

In short, it wasn't as easy to get the Japanese vaccine as the parents' groups, who bitterly denounced the Public Health Service's failure to import it, would have liked to believe. "It was very difficult to get efficacy data from Japan," said the FDA's Parkman. "We kind of want to have studies that show it's effective! I won't knock their system, but it's not ours." To provide the kind of experimental trials necessary to impress the FDA, it would be necessary to hold them in a country with low rates of pertussis vaccination, so that the vaccine's effectiveness could be tested where the germ circulated freely.[80] Sweden had not vaccinated against pertussis since 1979 and vaccination was not mandatory in Germany or Italy. The National Institutes of Health funded studies of 13 different vaccines from eight companies that contained varying amounts of pertussis toxin, filamentous hemagglutinin antigen, and two other ingredients, pertactin and fimbrae, which were cell-membrane proteins.[81]

In probably the best assessment of the trials, Plotkin and Michel Cadoz, of Aventis-Pasteur, concluded that there was a small but significant improvement in efficacy in the vaccines that had three or more proteins, compared to the single- or double-component vaccines.[82] They also found that all the vaccines' efficacy was limited. Most of them prevented severe pertussis, but they were less reliable at preventing infection and cough altogether. All the acellular vaccines were less effective against mild pertussis than the best whole-cell vaccines. "Public health authorities are thus faced with a difficult choice," wrote Plotkin. "Should the better efficacy of certain whole-cell vaccines be traded in for the better tolerance of acellular vaccines? The answer may vary in different parts of the world."

In the United States, of course, the answer was that a less-reactive vaccine was highly desired.[83] The trade-off, however, may have been less potency in preventing the disease itself. Perhaps most distressingly, despite

the $26 million the U.S. government poured into the acellular vaccine stud-
ies, the arrival of the best of the newer vaccines was significantly delayed.[84]
Acellular vaccines were introduced for older American children in 1992,
and for infants starting in 1996. But the vaccines that generally proved
strongest in the trials were not marketed in the United States until 1998 and
2002, respectively, partly because of a patent dispute with Chiron, whose
Italian branch owned the rights to refined pertactin protein.[85] "The sad fact,
despite the expenditure of millions of dollars," two pertussis experts wrote,
"is that the most efficacious vaccine available in the United States today is
one of the two generally available whole cell vaccines."[86]

Finally, a Compensation Program

While research on vaccines crept forward, the battle over the vaccine
compensation program continued in Congress for three years. A key
breakthrough came when Merck, which didn't make DTP, decided to
back the bill in order to protect its MMR vaccine, which produced a limited but consistent number of adverse events. Lederle actively opposed
the bill, but Connaught eventually came around, and that was enough to
satisfy most House Republicans. The measure was reintroduced repeatedly until finally being inserted into a "bulletproof" package—an omnibus
that included Senator Orrin Hatch's legislation for the export of unli-
censed pharmaceutical products and a measure to protect South Carolina
cotton mills, inserted by Reagan ally Strom Thurmond.[87] Medical pop-
ulists of both parties worked together. Al Gore, the Tennessee Democrat,
praised Hatch of Utah for "fighting to preserve some remarkable achievements on behalf of the American people."

The longtime Speaker of the House, Thomas "Tip" O'Neill was
retiring on October 18, 1986, the day the bill finally arrived in the
Speaker's lobby. It was missing a crucial sheet of paper, which sent staffers
racing back to the Senate to fetch it. As O'Neill made the rounds of the
floor, hugging his friends and weeping and preparing to gavel out the session, Waxman suddenly got up and said, "I send a bill to the House!" The
AMA and the administration were still opposed to the bill, so parents,
pediatricians, and even the pharmaceutical industry joined to push for
Reagan's signature. *The New York Times, The Chicago Tribune,* and other

newspapers editorialized in favor, and the bill brought as many phone calls and letters into the White House as any in the Ninety-ninth Congress. The American Academy of Pediatrics happened to be meeting that month at the Sheraton in Washington; Smith, its president, got 3,000 physicians to call Attorney General Ed Meese and Reagan.[88] Fisher, Williams, and 75 other parents held a candlelight vigil in front of the White House. On November 14, Reagan reluctantly signed.

The men and women who worked on the bill remember it with nostalgia as a great bipartisan effort for the common good, something that sounded almost unimaginable 20 years later in a sharp-elbowed, viciously partisan Congress. "I never thought this was the greatest thing in the world, but I thought that we were changing public consciousness," said Schwartz. "To keep whole new generations from being rushed roughshod into things like smallpox vaccination, regardless of the need and the risks."[89] The bill was not perfect, however, and some, like Schwartz, felt that important glitches were never worked out. Eventually, February 2, 1991, was set as a deadline for parents to file claims from injuries that occurred before 1988. A total of 4,000 were filed, many at the last minute; the chief clerk at the federal Court of Appeals got two or three FedEx trucks full of applications a day for a week. People with crippled kids in tow arrived at Kathi Williams's family business to use the Xerox machine. Some of the claims, naturally, were frivolous, or worse. There were cases of child abuse posing as vaccine injuries and unsubstantiated links to undocumented vaccinations years earlier. One lady came in with her dog, claiming it became stupider after getting a rabies shot.[90]

Initially, the court worked well for petitioning parents. Recognizing the extreme difficulty of determining whether a vaccine had caused an injury, Congress had called for a "benefit of the doubt" standard, which helped claimants. One attorney would later acknowledge that at least a third of the injuries presented were not caused by DTP.[91] More than 90 percent of infants whose deaths within a day of vaccination were defined as SIDS won $250,000 each. The special masters, federal judges who presided over the court, often found that crying, irritability, or sleepiness following DTP was evidence of encephalopathy—meaning that a child who later suffered ADHD might win thousands of dollars by associating it to early episodes of crankiness.

The Deathless Debate over DTP

In the years since the dispute over the pertussis vaccine first began, public and expert opinion have sharply diverged. While media coverage convinced more and more people that pertussis vaccine was unsafe, the growing consensus of pediatricians was that many bad events following vaccination were coincidental. This was largely based on several epidemiological studies showing that severe reactions to pertussis were rare, that SIDS occurred less often than usual in babies who'd recently gotten DTP, and that seizure disorders like the one that killed Julie Middlehurst were just as common in unvaccinated five month olds as in those who had recently been vaccinated.

Vincent Fulginiti's evolving position was perhaps typical of the shifting medical view of whole-cell pertussis vaccine. In 1976, Fulginiti raised the possibility that pertussis vaccination should be ended. The problem, he wrote, was the difficulty in knowing whether the vaccine was responsible for postvaccine events.[92] A six-month-old in his practice had suffered prolonged convulsions following her third DTP shot. A virus was recovered from the child's throat and feces. Was the episode viral or was it related to the pertussis shot? It was impossible to know. Although the common wisdom stated that serious complications were infrequent, "the absence of adequate field data makes such assumptions feeble. . . . it is inconceivable that we can steadfastedly recommend and employ pertussis vaccine without a parallel commitment to resolve the outstanding issues," Fulginiti lamented. Yet by 1983, he had become convinced that the whole-cell pertussis vaccine was responsible for no more than 1 in 100,000 severe complications.[93] And seven years later, as head of the American Academy of Pediatrics' infectious disease committee, known as the Red Book Committee, Fulginiti had come full circle. Now he agreed with Cherry and Mortimer that "so-called pertussis vaccine encephalopathy does not exist. . . . Why does belief persist?" he asked. "Because when neurological events occur in this age children, parents seek some explanation, and it's easier to blame the vaccine than their own genes or chance."[94]

The investigation of the DTP shot plunged researchers into the recondite field of childhood brain disorders. It was undisputed that pertussis disease caused brain damage, but some believed this was merely because coughing fits deprived the brain of oxygen. Toxins in both pertussis and whole-cell pertussis vaccine clearly excited brain cells and in doing so caused convulsions and shock. About 1 in 1,000 children who got the DTP shot suffered one or the other. But were such events, in the immediate aftermath of vaccination, capable of causing long-term damage? Vaccinated or not, something like 1 in 20 babies suffers a febrile seizure at some point in infancy. Such seizures are almost always benign. If they aren't benign—which is to say, if a child has a seizure and then goes on to have others, until being diagnosed as epileptic—they are called complex febrile seizures. Seizures without fever are often referred to as epileptic fits. Many seizures of all kinds are triggered by illness or fever. Using this logic, most vaccinologists came to the position that DTP could not be *blamed* for seizure disorders. Yet, a child who suffered a seizure after DTP might otherwise have been much older before a triggering event occurred. It was possible that the age of the child at the time of the triggering event affected his long-term health. Was it possible to admit that DTP triggered an event, and yet not to blame DTP for at least a share of a child's ultimate condition?

For leading pediatricians and public health officials who believed fervently in the importance of vaccination, the vaccine court—in particular, the injury table—did not reflect the established science, and they set out to change it. In searching for evidence to back their claims, both sides would rely heavily on the largest study of the pertussis vaccine safety question. This was the National Childhood Encephalopathy Study, or NCES, which recorded virtually every incident of childhood brain disease in England, Scotland, and Wales from 1976 to 1979, and eventually followed the children until 1993.[95] Whenever the authors of the NCES found a child who had suffered a neurological event, they established how recently the child had been vaccinated against pertussis and chose two random controls from the general public. The study design allowed the authors to determine what percentage of convulsions and more serious brain damage were linked, at least in time, to pertussis vaccination. The study found 1,182 children who had entered the hospital with neurological illness, 39 of them within seven days of the DTP shot. Of the 39, 7 died or were permanently damaged, while none of the controls suffered permanent harm. That established a risk factor for death or serious injury from pertussis vaccine at roughly 1 in 300,000 shots—which might translate into roughly 1 in 100,000 children in the United States, since they had three shots by the time they were 18 months old. While there was a large margin of error surrounding that number, it became the gold standard for risk from the whole-cell vaccine. DTP could cause brain damage, in other words, but such reactions were rare.[96]

Britain saw several highly publicized lawsuits against DTP, culminating in the case of Johnnie Kinnear, a 15-year-old with mental age of 20 months, allegedly due to pertussis vaccine. His case unraveled when contradictions were exposed between his parents' testimony and hospital records. Even more important was the Lovejoy case, heard in 1986 by Judge Stuart Smith. In a judgment that ran 14 chapters, Smith found that three of the seven DTP cases in the national encephalopathy study were not brain-damaged. Of the remaining four, two probably had viral origins, and a third was probably a case of Reye's syndrome. The Smith judgment officially ended the DTP controversy in England, where whole-cell pertussis vaccine continued to be used until 2004, with scant controversy.

In the United States, the Food and Drug Administration funded its own study of pertussis vaccine toxicity in the late 1970s.[97] It was far smaller than the British encephalopathy study and the vaccine being tested wasn't the same, since the U.S. pertussis shot had twice as many killed bacteria in it. In a study of 15,752 DTP immunizations, the authors, including Manclark at the FDA and Cherry and Larry Baraff at UCLA, found nine convulsions and nine cases of shock—and none in the control group, who received a diphtheria-tetanus shot that contained no pertussis vaccine. The study's methodology guaranteed that it would show less toxicity than the NCES because children who previously had experienced severe adverse reactions to DTP were excluded from the study and only deaths within 48 hours were counted. Thus, the death four days later of a child who became sleepy and developed diarrhea after getting her second DTP shot was not counted as associated with the vaccine. A follow-up of the 18 children with severe reactions found they were all "considered normal by their parents."[98] But psychometric tests found four with "minor neurological abnormalities," including a seizure disorder, speech problems, and language delay. Seven had low IQ. The authors attributed this to the fact that most were low-income Hispanics.

Later studies showed that children with family histories of seizures were at increased risk of neurological events, primarily febrile convulsions, after DTP.[99] But while this contraindication to vaccination was generally accepted, some doctors persisted in arguing that triggering did not mean the vaccine was responsible for a neurological event.[100] The Institute of Medicine (IOM), whose mission is the impartial resolution of medical controversies, relied heavily upon the British encephalopathy study in its 1991 report. The IOM committee was unable to determine whether pertussis vaccine caused permanent brain damage. But in a 1994 report it maintained that the balance of evidence suggested an *association* of whole-cell pertussis vaccine and long-term brain damage.[101] Based on its work, the committee came up with three scenarios for DTP-associated injury:

1. DTP caused brain damage.
2. DTP "triggered" an acute illness linked to an underlying problem; the harm could still be said to have been caused by DTP because the earlier such events occurred, many neurologists believed, the more damage they caused.
3. Children with underlying abnormalities might respond acutely to DTP without its being the cause of long-term brain damage. In other words, the underlying problem predisposed the child to react to DTP but that reaction did not trigger the long-term problem.

The committee determined that the data available "do not allow a distinction to be made among the three scenarios," and therefore "the balance of evidence is consistent with a causal relationship"—though a rare one—between DTP and brain damage.[102] Most of the scientists and doctors who followed the controversy accept the IOM's judgment as the bottom line. DTP, apparently, rarely caused brain damage.

Would it be possible to have a child who fretted all day after her DTP shot, then began to have seizures, and believe that the seizure disorder she suffered the rest of her short life was not a product of DTP? People tended to revert to the bedrock of their own direct experiences. But they also interpreted those experiences through a pre-existing mindset. My doctor's son suffered an episode nearly identical, in her description, to the one that Barbara Loe Fisher described in her son Christian. But while my physician's son, like Fisher's, grew up to be somewhat learning disabled, my physician did not associate this problem with the DTP reaction. A study of 215 children who suffered shocklike episodes after DTP vaccination found that a year later, one was autistic, a second was epilepetic, and a third had developmental delays.[103] It wasn't clear whether there was any connection to the shot in any of the three cases. The dry, clinical language of epidemiological studies left a huge vacuum into which individual parents poured their own interpretations. There was no way for an individual parent to know with certainty whether there was a causal relationship between a vaccination and the course of a child's mind. But with diagnosis of conditions like ADHD increasing, many parents were in the market for explanations.

"To say outright that it's not possible to have a brain injury after DTP, I just think that's foolhardy and reckless," an NIH scientist deeply involved in the controversy told me.[104] "The whole cell was a real dirty vaccine. The main reason it stayed on and was used was that it worked. But it was dirty enough that given the right situation—and there are a lot of vulnerable kids out there who we don't know who they are—I think it's possible it could induce severe fevers. How can you say it's not possible that somebody who had a fever of 106 didn't come down with an encephalopathy?"

In late 1991, the issue was coming to a boil as pediatricians and the government pressed to remove encephalopathies from the table of injuries that the vaccine court assumed were caused by DTP. Gerald Fenichel, a Vanderbilt pediatric neurologist and world expert on childhood seizures, resisted. "It is for the good of the program for the public to think they have been treated fairly," he said at a meeting where the injury table was being discussed. "There is certainly a public perception that the pertussis vaccine produces encephalopathy." Chairing an acerbic debate between Johns Hopkins professor Neal Halsey and Marie Rodee, the mother of an allegedy vaccine-injured child, Fenichel noted that "this program came to be because of the support of mothers like Mrs. Rodee and the American Academy of Pediatrics and requires public support for its continued existence and health. We have not and cannot show that pertussis vaccine does not produce chronic disability. It is not possible to show an event that may occur one in a million times. There is a clear feeling among a lot of people that this is something that does occur. . . . It is reasonable in such a system to compensate some number of people who perhaps have not been injured by the vaccine, to make the program work, and it is this program which I am supporting rather than a narrow scientific viewpoint."[105]

Despite Fenichel's argumentation, the Health and Human Services Department four years later removed almost all DTP injuries from the list that received automatic awards. Whereas in the past it had been too easy, perhaps, to win recognition of a vaccine-caused brain injury, after 1995 it was probably too hard. In the typical case, a clinician would testify that a child in perfect health got the DTP vaccine and became ill, and now was severely retarded. But more likely than not, the child's legal team would lack an accepted scientific theory for what happened. The government, on the other hand, had experts with résumés 50 pages long to testify that the epidemiological record did not support a finding of injury. The special master would be forced to turn down the claim.

By 1998, when I did a story on the court for *The Washington Post Magazine*, the government had "altered the game so that it's clearly in their favor," Gary Golkiewicz, the chief special master, told me. "This group has a vested interest in vaccines being good. It doesn't take a mental giant to see the fundamental unfairness in this." Roughly 30 percent of the children whose cases were adjudicated by the court from 1989 to 2004 won an award. Those who lost would end up qualifying for many of the same benefits under Medicare or Medicaid. But their parents would take home bitterness, rather than a feeling that justice had been done. Their child's soldierly sacrifice had not been recognized.

CHAPTER 8

—

NO GOOD DEED GOES
UNPUNISHED

I work on a virus that kills more people in a day than SARS ever killed.

— ROGER GLASS, ROTAVIRUS RESEARCHER, 2003

THE WINTER OF 1990–91 was a terrifying season for Dr. Robert Ross, deputy city health commissioner of Philadelphia. The biggest measles epidemic in two decades was sweeping the United States, most of its victims poor, unimmunized black and Hispanic babies and toddlers in big cities. The authorities made reasonably sure that schoolchildren were vaccinated, but vaccination rates were abysmal for the truly young, and if any virus could find the vulnerable child it was measles, an explosively contagious disease. Ross spent the winter seeking out sick children, isolating and quarantining their siblings and peers, and vaccinating all he could. And then in February, Ross and his colleagues stumbled upon a staggering epidemic within the epidemic—a Christian sect whose members would fight tooth and nail to prevent their children from being vaccinated or receiving medical treatment of any kind—even in the face of death.

Which is how Ross, a healthy six-foot-two man, found himself wrestling with a little girl's five-foot-one grandmother.

The nationwide outbreak had many flashpoints. In Philadelphia, a college student returned from Spain with a case of measles in late 1989, and it quickly spread. For most parents in the city, the epidemic was enough incentive to get their children vaccinated. That was not the case for the Faith Tabernacle Congregation, a church founded in 1875 by a doctor who had quit his profession when God's will was suddenly revealed to him.[1] Members of this church, based in North Philadelphia, shunned all medical care. That put Philadelphia health authorities in a difficult position. How were they to protect the health of these children without trampling on their parents' religious faith? There was a clear legal precedent for intervening to protect unvaccinated children during an epidemic. In *Prince v. Commonwealth of Massachusetts*, a 1944 case, the Supreme Court had said that "parents may be free to become martyrs themselves. But it does not follow that they are free . . . to make martyrs of their children." There were a variety of measures the health department could take, but it had no wish to stomp on a religious group.

Ross was tipped off to the measles outbreak in the Faith Tabernacle Congregation by an anonymous "Deep Throat," a woman worried about her unvaccinated two-year-old grandson, whose parents were congregants. Ross and a colleague visited Pastor Charles Reinert at the minister's office in Nicetown, which in spite of its name sounding like it had been lifted from a PBS kids' show was a drab, working-class neighborhood. Reinert was white, like most of his congregation, and in his 70s. During the meeting his office shades were tightly drawn. "We were in the dark," Ross recalled, "and that would also serve as a metaphor for this religious sect because they were so closed off." Reinert made it clear that the church believed in the power of prayer and that all medical care, including vaccination, was out of the question. The next morning, the coroner reported that a two-year-old had died of measles, and Ross traced the child back to Faith Tabernacle. He returned to Reinert and demanded the phone numbers of church members. "You won't change their minds," Reinert told him. And indeed, when a few days later measles claimed a second child in the sect, Ross realized that he had spoken with the girl's mother a few days before, but hadn't grasped the seriousness of the child's condition. "She told me the kid was fine, that all her kids were fine," Ross recalled. "Fine" clearly meant something different to people who believed everything was in God's hands. The health department ramped up its control measures, tactfully. Now it made home visits, promising not to "lay a hand" on the sick children unless their lives were in danger, in which case a judge was standing by to order hospitalization.

On a cold February day, Ross walked into a sparsely furnished house in Nicetown. The parents were out at work and the grandmother didn't want to let Ross in but he insisted. She lined up eight children for inspection on sofas and chairs in the living room. They looked all right, Ross thought. "Is that it?" he asked. The grandmother responded yes—except for Sarah, who was "sick but doing better," upstairs watching TV. Climbing the stairs and stepping into a bedroom, Ross encountered a ghostly pale eight- or nine-year-old girl taking deep, rasping breaths. "She looked like she was ready to code right there. Her eyes were sunken. It was clear she had a combination of pneumonia and dehydration." Ross said the girl needed to go to the hospital immediately, and he picked up the phone to call the judge, but the grandmother grabbed him. "This 90-pound grandmother was physically trying to restrain me from using the phone," he recalled. "She was shouting, 'Don't you believe in the power of prayer? Don't you believe the Lord Jesus takes care of all of us?'" Ross fended her off, bustled the girl into an ambulance and escorted her to the hospital, where by the next day she was recovering from pneumonia in both lungs. "I went to visit her, and I guess I was half expecting the family to say, 'Thank you, doctor.' Opposite reaction. They were furious. And there's no question in my mind that modern medical care saved that child."

Ross and other health officials made more than 60 visits to sect members' houses—a far cry from the vaccine raids made by Benjamin Lee and his enforcers during the smallpox epidemic 90 years earlier in Philadelphia. Then, a defensive medical establishment had maintained that its vaccine was perfectly safe and the only effective weapon against smallpox. Now, health officials acknowledged that imperfections in the measles vaccine were part of the cause of the outbreak. But they also knew that unvaccinated people were almost certain to catch measles if exposed to it, and they used quarantine, rather than forced vaccination, to protect them. In 1901, police and health workers had broken down doors to forcibly vaccinate recalcitrants. In 1991, health workers exercised their police functions gingerly. Armed with court orders, they eventually ordered nine dehydrated children to be hospitalized, and vaccinated 30 or so medically vulnerable kids while their parents wailed and moaned in the hallway outside. They closed two church schools but ordered no wholesale vaccinations.

Despite prayer and the health department's soft touch, the virus would eventually infect more than 480 Faith Tabernacle children, killing 5 of them. Four other Philadelphia-area children died, and a total of 1,400 people fell ill in the city's worst measles outbreak since 1954.[2] "We lost children we never should have in that epidemic," Ross said. "That still haunts me to this day."

The epidemic took authorities by surprise everywhere in the United States. Measles had been declining steadily since the first vaccine was licensed in 1963. In 1983, about 1,500 cases were reported, the lowest total up to then. But during the 1989–91 epidemic more than 55,000 cases were reported, and the deaths included 49 children who were 5 or under. Massachusetts, interestingly, had fewer than three dozen cases during the epidemic. True to a 270-year tradition that the Puritans had started during the great smallpox epidemics, Massachusetts was one of the few states with universal free vaccination.[3]

Though it wasn't widely noticed at the time, with the nation readying itself for the first invasion of Iraq, the return of measles marked the crest of a comeback for infectious diseases in the United States. Rubella also peaked, with more than 1,000 reported cases including 11 congenital rubella babies in 1990, up from a single case in 1988. Whooping cough was climbing—4,138 cases in 1990 compared to 2,823 in 1987. And tuberculosis was on the rise, its spread facilitated by immune systems weakened by its grim younger brother, AIDS.

The Faith Tabernacle Congregation was the locus of Philadelphia's measles story. But the epidemic hadn't started in the church, and once city officials looked beyond the religious exemptors—fewer than 200 of whom attended the city's public schools—a much larger immunization problem came into focus.[4]

Measles had traditionally been a disease of elementary-school-aged children. In the days when everyone got it, infants and to some extent toddlers were protected by maternal measles antibodies, passed along through the placenta and breast milk. In the 1989–91 measles epidemic, however, half the cases turned up among preschool and younger children. Little children no longer had the protection of maternal antibodies, because their mothers, unlike their grandmothers, for the most part had never gotten measles. The little children were also the most vulnerable to death, deafness, and brain damage from measles infection. Given this sobering change in the disease pattern, the nation's immunization program, it was clear, needed an overhaul.

The immunization campaigns of the 1970s had created school-entry laws around the country. "No shots, no school" meant that schoolkids tended not to catch measles. But the vaccine failure rate, estimated at 5 percent in the 1980s, was higher in those who'd been vaccinated before 1979, in the era of killed-measles vaccines and the first-generation live-measles vaccines, which tended to lose efficacy in storage. School-entry laws were of no help when it came to persuading parents of babies or preschoolers to vaccinate them. And when public health people did surveys it turned out that a shocking number of these children remained unvaccinated—out of parental ignorance, fear, disdain, poverty, or laziness. Although some states had vaccine requirements for entry into licensed daycare facilities, there were no other sticks, and few carrots, encouraging parents to vaccinate babies. In Philadelphia, in 1991, city officials estimated that 35,000 of the city's 115,000 preschoolers were unvaccinated. That was about par for the nation. The Reagan administration had stopped funding vaccination surveys in 1985,[5] but a Gallup report said only 28 percent of two-year-olds were vaccinated in 1991. The rates were extraordinarily low in some cities. Only about 10 percent of babies in Houston were vaccinated, for instance.[6]

Vaccines for Children

What accounted for this abysmal record? One cause was the DTP liability crisis and the ensuing jump in vaccine prices. Immunization budgets grew an average of 25 percent per year from 1981 to 1991, but the money was absorbed by rising vaccine costs. The price governments paid for the basic line of pediatric vaccines for a single child ballooned from $6.69 to $91.20 over the period. Because many insurers did not cover the costs, more and more physicians were sending their patients to public health clinics for their vaccinations. But public health clinics couldn't pick up the slack. They were closing or shrinking their hours in the new budget-cutting era, a fact that Henry Waxman, then chair of a House health and environment subcommittee, was quick to point out. "Parents want to vaccinate their children," Waxman said at a 1992 hearing. "But there are too many burdens." Long lines at understaffed clinics and inconsistent requirements for physical exams before kids got their shots dissuaded thousands of parents.[7] A 1991 survey by the Children's Defense Fund found that more than two-thirds of the nation's 540 community health centers had vaccine shortages.[8]

Sick, unvaccinated kids brought to emergency rooms were acquiring measles and passing it along at alarming rates. As many as 300 children were exposed to the virus by a single child waiting at an emergency room in southern California. As a measure of the severity of the disease and, perhaps, the failure of primary care for many children, perhaps 25 percent of California cases ended up in the hospital.[9] Pediatricians were incensed. "It is a national disgrace," said Dr. James Strain, president of the American Academy of Pediatrics. U.S. vaccination of its under-2 population was lower than in any country in the Western Hemisphere besides Haiti and Bolivia. "This isn't a measles problem. It's a systems problem," said D. A. Henderson, then a health aide to President Bush.[10]

Thinking about vaccination had taken a vacation during the Reagan presidency. As the measles data and deaths rolled in—and the 1992 presidential election approached—officials from Bush on down realized they had to do something about it. Within two years, the federal government would provide an enormous boost in funding and manpower for the program. This push would not only lead to the rapid eradication of homegrown measles in the United States but would also ensure that much more vaccine would be injected into children and would set the stage for the introduction of five new required immunizations in the 1990s.

In response to the epidemic, the Bush administration threatened to cut off benefits for welfare recipients who could not prove their children were vaccinated. CDC Deputy Director Walt Dowdle later clarified that the program would use welfare offices as a noncoercive point of access to kids.[11] Bush held a Rose Garden gathering to promote vaccination in June 1991, and senior public health officers visited six U.S. cities to help them develop vaccination plans. But Bush's promise of $91 million for additional vaccines failed to materialize, and when he hosted another

Rose Garden event a year later, Democrats derided him.[12] Even Sam Katz, the federally appointed chairman of the CDC's vaccine advisory committee, charged that the government had "finked out and did not follow through" after the first "Rose Garden show."[13]

At the CDC's urging, states began to create immunization action plans to meet the agency's goal of 90 percent vaccination of two-year-olds by the year 2000. Undoubtedly the most active official in these efforts was Walter Orenstein, who had become director of the CDC's immunization division in 1988. When the Clinton administration took over, Orenstein got his chance to right the nation's "grossly inadequate" vaccination levels. One of the administration's first initiatives was to promise that the government would purchase enough vaccines to immunize all the children in the United States free of charge. The pharmaceutical industry bitterly opposed the plan, as did some lawmakers and child health advocates, who feared that low government-set prices would drive more pharmaceutical companies out of the vaccine business. But unlike the rest of the Clinton healthcare agenda, a good chunk of the vaccine initiative survived as the Comprehensive Childhood Immunization Act, which passed in August 1993 as part of an omnibus spending bill. It included a $585 million federal entitlement and required federal purchase of vaccines for all children on Medicaid or lacking insurance coverage for immunization.[14] Though his sails had been trimmed a bit, the first Democratic president in 12 years would oversee the largest expansion of the U.S. vaccination program in its history.

Vaccination-based public health is an area in which market mechanisms alone are insufficient—or at least, there is no place in the world where the market alone has worked. Simply put, vaccine-induced immunity is an imperfect good; vaccines don't always work, and the duration of the immunity they provide is variable. An individual cannot count on lifelong immunity even if he or she is fully vaccinated. A critical mass of his neighbors have to be vaccinated, too, in order for immunity to hold. If the neighbors couldn't or wouldn't pay, someone else would have to pay for their vaccinations or the system wouldn't work. This principle was easier for Democrats to embrace than Republicans. But generally speaking, vaccines were a popular public health measure and Congress had been unwilling to abolish the programs entirely. So while vaccination programs languished somewhat during GOP terms, they were always hardy enough to outlast dry spells. The political dichotomy disappeared when disease outbreaks occurred. Fear could be counted upon to stimulate financing under administrations of any stripe, because epidemics, which struck a population's "readiness," were a matter of "national security."

Amid the outcry over Salk vaccine shortages, the Eisenhower administration had put up a few million dollars to purchase polio vaccine. When Kennedy signed Public Law 317 he expressed much loftier goals.[15] There was no reason, he said, that any American child should die of smallpox, diphtheria, polio, pertussis, or tetanus. Initially, the funding of PL 317 was supposed to be a one-off affair. But you couldn't expect to permanently whip any of these diseases just by sweeping in to vaccinate and then leaving, so the bill became a mechanism for purchasing vaccine for local health departments. But there was still not enough money to support the infrastructure of immunization, though, such as clinic staffing.[16]

In the 1980s, the DTP crisis made the need for government intervention clear, and the measles epidemic cemented that perception. Alan Hinman and other public health leaders who visited Philadelphia, Detroit, San Diego, Dallas, and Rapid City (North Dakota) during the epidemic heard complaints about the lack of funds for clinic hours and staff positions. Vaccination leaders ordered the state and local health departments to come up with comprehensive immunization plans. By the time these plans were ready, the Clinton administration had taken over, and it eagerly embarked on a major overhaul of the program. In 1994, Congress enacted the Vaccines for Children program, which obliged the federal government to finance vaccinations for children under 19 for all vaccines recommended by the Advisory Committee for Immunization Practices. States weren't required to implement the recommendations. But no state wanted to lag behind in disease reduction. Whether or not it was worth vaccinating against chickenpox, at the end of the day South Dakota would be embarrassed by chickenpox epidemics if North Dakota no longer had them. Mandating a recommended vaccine also brought money into a state's immunization program.

Laying Down the Law

The job of driving forward this suddenly vast, complicated structure fell to Orenstein, a slight, curly-haired New Yorker with a pebbles-in-the-mouth Bronx accent. Orenstein, who attended the Bronx High School for Science, followed Salk's footsteps into City College and graduated from Albert Einstein Medical School as a 23-year-old wunderkind. He was also among the youngest veterans of the elite lesson in vaccinology that was the global smallpox eradication campaign. Since the late eighteenth century, senior Public Health Service scientists have held military rank, and they are obliged to wear their uniforms at least once a week. For some CDC scientists, who generally dress with distracted schlumpiness, the crisp uniforms seem more of an embarrassment than an honor. One didn't get that sense from Rear Adm. Orenstein. An affable, good-humored fellow who liked to eat barbecue when his kosher-keeping wife wasn't watching, Orenstein seemed comfortable with imposing a soft-spoken yet firm authority born of precocity and conviction. He had a way of lifting his chin, il Duce–like, when making a pronouncement about measles or hepatitis. In Orenstein's office at Emory University, which he joined in 2004 after leaving the CDC, a big caricature, a departing gift from colleagues, depicts him sitting in uniform at his desk, a bright grin on his face and pen in hand. "I'm very concerned," he is saying; "This is a good start," and "Can you doublecheck that?" He ran a businesslike ship at the CDC, and by extension in pediatricians' offices around the nation.

Orenstein learned about the importance of tight ships in Uttar Pradesh, India, where he spent four months from December 1974 to March 1975. Working under Don Francis, a CDC official who would later create the world's first promising (though ultimately failed) AIDS vaccine as president of Vaxgen in San Francisco, Orenstein and his colleagues were able to choke off smallpox by strictly implementing control procedures. They stationed guards at the doors of smallpox patients and told them not to let anyone unvaccinated in or out. They conducted careful censuses and returned to villages to vaccinate the men who'd been working in the fields the first time they came through.[17]

It was a job of dotting I's and crossing T's, although the setting was

Polio, post vaccine. A boy on crutches is pictured at the threshold of his schoolhouse as he returns from the first time after spending eight months in therapy for polio, in the late 1950s. Sadly, the boy was not vaccinated against polio although the Salk vaccine was available. (Courtesy of the Library of Congress)

Albert Sabin, Jonas Salk, and Basil O'Connor, at a polio conference in 1961. Though making nice for the camera, Sabin was generally at odds with Salk and O'Connor. A more mature, well-respected scientist, Sabin patronized Salk, once joking at a congressional hearing that the latter's formaldehyde-killed polio shot was a "horse and buggy" compared to his race-car oral vaccine. (Courtesy of the March of Dimes)

Sabin and a true friend: Albert Sabin is pictured holding one of the 20,000 chimpanzees he personally injected with polio virus in experiments to develop the oral polio vaccine. Sabin's vaccine, which has been used billions of times, in 2007 was close to eliminating polio worldwide. (Courtesy of the Library of Congress)

The late Maurice Hilleman (left), here with Merck scientists Beverly Neff and Eugene Buynak, had a hand in developing most of the vaccines that have gone into use in the United States since the early 1960s. (Courtesy of Merck & Co.)

This half-empty Cleveland classroom, during a 1964 measles epidemic, shows the work that Hilleman's new vaccine had cut out for it. (Courtesy of the Library of Congress)

Harry Meyer (left) and Paul Parkman, government scientists in the vaccine regulatory division of the National Institutes of Health, in 1967. Meyer and Parkman's rubella vaccine was later adapted and sold by Merck. The arrangement raised some eyebrows, but the public health world was happy to have a means to prevent another outbreak of congenital rubella syndrome. (Courtesy of the National Library of Medicine)

Mechanical stars of *Star Wars* helped the CDC urge parents to vaccinate their infants in this 1977 poster. Despite widespread vaccination, public health officials battled sporadic measles epidemics in the United States until local transmission of the virus ended in 1998. (Courtesy of the Centers for Disease Control)

The twin on the left has a puffy eye, caused by inadvertent infection with virus from a vaccination on her sister's arm. Such side effects were not uncommon in the era of widespread smallpox vaccination. When the vaccine was brought back in bioterrorism preparedness in 2002, officials were much more careful to shield the vaccination wound—and children were generally not vaccinated. (Courtesy of the California Department of Health Services)

From right to left, Michael Lane, William Foege, and Donald Millar, three former directors of the CDC's smallpox eradication campaign, hold a magazine announcing the death of smallpox in 1980. (Courtesy of the Centers for Disease Control)

Walter Orenstein (third from right) in 1996 with then-CDC director David Satcher (third from left) and other colleagues. Orenstein cut his teeth as a vaccinologist in the smallpox eradication campaign. Later he would head the national immunization program for 15 years. (Courtesy of Walter Orenstein)

Jeffrey Schwartz and his daughter Julie Middlehurst-Schwartz. Schwartz and other parents, convinced their kids were hurt by DTP vaccine, organized to demand changes in the vaccination program after the airing of an investigative TV report in 1982. Julie, Schwartz's first-born child, died at age 3 while suffering her final epileptic fit. (Courtesy of Jeffrey Schwartz)

Whooping cough is back! Or did it ever leave? The waning protection of the vaccine and the difficulty of diagnosing whooping cough in adults allowed the bacterial disease to maintain a foothold throughout the country. In the new millennium, a combination of better diagnostic tools, improved surveillance, and pockets of unvaccinated children helped spawn a major comeback of the disease, leading the CDC to urge revaccination of teenagers and adults. (Courtesy of the National Library of Medicine)

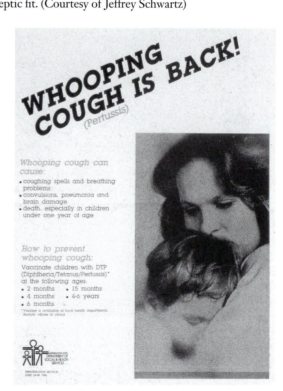

Barbara Loe Fisher (second from right) at Power of One rally, Washington DC, 2005. After creating Dissatisfied Parents Together (DPT) with Jeffrey Schwartz, Kathi Williams, and other parents in 1982, Fisher coauthored the book *DPT: A Shot in the Dark*, which described the well-known problems with the pertussis shot. Fisher and Williams later renamed their organization the National Vaccine Information Center, and continue to lobby for research into vaccine dangers. (Courtesy of author)

Power of One rally, Washington DC, 2005. Parents at a rally on the lawn of Congress in 2005 demanded that the government recognize the harm to their autistic children by vaccines containing thimerosal, a mercury-containing preservative. Thousands of parents had filed lawsuits claiming their children were left brain-damaged by thimerosal. (Courtesy of author)

Congressman Dan Burton speaking at Power of One rally, Washington DC, 2005. Vaccine critics were fortunate to find that the chairman of the Government Reform Committee from 1998 to 2004 was on their side. Burton, long a proponent of alternative medicine and a critic of the Food and Drug Administration, oversaw more than a dozen hearings at which vaccines were blamed for a variety of ills. In 2001, Burton's grandson was diagnosed as autistic, a condition his grandfather blamed on vaccines. (Courtesy of author)

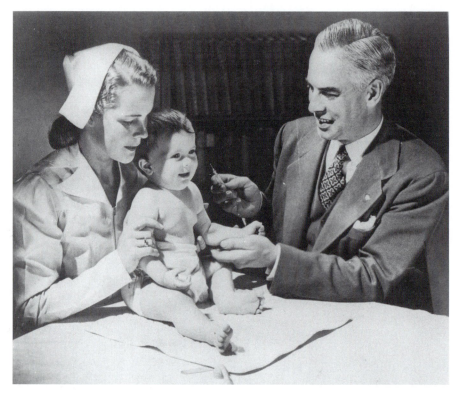

Undated, hubristic photo from the Golden Age of medicine. (Courtesy of the Library of Congress)

exotic. The World Health Organization (WHO) was a major employer in an area of high unemployment, but people didn't always do what they were being paid to do. So Orenstein spent a lot of time checking up on people who were supposed to be checking up on other people. "I would go into Indian villages where there were smallpox cases and I would spot-check houses and have people show me their arms to try to catch whether they'd been vaccinated or not," Orenstein recalled. "People would write on the house, mark it, to make it clear that the house hadn't been missed." Once, Orenstein was nearly lynched after being confronted by a woman whose two daughters went blind from smallpox keratitis shortly after being vaccinated. The mother blamed the vaccine, and "she was getting angrier and angrier. She only spoke Hindi and I didn't understand what she was saying, but finally they had to pull us out of there." His Indian colleagues hauled him out of the neighborhood in time to avoid a riot.

In February 1975, Orenstein witnessed the last case of smallpox in Uttar Pradesh state, and one of the last cases in India. The patient was a seven-month-old girl, Ashanti Priori Lal, and she died of hemorrhagic smallpox, the most virulent, contagious form of the disease, in a house by the main marketplace of Aligarh. Orenstein was called about two hours after her death and learned that the father had concealed an earlier case in the girl's sister. Orenstein took pictures, and when he returned 20 years later he brought them, along with a snapshot of his own son, and gave them to the girl's father. "He was very touched," Orenstein recalled. "He had no clue even then that he could have saved his daughter."

When Orenstein returned to the United States, the main focus of the CDC was measles. Although the WHO would end up focusing on the eradication of polio, doctors who'd battled smallpox in places like Africa and Pakistan knew that measles was a far greater threat than polio to children in the third world. Measles was killing nearly 3 million children a year as late as 2000, while polio's victims numbered in the thousands. In the United States, the drastic tightening of school laws during the Carter era showed that the public was willing to participate in eradicating measles. To do it, though, the system would need new resources and clever strategies. Once those tools were in place they could be used to fight a lot of other diseases. "Measles," Orenstein states, "was the primary driver for the immunization program in the United States for about 25 years."

Soon after his return from India, Orenstein took part in a trial to see whether ring vaccination—the systematic tracing, quarantine, and vaccination of a patient's contacts—could successfully control measles the way it had controlled smallpox. It didn't. The trial was conducted in several midwestern states, and was a complete failure—measles was too contagious and spread when the patient was asymptomatic, unlike smallpox. The experience fixed in the minds of Orenstein and his colleagues the need for a more rigorous vaccination program.

Throughout the 1970s and 1980s, the scientific consensus had shifted back and forth on the best times to vaccinate against measles. In 1963, the CDC's recommendation was for a single dose at nine months. A few years later, the age was bumped up to 12 months, then to 15 months in 1977, as it became clear that until they wore off, a mother's antibodies to measles effectively neutralized the vaccine virus, thereby leaving the child without the capacity to generate lasting antibodies of his or her own. In 1989, though, this had changed. Most mothers, not having suffered measles, lacked antibodies in placenta or breast milk sufficient to protect their babies. So pediatricians moved the recommendation for the first dose back down to 12 months. They also recommended a booster dose, to make up for an estimated 5 percent vaccine failure rate, as well as the ever-nebulous role of maternal antibody.[18] Public health officials began trials of vaccines that contained higher amounts of measles virus in them—high-titer vaccines, they were called[19]—to see whether they could vaccinate earlier in a child's life while overcoming maternal antibodies. Data from studies in Africa and Haiti published by 1992 indicated that girls who got the high-titer vaccine had an increased mortality rate. Although some scientists felt these results were a statistical fluke, the prevailing theory held that the high-titer vaccine caused transient lapses in immunity that made the children more susceptible to other infections. The CDC subsequently halted high-titer measles vaccine trials in the United States.

Orenstein took over the National Immunization Program in 1993 with a plan that he had helped develop two years earlier during the measles crisis. The blueprint, a White Paper by the National Vaccine Advisory Committee, stated that while "the measles epidemic is worrisome enough, measles, being the most contagious of the vaccine preventable diseases, is also an indicator that signals a failure in the system of vaccination."[20] It called for various reforms including more federal money for vaccine surveillance and delivery, mandatory HMO coverage of vaccines, and required vaccination of children entering daycare.

The Clinton administration moved quickly to enact most of these recommendations, and domestic transmission of measles in the United States suddenly stopped. The last cases of measles occurred in 1993; small outbreaks linked to imported cases continued, but ample resources tracked and squashed them. Shortly after Clinton's 1996 reelection, he and the First Lady attended a ceremony marking progress in the eradication of vaccine-preventable diseases. As gurgling and wailing babies were inoculated behind them, a folksy Clinton recalled getting the Salk shot as a young boy. "I screamed and squalled with the best of children," he said, but "I remember being very conscious that some enormous burden was being lifted off of my life."[21]

One little-noticed feature of the eradication of measles and polio (the last U.S. case of polio was in 1979, except for half a dozen vaccine-associated cases each year) was the role played by our neighbors. Cuba, which ridded itself of both diseases before the United States did, had followed Sabin's advice with a series of national immunization days in which every child in the country received vaccine. Following Cuba's success, Ciro de Quadros, for many years head of immunization programs at the Pan American Health Organization (PAHO), expanded eradication to Brazil in the mid-1980s and the rest of Latin America by 1991, when the Western Hemisphere's final case of polio appeared in Peru. "What that did was to knock the bottom out of polio," says Orenstein, "and that became the strategy to eradicate polio in the world." The strategy consisted of routine immunization, national vaccination days, and careful surveillance. Child-centric Latins viewed vaccination campaigns as sacrosanct. In El Salvador, left-wing guerrillas stopped their ambushes and army troops halted their campaigns during immunization days. In Mexico, Venezuela, and Colombia, hijackings and strikes were called off. During the 1989–91 outbreak, Orenstein was at a PAHO meeting where a Texas immunization official blamed the outbreak on Mexico. Indignant Mexican officials blamed their outbreak on the United States, and "they probably both were right," says Orenstein. As measles reports declined from 250,000 to 469 cases from 1990 to 2002 in Latin America—and 0 locally transmitted cases since then—immigrants and travelers from that region no longer seeded either disease in the United States. Most imported cases came from Japan, Korea, and Germany.

New Science, New Vaccines

Although the decade began with handwringing over measles, the early 1990s would be a promising time for the vaccine endeavor. While Orenstein and his expanding staff were working to remove obstacles that kept poor children from getting vaccinated, two promising, technically sophisticated new vaccines came on line. The first was a conjugate vaccine against *Haemophilus influenzae* type b (Hib), one of three major causes of bacterial meningitis in children, with an estimated 600 deaths and 3,000 cases of brain damage, deafness, or other serious aftereffects each year before vaccination began in 1985. The first vaccine against Hib was made from polysaccharide molecules purified from the outer membrane of the germ. The vaccine authorities recommended this polysaccharide vaccine for children 18 months and older only, though, because infants' immature immune systems were unable to build lasting protection from the vaccine. Since infants were the group hit hardest by Hib infections, three groups of U.S. investigators continued work on a new protein-polysaccharide conjugate vaccine.

Conjugate vaccines consist of polysaccharide molecules chemically bound to bacterial proteins. When I first started to grasp what they were about, I was reminded of an overgrown, abandoned garden near my childhood home where, as an eight-year-old, I used model airplane glue to stick hyacinth and daffodil stems together. How, I wondered, did anyone get the idea to fuse these two bacterial substances? Like many advances in immunology, the idea began with Karl Landsteiner, who discovered the polio virus and blood types, among other things, and who won the Nobel Prize in physiology in 1930. Landsteiner, a Vienna-born Rockefeller Institute scientist, noticed that there were substances that would bind with antibodies but that did not cause the body to create new antibodies. He called them haptens. After chemically combining a hapten with a protein, Landsteiner found that an injection of the larger molecule produced antibodies. Another Rockefeller biologist, Oswald Avery, discovered in the 1930s that he could combine the outer-membrane capsules of pneumococcal bacteria with proteins and elicit antibody. Avery's discovery was filed away for decades; like most bacterial vaccines, it became a post–World War II victim of the success of antibiotics. Rachel Schneerson, John Robbins's colleague at the NIH, first proposed the creation of conjugate *Haemophilus* vaccines in 1980; their vaccine was marketed by Merck in 1989.[22] They had an unusually collegial relationship with two other groups led by Porter Anderson at Rochester and David Smith of Praxis Pharmaceuticals, whose work eventually led to a Wyeth vaccine.

Late in 1990, the CDC recommended that infants receive the Hib conjugate vaccine at two, four, and six months, together with their first three DTP shots. Within five years, Hib meningitis had largely become a memory. While its defeat was not the kind of story that parents discussed on the playground, it was a remarkable triumph for medicine. Most pediatricians had seen children laid low with bacterial meningitis, and they enthusiastically applauded the Hib vaccine and its triumphs. The great unsung achievement of the vaccine was exemplified in the case of Heather Whitestone, a deaf woman who was crowned Miss America 1995. Whitestone had lost her hearing as an 18-month-old after suffering a severe Hib infection. Because the infection followed a DTP shot, however, her parents originally blamed the shot for her illness—and that was the account published in the newspapers at the time of her victory.[23]

With *Haemophilus influenzae* type b out of circulation, another strain of the bug—type a—has moved in to occupy part of the ecological niche. But type a is a less virulent form of the bacteria, so meningitis and other invasive *Haemophilus influenzae* diseases have declined precipitously.[24] A slightly different readjustment of the microbial world occurred a decade later after Wyeth, in 1999, introduced the conjugate pneumococcal vaccine to prevent the second most common form of childhood meningitis. The pneumococcal vaccine gained rapid acceptance among pediatricians who hoped it would help battle ear infections and reduce the need for the sweet pink antibiotic syrups that parents were constantly demanding for their sick kids. Antibiotic resistance among bacteria, including the *Streptococcus pneumoniae* strains that the vaccine targeted, was a rising concern, and the new vaccine contained antigens against seven of the dozens of strains of pneumonia. While it proved a capable fighter against the germs represented by its antigens, five years after its widespread introduction there was no evidence that the pneumococcal vaccine was reducing the rate of invasive pneumococcal infections overall.[25] However, one positive, unanticipated benefit of the vaccine was the reduction of the rate of pneumococcal disease in adults.[26] Even in a climate laced with paranoia about vaccines, these bacterial shots got largely positive reviews. A few studies had suggested that Hib might somehow trigger juvenile diabetes, but epidemiology showed no link. After a careful hearing of the evidence, the federal vaccine court threw out the diabetes hypothesis in 2003.[27]

Three viral vaccines put on the pediatric schedule in the 1990s—hepatitis B, chickenpox, and rotavirus—received a rougher ride than the conjugates had. Each of the three vaccines was in its way a vital product with tremendous potential to reduce disease. Each resulted from decades of research by dedicated pediatricians, immunologists, and epidemiologists. But these products would enter the corpus of childhood immunizations merely by dint of expert recommendation. They came about with little press attention, only scattered interest from pediatricians, and scant public enthusiasm. The experts' motivation was to make children healthier, but in selling the vaccines to their colleagues and patients, they at times obfuscated the complicated logic behind their choices. That would lead to criticism. None elicited more sustained and widespread skepticism than the recommendation of universal newborn vaccination against hepatitis B.

Hepatitis B—serum hepatitis, as it was known before the virus that caused it had been identified—showed a most unusual epidemiology. It was a virus that could cause a lot of harm, or none, depending upon the patient's age at infection, the mode and the dose or "inoculum," and traits of the individual humans it infected. Sometimes the virus caused an acute illness, then passed out of the body. Other times it caused only mild illness but provoked a chronic liver infection that could result in cirrhosis, cancer, and death. Hepatitis B was common in populations that came in intimate contact with blood products: drug addicts, hemophiliacs, people who have anal sex, and the babies of hepatitis B–infected mothers. It was, as we have seen, also common in soldiers injected or infused with contaminated vaccines or blood products. In places like military barracks, prisons, and mental asylums where hundreds of people might come into contact with the virus through sex or by deposits on toothbrushes, on bath towels and in hard coughs, it was a frequent visitor. Parts of Asia and Africa suffered more widespread infection, perhaps because the virus had turned up earlier in those continents, where injections with dirty needles were a common medical practice. A tiny amount of the virus could cause infection, although there was little documented transmission other than via blood and sex. In 30 percent of all cases, however, the source of infection was unknown.

In 1954, when Saul Krugman arrived at the Willowbrook State School in Staten Island, he found hepatitis to be endemic among the school's 5,200 residents.[28] Krugman established a special residence for his studies that was far more orderly and sanitary than the school's 25 other buildings. Thus, although he openly fed many of the children entering his ward with hepatitis B virus, parents readily sent their retarded children to him. The ethical argument for this practice was that most inmates at the school became infected with the disease anyway—and it was almost always a benign infection in children. Krugman's critics felt otherwise, of course. "We must ask who ends up at Willowbrook, why is hepatitis epidemic there, and what should a doctor do about it?" Ellen Isaacs, a left-wing New York University physician charged, at a 1972 meeting where the work was dissected. "The argument that bad conditions at Willowbrook justify experiments there becomes very insidious. These conditions are not inevitable."[29]

The CDC faced a similar conundrum in 1991 as it debated whether to recommend universal hepatitis B vaccination for infants in the United States. Over the past two decades the virus had made inroads into high-risk groups, despite the invention of two vaccines aimed at stopping it. Just as Krugman had to decide whether to go about his work in pursuit of an ultimate humanitarian goal, the CDC had to decide whether to stand and fight for a vulnerable population, or to aim for the eventual eradication of the disease. Like Krugman, it chose the further-off goal. Like him, it probably had no choice.

Krugman had created a crude hepatitis B vaccine out of boiled serum from former hepatitis patients in 1971 and used it to protect some of the inmates at Willowbrook. In 1982, Merck launched a vaccine made from hepatitis B surface antigen purified from human blood, but the vaccine did not sell well because of fears that it might contain AIDS virus. Four years later Merck came out with the world's first recombinant vaccine, which Hilleman developed with geneticist William Rutter at the University of Washington, in collaboration with scientists at Chiron, in San Francisco. Yeast was used as a microscopic factory to synthesize the antigen.

Chronic hepatitis B infection caused thousands of liver cancers each year, and Merck's new vaccine was thus greeted as the first anticancer vaccine. Alaska, where the native population had a particularly high rate of infection, began universal vaccination of teenagers and infants in 1982. In the rest of the country, most new infections were reported among intravenous drug users, prostitutes, and men who had sex with men. The Advisory Committee for Immunization Practices (ACIP) in 1984 targeted these high-risk individuals as well as the babies of pregnant women who tested positive for the surface antigen. These children would receive vaccine and passive immunization with immune gamma globulin shortly after birth.[30]

This approach to tackling hepatitis B proved disappointing. The groups in whom the virus was circulating were hard to reach, and the activities known to spread it were stigmatizing. In 1990, the ACIP recommended routine screening of all pregnant women, predicting that the policy would prevent the infection of 3,500 infants over the next decade.[31] But in the meantime, hepatitis B infections were getting more common, increasing by 37 percent from 1979 to 1989. An estimated 250,000 new infections occurred each year, with 5,000 to 6,000 deaths, mostly from liver cancer. And although only about 8 percent of infections in the United States occurred in newborns or other children, as many as 9 in 10 infants infected at birth would go on to become chronically ill with hepatitis B. Most healthy adults, on the other hand, could fight off an exposure to the virus without permanent infection.[32] In 1989, the CDC's immunization committee gave its first serious consideration to universal infantile vaccination against hepatitis B. The Reagan and first Bush administrations had never put any resources into vaccinating the most at-risk individuals. At the 1991 meeting that led to the universal vaccination of infants, ACIP members said that while they didn't want to treat high-risk adults as "second-class citizens" there weren't enough resources to recommend vaccinations for both groups. By 1992, Congress had provided money to vaccinate infants, but few states had the money to vaccinate high-risk individuals, too.[33] While there were good reasons to vaccinate kids, the CDC had downplayed the reasons for the earlier campaign's failure. Reading the political smoke signals—services for junkies and gay men were not a popular line item—public health officials had never really pushed the more obvious strategy. Eventually, there would be a price to be paid.

The old strategy had targeted the visible face of hepatitis B—"at risk," stigmatized groups. The new strategy was a piece of grand social engineering that would attack hepatitis B preemptively, before the humans who carried it had a chance to make the behavioral choices that spread it. The American Academy of Pediatrics sealed the deal by following the ACIP with its own universal recommendation in February 1992. Neal Halsey, a member of the committee, noted that safety was "really a non-issue with this vaccine. It's a very safe vaccine . . . safer than virtually all of the other vaccines that we have been using."[34] Pediatricians responded skeptically at first. In a survey conducted in late 1992 only 50 percent of pediatricians supported universal infant vaccination. Were the number of cases prevented in children worth the expense of the vaccine—estimated at about $100 for three doses? And more to the point, would immunity last into adulthood when people started engaging in the behaviors that put them most at risk?[35] Physicians had only just added Hib to DTP and polio.[36] Parents, they feared, would balk at yet another shot administered to children at two and six months of age. The CDC had never observed transmission of hepatitis B in a school, although an estimated 9,000 children a year somehow became infected with the virus.[37] Since infants born to hepatitis B–negative mothers rarely got the disease, why not vaccinate teenagers, some asked, or leave the decision up to parents?

The bottom line was that public health had to deal with society as it was, not as it ought to be. Under the circumstances, universal vaccination was good public health. Whatever objections parents had to the pumping of another vaccine into their innocent, pure babies, hepatitis B was a good shot. "We are notably poor as soothsayers in predicting which children will be at high risk by future behavior, environment or moving to a new locale," wrote Halsey and Carolyne Breese Hall, who chaired the ACIP at the time. "Pediatricians, therefore, must initiate an insurance policy for their young patients which matures in adulthood."[38]

Despite initial doubts, states moved quickly to include hepatitis B vaccination in their mandatory programs. By 1994, 82 percent of California pediatricians were immunizing infants,[39] and by 2002, laws had been enacted in 44 states mandating hepatitis B vaccination for kids entering elementary school and childcare centers. These laws passed, initially, with little mobilization of the antivaccine lobby. It was indisputable that vaccination was making inroads on the hepatitis B virus that would save thousands of lives and reduce the intense pressure on the nation's organ transplant system (sadly, latent hepatitis C infections, which also caused liver cancer, were *increasing* this pressure). Between 1982 and 2002, more than 40 million infants and children and 30 million adults received hepatitis B vaccine. In 2002, the CDC estimated that there were 79,000 new hepatitis B infections, compared to 200,000 to 300,000 in 1982. Part of the credit went to safe-sex behavior in the age of AIDS. But vaccination was having a major impact on infections in babies. There was less hepatitis B among teenagers who'd been vaccinated as children and among adults vaccinated as teenagers.[40]

Data from Alaska and Taiwan, where the vaccine had been in widespread use the longest, showed that the immunity it provided was long-lasting. The hepatitis B virus had a long incubation period, which apparently gave immune cells primed by the vaccine time to respond before new infections could take hold. Liver cancer and cirrhosis deaths continued to plague older populations, but as the vaccine cohort reached adulthood, the vaccine was sure to cut those numbers, too—and the lists of patients waiting for liver transplants.[41] The vaccine was doing what it was supposed to do.

No one denied that hepatitis B was a bad disease. But educated parents could legitimately weigh their own instinctual aversion to injecting a foreign substance into their child against a risk that in early life was very tiny. With the next vaccine introduced into the childhood vaccination schedule, against chickenpox, many wondered whether, on an individual basis at least, there was any health benefit at all. As these two vaccines were brought on line, resistance to vaccination began to grow from a tiny hard core of ideological opponents into a larger, more mainstream group of skeptics. In the case of hepatitis B and chickenpox, parents who found no fault with vaccination in general could still question whether the new injections were necessary. To be sure, these vaccines had come through the licensing process with excellent safety records. But that did not make them necessarily worth getting, from a parent's perspective.

When it came to new drugs, therapeutic nihilism—the philosophy of "the less drugs, the better"—was not a fringe concept. It had originated with William Osler, the Canadian physician who revolutionized medicine at the last turn of the century by picking through the list of traditional remedies and sorting out useless chaff. Therapeutic nihilism was a philosophy with legs; contemporary pediatricians often encouraged parents to wait out infections that were probably viral, rather than burdening their children's young bodies with useless, or even harmful, antibiotics. When it came to vaccination, guarding children against diseases like smallpox, polio, or even measles was one thing. But many baby boomer parents had been sick with chickenpox as children and remembered the experience as no big deal. Others felt, complacently perhaps, that they would not be raising children who would ever be putting themselves at risk of hepatitis B or, less complacently, that their children could get the shot later. Besides, adults who read the newspapers knew enough about unintended consequences to worry that the risks of these new vaccines might outweigh the known risks of the diseases they prevented. Among some educated and self-confident parents, the risk/benefit equation of certain vaccinations had now shifted, although there was little sign that the public health authorities recognized it.

To be sure, chickenpox was not always benign. Though it meant no more than a week of itchy discomfort for most children who contracted it, complications from the illness caused about 100 deaths and tens of thousands of hospitalizations each year. And if the vast majority of parents remembered it as a normal passage of childhood, in 1995, the year the chickenpox vaccine was licensed, most women with children worked and could not afford to stay home for a week with a sick child even if they wanted to. It would be useful, Halsey said, to think of the vaccine partly "in the realm of parents not having workdays lost, staying home with kids." Besides, children suffered, some more than others, when they had chickenpox. "It's not fun for anyone," he said.[42]

Pediatric vaccinologists had long sought this, the last jewel in the crown of childhood viral vaccines that Maurice Hilleman had set out to make in the late 1950s. Plotkin, Katz, and the other veterans of the era proselytized for making it a routine childhood shot. It had been a devilishly difficult vaccine to develop and was over 20 years in the making. Michiaki Takahashi and colleagues at Osaka University developed the vaccine strain from a boy named Oka in 1974, and it was licensed for general use in Japan and Korea in 1988, with 1.5 million children vaccinated by 1993.[43] Doubts about the vaccine's long-term effectiveness delayed its entry in the U.S. market. This was a serious problem to consider, because in a routinely vaccinated population, people made vulnerable by an ineffective vaccine would be more likely to catch chickenpox as adults, when it was more serious. Too, some scientists suspected that periodic exposures to chickenpox—when parents or grandparents, for example, got close to a child with the disease—provided a sort of booster that protected adults against shingles, a painful reactivation of the chickenpox virus in patients who had once had the disease. In the absence of chickenpox-y children, some wondered, would the rate of adults with shingles increase? People who were in no sense "antivaccine" were asking these questions. The concerns stretched back decades and at times pitted leading vaccinologists against each other.[44]

Susan Ellenberg, the director of biostatistics and epidemiology in the FDA's biologics center, had been taken aback at the paucity of data in many vaccine trials when she came to FDA in 1992. She noted that the final varicella vaccine was approved after exactly one randomized clinical trial on about 1,000 individuals, which was done 12 years before the vaccine was approved and with a slightly different formulation. To be sure, about 12,000 children had been vaccinated with the most recent formulation, but not in double-blind trials. Merck agreed to conduct postlicensure surveillance of 90,000 kids who got the vaccine, but Ellenberg wasn't convinced that was enough. She was "astounded" that the vaccine for a disease that rarely killed anyone "unless they were a cancer patient or with an immune deficiency . . . [would] be approved with almost no comparative data."[45]

The new vaccines stirred up what might be called philosophical objections among vaccinating doctors. Hepatitis B and chickenpox shots, wrote Arthur Lavin, a Cleveland pediatrician, "were recommended for a problem that the community and its physicians did not agree presented a danger sufficient to justify such an intervention. While varicella might cause 100 deaths and 1 in 600 hospitalizations, "How many children die of complications of the common cold? Probably more than 100—but I would still characterize the common cold as one of the most benign of infectious diseases," Lavin aruged.[46]

The possibility that the vaccine could transform disease patterns in unpredictable ways, perhaps converting chickenpox and shingles into major adult afflictions, added to Lavin's concern. "How can our profession accept a tremendous intervention in the epidemiology of an overwhelmingly benign disease, with little idea as to whether more harm than good will result in the long-term?" It came down to this: "Will the academic community and the Academy [of Pediatrics] ever meet a vaccine they don't like?" The counterargument was that with less virus circulating, fewer adults would get chickenpox. The rubella experience had given American vaccinologists the confidence that their virus elimination strategy was sound. "The most risky scenario," said Stan Plotkin, "would be sporadic vaccination of children, which could diminish the circulation of wild varicella without providing protection for all and thus augment the risk of disease later in life."[47]

Like most vaccines, the proof was in the pudding, and once a good share of children were vaccinated, the pudding was set. By 1999, when 60 percent of toddlers received the shot nationwide, only 1 kid in 10 was getting chickenpox—compared to virtually all children in the prevaccine era. And that was with only five states requiring the shot for daycare and school admissions.[48] A major epidemiological survey of Antelope Valley, California, Travis County, Texas, and West Philadelphia from January 1, 1995, to December 31, 2000, found the number of chickenpox cases declined by 71, 84, and 79 percent, respectively.[49] In one study in Philadelphia, 50 children and many mothers living in a homeless shelter where a parent and child had come down with the disease were vaccinated in the winter of 1998. Two siblings of the sick child became infected, but no one else did. At another shelter where no vaccinations occurred, an outbreak sickened 63 children and adults and lasted six weeks. Pediatricians might still be skeptical—about a third were, in 1998 surveys.[50] But once a vaccine was introduced, and worked (although by 2006 it was evident that a second dose of the vaccination would be required to provide full immunity), concrete results pushed future shock into the background. And for most of us, nostalgia for chickenpox proved to be a passing pang.

New Funds for Vaccine Safety

As the new vaccines were introduced in the early 1990s, the CDC vastly expanded its staff with new funding from the Clinton administration. The immunization program got its own center at the CDC, newly christened the National Immunization Program, and its staff grew from about 100 in 1993 to more than 500 a decade later. What's more, funds from the program increased the network of public health officials in cities and towns around the country who now worked on vaccination.

In 1988, a young Epidemic Intelligence Service graduate named Bob Chen took over the CDC's vaccine safety efforts. Chen, who had spent the previous two years working on hospice care for dying AIDS patients in San Francisco, had emigrated from Taiwan when he was 10 years old and spoke no English. He grew up in suburban Illinois and got his MD at the University of Chicago, then stayed for a masters in public policy under the philosopher Leon Kass, who questioned the utilitarian values that held sway in public health. Chen's father had grown up as a barefoot mountain farmer and risen to a senior post in the Taiwanese telecommunications monopoly before quitting rather than joining Chiang Kai-Shek's Kuomintang Party. He was a typical striving immigrant, but one with an acute understanding of the pressures that could be exerted by a large bureaucracy. His son would come to share that understanding.

Vaccine safety had never been a glamour position at the CDC. In 1978, Alan Hinman created the first system for monitoring adverse events, with one half-time epidemiologist and a health adviser assigned to it. Chen, armed with Kassian zeal for the value of individual lives, took to the field with relish. He would create, over the next 16 years, a vaccine safety program that employed more than 25 full-time scientists. The program's size and resources made it the envy of his colleagues in other countries and Chen a leading world authority on the subject. Chen's colleagues believed that he and Orenstein had done as much as anyone in the world to make sure shots were safe. But his success would put him in the hot seat, rather than on a throne. The more Chen's reputation grew among his peers, the more doubt and scorn grew among members of the public and even members of Congress. Most of the studies conducted by Chen and his scientists found that vaccines were safe. When he found problems, his superiors were skeptical and the antivaccine lobby jumped up and down. When he didn't find problems, the antivaccine lobby jumped on him, claiming a cover-up.

The 1986 act that created the vaccine court had required an Institute of Medicine panel of experts to examine controversial matters of vaccine safety, including the pertussis vaccine. In 1994, after an 18-month review of scientific and medical data, the panel issued a report that established certain infrequent but well-documented causal relationships between vaccines and medical conditions.[51] These included rare allergic reactions to hepatitis B and MMR vaccines; thrombocytopenia—a blood clotting disorder—in 1 out of 50,000 children vaccinated against measles; and 10 or so cases of vaccine-associated polio each year. The panel failed to find adequate data for or against other allegations of harm and recommended a number of specific studies to remedy this—including studies of Guillain-Barre syndrome following flu vaccination and of aseptic meningitis in children vaccinated with the Urabe strain of mumps. The panel also said there should be disease registries for rare conditions associated with vaccines.

Vaccine authorities, at the time, were showing a readiness to respond to specific safety complaints. In 1995, at the urging of John Salamone, the father of a boy injured by oral polio, the CDC and the American Academy of Pediatrics' policy-making committee recommended a switch from the Sabin vaccine to an improved Salk shot, to prevent vaccine-associated polio. Babies began receiving two doses of Salk and two of Sabin in 1996, and Salk alone in 1998. In 1996, the committees also recommended that babies receive acellular pertussis vaccine, instead of the dicier whole-cell shot.

Chen was building a program that could respond to these ideas. In the late 1980s he helped set up the Vaccine Adverse Events System, or VAERS, which took Hinman's system and added a more systematic method for pediatricians to report problems. He convinced several large West Coast HMOs with computerized patient records to link individual patients to vaccination data. The Vaccine Safety Datalink was up and running by 1991. Eight years later, after another new vaccine, against rotavirus, had been added to the vaccine schedule, Chen's system would pass its biggest test. But once again, there would be a vast gap between how the rotavirus story was seen by vaccine scientists and how the public, and even other doctors, understood it.

Rotashield—an Unwelcome Detection

On June 25, 1998, the Advisory Committee for Immunization Practices voted unanimously to recommend universal vaccination against rotavirus, a diarrhea-causing colonist of the gut. Though it was not a household word, and had only been discovered in 1973, rotavirus is among the most common acute infections of childhood; it infects the intestines of most of the world's mammals. Virtually every child in the world gets at least one rotavirus infection before the age of 5. In the United States, this translated into 500,000 office visits a year and an estimated 50,000 hospitalizations. Rotavirus infections are seasonal and sweep through the United States from the Southwest to the Northeast during the winter and early spring months. "When I finished my pediatrics training at Children's Hospital in Los Angeles, the admonition was 'don't get the infant ward during January,' " Walter Orenstein recalled. "Because that's when all these little kids were being hospitalized to get IVs for rehydration."[52] Deaths were infrequent in the United States, but in the poor areas of Africa, Asia, and Latin America, where sick children lacked medical attention, the resulting dehydration from vomiting and diarrhea killed as many as 870,000 children every year.[53]

In design, Rotashield and other successful rotavirus vaccines were something of a breakthrough into the past. The Australian microbiologist Ruth Bishop had discovered and named rotavirus for its wheel-like shape under an electron microscope. The NIH's Albert Kapikian conceived the design of the first rotavirus vaccines. He described it as "neo-Jennerian" because, just as Jenner had used a cowpox virus to protect against smallpox, Kapikian's vaccine employed a rotavirus native to the rhesus monkey to provide immunity to the virus that afflicted humans. Kapikian's vaccine, adopted by Wyeth, contained one gene each from three attenuated human rotavirus strains. Meanwhile, Merck was working with a chimeric virus derived from cows that had been created by Plotkin, Paul Offit, and Fred Clark of the University of Pennsylvania. GlaxoSmithKline's vaccine was of a simpler design; it was an attenuated version of a virus that had infected a Cincinnati toddler. All three were live oral vaccines designed, like Sabin's oral polio vaccine, to create a protective immune response in the gut.

The FDA approved Wyeth's Rotashield, but with a footnote. In the package insert, which alert pediatricians read so as to be on the lookout for potential side effects, the agency noted that the vaccine might be associated with high fevers as well as intussusception, a painful telescoping of the bowel that sometimes required surgery and could even be fatal.[54] In 1997, Margaret Rennels, a University of Maryland pediatrician who helped test the vaccine, had noted a case of intussusception in one of the children who received it; Carolyn Hardegree, a senior FDA official, drew attention to the problem and a chart review turned up four more intussusceptions among 10,000 vaccine recipients—compared to a single case among the 4,633 controls. Five cases of intussusception in the vaccine group failed to reach the threshold of statistical significance, but four of the cases had occurred within 15 days of vaccination. Intussusception was sometimes associated with viral gut infections, so Rennels's detection of the cases raised concern that more cases might emerge with widespread use of the vaccine.

In a literature search, Rennels and her colleagues found little evidence that routine rotavirus infection caused intussusception. Then again, intussusception was rare, which made it difficult to tell whether the association Rennels found was coincidental. Since most of the five cases had followed the second or third dose of Rotashield, and the intussusceptions had followed the administration of three different versions of the vaccine, "I concluded," Rennels told an FDA panel, "that they were probably due to chance temporal association." Still, Rennels and NIH scientist Roger Glass were concerned enough to publish her findings the next year and to call for postmarketing alertness to the condition.[55] Their two-paged article appeared in *Pediatric Infectious Disease Journal*, not a mainstay of family practitioners.

Glass, who has published more than 160 articles about rotavirus infections, was eager to get a vaccine on the market. As he noted a few years later, in the midst of another emerging disease scare, "I work on a virus that kills more people in a day than SARS has ever killed."[56] To be sure, it didn't kill more than about 60 of them each year in the United States, and while it did cause alarming, dehydrating infections, a cost-benefit analysis showed the vaccine would essentially cost $103 per case of diarrhea prevented.[57] But if the need in the United States was not pressing, the vaccine was a natural for the third world.[58] The WHO wanted a vaccine, and Wyeth had promised Glass and others that it would eventually discount it in poor countries. "Eventually" meant after it had earned back a portion of its costs in the United States, which it could only do if the vaccine was in universal use here.

In October 1998, several new people joined the ACIP, among them Paul Offit, a pediatric immunologist at the University of Pennsylvania who had spent much of his career working on the rotavirus vaccine now being developed at Merck. Offit was an outgoing, passionate scientist who spoke lovingly of his own children and wept when he talked about kids who had died. His wife, Bonnie, was a pediatrician whose warmth and patience attracted many well-heeled, vaccine-skeptical parents to her practice. Unlike a lot of people in the vaccine world, Offit himself did not humor vaccine skeptics. He thought they were misinformed and said so. "Science isn't very politically correct," he would say, "Something is either proven or it isn't." Offit's forceful defense of vaccines, published in places like *Pediatrics*, made him a lightning rod for the antis. He even got death threats from people who blamed vaccines for their children's problems.[59]

After growing up in a Baltimore suburb, studying at University of Maryland Medical School, and doing a pediatric residency in Pittsburgh, Offit had gone on to a fellowship at Children's Hospital in Philadelphia. Stan Plotkin, who had tested oral polio vaccines in the Congo in 1957 and later developed the definitive rubella vaccine, was Offit's mentor, and it was Plotkin who lured Offit into the less-than-glamorous path of vaccinology. "I remember walking into Stan's office and he'd be leafing through the *MMWR* looking at the latest statistics on congenital rubella," says Offit. "That was his report card, and the numbers kept going down and down. To have that kind of direct, tangible impact on the health of children just seemed fantastic." With Fred Clark, a veterinarian, Offit took up the development of a rotavirus vaccine that Plotkin started. Offit's participation on the ACIP theoretically gave him a conflict-of-interest problem, since his vaccine was in competition with Wyeth's product. Vaccine critics would claim that Offit's backing of the Wyeth vaccine was self-serving since its approval would theoretically open the way for Merck's product to enter the market. On the other hand, vaccine manufacturers preferred markets in which they were the sole providers, so you could have argued the opposite. Besides, anyone who thought vaccine developers and pharmaceutical companies worked hand in glove would be mistaken. Plotkin had tried to interest his employer, Aventis-Pasteur, in the rotavirus vaccine, but Offit and Clark ended up shopping it to Merck after Aventis declined. Offit had spent an average of one day a week for two years lobbying Merck officials to make sure they continued with their trials of his vaccine. He had an acute sense that vaccines were the runt stepchild of the pharmaceutical industry.

"Yes, it's true that we want our vaccine to be used," Offit would say, sarcastically. "It would be nice if we could make it ourselves, but we can't. So we have to work with the evil drug industy." It was a sensible bias to have vaccine experts recommending vaccines, Offit felt—who else could evaluate whether a vaccine was any good? In any case, by the time Offit got on the ACIP, the Food and Drug Administration had licensed the Wyeth vaccine and the ACIP had already made its formal recommendation for universal vaccination.

Pediatricians vaccinated an estimated 1.2 million children with Rotashield starting in late 1998. Vaccination had for the most part started too late to forestall the annual rotavirus season, but scientists who'd spent their lives preparing for this moment were thrilled. "I think it's going to have a major impact," said Kapikian, who had worked on the vaccine for 24 years.[60] It was a short-lived triumph. At the CDC, the vaccine safety system that Chen had been carefully building was doing what it was supposed to be doing, which was finding signals of vaccine damage. Confirming the worst fears of Glass and Rennels, Rotashield seemed to be associated with a larger-than-expected number of intussusceptions. Scientists at FDA and in Chen's office started culling reports out of the VAERS database in December. By June 2 they had accumulated 10. Intussusception had only been reported four times previously since the VAERS system came into being in 1990.[61] Since the FDA had asked doctors to be on the lookout for it, there was always the possibility of an ascertainment bias—the tendency for people to find things more frequently once they start searching for them. Still, it didn't look good for Rotashield.

When the number of intussusceptions reported on VAERS reached 15, the CDC, at Orenstein's urging, pulled the plug. On July 16, 1999, CDC director Jeffrey Koplan urged healthcare providers to suspend Rotashield vaccination. By the year's end, 112 cases of intussusception had been reported to VAERS in association with Rotashield. In the meantime, the CDC had withdrawn its recommendations, and the manufacturer recalled the vaccine.

The well-publicized suspension of Rotashield, and its subsequent recall, marked the first serious setback for the federal vaccine program since the resolution of the whooping cough crisis. It contributed to new attacks on the program by the usual critics and by some new ones, including members of Congress, who would claim that it exposed conflicts of interest in the program's oversight committees.[62] A report published by Congressman Dan Burton of Indiana in 2001 stressed that Offit and other members of the FDA and CDC committees had either worked on rotavirus vaccines or received research grants or other money from the pharmaceutical industry.[63] That was news to no one who knew anything about vaccines. Rotavirus experts, to be sure, were more likely than the average pediatrician to be advocates for a rotavirus vaccine. There was nothing in the report to indicate that drug company ties had forced anyone to make a decision he or she later regretted.

Not long after the CDC published its first studies, a group at the NIH responded with studies that seemed to vindicate the vaccine. Lone Simonsen, a Danish researcher at the National Institute for Allergy and Infectious Diseases (NIAID), surveyed intussusception rates in 10 states and found that the overall rate was *lower* in the six months during which Rotashield was employed. The trick was in the timing, she believed.[64] The risk of intussusception following the first dose of Rotashield was indeed higher, her data showed—though she calculated a rate of only 1 in 36,000 shots compared to the CDC rate of 1 in 10,000. But the decline in intussusception rates over the full six months in those 10 states—she later added 10 more with essentially the same result—suggested that the vaccine merely triggered an event that a baby would have suffered later in life. In other words, it was a phenomenon not unlike the seizures that followed DTP vaccination. But Simonsen's study had many weaknesses—the main one being its passive reliance on hospital discharge data, which was unreliable for a condition such as intussusception. At an October 2001 meeting of the ACIP, the vaccine's advocates pled their case. Glass argued in favor of reinstating the vaccine, noting that even accepting the CDC's data, for each intussusception the country could avoid 150 hospitalizations and 1,000 office visits for rotavirus. Moreover, if America could do without the vaccine, the rest of the world could not. "Hundreds of thousands of young children die from rotavirus diarrhea annually world wide. Completely safe vaccines are ideal, but how long will we have to wait?" asked a plaintive Kapikian.[65] Without a full-throated universal recommendation in the United States, Wyeth was not inclined to manufacture any more vaccine.

Orenstein had the last word, and it was no. The CDC had on its hands a vaccine that obviously caused—or at least triggered—a serious problem. Furthermore, Chen and his colleagues had dug up more suspicions about Rotashield in the form of additional VAERS reports, some quite serious, of rectal bleeding following vaccination. Only a small minority of doctors surveyed said they would use Rotashield again.[66] In fact, many were furious that the warnings about intussusception hadn't been broadcast more loudly before the vaccine came out. "I don't expect perfection. I think a vaccine could be released without any evidence of a problem in the early trials, and a problem might later be discovered. I accept that," said Al Mehl, a physician at a large Colorado clinic. "I am less accepting of discovering in the early trials that there might be risk for a potentially dangerous complication, and yet pushing ahead with an incomplete answer, even knowing the vaccine wouldn't save many lives and probably wouldn't ever save money."[67]

Public Health officials hoped to get Wyeth to introduce the vaccine in the third world, but at a WHO meeting in 2000, it became clear that health ministers in poor countries weren't willing to take the political risk of adopting a vaccine nixed by the United States.[68] "I'd love to use it, we have 100,000 deaths each year of this disease," a senior Indian health official told Roger Glass. "But when I have the first case of intussusception I will be tarnished in the press for having accepted a vaccine that was rejected in the U.S." Despite the fact that in some countries 1 in 300 children died of rotavirus infection before age 5, added John Montaigne, then NIAID's deputy director, "there's a sense that the vaccine has to be approved for use here or it won't be touched by anyone overseas."[69] Most of the health ministers in countries that could afford Rotashield had been educated at U.S. medical schools and followed the American Academy of Pediatrics' assessments of vaccines. "The WHO urged acceptance but the offer was met with silence," wrote Stan Plotkin. "No country was willing to place public health above possible criticism for using a vaccine rejected by the United States. This was not exactly a profile in courage."

The ACIP did recognize, as Simonsen said, that in the third world Rotashield could do a world of good. But the committee was preoccupied with a new set of worries. And it would be five years before a new and improved rotavirus vaccine saw the light of day. In the fall of 2001, a terrorist or terrorists of unknown origin and motivation had mailed several letters full of anthrax to Congress, killing five people and setting off a massive, multibillion-dollar bioterrorism "preparedness" project. Rotavirus and the millions of poor children who died of it receded to a blur on the passing landscape. "Ours," Simonsen noted sadly, "was only a minor issue." The vaccine, the ACIP decided at a February 2002 meeting, was dead. Like Brigid O'Shaughnessey in *The Maltese Falcon*, it had to take the fall. "At a time when many parents express concerns about the safety of vaccines and vaccine adverse events are the focus of increasing attention by the public, media and the U.S. Congress, the wisdom of recommending a vaccine that causes a severe adverse reaction in an estimated 1 in 10,000 infants must be considered," the CDC's investigators argued. "Public confidence and the support of vaccine providers for vaccination recommendations, although difficult to quantify, are important factors in the decision making process."[70]

The Rotashield experience taught some lessons. One was that you didn't put all your hopes on one company's vaccine. Another was that multinational firms alone couldn't be counted on to provide a vaccine for the third world. The State Serum Institute in India was producing a hepatitis B vaccine and wanted to start making its own rotavirus vaccine. GlaxoSmithKline, which had tested its rotavirus vaccine on more than 110,000 subjects by early 2005, introduced it in Mexico, rather than Europe or the United States. It hoped to gain the interest of the Global Alliance for Vaccine Initiatives, which partnered with Gates to buy vaccines for the third world. Merck planned to market the vaccine in the United States and 11 other countries, including Haiti and Colombia.

The Rotashield trial seemed to vindicate the FDA's Susan Ellenberg, who had been quietly advocating that pharmaceutical companies hold larger trials before licensing. "Before Rotashield, people thought I was both crazy and dangerous," she said. "Afterward I think maybe they thought I was just dangerous." But if Rotashield caused intussusception in 1 of every 10,000 children vaccinated, a trial of 100,000 or more children might have been required to find clear statistical evidence of it. In a postmortem, Rennels asked, "How many children are required to satisfy us that a vaccine is safe to license and recommend? This is somewhat of a 'Catch-22' question because to accurately perform the calculation you must already know what side effect you are looking for."[71]

America was a risk-averse culture, and for a vaccine like Rotashield, designed to battle a disease no one had heard about, 1 in 10,000 side effects was too many. The CDC could not endanger its reputation among pediatricians and the public—it could not endanger the rest of the immunization program—by ignoring qualms about this product.

But where did this logic end? After all, there was no vaccine that had zero side effects. MMR caused thrombocytopenia, a serious blood-clotting disorder, with a regularity that approached that of Rotashield-linked intussusception. Acellular pertussis, though milder than the whole-cell shot, caused seizures in about 1 in 14,000 children each year. Oral polio crippled about 1 in 750,000 children. Would it be impossible, now, to introduce a new vaccine capable of achieving the benefits of an MMR, a polio, a DTP? The implications for the future of new vaccines were staggering. In an age of informed consent, expensive patents, and skyrocketing medical liability, 60,000 was a huge number of people to enter in a vaccine trial. To recruit and process a single patient for a vaccine trial cost in the range of $8,000 to $20,000 in 2005.[72] That came out to around $500 million minimum. The Rotashield episode forced vaccinologists to face the increasingly harsh environment for their wares. "We all want 100 percent effectiveness and 100 percent safety," said Orenstein. "But the kinds of regulation in place will make that more and more cost-prohibitive. I don't see us rolling back the regulations or compromising on the standards. But each new regulation, each new effort . . . has unintended consequences in terms of the cost to society."[73]

The vaccine safety system established by Chen and his colleagues was a wonderful tool, but it was a dangerous one, too, a sorcerer's apprentice that cranked out the data in the absence of a social agreement about how to assess the answers it produced. If the Vaccine Safety Datalink spat out an equation of risk for a vaccine, what did you do with that? If no amount of risk is acceptable, how could we possibly convince drug companies to sink millions into developing vaccines that were almost sure to have at least *some* risk? Who decided what level of risk was acceptable? Paul Offit tried to get the ACIP involved in a discussion of risk and benefit at a meeting in 2000, but "people were too scared to talk about it." Someone at the meeting described driving to work and hearing a mother interviewed on the radio discussing her child's case of intussusception. "We have to think about the mother whose kid gets intussusception," Offit acknowledged, "but we also ought to think about the 40 mothers who lose their kids to rotavirus. They are less likely to get on NPR; they're more likely to be poor African-Americans. But we need to represent them too."[74]

Six years later, in February 2006, the ACIP and the Academy of Pediatrics would recommend that all infants receive another rotavirus vaccine, the Merck Rotateq formulation that had been developed by Offit, Clark, and Plotkin. After Merck tested it on 70,000 children in 11 countries, in one of the biggest vaccine trials in history, the company was reasonably certain it didn't cause intussusception—although who knew what other rare condition it might be eventually linked to, rightly or wrongly?[75]

In 2002, in any case, the CDC was up to its ears in alligators and in no position to make a controversial decision. "There were so many issues around vaccine safety," said Ellenberg, "that they felt they really couldn't delay on this." The CDC's researchers were being sued, subpoenaed, and even threatened with death, and the chairman of the Government Reform Committee in Congress was seemingly intent on destroying the nation's immunization system. The accusations of Congressman Burton and hundreds of parents who supported him went far beyond a few rare side effects. They were claiming that vaccines had caused several generations of U.S. children to lose their minds.

CHAPTER 9

—

PEOPLE WHO PREFER WHOOPING COUGH

A nature reserve preserves its original state which everywhere else has to our regret been sacrificed to necessity. Everything, including what is useless, and even what is noxious, can grow and proliferate there as it pleases.

— SIGMUND FREUD,
Introductory Lectures on Psychoanalysis

SHINING MOUNTAIN WALDORF SCHOOL lies on the northwest outskirts of Boulder, Colorado, under the craggy shadows of the Rocky Mountains. The 315 children attending the K–12 school's nine-acre campus spend their days in an idyllic setting. A creek winds through cottonwood and willow trees whose shade drapes neat bungalows linked by flower-lined paths and lawns. In wintertime, thick snows close the road leading toward mountain hamlets like Gold Hill and Sunshine; in the spring, sage and alfalfa blossom in nearby pastures and scrub land. Beauty is everywhere and quite intentionally so, for aesthetics are central to the idea of a Waldorf education.

The Waldorf movement was the creation of the Austrian mystic philosopher Rudolf Steiner, who founded his first school in 1919 for the children of the workers at the Waldorf-Astoria cigarette factory in Stuttgart. Steiner saw many spiritual connections to the physical world, and he felt that particular colors and textures stimulated a child's emotional and spiritual development. "Real aesthetic life in human beings consists in this, that the sense-organs are brought to life, and the life-processes filled with soul," he said in one of his innumerable lectures. The colors inside Waldorf schools lean heavily to pastels, and the instructors discourage television and reading matter of any kind for children under age 7 because such media are thought to interfere with the child's natural inclinations and unfolding imagination. The bundling of children is another tenet of Waldorf education handed down by Steiner, who believed that physical and spiritual warmth were related. Waldorf kids are instructed to dress warmly year-round in wool hats and layers of cotton sweaters.

Some parents are drawn to Waldorf schools for their wholesome approach, dedication to play, and vigorous eschewal of the corruptions of mass media. Another, less advertised Waldorf belief, one that comes as a surprise to many parents, is that children need to become quite ill with infectious diseases in order to develop into spiritually whole beings. The same aesthetic view that favors wooden toys and cotton clothes prefers the natural trajectory of strong fevers, coughing, vomiting, and (hopefully) recovery to the sterile efficacy of antibiotics and other medications. As a result, children at Shining Mountain and 700 or so other Waldorf schools around the world are disproportionately unvaccinated, and these schools have often been epicenters of vaccine-preventable illnesses, particularly in Germany and the United States.[1]

When I visited Shining Mountain in 2001, vaccines were newly under attack around the country. The DTP controversy had been a serious but rather simple debate, revolving around the question of whether the whole-cell pertussis shot was as destructive as some claimed. The controversy had sparked the first significant consumer movement against vaccines in nearly a century and opened the doors to latent skepticism that went well beyond fears of the pertussis shot. By the end of the 1990s, polio, measles, mumps, rubella, diphtheria, and tetanus were vanishingly rare diseases, as was *Haemophilus influenzae* type b bacterial infection. Few American parents had any experience with the diseases their children were being vaccinated against. Even for grandparents, these ailments were becoming vague memories. Vaccination had done its job. Now, many thoughtful parents were questioning the value of vaccines.

Once such couple was businessman John Hickenlooper and his wife

Helen Thorpe, a journalist, who gave birth to a son in 2002, several months before Hickenlooper was elected mayor of Denver. "We were doing all the vaccines," he said, "But there was part of us that was saying, 'Is this really a good thing? Why do you get them all in one day?' There is never a time in your adult life when you feel more vulnerable than when you have your first child. There's so much you don't know about, and you don't have time to research it all. Your relationship with the pediatrician is incredibly powerful."[2]

The family pediatrician has limited time for an individual visit, and the logic behind collective vaccination is complex. It is the nature of our extremely complex, technology-dependent modern world, as George Orwell once wrote, that "much the greater part of our knowledge . . . does not rest on reasoning or on experiment, but on authority. And how can it be otherwise when the range of knowledge is so vast that the expert himself is an ignoramus as soon as he strays away from his own specialty?"[3] To one extent or another we all accept on faith the advice of those we consider experts. By 2002, when most American children had been receiving the acellular pertussis vaccine for five years, a new generation of vaccine critics were claiming new forms of harm from vaccination—autism, diabetes, ADHD, and autoimmune disorders. Parents who believed their children were vaccine-damaged were starting to become more audible and to find one another in far-flung places, because of the Internet. The Internet is a fabulous way for people around the world to communicate. It also has the potential to amplify untruths, to present an opinion as authoritative when it is not. Before the Internet, the parents of a child with autism would rarely meet other parents except at special psychiatric centers or conferences. Now you can reach out and touch thousands by typing a phrase. This is wonderful. But does it mean there are more autistics? Or has the Internet, by making instant connections among the disparate, simply taken away the denominator, the baseline of normality? After all, people don't create e-mail forums to talk about how normal their children are.

In 2001, Representative Dan Burton, a populist Republican from Indiana, began a series of contentious hearings to investigate his belief that the Public Health Service was allowing unsafe vaccines on the market. Burton was convinced that vaccines had caused his grandson to become autistic. He called FDA, CDC, and NIH officials before his committee and accused them of covering up associations of vaccines to disease. He also invited alternative and maverick practitioners to present their perspectives on infectious disease. When threats of bioterrorism appeared in late 2001, Burton recommended that the NIH look into using homeopathy to treat tularemia, one of the bacteria listed as a possible bioterrorist agent.

Of course, vaccines were not the only pharmaceutical products under attack. Starting in the late 1990s manufacturers pulled a series of popular drugs off the market because of safety worries. Well-publicized studies showed that medication errors were a major contributor to mortality, both in and outside of hospitals.[4] Antidepressants were linked to teenage suicide, pain relievers to heart attack, hormone replacement therapy to breast cancer. Each time a new drug was recalled new questions arose over whether the Food and Drug Administration was adequately monitoring the products it regulated. Some questioned whether the country was overmedicating itself. Pain medications like the Cox-2 inhibitor Vioxx, former *New England Journal of Medicine* editor Marcia Angell said, "are a testament to the fact that you can persuade people to spend more money for almost anything if you market it right."[5]

Attacks on vaccines, however, had a particular bent because of who the critics were. Drug industry watchdogs were for the most part whistle-blowers or experienced physicians with jaundiced views of the drug industry, people like Angell or Dr. Sidney Wolfe of the consumer rights group Public Citizen. These critics weren't questioning the value and necessity of pharmaceutical products per se, even when they accused the industry of pushing bad or unnecessary pills. The vaccine safety movement, on the other hand, was led by people who were skeptical of the value of vaccination, if not downright opposed to it. They were part of a tradition that reached well beyond the muckrakers of the turn of the century, back to the earliest battles between established medicine and its foes. To be sure, Barbara Loe Fisher and Kathi Williams defined themselves as consumer safety advocates, denying they were "antivaccine." But this self-definition was problematic, because they viewed mass vaccination itself as a dangerous process of questionable value. They felt that vaccines were a small or negative contribution to health and should be optional. When I asked Fisher whether she felt there were any "good" vaccines, she declined to mention any specifically, but responded that she supported "the availability of the safest, least toxic, most advanced vaccines that can be produced as a health care option for anybody who wants to use them."[6]

In the early 1990s, Dissatisfied Parents Together had been on the verge of folding, with only $2,000 in the bank, when a chiropractors' organization came through with a large donation that helped keep the group alive (it subsequently changed its name to the National Vaccine Information Center).[7] Although Fisher and Williams welcomed mainstream science when its findings supported a critical view of vaccines, they never broadcast good news from the vaccine front. Alternative views of infectious disease permeated their communications. In Fisher's worldview, as represented in her speeches and e-mail messages, official worries about infectious agents—whether anthrax, whooping cough, or pandemic flu—were always part of a campaign or plot to make money for the pharmaceutical industry by railroading parents to vaccinate their children. A successful trial of a new vaccine was typically a cause for breast-beating, because it meant, inevitably, that public health would force everyone to take the vaccine. It seemed to an outsider that in Fisher's universe, parents who claimed a vaccine injury were always right, the authorities constantly suppressed the truth, and the good studies showing harm from vaccines were only lacking because of the establishment's refusal to fund them.

Nationwide, fewer than 2 percent of parents refused to vaccinate their children in the early 2000s. But nonvaccinators, or partial vaccinators, tended to be concentrated in certain places. Ashland, Oregon, a town of 20,000 near the California border and home to a famous summer Shakespearen festival, was a typical antivaccine community. In Ashland, where a large hippie community settled in the 1970s, "the normative standard is alternative care," explained a public health official. "If you get your kids vaccinated it means that you are a dupe, that you haven't done your research."[8] A third of the children in towns like Ashland weren't fully vaccinated, leaving large gaps that would allow vaccine-preventable illnesses to spread easily in schools. Luckily, polio, diphtheria, and invasive *Haemophilus influenzae* type b infections were no longer circulating in the United States, so fluctuations in the percentage of vaccinated children caused no outbreaks of these diseases. Even measles and rubella were rarely transmitted within the United States. There was, however, one vaccine-preventable germ—pertussis, or whooping cough—that continued to be endemic in the United States, with thousands of cases reported each year. Although vaccinated people, especially older children and adults, frequently caught the disease because of the waning immunity of the vaccine, unvaccinated people were at a greater risk for catching and spreading it. Like AIDS, pertussis was in a sense a behavioral disease. There were known means of prevention that some people avoided, despite the best efforts of public health.

In 2001, I went to Boulder to find out whether vaccine shunners were contributing to the spread of the disease outside their narrow group. Whooping cough had been endemic to Boulder County since 1993, with an average of about 80 cases reported each year. Public health officials traced many of the outbreaks to the Shining Mountain school, although as time wore on the disease was popping up more and more widely along the eastern edge of the Rocky Mountains in towns like Fort Collins, Colorado Springs, and Golden—and in other parts of the state. Colorado's two-year-olds were the least vaccinated in the country, according to a CDC survey in 2003.

As it happened, Alison Kempe, Henry Kempe's third daughter, was a pediatrician at Denver's Children Hospital who did research on how public policy affected the health of children. She had noticed that the plunging immunization rates coincided with thousands of poor children on Medicaid not being able to find doctors who would take care of them. These children had been forced out of group practices that no longer wanted low-paying patients. It seemed to Kempe that there were really two sources of unvaccinated children in Colorado: "The privileged pockets of non-immunization because of ideology and paranoia, and the underprivileged not obtaining vaccines." Although the pockets of underimmunization were small, that didn't mean they had no impact, she said. "Once you get below a certain level of herd immunity, individuals can have a lot of impact."[9]

In Colorado, all a parent needed to do to avoid immunization was to sign an exemption statement on the back of a child's immunization form stating that his or her parents were philosophically opposed to vaccination. Twice as many parents were doing that in 1998 as had in 1987.[10] If the 2 percent of Colorado exemptors were spread across the state, this might not have been a major concern. But they tended to cluster in certain areas, particularly counties with high percentages of relatively educated people. In fact, a high concentration of PhDs was a risk factor for pertussis in Colorado. The parents of roughly half of Shining Mountain's children had exempted their children in 2001. With so many children unvaccinated, the school was highly susceptible. When the germ arrived—epidemics usually run in three-year cycles—unvaccinated students tended to get sick.

Vaccines as Spiritual Pollution

While the history of vaccines has its share of mistakes and tragedies, the antivaccine movement follows its own rhythms. In the 1930s through the 1970s, scientists, rather than consumer rights advocates, carried the weight of vaccine safety on their shoulders. By the time safety concerns emerged before the broad public for a second time, vaccines were safer than they had ever been before. Antivaccine movements conform to a social dynamic that has to do partly with changes in society, partly with the internal dynamics of the neo-Luddite movement, partly with the very success of vaccines in eliminating or reducing feared diseases. Fear of science, dread of change, mistrust of anonymous corporate and government entities, and a nostalgia for a simpler real or imagined past all fed these movements in the past and help explain their resilience. The viewpoints of third millenium critics of vaccination were remarkably similar to the antivaccinationists of a century earlier. But the original message had become subliminal, to the extent that even some of the most dedicated foes of vaccination were unfamiliar with the ideological and religious roots of their beliefs.

The principles of health that influence Waldorf schools and communities are vestiges—like Christian Science, Rosicrucianism, and Swedenborgianism—of spiritual philosophies that view the body as a temple, the blood as a divine fluid, and vaccines as spiritual pollution. The original Christian scriptures contain no such explicit concerns or strictures, although pollution and defilement are perhaps no more than vivid expressions of the idea of original sin. But each of these postindustrial religious sects arose at a time when society was visibly mucking up nature and humankind. They were a millenarian response to the despoiling of traditional life. And although arguably our air, water, and food are less polluted than they were 50 years ago, the belief that we have defiled ourselves with pollutants and toxins remains just as intense if not more so. Spiritual holists are unimpressed by the progress achieved through secular achievements such as the Clean Air Act. Laws cannot alter the fact that we are a fallen civilization. The government agencies and corporations that maintain otherwise are merely the agents of a worldview that is dishonest and impure. Just by dint of industrial processing, our air, water, and food are spiritually dead.

Because of their ubiquitousness and the government's involvement in administering them, vaccines are a natural target for paranoid caricatures. (In a December 2000 episode of *The Simpsons*, Homer is kidnapped to keep him from revealing that flu vaccine is administered just before Christmas because it contains a serum that heightens the impulse to shop.)[11] In a recent best-selling airport novel by the parent of a child with autism, a thinly fictionalized Barbara Loe Fisher character intervenes at the last moment to foil a plot by the pharmaceutical industry (and a thinly veiled Hillary Clinton) to foist a deadly, 30-component "super vaccine" on the public.[12] Such depictions of evil masterminds aren't that far off from the actual fears of vaccine critics like Fisher and Phyllis Schlafly, who have battled state immunization tracking programs—which aim to assure that kids are vaccinated on schedule and that they aren't overvaccinated—on the grounds that they are a step toward tracking Americans "to deny admission to daycare, kindergarten, school or college, or even access to medical care for any child who has not had all government-mandated shots."[13]

Elements of political paranoia characterize the perspectives of both the Christian and New Age alternative health activists who oppose vaccination. In general, paranoia is more acute in those who feel wronged—the chronically ill, the adult abused child, the parents of terribly ill or prematurely dead children. Paranoia bridges the gap between conservative and radical left critiques of modern medicine.[14] When it comes to vaccines, the paranoid contingent feeds heartily on the writings of Leonard Horowitz, a former Gloucester, Massachusetts, dentist who theorized that the Ebola and HIV viruses were created in experiments conducted by Hilleman, Krugman, and scientist Robert Gallo with the backing of Henry Kissinger and various ex-Nazis, CIA agents, cold warriors, and corporations.[15] Such florid fantasies, which circulated in books and pamphlets sold in health food stores and over the Internet, earned Horowitz an audience in 1998 with Louis Farrakhan, leader of the militant Nation of Islam. The Nation of Islam subsequently declared a moratorium on vaccinating its children, lifting it only after Farrakhan met with then-CDC director David Satcher.[16] Horowitz continued to spread his theories at gun shows and traveling holistic medical conferences. In 2003, Republican Congressman Chris Shays invited him to testify from the audience at a hearing critical of the vaccine establishment's response to autism.[17]

Notions of authenticity and autonomy that surfaced in the late 1960s also underlie the antivaccine philosophy. Ivan Illich, a Vienna-born philosopher and theologian who founded a school for nonviolent change in Cuernavaca, Mexico, stimulated these ideas in *Medical Nemesis*, a book in which he argued that modern medicine had "undermined the ability of individuals to face their reality, to express their own values, and to accept inevitable and often irremediable pain, impairment, decline and death."[18] One prominent foe of vaccines, the homeopathic physician Richard Moskowitz of Massachusetts, contended that vaccine-mediated immunity is "inherently counterfeit." Health comes from Nature, not the physician, Moskowitz wrote, and "healing applies only to individuals in unique here and now situations rather than to abstract disease, principles or categories. It is an art and can never be reduced to a technique or procedure however scientific its foundation. . . . health, illness, birth and death are inalienable life experiences belonging wholly to the people undergoing them. Nobody else has the right to manipulate or control them, or any part of the body involved in them, without their explicit request."[19] Not only healthcare is an inalienable right, so is the right to get sick.

Another group that has contributed, in a small way, to vaccine-preventable disease is the Maharishi organization, a Hindu cult with a considerable following in the United States. The Maharishis shun vaccination, preferring to "reestablish balance between the body and its own inner intelligence through Vedic knowledge." *Veda* is a Sanskrit word that means "knowledge," and Vedic medicine is a form of traditional Indian medicine. In 2002, six students from Maharishi University in Fairfield, Iowa, flew to India during spring break and got measles (a seventh became ill returning home). Either on the plane or in Iowa, they exposed about 1,000 people to measles, forcing the state's Health Department to organize a large quarantine effort. A measles outbreak tied to a Christian community in Indiana in 2005 had similar results.[20]

While early-twentieth-century religious sects frequently attacked vaccination on religious or "commonsense" grounds, contemporary vaccine critics incorporate "science" into their arguments. By invoking science while implying that government and industrial science are corrupted by power and money, holistic medical practitioners gloss over the extent to which alternative philosophies, rather than science, shape their critique of vaccination. This may reflect a philosophical insecurity, but it also shows how powerful established medicine's authority has become over the decades. As Harvard historian Charles Rosenberg has pointed out, mainstream medicine is now confident enough of itself to incorporate—following the suitable scientific trials—any "alternative" method that works.[21] Thus to reach a larger audience, alternative medical practitioners—and those who oppose vaccination—have adopted the language of science. (This is not unlike Christian evangelicals who, believing in the literal truth of the Bible, arm themselves with "creation science" or "intelligent design" to do battle against Godless evolution.) The alternative health worldview—a set of convictions and emotions, a way of life, a source of income—now work synergistically with straightforward concerns about the health effects of vaccines.

Like any other service in which the consumer has a choice, medicine asserts its authority partly on the basis of trust and partly on the basis of tradition. Most of us trust the family doctor and the specialized machines and experts to whom he refers us. The bedrock underlying that trust is the scientific proof of the success of medicine's offerings. Reductionist science, as the cumulative wisdom of small experiments is known to both critics and proponents, undergirds medical authority. By its nature, though, reductionist science does not offer all the answers to sickness and health. By contrast, holistic medicine—whether it is anthroposophic, homeopathic, Vedic, or just plain fruitcake—answers any question on the basis of absolute principles. Thus, trust in medicine can fail either because there is no tradition for it, as in Christian Scientist families, or because medicine fails to relieve a particular ailment. Modern medicine excels at treating acute illness and its track record with chronic diseases like cancer, diabetes, and depression is improving. For diseases where there is neither a cure nor an explanation—for example, autism—"holism" holds its ground. When it comes to gray zones of unknown risk and cause, many people seem to find holistic approaches more comforting than the unripe fruits of reductionism. Meaning is central to the holistic approach—the endowing of meaning to the illness and, almost always, of meaning to the cure. Meaning of this sort is something that mainstream medicine fails to provide.

Homeopathy, Rosicrucianism, and the Battered Child

"What makes homeopathy scientific, in the opinion of its practitioners, is the rigor of its method," wrote medical historian Harris Coulter, who was a foe of vaccination and a cheerleader for homeopathy. "Homeopathic method is very precise, and the practice of this form of therapeutics is a demanding discipline." Whereas traditional medical practitioners feel that "the basic laws of health and disease" have not been disclosed, Coulter continued, homeopathy "always insisted on the necessity of practicing medicine guided by a set of principles."[22]

Few outside its discipline would call homeopathy a science. Working from the principle that "like cures like," homeopathic physicians use microdilutions of deadly substances—millionths or even trillionths of a gram mixed in water—to heal illness. Most rigorous trials of homeopathy have shown that it does little but provide a placebo benefit. It does, of course, have method and principles. But while "rigor" and "precision" are important to science, they are not science. The "rigorous" eighteenth-century application of mercury, sulfur, bleeding, and leeches was not scientific, despite what its practitioners pretended. In fact, it was the unscientific character of mainstream medical practice at the time that enabled homeopathy and other light-handed cures to grow.

Many of the most widely distributed antivaccine books—typically written by homeopaths or naturopathic practitioners—contain ample footnotes, graphs, and tables intended to scientifically discredit mainstream science. As I traveled around Colorado in the fall of 2001, and again three years later, I frequently came across a book in health food stores titled *Vaccines: Are They Really Safe and Effective?* The author, Neil Z. Miller of New Mexico, claimed in an e-mail communication in January 2005 that he had sold 125,000 copies of the book. In one of his more egregious manipulations of data, Miller argues that rubella vaccination actually *caused* an increase in congenital rubella syndrome. In another instance, Miller takes a typographical error in a published report as the basis to claim that the smallpox fatality rate in Europe in the eighteenth century ranged from 1 per 250 to 1 per 5,000 cases, when the true figure is closer to 1 in 4.[23] He also claims to have discovered by diligent research that polio vaccine was not responsible for the decline in polio. To believe this, one would have to assume a massive cover-up involving thousands of scientists and doctors.

Miller and the homeopathic practitioner Randall Neustaedter of California, whose *The Vaccine Guide: Making an Informed Choice* is similarly vituperative, frequently cite Harris Coulter's work. Coulter wrote a multi-volume history of homeopathy in America and also coauthored *DPT: A Shot in the Dark* with Fisher. He developed a view of vaccination that was very dark indeed, as reflected in the title of his 1990 book, *Vaccination, Social Violence, and Criminality*. Coulter's thesis is that brain inflammation caused by whole-cell pertussis vaccine causes most serious pathologies in American life—everything from blindness, asthma, and chronic diarrhea to rape, autism, and ADHD, not to mention obesity, innumeracy, serial murder, and crossed eyes. Pertussis brain damage is "organically linked to the rock group," he adds— clearly, Coulter doesn't appreciate headbanging music.[24] Sociopaths whose medical records indicate no reaction to DTP, Coulter asserts, undoubtedly sustained hidden damage.[25] In short, "vaccination has altered the very tone and atmosphere of modern society. Because the changes are so insidious and widespread, and because we lack perspective, they have been largely overlooked."[26]

Coulter was in a nursing home by the time I started writing this book, and I had no luck contacting him. I did locate Dr. Harold Buttram, another prominent vaccine foe who had blurbed many of Coulter's books. Unique in the antivaccine world, as far as I determined, Buttram is directly linked to an antivaccine movement from a century ago. He is head of a Quakertown, Pennsylvania, clinic that specializes in healing diseases caused by the hidden toxins of modernity. He is also a high priest in the Church of Illumination, the "outer door" of the Rosicrucian Fraternity, established by Reuben Swinbourne Clymer, who was vice-president of the Anti-Vaccination Society of America in 1902. Buttram's clinic is a short stroll through the woods from the Rosicrucian meeting house on Clymer's estate north of Philadelphia.

I called Buttram in the fall of 2004 and he invited me to meet him the following week in Brooklyn, New York, where he had been hired to testify in family court on behalf of a couple who were attempting to regain custody of their 15-month-old daughter. The authorities had placed the girl in foster care in September 2003 after she was taken to the hospital with a swollen arm. On examination, she was discovered to have suffered rib fractures, a broken arm, two broken legs, permanent brain damage, and retinal damage during her first three months of life. Buttram was to testify that vaccinations and vitamin deficiencies were responsible for the girl's gruesome condition—an argument he had already made in about 60 other cases. Buttram worked closely with a San Diego attorney who represented parents and caretakers claiming that their charges' battered condition resulted from vaccination. This sort of defense against child abuse allegations was directed, as it happened, at the legacy of virologist C. Henry Kempe, whose "battered child" hypothesis began to open eyes to the abuse of children by their parents and caretakers. In the mid-1960s, after colleagues ousted Kempe from the smallpox world for his vaccination heresy, he had turned to diagnosing the hidden epidemic of child abuse. The irony was rich, but in the small world of pediatrics it often seemed that controversies were separated by only a few degrees.

Buttram had made a name for himself in this world by helping to spring Alan Yurko, a convicted felon who was arrested and charged with murder in the death of his infant son in 1997. The baby, Alan Jr., had suffered multiple fractures, retinal hemorrhage, and brain swelling. Yurko, who had just finished a seven-year prison term for armed robbery, was charged with battery on an officer and resisting arrest when police came to detain him for the baby's death at his house in Orlando, Florida.[27] Once in prison, he connected with vaccine opponents on the outside through Internet research. Despite his criminal past, much of the antivaccine community viewed Yurko as a martyr for the cause. Buttram convinced a group called the International Chiropractors' Association to foot the bill for Yurko's defense, and a group of 50 supporters, wearing matching "Free Yurko!" T-shirts, attended his retrial near Disneyworld in the summer of 2004. Buttram testified, as did Dr. Jane Orient, a survivalist who led a libertarian group called the American Association of Physicians and Surgeons. Although the judge said there was little evidence that vaccines had harmed his son, Yurko was freed because a medical examiner had botched the autopsy. Florida authorities immediately deported Yurko to Ohio to serve out a parole violation. From prison he coauthored articles about shaken baby syndrome for Buttram's website.

Buttram, like Coulter, has written that "much of the crime, personality and nervous disorders, mental illness, drug addiction . . . have their basis in subtle immune alterations taking place, largely unrecognized," because of "vaccines, denatured foods, formula feeding of infants and antibiotics."[28] Buttram has claimed that smallpox vaccine causes diphtheria and vitamin C prevents polio. Modern childhood vaccines, rather than offering protection, in fact deplete the immune system, he insists. This is a common belief in antivaccination circles.[29]

During a break in the Brooklyn shaken baby hearing, Buttram and I left the putty-brown family courtroom and had lunch at a nearby hotel. A lean, thin-lipped 79-year-old, he wore tortoise-shell bifocals and an old herringbone jacket. Buttram had grown up in Oklahoma, he told me, in a family that was suspicious of medicine. At age 17 he suffered a nervous breakdown while at boarding school. Poor nutrition and vaccinations he got in the military and in medical school subtly poisoned him, accounting for his "subpar performance" in these institutions. Briefly diagnosed with schizophrenia, he found solace and recovery in the Rosicrucian order. Around the time he graduated from the University of Oklahoma Medical School in 1958, Buttram had the good fortune to meet Clymer, "an aloof Victorian of the old school," as Buttram remembered him. Unlike Rudolf Steiner, Clymer gave no public lectures, confining himself to the written word because "he did not wish to be deified like Jesus."[30] But to Buttram, Clymer was a visionary. "I believed him to be infallible," Buttram explained simply. That is also the official belief of the Rosicrucian Fraternity. Clymer died in 1966 at age 87, but the fraternity has faith that he will eventually return as its master. After the terrorist attacks of 9/11, Clymer was said by the fraternity leadership to have sent an apocalyptic message from the afterlife that prophesized the start of a new world war.[31]

Buttram told me that his health philosophy was deeply influenced by Clymer's 1957 book, *The Age of Treason*. After reading this jeremiad against vaccination, food additives, and artificial hormones, Buttram spent a career in the ebb and flow of the medical antiestablishment. He has written books about herbal therapies and marital sex, prescribed megavitamins, treated autistic children with porcine hormones and heart patients with chelating agents. If it's nontraditional and controversial, chances are good that Buttram will try it. For all that, Buttram's enthusiasm for *The Age of Treason* is quite extraordinary. True, the book's dark meanderings about our toxic universe are commonplace in the alternative medical literature. What makes this book a classic, in the paranoid genre, is that it blames these ills on Communists and Jews—an update on the old well-poisoning myth. The book's thesis is that birth control, polio vaccine, mood-altering drugs, hormones, and racial miscegenation are the tools of a Satanic plot to undermine Anglo-Saxon America.[32] Vaccinations are especially diabolical, in Clymer's view: "The enemies of God and mankind . . . have used or plan to employ inoculations for the purpose of destroying mental balance and making it impossible for the minds of children to develop beyond a more or less moronic or robot degree." He fears that vaccines will be used to introduce a "Marxist agent to degenerate human reason and make of man a robot, a human monstrosity." (Racial desegregation is another "means to bring about demoralization and degeneration of both the white and the colored races" that he blames upon "militant socialists and the enemies of God and Man, many of whom are admittedly Jews.")[33]

The Age of Treason makes a rather exceptional bible for Buttram, given the fact that when I met him he was being paid $1,000 to testify on behalf of an African American family defended by Nigerian-born lawyers in front of an African American judge. The battered child's mother was a well-spoken financial analyst, and her semi-employed husband came across as gentle and solicitous. They seemed like nice people. The state had no specific theory of how the child was harmed. But it was not impossible, given the evidence, to imagine some truly terrifying scenarios. Buttram presented his theories of vitamin deficiencies and free-radical attacks on brain fats. Shaking couldn't have caused the little girl's multiple fractures, her brain damage, her blasted retina—the baby must have been born with weak bones. The court officers were courteous. Only when Buttram stated that he didn't believe there was such a thing as shaken baby syndrome did the women in the courtroom exchange raised eyebrows. Do you believe in vaccination at all? the judge asked. "None of the current vaccines have been tested for safety," he responded.

Somebody in this sad, claustrophobic room was concealing a secret, but it didn't seem to be Dr. Buttram, who tenaciously clung to his thin raft of a narrative. At a break, the mother's lawyer, Omotayo Orederu, complimented Buttram for his testimony. But Orederu, who grew up near Lagos, couldn't help mentioning a nagging doubt of his own. "What about polio—isn't that a good vaccine?" he asked. "In my village there was a guy who still can't walk because of polio, but the disease is almost gone now." Buttram didn't respond. He could speak quite a mouthful about polio. But one has to choose one's moments carefully.

Anthroposophy: Smallpox on the Road to Salvation

In his own day, Rudolf Steiner was called a "Rosicrucian theosophist" by people who were interested in such matters. What they meant was that he blended the mystical Christianity of Rosicrucianism with the exoticism of theosophy, a cult made up of mystical, neo-Buddhist intellectuals who liked to hold hands under blankets in dark rooms and look for multi-hued auras around each other's heads. Steiner said that his revelatory work in the mystical sphere had enabled him to dissolve the dualism of science and religion and would help humankind re-create the spirit world that existed earlier in our "evolution." *Anthroposophy*, a term Steiner coined himself, was meant to be a sort of occult science of humankind. It is a long way from the séance parlors of Mitteleuropa to the thin air of Boulder, yet the distinctive Waldorf pedagogy of the Shining Mountain School fits quite well into the local culture. Boulder can seem very far away, too, from the East Coast and its professional anxieties. Boulderites tend to be well-educated, wealthy, and outdoorsy. They call their city the "Athens of the West," and like to think of themselves as rugged individualists with an intellectual bent; their state was one of the last to have a mandatory vaccination law.

Steiner was influenced successively by the works of Goethe, Nietzsche, and the theosophist Madame Blavatsky before turning back to his own form of Christianity. From around 1900 until his death in 1925, Steiner espoused his beliefs in more than 6,000 lectures on everything from dance therapy to cooking. Though he had limited scientific learning, Steiner did not step hesitantly into any subject. Among his many footnote-free works were dozens in which he spoke authoritatively on complex matters of education and health. His ideas are fantastic, bizarre even, but not maniacal. Though baffling there is much beauty in his writing, in the manner of a mellow trip down a quiet stream in a state of mild hallucination. But Steiner, who in photographs bears an uncanny likeness to the actor Jeremy Irons, declared that his gospel represented the fruit of a spiritual science. His findings were achieved by "supersensory" observations of hidden, prelapsarian worlds and dimensions accessible only to those with cleansed powers of perception. Since most if not all of his followers presumably lacked the ability to carry out such esoteric research, they had to take his word on it. Anthroposophy was in effect a cult.[34] Franz Kafka dallied with anthroposophism, resonating with Steiner's obsessive advice on diet and color schemes. "The efforts of Dr. Steiner will only succeed if the Ahriminian forces do not get the upper hand," Kafka wrote in his diary after attending a lecture in Berlin in 1911. "He eats two liters of almond and fruit emulsion that grow in the air. He communicates with absent disciples by means of thought forms." Kafka sought Steiner out and asked whether he should become a disciple. Steiner picked his nose constantly during the interview, which appears to have been inconclusive.[35]

Like many turn-of-the-century sectarians, Steiner gave great significance to the blood, including its racial connotations, and vaccination was at times equated with blood pollution. He linked the biblical Fall with race mixing, which he believed had caused humans to lose their biological connection to an ancestral world in which spirit was visible and omnipresent.[36] In one lecture, he stated that blond, blue-eyed people were disappearing from the world because they were "weaker physically and mentally stronger" than "dark people." This, according to Steiner's occult science, was because a "blue-eyed person's organism does not drive the nourishing substances far enough into the eyes to fill the front part of the iris with them."[37] The vital substance that "the fair people retain in their brains" was in dark people driven into the hair and eyes. That allowed them to "last longer, for they possess greater driving force" while "blonds and blue-eyed people are already marked for extinction"—an unfortunate development because "if the blonds and blue-eyed people die out, the human race will become increasingly dense." Our only hope was that people might evolve a new level of intelligence that was "independent of blondes."

When it came to disease, Steiner had a hailing familiarity with the emerging discipline of bacteriology, but he endowed germs with a spiritual significance that probably didn't impress his contemporaries Pasteur and Koch, if they happened to notice. "There is a certain species of bacilli who are the carriers of infectious disease," he stated in one lecture; "these beings are the progeny of the lies told by human beings; they are nothing else than physically embodied demons generated by lies."[38] In a series of lectures to the workers building his new headquarters in Switzerland in 1922–23, he stated that overfeeding infants caused measles and scarlet fever, that smallpox resulted from heavy breathing or sweating, and that excessive bathing caused diphtheria. Influenza was contagious not through viruses but as a result of the garlic-and-onion odor emanating from people with flu, which "disrupts the astral body" of the people who smelled it.[39]

Elsewhere, Steiner fell back upon second-hand Eastern spiritualism to define illness. Diseases of childhood were the result of spiritual impurity. Epidemics occurred when "a great number of people felt impelled due to their unloving attitude to their fellow human beings" to "absorb certain infectious substances in order to succumb."

Vaccination was in general a bad idea because it "would prevent the outer physical nature from expressing the unloving disposition while failing to remove the inner inclination to unlovingness." That is, vaccination might prevent epidemics but at the cost of leaving spiritual work undone: "If we destroy the susceptibility to smallpox we are concentrating only on the external side of karmic activity." Since "man cannot escape his karma" and smallpox was the "organ of unlovingness," vaccination might prevent the disease but the "cause of unlovingness would still remain, and souls in question would then be forced to seek another way for karmic compensation either in this or in another incarnation."[40] (This is similar to the Swedenborgian dismissal of vaccination as a spiritual trick [chapter 3].) Managing your karma was a matter of grim, Nordic hard work—you couldn't fight it with some cheap gimmick like a vaccine.

Anthroposophical medicine in practice was similar to homeopathy, the oldest Western system of holistic therapy. The system of homeopathy, attributed to the nineteenth-century physician Samuel Hahnemann, rested upon the ancient belief that there were four humors in the body that needed to be kept in balance. Homeopathy sought to balance the humors by administering tiny therapeutic doses of medicine. To treat smallpox, a homeopathic physician might administer a solution containing smallpox virus that had been diluted until there was no more than one smallpox particle per trillion parts water. Like the Rosicrucians, the anthroposophists stressed that hardening (sclerotic) and softening (inflammatory) forces were at work in the body; inflammation was characterized by "calor, dolor, tumor, and rubor." To treat an infectious disease, you rubbed an extremity until it turned red, because this "rubor" would encourage the eruption of rash that helped rid the body of the infection.[41]

Steiner was also a bit of a numerologist. He divided the child's development into seven-year periods. Before the age of seven, the child's "etheric body" remained linked to that of its mother. It made no sense to teach children under this age because that used up forces "which should be conserved for a latter state of development." Vaccination, like reading and writing, was an extraneous influence that disrupted the mother's spiritual dominance of her child. Steiner warned of a day when medicine would create a vaccine "which will be injected into the human organism in earliest infancy, if possible immediately after birth, to ensure that this human body never has the idea that a soul and a spirit exist," a day when "materialistic doctors will be entrusted with the task of driving souls out of human beings."[42] These Steinerisms, remarkably similar to Clymer's paranoid vision, were recently published by an anthroposophist physician in an article distributed at a Waldorf school in San Francisco. Dr. Philip Incao, who assembled them, concluded that Steiner's comments "leave no doubt about the 'hidden agenda' behind the plan to vaccinate all the world's children with as many vaccines as possible, thus devastating their spiritual development."[43]

When I inquired about the Waldorf vaccination policy with the schools' licensing body, the Association of Waldorf Schools of North America, a spokesman said the group took no position on vaccination, but referred me to Dr. Incao, then living in Denver.[44] Eventually I spoke with him after his address to a November 2002 conference on vaccination organized by Barbara Loe Fisher and Kathi Williams. I found that Incao faithfully recapitulated Steiner, with a light scientific coating. A bearded man with imposing eyebrows who would have looked right at home in a North Beach café, Incao wore a gray beret and walked with a cane. The 300 or so people attending the conference, most of them vaccination opponents with naturopathic perspectives, greeted his speech with frequent applause. Incao said that medicine was on the brink of a holistic revolution, a turning away from the "fear-based paradigm" of modern science. Injuries caused by vaccines, he stated, were "symptoms of a dying paradigm." Illness and healing were fated events that helped develop human consciousness, and vaccines were part of our fearful suppression of the natural way. "Illness happens unexpectedly to people just as accidents happen to cars," he said, but "in both cases the cause is usually the driver, not mechanical malfunction. The new paradigm will have to rediscover the human spirit."

Trained in both traditional and anthroposophic medicine, Incao established a practice in Harlemville, New York, in 1973, and was also the medical director of a nearby residential home for disabled students. Incao saw hundreds of cases of whooping cough in unvaccinated children during his 23 years in that upstate Waldorf community. "It would sweep through, infecting 20 or 30 people at a time," he told me. No one ever died, and although the youngest was only four months old Incao never hospitalized any of the children. He treated them homeopathically. "The ones who did best came from the stablest families," Incao said. "The ones from difficult upbringings got the sickest." In 1988, the state medical board pulled his records, presumably over vaccination, but no charges came of it. In 1996, after his first wife died of breast cancer, Incao moved his practice to Boulder and Denver, where most of his patients were adults with what he called "mystery ailments."

Incao viewed himself and other alternative healers as modern-day Galileos, cast out by the chauvinism of the ruling paradigm. "Dissent from the prevailing wisdom is never well received, yet it is the only way progress in medicine or any field, has ever been made," he writes.[45] It is pretty hard to imagine a universe in which the mainstream current would be swimming around a doctor who advocated the benefits of serious infectious diseases for children. Yet that is most unabashedly Dr. Incao's position. "One of the best ways to ensure your children's health is to allow them to get sick," he has written.[46] "Standard childhood illnesses, such as measles, mumps and even whooping cough, may be of key benefit to a child's developing immune system and it may be inadvisable to suppress these illnesses with immunization."

In keeping with alternative medicine's need for a scientific veneer, Incao often cited the "hygiene hypothesis" as the modern-day version of anthroposophical beliefs about the immune system. The theory is an intriguing one, with a certain amount of data behind it, and it is worth some examination.

The hygiene hypothesis attempts to explain the apparent increase, in the developed nations, of allergic disorders such as asthma, as well as certain other immune-mediated conditions such as multiple sclerosis and diabetes. The basic idea is that it is unhealthy to grow up in an environment that is too "clean" in terms of microbial or viral exposures. The absence of such exposures may cause a shift in the predominance of two types of white blood cells, the T-helper cells. The two types, known as Th1 and Th2, trigger other immune cells to release different mixtures of inflammatory proteins called cytokines. The right kind and quantity of microbial exposures, so the theory goes, will produce mostly Th1 cells in the respiratory system, resulting in a cascade of cytokine products designed to fight microbes—which is what you want. In the absence of Th1-stimulating infections the Th2 cascade predominates, which causes the overproduction of IgE, a type of antibody that binds to things like cat dander and peanuts—in the process creating an allergic, or atopic, response.

In its modern scientific form, the hygiene hypothesis originated with a 1997 paper in *Science* by Julian Hopkins and Taro Shirakawa of Churchill Hospital in Oxford, England.[47] They found that Japanese children who had been vaccinated against tuberculosis were less likely to develop allergies. Since the TB vaccine was a live attenuated bacteria, the study suggested that exposure to some microbes, such as the mycobacterium tuberculosis, might help protect against allergic disorders. Since then, more than 200 studies have been conducted to assess the theory, with mixed and complex results. There are good data demonstrating that children who grow up on farms and children in large families get fewer allergic disorders. Studies in England have also suggested that juvenile diabetes and acute lymphoblastic leukemia, the most common childhood cancer at about 30 cases per million, are also higher in children who have fewer early infections.[48] First-born children, by contrast, are consistently more at risk for allergic disorders and multiple sclerosis.[49]

But the evidence does not consistently show that exposure to germs prevents allergic diseases. Children exposed to particular viruses at an early age, particularly the respiratory syncytial virus, for example, are *more* rather than less likely to become asthmatic. Exposure to endotoxin, a fever-inducing component of pertussis as well as the whole-cell pertussis vaccine, seemed to protect against allergy in some studies, while in others it provoked it. So there may be specific microbial proteins that protect some subset of children from allergic illness, but there's little more definitive to be said than that. As for vaccination, a few small studies suggested a link to childhood asthma and allergy but the larger studies were negative.[50] Still, working from the evidence that something in modern life was screwing up babies' immune systems, scientists are experimenting with vaccines that contain tiny fragments of recombinant bacterial proteins known to stimulate the Th1 leg of the immune system. The peptides would be added to vaccines to push the immune response in a healthy direction.[51]

"The hygiene hypothesis is a nice tidy hypothesis and there are data to support it and data not to support it," said a senior NIH scientist. "Like every grand hypothesis there's some truth to it, and some holes in it."[52] Based on the available evidence, the hygiene hypothesis, though intriguing, was hardly a basis upon which to decide whether to vaccinate. But on a philosophical level, the hypothesis meshed well with the cultural pessimism that predominated in the antivaccine scene. It harkened back to a time when health was theoretically better—when we were happier, grubbier, and more robust. It employed science to rehabilitate earlier forms of nonscientific "wisdom." It reminded some of us, perhaps, of our grand-parents' old saws about how a peck of dirt a year made us healthier. It took the priming of the immune response out of the hands of medical experts and into our own humble, soiled ones.

To be sure, this type of nostalgia and cultural pessimism was not con-fined to healthcare Luddites. Many scientists have speculated on the uncer-tain meaning of artificial immunity. The Nobel laureate Alexis Carrel, whose pathbreaking experiments on viral cultures at the Rockefeller Institute did a great deal to advance the creation of viral vaccines, ended his career on such a note of pessimism. "Medicine is far from having decreased human sufferings as much as it endeavors to make us believe," he wrote.[53] "Indeed the number of deaths from infectious diseases has greatly dimin-ished. But we still must die, and we die in a much larger proportion from degenerative diseases. The years of life we have gained by the suppression of diphtheria, smallpox, typhoid fever, etc., are paid for by the long suffer-ings and lingering deaths caused by chronic afflictions."

A vague discomfort with modernity is part and parcel of the antivac-cine worldview. But while asthma and a few other childhood illnesses have increased over the past few decades, the gloom and doom that were the mainstay of Barbara Loe Fisher's message seemed to have little basis in fact. Fisher maintained that chronic medical conditions, including autism and ADHD, had skyrocketed since Public Health began requiring more vaccinations. But the rates of disability—7 percent of children aged 5 to 17—remained basically unchanged from 1984 to 1998. With infec-tious diseases under control, cancer had become the leading cause of death for children under 15, but it was still pretty rare—one to two cases each year for 10,000 children—and less common than 20 years earlier.

The hygiene hypothesis provided a handy scientific gloss for the views of anthroposophy, but the philosophy drew on ancient medical ideas. In a recent antivaccine anthroposophist collection that included an essay by Fisher, Incao extolled a theory of Constantine Hering, who brought homeopathy to the United States in the early nineteenth century.[54] Hering believed that the rashes characteristic of smallpox and measles were a sign of health, in that "as an illness resolves, its manifest signs and symptoms travel from inner vital organs and blood to the outer surface, often visible as a rash or discharge to 'throw off illness.' " Suppressing infections by vaccination would only keep the diseases inside the body, where they could do more harm, Incao concurred. Therefore it was healthier to get measles than to be vaccinated against it. The problem today, Incao wrote, was that "we don't fight our battles vigorously enough" and therefore are "liable to be infiltrated by the enemy in disguise and suffer from chronic allergic or autoimmune disorders."

In the anthroposophist interpretation of illness, with nuggets of immunology plucked blithely out of context to dignify the argument, "illness was part of a child's destiny" and it was not good to "do the work for them."[55] Every childhood inflammation," wrote Incao, "every cold, sore throat, earache, fever and rash is a healing crisis and a cleansing process, a strong effort by the human spirit to remodel the body, to make it a more suitable dwelling."[56] The stirring metaphors could almost make one nostalgic for smallpox. Almost.

The Athens of the West

In October 2001, snow already dusted the higher peaks above Boulder and the aspens were changing to yellow and orange. Although pertussis was a familiar visitor to the school, there hadn't been an outbreak at Shining Mountain in a few years and the school's principal, a youthful-looking, bearded 47-year-old named Robert Schiappacasse, found my interest in this "very uninteresting subject" to be droll, though he was willing to discuss it briefly between phone calls and paperwork. Parents at Shining Mountain, he said, are "more likely to be concerned about fumes from a new carpet than they are about infectious disease." The school didn't give parents advice one way or another. Schiappacasse's own daughter had suffered whooping cough with no ill effects, though he admitted to some concern when a secretary's infant "coughed itself into a hernia" after being exposed at the school.

A group of parents I found congregating in the school parking lot, on the other hand, were eager to discuss their opposition to vaccination. They were full-time moms who felt that mothers, not public health officials, should be in charge of deciding how to handle their child's health. They harbored a nostalgia for illness that was touching, evoking a past like *Little House on the Prairie* in which everyone had time for each other, children were brave and responsible and content with their simple wooden toys, and parents feared epidemics but were sure that if they stuck to their beliefs and used time-honored remedies they would pull through. "What I've heard that resonates with me is there's a little bit of soulfulness with getting ill," said Christine Anderson, whose 13-year-old daughter Monika, she regretfully noted, was vaccinated before Christine had become aware of how bad vaccination was. "Sometimes people say that after a fever you see a difference in a child's being. It really strengthens them. No one wants their kid to have polio but it almost seems like we've gone to the other extreme. We treat our children like machines that are never supposed to slow down or let us miss a day of work or function right on schedule all the time. To me that's an opposite extreme. We never allow them the soulfulness of being ill. That is a human experience, and hand in hand with that is nurturing and nursing people, which was an art we all engaged in."

Christine's friend Johnnie Egars agreed. "Some of the most bonding soulful times with my daughter have been when she's had 'inconveniences' that go on for five or six days. They weren't polio or anything life-threatening but you know, chicken soup." Johnnie recalled that as the fourth of four children, she seldom got any attention from her mother unless she was sick. She didn't want to deny that pleasure to her children, and didn't vaccinate them, making an exception for tetanus because the vaccine in that case was similar to the puncture wound of "natural" tetanus infection.

As a parent of small children at the time I could see, at first, what she meant. My wife and I were classical therapeutic nihilists—we avoided medicine when it wasn't necessary—and, back before our school had required it, had declined the first opportunity to vaccinate our son against chickenpox. Our pediatrician had been indifferent to the vaccine, and the CDC information sheet mainly emphasized the benefit of fewer missed days of work and school. But I wondered if it was really necessary to be close to death to spend extra time with your parents. Wasn't an ordinary ear infection excuse enough to lay on the chicken soup, saltines, and picture books and stay in pajamas all day long, good enough reason for mom or dad to skip work once in a while?

The rest of Egars's story raised my eyebrows a little further. All three of her children had had whooping cough, including the youngest, Elise, who was 2 and undergoing chemotherapy for a congenital Willms tumor of the kidney. "It's intense," Egars noted. "a loud cough that goes down to their toes. The whoop is a sharp intake of breath. They cough and cough until they throw up, then they sleep for an hour or two, then they wake up and start all over. That went on with Elise for about three weeks." Her daughter was hospitalized and put on steroids with a nebulizer for three days in the pediatric infectious disease unit. "Antibiotics probably saved her life," Egars admitted. Hundreds of thousands' of dollars worth of the world's best medical care also helped. Despite all that, Johnnie claimed she had no second thoughts about her decision not to immunize. Convinced as she was that vaccines weaken the immune system, and that illness strengthens it, she was happy to have her children endure the disease. Elise was now 10 and ponytailed, passing out leaflets for a school event by her mom's side and looking very healthy. Thanks to modern medicine, it seemed to me.

I wrote an article about whooping cough in Boulder that appeared in *The Atlantic Monthly* in 2002 and gave the concluding quote to a holistic therapist who compared some Boulder parents unfavorably to the firemen at the World Trade Center: "Their view is, 'I'm going to let everyone else's child take a risk but not my own.' That's not avant-garde. That's not enlightened. It's pretty primitive." The vaccine critics understandably hated that sentence, but I felt it was fair enough. Herd immunity was real: the more members of a community failed to vaccinate their children, the more risk there was for everyone, including those who were unvaccinated because their parents were poor or neglectful or didn't speak English, or even those who were vaccinated—because heaven knew, the pertussis vaccine didn't work perfectly. Millions of dollars had been spent to develop acellular pertussis vaccines to convince Americans like these to keep vaccinating their children. And whooping cough was a serious disease. It killed three babies in Colorado between 2000 and 2002, and scores of others were hospitalized with severe breathing problems. But many people in Boulder did not care to run the very slight risks of vaccination to serve the common good, or even the good of their own kids. When I returned to Colorado three years later I found the vaccination rate even lower, and the whooping cough epidemic much worse. By the end of 2004 there were

1,200 cases of the disease in the state. Coloradans were getting sick with pertussis at a higher rate than any year since 1964.

To be sure, this was not a return to the 1930s, when there were 20,000 whooping cough deaths each year. With healthier children and better intensive-care medicine, perhaps 20 children died of pertussis in the United States. But there were an estimated 2,000 hospitalizations, and they weren't pretty affairs. I spoke with Marty and Helena Moran of Longmont, Colorado, whose infant daughter Evelina had caught the disease from her coughing mother during labor. Helena had probably picked it up from an unvaccinated teenager at the dental office where she worked as a hygienist. For several nights, they cared as best they could for Evelina while she went through coughing spells that left her blue in the face. "They were horrible days and nights, taking shifts, watching her," recalled Marty, who worked at a telecommunications firm. "When she went into her spells you'd pick her up, put her on her stomach, pat her on her back, she'd curl up and turn blue. When she stopped coughing she turned blue. It was actually better when she was coughing." Eventually they got her into a neonatal intensive care unit, where she stayed for five weeks. For months afterward Evelina coughed badly each time she got sick, and her lungs would remain scarred for years. The medical bill was about $200,000.

There were many pockets of the country where antivaccine sentiment was entrenched enough to constitute the conventional wisdom, almost to become the local "standard of care." In 2003, Michigan had an exemption rate of about 5 percent, tops in the nation—and the rate was 20 percent in some counties. Upstate New York had skyrocketing exemption rates, reminiscent of Niagara Falls in the 1920s.[57] Exemptions doubled or tripled in many states where it was easy to get them in the early years of the new century. A study of the influence of nonvaccinators on disease found that unvaccinated children were six times as likely to get pertussis during outbreaks.[58] Despite the risks, part of the public was thumbing its nose at public health.[59] During interviews conducted in 2002, Johns Hopkins researchers found that about a quarter of parents were missing scheduled vaccination of their infants, though most caught up in time to make school requirements. Mistrust of vaccines was even starting to creep into the class of professionals responsible for promoting them. In a survey of attitudes about vaccination among school nurses and others responsible for vaccination records at 1,000 schools in Colorado, Massachusetts, Missouri, and Washington, 19 percent of respondents were worried that children's immune systems could be weakened by too many shots.[60] Fourteen percent agreed with the statement, "Children get more immunizations than are good for them."

Exemption rates were highest among people who had read reports put out by an Institute of Medicine committee that was conducting a review of vaccine safety. This seemed paradoxical, since the institute's reports indicated vaccines were generally safe. But the correlation showed that exemptors were digging deeply into the literature that focused on possible causal association between vaccines and adverse events.[61] Exemptors of conscience were not slovenly, poorly informed parents. They were "people who know too much but not enough," in the disrespectful but probably truthful words of James Cherry, the pertussis expert. Since vaccination was a conventional, government-promoted practice, people who mistrusted the government tended to transfer their allegiance to the rejectionist position. They preferred to believe that parents who had researched vaccines by looking up negative articles in the Internet conveyed more trustworthy information than the experts who spent their lives studying vaccines.

Sadly for the unvaccinated, however, vaccine-preventable illnesses continued to circulate. By the end of 2004, obscured by news coverage of the frantic search for flu vaccine, health authorities around the United States were reckoning with one of the largest outbreaks of vaccine-preventable disease since the measles outbreaks of 1991. The year saw more than 19,000 cases of pertussis, compared to 11,647 in 2003. The numbers had been steadily growing from the late 1980s, then exploded in the early part of the new century. How would vaccine opponents react, I wondered, knowing that their actions were contributing to an epidemic of disease that could kill and maim small babies?

Life on the Western Slope

To find out, in the winter of 2004 I started hunting around for a pertussis outbreak. Barbara Loe Fisher's website ran an antivaccine screed written by a woman named Dawn Winkler, so I set out to interview Winkler in

Gunnison, on the western slope of the Colorado Rockies, where she and a few thousand other hearty souls lived in one of the coldest spots in the continental United States. As I drove into town along an icy state highway, the local radio station was running an unpleasantly vivid ad for a new cough medicine. The thermometer stood at 16 below zero that late November morning. I met Winkler at The Beanery, a café on the main drag. She was 33, with a square jaw and high cheekbones, shoulder-length blondish hair and steel-blue eyes behind wire-rim glasses. Like everybody else in town she wore a down vest with layers of sweaters and vests underneath. She had a slightly metallic, western voice, and did not laugh easily.

Winkler and her husband and eight-year-old son, Levi, had lived an itinerant existence for most of Levi's life. Her husband worked for the Bureau of Land Management and had been subject to frequent relocations around California, Nevada, Colorado, and Washington State. Winkler did not particularly like winters in Gunnison but her husband's current BLM job wasn't bad. He had been setting controlled burns as part of a wildlife restoration program. The idea was that burning unused ranching lands would encourage certain grasses to come back. The snow hares would return to eat the grasses, and then the state would reintroduce lynxes to eat the snow hares. The lynx, which had disappeared from Colorado a few decades earlier, would thereby become fruitful, multiply, and cease to be an endangered species.

While her husband stewarded the public lands, Winkler agitated the public health. She was sort of an antivaccine slacker, fighting mandatory vaccination at each stop the family made while taking odd jobs to supplement their income. Before I met Winkler, a public health epidemiologist in nearby Durango had already told me about her. "This one parent has made it her mission to stop people from vaccinating and has been successful to some degree," said the official, who was worried because whooping cough had already popped up in three counties adjoining Gunnison in 2004. It seemed like only a matter of time, and there were a lot of infants being born.[62]

After reading Winkler's writings, I was not surprised to find that she had chronic medical problems nor that her life had a tragedy at its core—the unexplained death of a baby.[63] Winkler's personal history met the profile for ardent vaccine opponents that I had unscientifically developed over years of talking to them. Her daughter, Haley, was born in Keokuk, Iowa, where Winkler had grown up. A lovely, brown-haired baby, she had died unexplainedly at the age of five months, in 1995. Winkler was studying for a tax accounting class in her mother's kitchen the night Haley died in a room upstairs. No one living in the house would ever forget that night. Winkler had gone to SIDS support group meetings but found little solace there. "You're told over and over again it's a mystery," she says. "That never sat right with me. If someone dies, there's a reason." She heard some of the other mothers blame DTP for their babies' deaths. But Haley hadn't suffered a seizure or shock after vaccination— her last vaccines were administered six weeks before she died. Several years later, after allegations about thimerosal, a mercury-containing vaccine preservative, had gotten into the news, Winkler decided that Haley had died of mercury poisoning. By then she had remarried and moved out west.

After arriving at this explanation, Winkler began to see many difficult facets of life as by-products of vaccination. She blamed childhood smallpox and typhoid vaccines on her poor immune system. She'd had severe endometriosis for years, and "if I eat even one bite of fish that has high mercury content, like tuna, I am in the hospital. I can't breath, I vomit to the point where I strain muscles in my back." This had to be explained by the high mercury content in fish, she said, a mercury sensitivity she'd no doubt passed on to her daughter.

Communicating over the Web, Winkler found a few doctors who concurred with her thesis. She confidently slung medical buzzwords— T-cell differentiation and hypoxia and the apoe4 gene variant—to explain why vaccines made children sick. She drove around the valley in a dusty Toyota SUV with antivaccine bumper stickers. Although she saw the vaccinated more as dupes than as devils, it was hard for her not to view the world through an angry lens: Kids threw tantrums because they were vaccinated; calm children were unvaccinated. The neighbors who shared her terrific view of the mountains were teachers at the local college and had an adopted Russian kid. "She must be vaccinated because every time I see her she has a huge amount of snot on her face," Winkler said.

Winkler was a soldier in a small army of antivaccine activists loosely affiliated with Fisher and Williams, with computerized outposts around the country. Her grain of sand was to raise thousands of dollars to pay a chemistry firm to test various vaccines to show that they still contained traces of thimerosal. The money for those tests had been donated by J. B. Handley, a big San Francisco venture capitalist who had a child with symptoms of autism that he blamed on thimerosal. Winkler was writing a book that would show that mercury caused SIDS. She had given seminars to "educate" locals about vaccines, and she helped connect activist lawyers with parents seeking advice on their vaccination requirements. Sometimes, during the nice weather, Winkler would approach a mother right in Gunnison's little municipal park and hand her a card directing her to antivaccine websites. As a result of her efforts, the vaccination rate in Gunnison schools had fallen about 10 percent in three years. At Levi's tiny private school in Crested Butte, a ski resort town 20 minutes up the valley, only a few of the children had been vaccinated.

There was really no one to counter the message. Gunnison had no pediatrician, and the two general practitioners found it cumbersome to purchase and administer children's vaccines on top of all their other workload. They referred children to the county public health nurse, who ran an immunization clinic once a week. Since 1999 the nurse had been a pleasant California transplant named Carol Worrall. The parents who came to Worrall had already decided to get at least one vaccine, and she did her best to fully vaccinate them. But parents could do what they wanted. Winkler was a shadowy nemesis in Worrall's life; they ran into each other at the supermarket and didn't say hello. In effect, Gunnison valley had a part-time vaccinator and a full-time antivaccinator. When we spoke, Worrall declined to name Winkler—referring to her as "that individual," as if she were Voldemort in *Harry Potter*. Worrall tried to explain to parents that the pertussis vaccine was safer than it used to be and that it was now virtually mercury-free, but mistrust was high. In her experience, parents often vaccinated their children in the first year of life, but once they entered playgroups and started hearing the vaccine rap from other parents, they felt they'd been duped. As a result, there were many partially vaccinated children in Gunnison. The main playgroup in town was run by mothers who were viscerally antivaccine. "I know people who vaccinated their kids who were sort of verbally abused; 'Your kids are going to die, you're a bad parent,' " reported Worrall. "The hard core members of that playgroup are very against it. Young, highly educated. College grads. Plugged into the Internet. Wanting everything organic."

The most reliably vaccinated children in Gunnison were the children of Kora Indians from northern Mexico who came to the Gunnison Valley to work construction, pick fruit, and clean the condominiums of rich skiers. "Any vaccine you offer they are more than grateful to have," says Worrall. "And no matter how poor they are, when they show up at the clinic they always have their yellow immunization card from Mexico."

Worrall, who had two teenage children of her own, had worked in public health much of her life but never in a place like Gunnison. "It's really frustrating. You don't want people to get sick, especially a little baby, or a death, but you have this feeling like, unless that happens people will not take this seriously. And that's a bad feeling. But as long as the column under deaths is at or near zero—which hopefully it will be with today's medicine—all the kids who were in the ICU, or on a respirator, or have damaged immune systems from whooping cough, well they don't really register."

Winkler and I had been sitting at the Beanery a while when Sharalee Pederson, a perky brown-eyed woman in her twenties, turned up to chat. Pederson had an 18-month-old boy, Griffin, whom she had decided not to vaccinate after attending one of Winkler's seminars. She looked at Winkler when I asked her whether pertussis was a concern. "Maybe I'm naïve, I should research more about it. But what I've heard is that it doesn't protect you from the disease anyway. . . . I'm just very vigilant and picky with my child and I believe I do everything I can to support his immune system." If her son got sick, Pederson explained, she took him to a homeopath, but if that didn't work she took him to a real doctor, and might even administer antibiotics. But if he died of a vaccine-preventable illness she'd consider it all part of God's plan. "I could walk out the door and get hit by a car and there's nothing that anybody could have done to prevent that either. We can't micromanage our lives to the extent of extracting all the variables of danger." She would never vaccinate her own children, she said.

Later Winkler drove me to her house and showed me a photograph of Levi on skis, negotiating slalom poles, looking robust and unvaccinated. She told me that America's health had declined since World War II because of all the vaccines, that polio had disappeared mainly due to better diet and banning DDT. It occurred to me that it was awfully hard to change someone's mind in the course of a discussion.

Finally, Winkler showed me the sad baby videos of her child, who was born, with a quite normal weight and size, in February 1995. In one sequence the baby looked solemn, in another very animated, lying on her back in a crib batting at hanging stuffed animals. On May 27 she was in her Johnny Jump-up, giggling at the camera; on June 8, a few days after her four-month shots, she could be seen alertly responding to her mother's voice—amused and fussy, gurgling and smiling and drooling and whimpering. Winkler said: "Her pupils are dilated." On July 14, she sat with her back against cushions, playing with Sesame Street characters, smiling, exploring with her feet. A typical baby. Three days later, she died. I told Winkler that her daughter didn't look developmentally delayed to me. "Maybe you can't see it," she said. "But I felt a very distinct mother's intuition." Each time she'd had the baby vaccinated, she said, when the needle went in "it just seemed so wrong."

A FEW DAYS after I met Winkler, I discovered that pertussis was just breaking out in the town of Hotchkiss, about a 45-minute drive—when the road isn't closed by snows—from Winkler's house in Gunnison. The Hotchkiss pertussis outbreak was about a week old when I pulled up at South Fork Elementary School, a small, brightly decorated building at the foot of a big mountain. Bill Eyler, the principal, was a sandy-haired man in his forties with a neat Van Dyke beard and friendly blue eyes. A veteran of Montessori education, he had come to Hotchkiss from Denver in 1999 to help start the public charter school. The parents of his students included poor rural families, back-to-the-land types, and professionals who could do their jobs online in a mountain hamlet as easily as in a Denver high-rise.

The public health department had ordered children exposed to the germ to stay home or go on antibiotics. Some of the teachers were home coughing, and Eyler invited me to share lunch with the few remaining staff. Although I was hungry and lunch was delicious—spaghetti, salad, and garlic bread prepared in the high school cafeteria across the road—I couldn't help picturing a flabby-wattled Typhoid Mary hacking away back in the kitchen. I ate quickly. Everyone in the room jumped when I cleared my throat. "Was that a deep constricting cough?" someone asked. She was joking, sort of.

"It's awful," Eyler said, shaking his head. "At least it isn't smallpox."

A case of adult pertussis had been diagnosed in Hotchkiss the second week of November, a second case in a senior at the high school, at which point the state declared an outbreak and started investigating. Eyler saw the third official case, three-year-old Noah Stone, on November 22. There was a bake sale fund-raiser for the annual field trip that day, and Eyler ran into Noah and his mother, Valerie, at the grocery store. "I walked them over to their car and saw him go through this convulsive coughing fit," Eyler recalled. "He just had the most horrific cough." Noah had not been vaccinated, nor had his friend Blaise Keenan, who was diagnosed a short time later. Three or four other preschoolers also became ill, including Aki Blake's partially vaccinated son, Finn. He hadn't seemed very sick at first, and the family practitioners in town weren't experienced with pertussis, and told Aki that Finn didn't have it. So she and her family went away for Thanksgiving and exposed all of her relatives, including a niece with an unvaccinated baby.

When it became clear to everyone how many of the students at South Fork Montessori were children of nonvaccinators—more than 40 percent—it created a lot of polarization. For the nonvaccinators, the sickness of their children was a political choice, and the others didn't know quite what to make of that. Vaccination had become a cultural line in the sand, with the community evenly lined up on each side. Most of the vaccination opponents were women who didn't have jobs. The working women felt the nonvaccinators were endangering their kids and making life unnecessarily difficult. But when the nonvaccinating women felt under attack, they pulled out little knives of their own. Vaccine-preventable diseases like pertussis and chickenpox were just nuisances. They could be positive experiences if you approached them with the right attitude. The working women were negligent in their own way, went the thinking of the nonvaccinators. They rushed around too much, stressing out their kids. If they felt so strongly about vaccination they should live someplace else.

Aki Blake was a fence-straddler. Like many parents at South Fork she had entered her child-bearing years with a nebulous belief that vaccinations were harmful. But she knew there were also risks to not vaccinating, so she compromised, starting her children's vaccines late and then staggering the shots at separate doctor visits. Blake, who ran a company with her husband that removed a thirsty invasive plant called tamarisk from desert park lands, was a little vague when it came to what she'd read about vaccines or how it influenced her. "I probably didn't do enough research on it," she said. "I couldn't find a doctor who would sit and discuss it in a nice manner." Blake ended up delaying the children's first shots until they were a year old, but then she sort of lost track of them, and her children were each missing several pertussis boosters. After a week of listening to Finn's incessant cough, with her daughter Serena also beginning to sicken, Blake was fed up with the disease. "I guess I'm a good example of why the doctors want you to get the shots on schedule," she told me. She told a friend: "I'm a convert. If anyone asked me now, I'd say vaccinate."

But if pertussis was something of a transforming experience for the Blakes, it really wasn't for most of the community. Noah probably caught the disease from a younger boy, Ethan Bartlett, who been "coughing his guts out," as his mother, Melinda Doden, put it, since just before Halloween, although he'd never been officially diagnosed. Doden was a massage therapist who generally shunned the doctor's office. For three weeks she treated Ethan with herbs, although he woke up 10 or 12 times a night with a terrible cough and vomiting. Eventually she took him to the family practice, but the doctors didn't recognize the cough either, so he returned to school for three days. Then she took him to a pregnant acupuncturist friend, whose own two-year-old daughter had turned blue during her frightening bout with the disease. The acupuncturist did "body work" on Ethan. It didn't do much good and the acupuncturist herself was later admitted to the county hospital with an extruded cervix, a complication of prolonged forceful coughing—leaving her sick daughter in the care of her sick husband. By the time Doden isolated her son, he had unintentionally exposed other children at a Halloween party and spent a day at a clothing-optional spa in the mountains. At Christmas he was still coughing, but hardly vomiting at all, which Doden viewed with satisfaction. She continued to believe that vaccination was an evil to be avoided.

I asked Eyler whether he was bothered by the fact that 41 percent of the children at his school were unvaccinated. Like all the public servants of Colorado whom I asked this sort of question, he answered sheepishly. "The way the state law is written parents have the right to follow their own ideologies," he said. "Do I wish that the state of Colorado would revisit their law? Maybe. There's not a whole lot we can do.

"Some people believe in science," he added. "Some believe in witchcraft."

Hotchkiss and its neighbor Paonia have a combined population of about 2,500. They form an unusual socioeconomic patchwork created by the layering of successive transplants. The process began in the 1880s with the arrival of the first white ranchers and fruit growers. Then came a coal mine, and Christian fundamentalists, and hippies. Then a few years back, young professionals freed from their cubicles by the Internet began to arrive from places like Denver in hopes of raising kids in a low-key, wholesome setting. You couldn't beat the look of the place, high-desert mesa studded by Daliesque reefs and buttes with sweeping, ever-changing views of the mountains. Since Hotchkiss and Paonia were basically just one big town, people who might have run in different circles in the city were all thrown together. The discussions and backtalk over pertussis at South Fork Montessori were a miniature, face-to-face version of the national debate on vaccination.

After meeting Eyler I drove up the Gunnison River to Paonia and took a stroll on the boardwalk of the wood-facaded Main Street , which was bristling with expensive boutiques selling tasteful Christmas gifts. Hotchkiss had a somewhat scruffy, commercial appearance, but Paonia was blessed with all the charm and none of the tourists of a traditional western town. It had even been featured in a 2005 *U.S. News and World Report* feature on idyllic second-home communities. A block off the main drag I ran into Ron Edmondsen and Moni Slater outside a resource center for parents. Moni and Ron were vaccinators who had just realized that some of their neighbors weren't overly concerned about the spread of a disease they viewed as dangerous. That worried them. Edmondsen, a compact, bifocaled 49-year-old businessman, had two nonvaccinating sisters in Denver and was aware of the vaccines-cause-autism argument. "Our thought was, you're playing Russian roulette. Which one has more chambers in it?" On balance he figured it was riskier not to vaccinate. He was right. His daughter ended up being asthmatic, which put her at risk of serious respiratory compromise when there was a cold virus going around, let alone whooping cough. Because of her delicate condition, Edmondsen was keeping her at home even though she was on antibiotics. He wasn't bothered by unvaccinated children per se. "What's frustrating is that not everyone is taking this seriously. A bunch of those who are not immunized jumped right on board and got the antibiotics. There are some who didn't." Moni agreed. "I'm having a hard time being as sympathetic as maybe I should be with some of these people," she said.

Another friend, Gingy Molacek, who taught at a charter school for unwed teenage mothers near Hotchkiss, had even less patience for the nonvaccinators, because her school had a nursery for the babies, an especially vulnerable group. She and her husband, who had moved from Seattle, considered themselves very liberal. But they had their kids in the hospital and shopped at City Market (a regional grocery chain) in addition to the health food store, which made them conservatives within the local milieu. "I heard friends describing us as 'people who shop at City Market.' There's that kind of line drawn." Molacek's father and brother-in-law were pediatricians and she knew that if whooping cough hit her kids, who were 4 and 6, "it wouldn't be fatal but that it would be really awful, and that I'd miss a couple months of work. I can't do that. I listen to some of my friends with astonishment saying, 'If they get sick we'll just stay home with them.' I also think that they truly believe if their kid gets sick it will make them stronger and resistant. I honestly don't think that most of the non-immunizers understand risk analysis. They don't understand, sure there's a risk to immunizing but how does it compare to not immunizing?"

One of Molacek's nonvaccinating friends was Valerie Stone, mother of Noah. Stone, a Long Island–born organic farmer, said her son's illness had made her revisit her decision not to vaccinate, but she wasn't apologetic about her choice. "Part of why people live here is to do something a little different and not just have the 40-hour job," she says. "If you have that kind of contact, then you have to vaccinate your kids. But being a stay-at-home mom and farmer I'm not really at high risk for getting other people sick." At Thanksgiving her son couldn't keep much food down and was very tired. "When he was in the middle of a coughing fit it was terrible. He'd be gripping the furniture with his leg and stomach muscles clenched. Just horrible looking. Then when he was done coughing he'd say, 'I'm never going to stop coughing. Why do I have this sickness?' " But rather than second-guessing herself, Stone focused on the fact that the local hospital had allowed her and Noah to sit in a waiting room for two hours with two vulnerable infants.

Molacek found it frustrating to talk to Stone about vaccination. "Valerie is always saying, 'if we were in another country I'd get them immunized but I don't think it's something we're worrying about.' And I was like, 'Yeah, because every one else is carrying the burden.' It's like paying your taxes. That's what makes things work. I don't want to pay taxes but I drive on the roads and reap all the benefits, and immunizing your kids is the same way. All these guys have the privilege of not immunizing their kids cuz the rest of us do. They're riding on our shoulders."

"Yes," Stone responded, "I'm getting a free ride on the fact that this country has had a good vaccine program and many diseases have been eradicated. But I don't feel like I'm doing anything wrong that's endangering people. I've chosen an alternative lifestyle all around and am used to making choices that make people feel offended. Not wanting to buy junky toys made in China, or not using paper plates. . . . There are parents who are full time, rushing their kids around, and a lot of those kids are exhausted and rundown. I could say, that's being neglectful of your kids. But it's just different people choosing what their priorities are."

The many nonvaccinators I interviewed for this book spoke of being able to withstand diseases like whooping cough because of the healthy lifestyles they lived. In the alternative health culture, seeking alternative cures was itself a sign of health, as were eating whole-grain foods, getting enough rest, and shopping at Whole Foods or Alfalfas, instead of Safeway and Kroger's. These people didn't seem particularly healthy, just health-obsessed. They seemed to forget that natural healthcare could be an oxymoron. Natural wasn't just fresh vegetables and wooden toys and Tom's toothpaste. It was pathogenic bacteria and rabid raccoons. Nature could kill you, and smell and sound and look really ugly in the process. In

Hotchkiss as anywhere else, the unvaccinated children got severely ill, while most of the vaccinated children either got less ill or remained well, depending on their level of vaccination and exposure. This was pretty much what the medical establishment expected and had shown in dozens of vaccine trials. The difference was, the nonvaccinating parents were banking on some future benefit from their children's illness. And since the mothers typically didn't have jobs outside the home, sickness was less obtrusive in their lives. Resorting to folk remedies to deal with an awful yet certainly nonfatal illness had only enhanced the feeling of being a people apart from the mainstream.

"I think a death might change things," said Molacek, the provaccine teacher. "It would be easier to point fingers if there was a death." Barring a death in their community—there were, after all, pertussis deaths almost every year within the boundaries of their state—it seemed that many parents clung to the belief that what hadn't killed their children had made them stronger. "I don't think it changed anybody's pre-outbreak notions," Edmondsen told me several months later. "The people who made the decision not to vaccinate—it's so ingrained and deeply embedded, they're just saying, 'See, we didn't have anything to worry about.' "

Bonnie Koehler, who had been the county public health officer for 20 years, based in the nearby town of Delta, Colorado, was quite familiar with the arguments over pertussis and vaccination. When an outbreak occurred, it was her job to order antibiotics and school exclusions, and to warn parents with infants to keep them away from the movie theater and shopping mall. "My role is to plug the holes that exist but not to tell people how to manage their children and their immune system," she said. A public health nurse who worked for Koehler had gotten pertussis in 2004 and passed it to her children. The previous year, a colleague based in Grand Junction caught whooping cough from her husband, a Walmart employee who broke a rib coughing. Despite these sacrifices, the public health workers accepted the nonvaccinators and focused on saving the lives of babies. They were committed to the law of the land, which included the right of the people in that state to buck the immunologic herd. "I've volunteered more than once to be the cosmic hall monitor," Koehler told me. "But Colorado is one of those states—this is still the frontier. We believe in mending your own fences, caring for your own livestock and being responsible for choices you make."

Increasingly, though, it seemed to me that Koehler was not talking just about Colorado. Americans had ditched the World War II mindset of doing what you were told, hoping for the best, and taking your lumps and sore arms on the assumption it was all for the greater good. Too many people seemed more likely to trust what they read on some website than what their pediatricians told them. Too many families took a libertarian, individualist view of child-rearing, making their own way and not worrying much about anyone else's. And as a result, a piece of the national commons, our collective resistance to pertussis, was withering away in some places. As vaccination rates crept slowly down, the loss of protection invited disastrous consequences, like a slow oil leak from an automobile that the driver noticed, vaguely, but did nothing to stop.

Whooping Cough Nation

At the turn of the third millenium, for reasons that had partly but not entirely to do with exemptors, whooping cough was sweeping large areas of the country, striking cities and towns, suburbs and rural outposts. Arkansas, with more than 1,300 confirmed cases in 2001 and 2002, first realized it had trouble on its hands when a doctor observed that athletes were coming to the sidelines to gasp and vomit between plays at a high school football game.[64] New Jersey, Ohio, and Virginia had major upticks in pertussis, as did Chicago and its suburbs. Wisconsin had 5,000 cases in 2004. Nashville's Children's Hospital reported four deaths in the 2002–3 season, prompting a pertussis expert at Vanderbilt University to recommend universal pertussis vaccination at birth, rather than the current practice of beginning at two months.[65] Even North Dakota suffered a record 800 cases. The disease had an impact on the state's 2004 senatorial race: Republican Mike Liffrig left the campaign trail for several weeks after his wife and two children were diagnosed with pertussis. He narrowly lost to the incumbent Democrat, Senator Byron Dorgan. In Philadelphia, where the vaccination rate had dipped as low as 34 percent in the late 1980s, an outbreak among nurses at the Temple University Hospital neonatal ward led to the death of a newborn and sickened 27 others. Barbara Watson, a public health officer with broad and deep experience with pertussis—she had suffered the disease and a whole-cell vaccine had given one of her children a fever of 106 degrees—was incensed. She saw sick children from disadvantaged families, and "a lot of the antivaccine groups are rich, white, opinionated and don't care. That really aggravates me. They've never seen a child die. They don't understand the consequences of their actions."

Even wealthy Westchester County, New York, was not immune. By April 14, 2004, the county had identified 98 cases in an epidemic that began among four intentionally unvaccinated children. In Denver, John Hickenlooper's four-month-old son Teddy was infected by an underimmunized playmate during Hickenlooper's campaign for mayor. Teddy stayed in the Intensive Care Unit at Children's Hospital for two days and coughed for four months. The experience made vaccine believers out of Hickenlooper and his wife. After he became mayor the following July, Hickenlooper got the state to pump another $400,000 into vaccination programs. "We had never realized that this disease still existed," he said.

And yet not all of the resurgence could be blamed upon vaccine shunners. Boulder County had another big upswing in pertussis, with more than 150 cases in 2004. But unlike some past years, most of the cases were in teenagers, roughly 90 percent of them fully vaccinated. Only 10 percent of the cases were in children under 4, and authorities were also seeing an increase in adult pertussis. Officials in the county were keeping their eyes on Shining Mountain, but by December the school hadn't reported anything. "Based on what we're seeing this year," said Heath Harmon, an epidemiologist for the county, "the data doesn't support that exemptors are playing a major role in Boulder." While exemptors were a reservoir for the disease in Boulder, it also spread in the vaccinated population.

That was also true for many other areas of the country. Why, at a time when American children were being vaccinated at record levels, was pertussis returning with such vigor? Could all the blame be laid at the feet of the Dawn Winklers, the Philip Incaos, and the Melinda Dodens, spreading the false gospel that pertussis vaccine was a dangerous toxin, pertussis a soul-stirring adventure of life? Or to the contrary, was the vaccine itself to blame?

It turned out that pertussis had never really left us. Research in the 1990s showed that many adults with coughs had undiagnosed, but still contagious, whooping cough. It was not until a decade later, however, that scientists were able to generalize this finding with improved diagnostic techniques.[66] In 2004, UCLA pediatrician Joel Ward estimated that 700,000 to 1.5 million people become ill with pertussis each year in the United States. As many as 6 million more seroconverted, meaning they were exposed to pertussis but did not fall ill. The increased pertussis reports of recent years, in Ward's view, were mainly "an artifact of detection."

Until recently it was nearly impossible to diagnose pertussis in adults or older teenagers. Because their coughs were not nearly as serious as those in children, older people rarely sought medical help until they had been coughing for at least a few weeks. By then there were not enough bacteria in their air passages to make a good culture, so tests for pertussis proved negative. In the process of developing a testable, quantifiable acellular whooping cough vaccine, scientists developed antibody measures that allowed them to compare patients with pertussis disease against healthy controls. The state laboratory of Massachusetts, which until recently made its own pertussis vaccine, was the first in the nation to standardize the system for comparing pertussis antibody titers. Within a few years of establishing the standard, Massachusetts began to report levels of pertussis far above the national average. There was no reason to think Massachusetts had more pertussis than any other state. It just had better detection.

In the mid-1990s, Colorado started using polymerase chain reaction, or PCR, a highly sensitive testing method. Looking for pertussis used to involve sticking a long swab through the nasal passage and rubbing it against the wall of the upper throat—a nasty business for a coughing, vomiting child—and hoping that the notoriously fickle pertussis organisms would grow in the culture medium into which the swabs were placed. With PCR, doctors could squirt saline solution in a child's nose and collect the fluid in a little tub. A bit of bacterial DNA would produce a quick, positive result. Not only had the tools of detection improved but there were also more people looking. Pediatricians were growing more aware of the disease and increasingly testing for it. Congress, after the anthrax attacks on Senate office buildings in October 2001, boosted federal cash for epidemiological surveillance. Colorado now had a bioterror epidemiologist, as well as a network of 11 health districts with an epidemiologist in each one. More eyes meant more detection, and more detection meant higher numbers. So did something as simple as elementary school e-mail listservs. "With pertussis," said Heath Harmon, "the more you look, the more you find."

Yet better diagnosis alone did not explain the whole story. Colorado had been on the outlook for pertussis for many years and had used PCR testing since 1995. The year 2004 still produced the highest rate in 40 years. John B. Robbins, the NIH vaccine inventor, believed that the children whose parents were scared away from vaccination prior to the introduction of the acellular vaccine continue to play a role in spreading pertussis. James Cherry, the UCLA pediatrician, had a more controversial hypothesis. "Until recently," he told me, "the vaccines we had been using against pertussis were not as efficacious against the disease as the whole-cell vaccines we used up until about 1985."[67] The first-generation acellular vaccines used until recently in the United States seemed to provide less lasting immunity than did vaccines used in Europe and Canada.[68]

The failure of pertussis vaccines to protect for much more than a few years has had a ricochet effect. Older people who were protected by pertussis vaccines as babies were now getting the disease if exposed to enough of the germs. They, in return, threatened the youngest, unvaccinated babies. In recognition of this fact, the FDA in 2005 licensed new DTP formulations, and the CDC and pediatricians recommended them for teenager and adult booster shots.

However disappointing the newer acellular vaccines were, they did protect the young against deadly pertussis disease—hospitalization was necessary mainly for unvaccinated or undervaccinated babies.[69] From 1990 to 1999, U.S. authorities reported 18,500 infant pertussis cases. Of the 93 pertussis deaths reported, 90 percent were in entirely unvaccinated children. In 2003 there were 2,000 cases of pertussis in children under six months—and two-thirds were hospitalized.[70] One study showed evidence of undiagnosed pertussis infection in 6 percent of babies whose deaths were reported as SIDS.[71] Could pertussis, rather than pertussis vaccine, be a major cause of SIDS? It was an intriguing possibility, since babies with pertussis didn't whoop, but often stopped breathing.

Not vaccinating against pertussis was truly a form of roulette. But such speculation had little impact on thousands of Boulderites, who hacked through the season but still insisted that they were leading healthy lifestyles. "I don't fault people for not vaccinating their kids, but I don't think they know the truth about vaccines," said Stephen M. Fries, a busy family practitioner in Boulder. "My partner took care of two kids who died of whooping cough. But it makes no press. You're not going to go to the papers and say, 'Make a big deal of this, my kid died. I'm an idiot and didn't get them immunized.' You try to make people understand that pertussis is no fun, but they don't believe you."

CHAPTER 10

—

VACCINES AND AUTISM?

Life is unfair. The unfairness of the life dealt to autistic children, however, is so unfair that it defies description.

— MICHAEL D. GERSHON

WHEN WILLIE MEAD was about $2\frac{1}{2}$ years old, his parents realized that something was wrong with their freckled, brown-haired boy. His speech had become nothing more than a series of squeaks and grimaces. "He just wasn't there anymore," said George Mead, Willie's father. "He spent all of his time spinning in circles, looking out of the corner of his eye and going, 'ticka-ticka-ticka-ticka.'" George and his wife Tory made the rounds from pediatrician to developmental neurologist to special education counselor until they confirmed the devastating diagnosis: their son had autism, an incurable condition marked by a profound incapacity to connect with other people.

Autism was only recognized as a disease in 1943, when Leo Kanner, the Viennese emigrant who founded the field of child psychiatry in the United States, pronounced its arrival in a study of 11 cases he had seen in his office at Johns Hopkins University in Baltimore. "Since 1938," Kanner wrote "there have come to our attention a number of children whose condition differs so markedly from anything reported so far that each case merits—and I hope, will eventually receive—a detailed consideration of its fascinating peculiarities."[1] The outstanding feature of these children was their "inability to relate themselves in the ordinary way to people and situations from the beginning of life"—although a few of them had seemingly normal infancies, only to regress. The children shunned the familial embrace and occupied walled-off worlds of repetitive, often self-injurious behaviors. "Their parents referred to them as having always been self-sufficient . . . 'like in a shell;' 'happiest when left alone;' 'acting as if people weren't there' . . . There is from the start an extreme autistic aloneness that whenever possible disregards, ignores, shuts out anything that comes to the child from the outside." A few of the children had remarkable memories or mathematical abilities, but others were severely retarded. All were obsessed with preserving sameness, order, patterns.

Thirty years later, autism was still considered a rare diagnosis, with a prevalence estimated at between 3 and 5 in every 10,000 children. In 1971, in the first volume of the new *Journal of Autism and Child Schizophrenia*, Kanner went in search of his first 11 children and found 8 of them in institutions, while 3 were working, relatively happily—as a bank teller, a duplicating machine operator, and a tractor driver on a farm. The visits with the other children were sad. "State hospital admission was tantamount to a life sentence," he wrote, "with evanescence of the astounding facts of rote memory, abandonment of the earlier pathological yet active struggle for maintenance of sameness, and loss of interest in objects added to the basically poor relation to people—*in other words, a total retreat to near-nothingness.* [Kanner's emphasis] . . . After its nearly 30-year history and many bona fide efforts, no one as yet has succeeded in finding a therapeutic setting, drug, method or technique that has yielded the same or similar ameliorative and lasting results for all children subjected to it."

By 2000, when Willie Mead was diagnosed, the prevalence of autism was estimated at something more like 3 to 5 children in 1,000. There were still few options for improving the behavior of autistics, the doctors and counselors told the Meads. An attractive professional couple from Portland, Oregon, the Meads did what people confronted with horrible medical news do these days—they got on the Internet and researched it. They quickly latched onto stories from parents convinced that government-mandated vaccines had caused their children's autism.

American children in the year 2000 were receiving 16 or more vaccine injections during the first two years of life. The symptoms of autism often made their initial appearance when a child was between the ages of one and two years. Most autism experts believed these two facts were coincidental, but a small group of doctors and a vocal portion of parents, which the Meads joined, disagreed. The fact that the mainstream medical community rejected this theory would not bother them much, for established medicine had so little to offer the parents of autistics that turning one's back on the advice of the American Academy of Pediatrics was almost no sacrifice at all. There seemed to be an epidemic of autism, and there were no drugs that consistently treated the disease. There was little money to meet the overwhelming demands of educating autistic children and, perhaps worst of all, there was no meaningful light to shed on the grievous mystery of autism itself. Who was going to take care of all these kids? Who was going to take care of their son when they were gone? Why should the Meads care what people thought?

Before the tragedy of their son's illness commandeered the couple's life in the summer of 2000, George Mead had been a medical malpractice lawyer who defended hospitals and drug manufacturers. His wife, Tory, was a peppy junior leaguer with a masters in journalism from New York University who'd held a series of fun jobs in journalism and business before giving birth to Eleanor, Willie's older sister. After that she became an overeducated, energetic, and pretty fulfilled stay-at-home mom, like a lot of women of her postfeminist generation. One day Tory called George at his new law firm. Willie was lying on the stairs, staring at the wall, laughing maniacally. She was frightened. There was something so cold about a diagnosis of autism. "I went and sat in Willie's room and felt like the winter walls were moving in, I felt like my life was over," recalled Tory. "When you get the diagnosis you fall off the map. They have nothing for you."

Activism offered a way out. The vaccine theory made sense to the Meads, and it gave them something to fight against, a focus for some of their rage and grief and frustration. In addition, a whole cottage industry of alternative practitioners was offering cures for autism that were premised upon the disease being a product of vaccine damage. Some blamed the MMR shot, and some DTP, some thimerosal, a vaccine preservative that contained ethyl mercury, while some blamed vaccines in general as part of the "toxic" universe in which we were alleged to live. To the extent that the treatments offered by people who blamed vaccines seemed to improve the condition of the children—and there were a few "miracle cures"—the small improvements in Willie's mental and physical health confirmed the Meads's belief that vaccines were to blame for the problem in the first place.

The vaccines-cause-autism mindset was the product of a set of assumptions that were impossible to completely prove or disprove. It was true that more cases of autism were diagnosed in the 1990s than ever before. But to what extent was the increase based on more elastic definitions of the condition and more funding for detecting it? If there were more autistic children, what was causing the increase? Pediatric vaccinations had increased significantly in the 1990s. What was in those vaccines? Well, it turned out that a few of them contained mercury, though not very much of it. Mercury was a known neurotoxin—"Mad as a hatter" was a phrase that originated in eighteenth-century hatters' use of mercury to block felt hats. If you treated the child as though mercury caused the problem, and the child seemed to get better, then it proved your point, didn't it? This wasn't science, but there was a certain seductive deductiveness about it.

After Willie's diagnosis his parents mourned for a while and then, almost overnight, they metamorphosed. Convinced that thimerosal had damaged Willie, they started sending him to an environmental health specialist who wore Hawaiian shirts to work, drove a Volvo graffitied with green and red spray paint, and prescribed a cabinet full of dietary supplements. The Meads sued their pediatrician and the vaccine manufacturers. They were now part of a political movement, a coalition that included Barbara Loe Fisher and Dan Burton, who was devoting the considerable resources of his congressional committee to finding a link between vaccines and autism. By way of a medical problem the Meads had crossed a psychic divide, leaving behind the world of prosperous, reasonably contented professional people for the spooky realm of herbalists and populist mavericks and—not to put too fine a point on it, conspiracy kooks—who viewed America as a toxic hell. The Meads called it "going down the rabbit hole." In their world—dealing with a child who sometimes rocked for hours, banging his head on the wall, who chewed dirt and didn't speak but obsessively scratched his enormous welty mosquito bites—white was black and up was down. Unlike Alice they were not dreaming of Wonderland. They and thousands of other parents had become convinced that vaccines, which most of the world viewed as safe and wholesome and life-giving, were poison. And, like thousands of others, they would seek an antidote by trading experiences and ideas through the Internet.

The Loneliness of the Long-Distinguished Vaccinologist

Thimerosal might never have become such a lightning rod for the vaccine skeptics, parents of autistics, and trial attorneys were it not for Neal Halsey. There was supreme irony in that. Dr. Halsey, a distant relative of Adm. William F. Halsey, the famous World War II naval commander, took his responsibilities within the American vaccination world extremely seriously. In June 1999, Halsey had just finished a four-year term as chairman of the Infectious Disease Committee of the AAP, a position that had been the cherry atop a busy, productive career in vaccines. Halsey had been a pediatric resident under Henry Kempe at Colorado and spent his early career at the CDC working on measles and other vaccines. In 1997 he'd set up the Institute for Vaccine Safety at Hopkins.

Vaccine safety was a matter dear to Halsey's heart for many reasons. He had seen three cases of vaccine-associated polio and conducted trials with inactivated polio vaccine to help establish the viability of moving from the Sabin to Salk vaccine. He had also joined others in pressing for the adoption of the acellular pertussis vaccine. Halsey understood that the benefit of vaccines relied upon parental confidence in the public health authorities' recommendations. At 56, with a full beard, thick glasses, and a wrestler's build, Halsey had a curmudgeonly air that conveyed a respect for facts, training, and medical authority. But Halsey cared for children, and there was no one in the vaccine world who felt more deeply about the bond of trust required for the vaccination enterprise.

Halsey became privy to an alarming bit of news while sitting in on a luncheon meeting at the FDA's Center for Biological Evaluation and Research in suburban Maryland.[2] At the meeting, FDA scientists Leslie

Ball, her husband, Robert, and Douglas Pratt presented data indicating that millions of American children had been exposed to levels of thimerosal that were potentially far above the EPA-recommended limits on mercury. Thimerosal was an ingredient in three of the childhood vaccines administered during well-baby visits—DTP, which had been given routinely for decades, *Haemophilus influenzae* type b, and hepatitis B, which had been added to the universal vaccine schedule in 1991 and 1993, respectively. Thimerosal, used to prevent bacterial contamination in certain vaccines, was 49.5 percent mercury by weight. As a member of the ACIP and the Red Book committees during the 1990s, Halsey shared responsibility for the implementation of universal vaccination of infants against hepatitis B and Hib. Most DTP and Hib vaccines had 25 micrograms of ethyl mercury each in them, while the hepatitis B shot, given at birth, two, and four months, contained 12.5 micrograms. How these levels related to the EPA's recommended limits was not easy to determine.

The EPA had suggested limits for mercury ingestion aimed at keeping pregnant women from eating large amounts of fish containing methyl mercury, a related compound. The EPA felt that 0.1 micrograms per liter was a safe amount of mercury in the cord blood of newborns, based on chronic exposures of mercury that mothers passed along in small amounts to their fetuses—not the single pop of mercury a child got in a vaccine. The EPA limits did *not* mean that blood levels above that limit were unsafe, or dangerous, and there was certainly no mention of a risk of autism. Still, in the first six months, a fully vaccinated child might receive as many as 187.5 micrograms of ethyl mercury. Leslie Ball and her colleagues had done the math and found that if you averaged it out, the babies' blood would theoretically be far above the 0.1 microgram per kilo level. But how much above, and was it significant?

There were a number of unanswered questions. How much of the mercury from thimerosal was absorbed into the blood, and from there into the brain? How much of it remained in the body, and the brain, between the doses given—birth, two, four, and six months? How comparable, in its toxic effects, was ethyl mercury to the much more widely studied methyl mercury? Halsey worried that the single, relatively large dose from a vaccine might be more dangerous than the gradual intake of methyl mercury. He found the numbers frightening.

"My first reaction was simply disbelief, which was the reaction of almost everybody involved in vaccines," he said a few years later.[3] "In most vaccine containers, thimerosal is listed as a mercury derivative, a hundredth of a percent by weight. And what I believed, and what everybody else believed, was that this was a truly trace, biologically insignificant amount. My honest belief is that if the label had had the mercury content in micrograms, this would have been uncovered years ago. But no one before Leslie did the calculation."

The inquiry into thimerosal had begun as a result of an amendment introduced into an FDA reauthorization bill by Frank Pallone, a New Jersey Democrat, who had a strong interest in environmental issues.[4] Pallone's amendment gave the FDA two years to "compile a list of drugs and foods that contain intentionally introduced mercury compounds and . . . provide a quantitative and qualitative analysis of the mercury compounds in the list." The bill evolved into the FDA Modernization Act and became law on November 21, 1997. In April 1999, the FDA published a notice in the *Federal Register* requesting manufacturers to provide data on mercury. Meanwhile, the agency had its attention drawn to thimerosal by its European sister, the Agency for the Evaluation of Medicinal Products, which was just completing an 18-month risk assessment. The European agency found in June 1999 that "although there is no evidence of harm caused by the level of exposure from vaccines, it would be prudent to promote the general use of vaccines without thimerosal."[5]

After confirming Ball's equations with a colleague at Hopkins, Halsey started calling mercury experts all over the world to get a sense of how serious the problem was. He eventually reached Orenstein and convinced him something had to be done. "There were many pediatricians involved in the vaccine issue who truthfully didn't know the story about mercury. . . . I was one of them," Halsey told me later. "But I did my best to educate myself."

Halsey had been a vaccine researcher in the United States and abroad for three decades, and he was not shy about speaking up on potential dangers. He had helped bring to light dangers from high-titer measles vaccine trials in the early 1990s; in 1996, he alerted the CDC to an epidemic of diethylene glycol poisoning that had caused 30 deaths in Haiti, where he worked for many years.[6] Halsey took this step although blame rested with a company owned by the powerful Haitian family under whose auspices Halsey was conducting his work. At the same time, Halsey was firm about debunking spurious theories of harm from vaccines. "Neal is a man who does what he thinks is right, bordering on zealotry," said a colleague, Paul Offit. But thimerosal worried him. It was almost too much to accept that vaccines, which he and other public health authorities had urged upon American children, might have poisoned their minds. But Halsey, unlike many of his colleagues in coming months, was stout-hearted enough to accept the possibility, and its consequences.

"Most of the concerns about vaccines are based on hypotheticals and temporal relations with little evidence," Halsey told me in a 2002 interview in his office at the Bloomsberg School of Public Health, a hilltop building that overlooks vast Baltimore slums. Thimerosal, he suspected, might be different from other vaccine scares. He showed me a graph, produced by CDC researchers, that showed a positive correlation between increasing thimerosal ingestion and neurodevelopmental problems like tics and speech delay. Even at the time Halsey was far from convinced that thimerosal had done any damage, and he was certain it didn't cause autism. But the Institute of Medicine in 2001 had said there was "biological plausibility" of some harm resulting from the substance, "and I agree with that," he told me, in an interview for a *New York Times Magazine* article.

In the face of uncertainty, Halsey felt it would undercut the credibility of the immunization program not to take steps to remove mercury from vaccines. A scientific review conducted by the National Academy of Sciences in 1998 had found that 8 percent of childbearing women in the United States had levels of mercury in their blood above the EPA recommended level. A single 5.6 ounce can of tuna on average contains about the same amount of organic mercury as the hepatitis B shot;[7] the average breastfed child would get roughly twice as much methyl mercury in mother's milk as ethyl mercury from vaccines in its first six months of life.[8] That statistic would cut both ways in the coming debate about harms from thimerosal, but Halsey worried about the additive effect of ethyl and methyl mercury in babies whose mothers were big fish eaters. Too, some pregnant women had received ethyl mercury in Rhogam shots, which were given to prevent the placental rejection of a fetus with a different blood type.

Thimerosal had first been introduced as a preservative in vaccines in 1942 in response to 11 deaths that resulted from staphylococcal contamination of a lot of tetanus antitoxin. The substance was used to deactivate whole-cell pertussis bugs in DTP and added to vaccine lots during production to keep them aseptic. Most of that thimerosal was subsequently removed, but a small amount was kept in multidose vaccine vials in order to prevent contamination when a doctor or nurse stuck in a needle to extract a dose. Thimerosal was not used in polio or measles vaccines. When thimerosal was introduced it must have seemed an innocuous product indeed, especially since it was being added to vaccines in doses that appeared homeopathic by comparison to other applications of mercury in medicine. Massive doses of mercury had been used as a purgative in the blood-and-guts medical world of the late eighteenth century. Even into the 1960s, teething powders marketed for colic could easily expose an infant to a gram of mercury in a few months. Wound care often called for ethyl mercury slathered on as a disinfectant. There was mercury in thermometers, of course, and many a baby boomer could recall breaking one open to roll the gleaming silvery puddle around in the palm of one's hand—inhaling its toxic fumes all the while.

Halsey found that environmental health and infectious disease experts had a different perspective on thimerosal. Scientists who studied the effects of minute amounts of metals on the brain worked with imprecise data and multiple variables; they tended to err on the side of caution. The vaccine world was used to evaluating concrete harms and benefits from new vaccines. "We have the luxury of good numbers," was how Bob Chen put it.[9] Halsey was starting to see things from the toxicologist's perspective. "It became apparent to me that what has happened with mercury exposures," he said, "has a parallel to what happened with lead." Many of the studies of lead involved children living in crumbling apartment blocks not far from the Hopkins campus. The more scientists studied lead, the more dangerous it appeared to be even at low levels.

Most of the mercury poisoning in the medical literature resulted from doses thousands of times higher than what infants received in thimerosal. The science on smaller exposures was ambiguous, as was often the case with attempts to correlate specific environmental exposures to medical problems. Not only that, there was precious little data on ethyl mercury, even from animal studies. This was a regulatory lapse that had to do with the bureaucratic history of vaccines. When vaccine regulation switched to the Food and Drug Administration from the NIH in 1972, many aspects of production were grandfathered in. Thimerosal was one of them. "FDA folks focus on one single product at a time," Chen said. "It wasn't anyone's job to look at the toxicology overall."

EPA guidelines were based mainly on a set of studies involving 917 children born in 1987 in the Faeroe Islands, a windswept archipelago in the North Atlantic. The children had been exposed to varying levels of mercury as fetuses by mothers who ate pilot whale meat. Some of the Faeroese children whose umbilical cord blood showed four times the EPA reference dose exhibited "neuropsychological" effects seven years later. They tended to have slower reaction times, diminished attention spans, less keen vocabulary and memorization ability than their mercury-free classmates, according to Philippe Grandjean, a Dane who led the ongoing Faeroes study and taught environmental health and neurology at Boston University.

Thomas Clarkson, a mercury toxicologist at Rochester University, looked at the thimerosal evidence initially and felt that it was reassuring. Methyl mercury lingered in the body longer than ethyl mercury. Most of the single "bolus" dose of mercury delivered by a vaccine would pass through the body without crossing the blood/brain barrier, he believed. He pointed out, at an August 1999 meeting of vaccine experts, that moms had to eat a lot of whale meat before mercury even showed up in their blood. Mercury in a vaccine did not translate immediately to mercury in the blood, much less the brain. Grandjean was more worried than Clarkson, but even he doubted the exposures could have caused major brain damage. To the experts, with a few exceptions, the notion that thimerosal was behind the growing diagnosis of autism in American children seemed ludicrous. But the tendency of vaccinologists to brush off the mercury problem bothered scientists who were starting to try to assess whether there had been any harm. As one CDC scientist put it in an e-mail to a colleague, the Faeroes and thimerosal exposures were "comparable as apples and pears. Unfortunately, too many vaccine experts seem unwilling to compare apples and pears."[10]

Mercury experts thought it wouldn't be a bad idea to remove thimerosal preventively. There were vaccines on the market that did not contain thimerosal. In Europe, the practice was to use preloaded, single-dose syringes of vaccines, and those shots didn't need or contain thimerosal.[11] Although no one at FDA acknowledged it at the time, the agency had been aware of the thimerosal issue at least since early 1991, when Maurice Hilleman had pointed out the potential concern in a memo to Merck management. Hilleman noted that by six months, children receiving the Hib and hepatitis B vaccines on the anticipated schedule would get doses up to 87 times higher than FDA guidelines for the maximum consumption of mercury from fish. "When viewed in this way, the mercury load appears rather large," he wrote. "The key issue is whether thimerosal in the amount given with the vaccine, does or does not constitute a safety hazard. However, perception of hazard may be equally important." The FDA didn't seem worried about it, Hilleman said. But he considered it "reasonable" to remove it.[12]

Switching over to an exclusively thimerosal-free schedule would cost money and create temporary shortages, but it was the prudent thing to do in Halsey's view as well. On June 30, 1999, after a whirlwind of communications among academic, industry, and government scientists, the Academy of Pediatrics convened an emergency meeting at its 13th Street office in Washington. The debate was furious. In e-mails later made public by an advocacy group, one FDA official was quoted as saying that a decision to move against thimerosal would make them all look bad, since it boiled down to a ninth-grade algebra equation being used to uncover a risk missed by hundreds of scientists. The implication, as he put it, was that the FDA and CDC had been "asleep at the switch" for nearly a decade. Halsey, Wake Forest University pediatrician Jon Abramson, and members of the academy's environmental health committee wanted pediatricians to stop using thimerosal-containing vaccines altogther if alternatives were available, and thought the first dose of hepatitis B vaccine should be pushed to six months as long as the mother was hepatitis B–negative. The CDC leadership and many other academy members felt thimerosal was a theoretical risk and that Halsey's hastiness could threaten a cornerstone of public health. In the end, a compromise was struck. Manufacturers were urged to remove the substance from all vaccines as quickly as possible. Pediatricians were told they could postpone the birth dose for two to six months.

The decision was taken scarcely a week after Halsey had first learned about thimerosal. The Academy of Pediatrics had acted with remarkable— perhaps excessive—speed. Some admired Halsey for his urgency but others were disgusted. A few days later, Halsey, who hadn't slept much in the previous two weeks, left for a vacation in Maine. There, canoeing and fishing on a beautiful lake, he came across signs warning pregnant women to avoid eating fish. "I thought, 'How can we say don't eat fish but inject this stuff right into babies?' " At a National Vaccine Advisory Committee–sponsored thimerosal workshop in August, Halsey was practically alone in a hall full of hostile colleagues. Stan Plotkin took the role of prosecutor. Ethyl mercury wasn't methyl mercury, he said, and the fish studies were contradictory. In a large study done among Seychelles Islanders, the babies of pregnant women who ate an average of 78 micrograms of methyl mercury per day per kilo—that is, a typical 100-pound woman had consumed 3,500 micrograms a day—suffered no harm. Even in the Faeroes, vaccinated schoolkids, and kids with higher mercury content in their hair, were doing better in school than their peers. A ban on thimerosal would cripple vaccination programs in the third world, which relied on multidose vials in order to be affordable. "If physicians or state public health services insist on access to thimerosal-free vaccines," Plotkin predicted, "chaos will ensue." The academy had panicked, rushed to judgment, lacked *sangfroid*. He compared the decision to the Red Queen's court in Alice in Wonderland: "First the sentence, then the trial!" As for the public perception, Plotkin railed, "If antivaccinators didn't have mercury they'd have another issue, and one cannot prevent them from making hay regardless of whether the sun is shining or not."

A few of those present disagreed. Dixie Snider, a veteran CDC scientific director, said that environmental scientists as well as vaccine foes were concerned about thimerosal. While the limited data showed that ethyl mercury left the blood faster than methyl mercury, that didn't mean it didn't cross the blood-brain barrier. High single doses might be a bigger problem than the gradual accumulation of fish consumption. Public concerns, like it or not, were more and more important to the future of the vaccine program. "The time has passed when we can push these aside and say they aren't relevant," agreed Sam Katz. Walter Orenstein suggested that the CDC begin a study using data from the Vaccine Safety Database.[13]

Halsey strongly supported Orenstein, but the study would prove more controversial than anyone had anticipated. The audience that took Halsey's concerns most to heart, it turned out, was the one he had the least interest in reaching. That was the trial lawyers, who began signing up parents of autistic children convinced that thimerosal was responsible for their problems. Although Halsey believed that the antivaccine lobby would eventually have fastened onto thimerosal without his calling attention to it, the apparent fruit of his blow for prevention was a bitter one. A flood of lawsuits had cost the vaccine industry more than $200 million even before any cases reached trial. Vaccine skepticism was growing in the American public, and the antis had been given their strongest tool yet to chip away at the public health system.

Down the Rabbit Hole, into Rim-Land

A survey published in 2004 showed that a third of autism-affected families relied at least partially upon alternative healthcare practitioners.[14] Given that many parents of autistics have given up on their neurologists altogether, the number who have tried alternative diets and supplements was probably much higher. Many autistic children have a baffling array of symptoms, medical as well as mental. "Our attitude," said Eric Colman, an MD who worked as an obesity drug reviewer at the Food and Drug Administration and who has a mildly autistic son, "is that if it doesn't seem likely to hurt and there's some decent anecdotal evidence of help, we'll try it."[15]

The I'll-try-anything-once approach has been a staple of the autism scene for decades, and autism has been "cured" dozens of times. It can be difficult to gainsay the absolute certainty with which some of the cure peddlers—many of them parents of autistics—have asserted the supremacy of their approaches. "Various groups have exhibited an attitude of monopolistic ownership of anything pertaining to autism, and from time to time have awed the lay press with 'evidence' of miraculous-sounding 'cures,'" Leo Kanner wrote in 1968.[16] That statement was even truer four decades later. Anyone who has looked into the thimerosal controversy is struck by how bitter and defiant the parents of autistics are, and who can blame them? What could be more unacceptable than to lose a child, and to be reminded of it every day? "Parents of children with autism are more depressed than parents of children with cancer," says Tory Mead. "It's a free fall all the way to the bottom. Most of the moms are on anti-depressants. These were supposed to be the best years of my life. Our houses used to be full of children, and now they're empty."

Some of the bitterness can be traced back to the original bitter parent, a man by the name of Bernard Rimland, and the steps he took to improve the lot of autistics. Rimland resembles the *Mad* magazine founder William Gaines, with a full gray beard and clothes that look as if they were purchased around the time of the last flu pandemic. Amiable but tenacious, Rimland was a navy psychologist in San Diego whose job was to screen sailors for IQ and other abilities that would determine their placement in the service. In 1955, his wife Gloria gave birth to a son, Mark, who before long began to show symptoms of autism. The boy spent most of his days staring into space, banging his head on the wall, and tearing his crib apart with wild rocking. He became apoplectic at the slightest change in his environment, to the extent that Gloria Rimland had to buy identical flower-print dresses for herself, her mother, and mother-in-law, to keep him from screaming.[17]

Mark had an uncanny mathematical ability—given a person's date of birth, he could instantly calculate the day of the week he or she was born on—and he eventually developed into a talented artist. But his childhood was hell. What made it worse for the Rimlands was the reigning psychiatric paradigm, which blamed them for their child's condition. The Freudianism of the day held that whatever brain abnormalities psychotic children might have were dwarfed by the importance of their relationships with parents. Kanner himself was ambivalent about this idea. In his first paper he wrote that "we must assume these children have come into the world with *innate inability* [Kanner's emphasis]" to form emotional ties with other people "just as other children come into the world with innate physical or intellectual handicaps." But by 1949, by which time he had seen 55 patients, Kanner wrote that the parents of autistics were generally "mechanistic" about human relationships, and their attitudes had affected the children.[18] "Most of the parents declare outright that they are not comfortable in the company of people; they prefer reading, writing, painting, making music, or just ' thinking.' . . . Maternal lack of genuine warmth is often conspicuous in the first visit to the clinic. As they come up the stairs, the child trails forlornly behind the mother, who does not bother to look back." The kids, "reared sternly in emotional refrigerators, have found at an early age they could gain approval only through unconditional surrender to standards of perfection." It did not seem to occur to Kanner at the time that the mothers might be holding these children at arm's length because that was the only proximity the children could tolerate.

Kanner would switch directions again, absolving parents of guilt by the late 1960s. In the meantime, however, Bruno Bettelheim, a glib and gloomy concentration camp survivor, had emerged on the scene of child psychiatry. Bettelheim compared the autistic child's experience to his own at Dachau and Buchenwald. "The precipitating factor in infantile autism is the parent's wish that his child should not exist," Bettelheim wrote. "All my life I have been working with children whose lives were destroyed because their mothers hated them." The fact that many autistic children seemed to regress at age 1 or 2 helped sustain Bettelheim's theory of a postnatal, psychological insult to the child. It is Bettelheim who is most often associated with the phrase "refrigerator mother."

Rimland was not the type to take such nonsense sitting down. The original inspired amateur, he investigated autism at university libraries while on business trips for the navy over several years and found no evidence that the parents of autistics were colder than other parents. When he decided to put his findings together, Kanner was encouraging, writing a forward to the book. Bettelheim, however, rebuffed Rimland's requests for information, perhaps because he was preparing his own *Empty Fortress*, published a few years later, to great acclaim. Writing at home until 2 or 3 each morning, Rimland would shake his fist. "You son of a bitch Bruno Bettelheim," he would say to himself. "I'm going to show you."[19]

Rimland's book, *Infantile Autism*, which debunked Bettelheim under a deluge of research, was published in 1964. Afterward Rimland set up a research institute and began reaching out to autistic parents around the country who like him were seeking contacts, reassurance, and information about the causes of their children's illness and hopeful approaches to treating it. Psychotherapy, it was clear, was not doing the trick. The lives of these parents were profoundly difficult. Should they use corporal punishment, or aversion therapy, to control the behavior of children who sometimes ran out into traffic or attacked other kids, or picked up dog turds and ate them? How could they stay sane when their children refused to sleep? Were Thorazine and Haldol helpful, or would they simply zombify their children?

In addition to their behavioral challenges, many autistic children seemed to be plagued with odd gastrointestinal problems, food intolerances, and allergies. Parents were desperate for some way to help their children with all these problems and willing to try extreme measures. Rimland led the way, clashing all the while with mainstream scientists who were skeptical of therapies that hadn't withstood double-blind placebo trials to test their efficacy. After his book came out Rimland started pushing the efficacy of megadoses of B vitamins and magnesium in improving the symptoms of autism. He worked with a supplement maker, Kirkman Labs of Portland, Oregon, to create soluble forms of vitamins and other supplements for autistic children, which he sold on his institute's website.[20] These and other alternative methods still hadn't been subjected to rigorous testing by the late 1990s. Mainstream and alternative practitioners were way out of step with one another.

The gluten-free, casein-free diet was one of the favorite alternative practices employed by the parents of autistics. Gastroenterologists had used it for decades to treat celiac disease, a genetic disorder in which people are unable to tolerate gluten-containing grains. But the association of diet and madness extends well into the past—nineteenth-century scientists investigated whether *Selbstvergiftungen*, "ferments of the gut," provoked mental illnesses.[21] In the mid-1960s, a Philadelphia clinician named F. C. Dohan theorized that nutritional deficits were behind mental illnesses like schizophrenia and autism. Nutritional scientists had noticed that celiac patients, besides their inability to tolerate gluten, frequently suffered depression. Why not treat the mentally ill with the gluten-free diet of the celiac disease patient and see if the neurological problems fell away? Three New York psychiatrists published a case report that found that 7 of 21 children randomly selected from a group of

65 autistic children under their care had severe gastrointestinal problems including bowel obstructions, chronic diarrhea, and dehydration. Earlier, one of the therapists had treated an autistic child with celiac disease whose behavior improved notably on a gluten-free diet.[22] The study found that the children responded negatively to gluten in their diet. Paul Shattuck, a British autism researcher, and Karl Reichelt, a Norwegian biochemist with an autistic child, both picked up on Dohan's research. Reichelt's group at the Pediatric Hospital of Oslo believed that the children's intolerance for foods was what *provoked* autism. Other autism experts tended to believe these problems were co-inherited neurological and immunological traits.

By the late 1990s, no one doubted that autism was biological in origin. But how much of it was inherited, how much the result of environmental insult, in or out of the womb? To get a handle on these questions it was important to establish whether the prevalence of autism had increased and, if it had increased, when. More and more autistic children were flooding state clinics. Many people who tried to help these children described the growth in autism as an "epidemic." That word was loaded, because epidemics were not, presumably, genetic. If there was an epidemic, there had to be an environmental cause. No one could deny that autism diagnoses had gone up during a period in which vaccine use was going up. For many, it was a handy correlation, whether coincidental or not.

The claim that vaccines caused autism had been bruited about in shadowy corners of the alternative medical universe for decades. Rimland, who came to believe that any "natural" substance, such as a B vitamin, could be taken in unlimited amounts, while any pharmaceutical drug was intrinsically dangerous, concluded in the 1960s that DTP vaccinations played a role in autism.[23] Rimland blurbed Coulter's 1990 book on the social pathology of DTP, calling it a "thoroughly documented exposé of the dangers of childhood vaccinations" that would "no doubt start an acrimonious but timely debate about the known benefits and hidden costs of childhood vaccination programs." In 1998, *The Lancet* published the case reports of 12 autistic children suffering from severe constipation, diarrhea, and immunological deficiencies.[24] Its lead author was a polished and respectable London gastroenterologist named Andrew Wakefield, who earlier had speculated that measles virus was the causative agent of Crohn's disease. Wakefield noted that eight of the parents associated their child's regression with administration of the MMR shot at 15 months. At a news conference at the prestigious Royal Free Hospital, he stated that it might be a good idea to offer separate measles and rubella shots.

Wakefield described the children's condition as "new variant inflammatory bowel disease," suggesting that he had discovered a new causative agent. In fact, a major report on bowel disorders in autistics had been reported in 1971, and even some of Kanner's children reported GI problems—all years before MMR existed. The Legal Aid Board, the British agency that represents parents who believe their children have been harmed by vaccines, had paid Wakefield about $100,000 to conduct his research, and in 1997 he had applied for two patents, one on a single-measles shot, the other an alleged cure for MMR-related autism. When a reporter turned up these facts in 2004, *The Lancet* retracted the original Wakefield paper, and most of his 12 coauthors denounced it.[25]

In March 1998, a week after Wakefield's study appeared, Barbara Loe Fisher's group issued a news release together with Rimland and Cure Autism Now, an activist group funded in part by contributions from Hollywood parents with autistic children and their friends. Portia Iverson, the president of the group, said that about half of the hundreds of parents who called her office each month reported that their child had become autistic shortly after receiving a vaccination. Acknowledging that the timing could be coincidental, Iverson was troubled by the CDC-produced editorial that *The Lancet* had run alongside Wakefield's article, warning that poorly substantiated research could lead parents to abandon immunization. "Isn't it the responsibility of the government to take a proactive position on behalf of these children rather than a defensive one?" she asked.[26]

That news release would herald a tidal wave of antivaccine activism in the years to come. Later that year, a Vermont woman named Victoria Beck provided a nice complement to the Wakefield story by announcing on an ABC News program that her son Parker's autism—which she blamed on the MMR vaccine—had been virtually healed by an injection of a pig hormone called secretin. Her theory was that secretin, used to diagnose pancreatic problems, had healed Parker's vaccine-damaged gut, thus staunching the flow of opioid-like peptides to his brain. Beck patented the use of secretin for autism and transferred the title to Rimland, who sold it for an estimated $1 million to Repligen, a biotech company that planned to produce the hormone artificially. Repligen's president, Walter Herlihy, had an autistic daughter who had apparently benefited from secretin. Rimland told me at the time that he expected 70 percent of autistics to benefit from secretin. Only the "pathologically skeptical" could deny the benefits of the substance, he said. As for his critics, "I pity them."

Rimland had inspired a great deal of controversy over the years. Many people admired—even revered—him for demolishing the ill-founded "refrigerator mother" thesis. But many of those who treated autistics or raised money for research felt that some of his assertions hindered progress in the field as much as they helped it. "If you read Dr. Rimland's newsletter, in every issue there will be five or six treatments for autism," a developmental pediatrician in Texas told me. "If you've got five or six treatments every month, there must be something not exactly right about the treatments."

In May 1999, the secretin craze was in full swing—thousands of children had been injected with the substance—when Representative John L. Mica of Florida held a hearing in his Government Reform subcommittee into claims that the hepatitis B shot caused multiple sclerosis and other autoimmune diseases. This was the first major congressional investigation of vaccine damage since Mica's brother Dan had helped organize the DTP hearings more than a decade earlier. It was the first of more than a dozen hearings on vaccine safety in the Government Reform Committee, chaired by Dan Burton through the end of 2002.

In the substantive part of this hearing, several prominent scientists testified about reports of autoimmune disease.[27] In 1998 the French government had temporarily suspended its hepatitis B program when a French researcher collected 600 cases of autoimmune disease allegedly caused by the vaccine. Bonnie Dunbar, an animal vaccines researcher at Baylor University, had seen the health of her brother and a laboratory mate deteriorate following hepatitis B vaccination. She had gathered more than 100 cases of multiple sclerosis or other autoimmune reactions following the vaccine. Dunbar voiced the theory that parts of the hepatitis B vaccine mimicked human proteins, triggering an immune response against the protein sheath that surrounded nerve connections. Dunbar's hypothesis was biologically plausible, and the Institute of Medicine safety board set to work examining epidemiological data.

Mica noted that in 1996 the CDC had reported 10,637 cases of acute hepatitis B, of which 279 were in children under age 14. Given the claims of harm from the vaccine, he asked, why was it necessary for infants? Mica clearly hadn't been briefed on the complex but sturdy reasons for the infantile hepatitis B vaccine recommendation. Judy Lafler Converse of Cape Cod testified that day in May of 1999. Her baby was vaccinated against hepatitis B after birth, suffered seizures starting four days later, and was eventually diagnosed with autism and a seizure disorder. No evidence was put forth linking the two, but her belief in a link was strong.

Because the public health justification for the vaccination of newborns was difficult to explain and because hepatitis B was a disease with a stigma, the reports of harm from the vaccine provoked outrage. Many Americans, it seemed, didn't like being asked to take a risk of any kind to prevent a disease they believed they could prevent perfectly well themselves. They instinctively mistrusted anyone who forced them to belong to a group they hadn't joined, especially when that group included people they regarded as morally suspect. Middle-class moms who had never seen the disease were outraged that their children had been roped into the herd. Kathleen Rothschild, a suburban Chicago housewife, got involved in fighting for the right to a philosophical exemption from vaccines when a school nurse sent home a memo informing her of a new law requiring her daughter to be vaccinated against hepatitis B before entering fifth grade. Rothschild had "started investigating and was alarmed" by a package insert stating that there was no information on whether the vaccine "can affect reproductive capacity," she told an Illinois legislative committee. This was standard wording for any medication that had not been tested on pregnant women, but Rothschild felt the government was threatening her 10-year-old's future ability to have a family.

The issue mobilized a portion of the religious right. Catholics had long debated whether it was a sin to be vaccinated with a shot like rubella that was grown in a cell line originally obtained from an aborted fetus. Despite papal absolution for such vaccines, the hepatitis B vaccine provoked a new variety of religious resistance. Susan Brock, who brought suit against the requirement in Arkansas, wrote in a legal brief that the vaccine prevented a disease that (she believed) spread only through sex and illegal drug use. "The only protection my children need" she wrote, "is in Ephesians 6:13–14 which tells Christians to 'take up the full armor of God' and 'stand firm.' . . . I believe that immunizing my children against hepatitis B gives the appearance that my children will be sexually promiscuous or drug users." Another Arkansas parent, Cynthia Boone, charged in her lawsuit that the vaccine "supports the devil in his effort to encourage her daughter to engage in sex and intravenous drug use."[28] The conservative antifeminist Phyllis Schlafly and her son, Andrew, charged that mandatory vaccines were part of a government plot to control Americans.[29]

Three years after Mica's hearing, the Institute of Medicine panel issued its verdict on hepatitis B and autoimmune disorders.[30] It found no link—although the special masters of the Vaccine Court later did award patients for hepatitis B vaccine–related nerve damage. But this was only one of the many vaccine injury theories aired in the committee's hearing room over the next several years.

Burton's hearings were carefully choreographed to generate as much negative feeling toward the vaccination system as possible. Parents, most of them just plain folks whose honesty seemed unimpeachable, would appear one after another before the committee to share their stories of woe. Often, the parents brought their crippled, mentally ill children into the hearing room. A few scientists or alternative medicine advocates would then offer endorsements of the parent's theory and demand more research into the causes of autism and the therapies they were pursuing. Next, a panel of government scientists who spouted statistics and lacked heart-tugging stories would try, while remaining respectful toward the parents, to debunk the maverick theories and offer reassurance about vaccines.

Burton, who made little secret of his conviction that vaccines were behind autism and other problems, subjected government scientists to harsh cross-examination. He frequently accused them of bad faith and conflicts of interest, and sometimes threatened to cut their funding if they didn't find what he wanted them to find. "If you think these problems are going to go away, you're blowing smoke," he said at a hearing in 2001. "If the health agencies don't deal with this and deal with it quickly, you're going to have a big problem over there."[31]

To anyone who knew the history of Burton's stands on traditional and alternative medicine, his attacks on the vaccine establishment were not surprising. In the Indiana Senate in the 1970s, he had spearheaded a successful, John Birch Society–funded campaign to legalize the sale in Indiana of laetrile, a bogus cancer remedy made from apricot pits. Burton proudly stood on the steps of the State Capitol Building in Indianapolis in 1977 to announce that he had won the right of Hoosiers to get the treatment, though it was proven to be ineffective and to cause potentially fatal side effects.[32] In the early 1990s, Burton's wife was diagnosed with cancer and sought treatment from Georg Springer, a German practitioner then living in Illinois. But the FDA had blocked Springer's cancer vaccine trial from proceeding; the action infuriated Burton and may have motivated his many campaigns against the FDA. Burton felt the government had no place telling people what cures to seek.[33]

Burton joined Congress in 1983 and was known for his vituperative attacks on Democratic opponents. The Helms-Burton Act on Cuba, one of many draconian crackdowns that failed to topple Fidel Castro, was his best-known piece of legislation. At one point he theatrically fired bullets into a watermelon in his backyard to support his theory that Clinton aide Vince Foster had been assassinated. More to the point, Burton consistently opposed closer FDA regulation of companies that produced dietary supplements and alternative treatments. One of his major campaign backers was Metabolife, the leading manufacturer of the controversial diet pill Ephedra.[34] Burton had no scientific education; his college degree was from the Cincinnati Bible Seminary. In 2000, according to his website, he received an honorary doctorate from the Capitol University of Integrative Medicine. His staffers were alternative healthcare advocates. His Democratic foe, Henry Waxman, had a doctor and a longtime FDA official on his staff.

Burton's legislative jihad against vaccines was sparked by the conviction that his grandson Christopher, who was diagnosed with autism in 1998 at age 2, was a victim of vaccines. Burton's daughter, Danielle Burton-Sarkine, said that Christopher had received DtaP, MMR, HepB-

Hib, and OPV on a single visit to the doctor.[35] Those vaccines, she was quoted as saying, gave her child 62.5 micrograms of mercury in one day. But of the vaccines she mentioned, only one—the DtaP—contained thimerosal, a total of 25 micrograms.[36] If the four vaccines were given at a catch-up appointment, chances are that the boy had received even less thimerosal as an infant, when thimerosal was most likely to do harm.

According to Beth Clay, a former staffer at the NIH's Center for Alternative and Complementary Medicine who came to Burton's staff in 1998, she got a call from Burton about his grandson's autism diagnosis just as Barbara Loe Fisher walked into her office to talk about vaccine-related autism. "I'm the kind of person who believes in coincidence but I never see them," Clay says. "When things like this happen, you just need to pay attention. I put down the phone, turned to Barbara and said, 'You're not going to believe what just happened.' " Clay believed that the CDC's vaccine division was a corrupt agency in the pocket of the pharmaceutical industry. Obviously, her boss came to believe the same thing.

"I'm so ticked off about my grandson, and to think that the public-health people have been circling the wagons to cover up the facts!" Burton fumed at a June 2002 hearing, when Waxman pointed out that Burton had produced more heat than light at the hearings. "Why, it just makes me want to vomit!" Waxman, as the coauthor of the National Vaccine Injury Compensation Program, had done far more than Burton to help children truly harmed by vaccines. His role during the Burton hearings was to carefully reexamine the witnesses whom Burton had savaged, to make sure that the real, conjectural, and downright false were separately reflected in the record. For this, Waxman was sometimes booed in the hearing room. His careful assessment of the benefit and risk of vaccines was not welcome to those who were certain.

Ironically, the pharmaceutical industry had long targeted Waxman because of his support for cheaper and safer drugs. Among his drug-safety accomplishments was bringing to light the allegation that Merck quashed data on the risks of heart attack following the use of the anti-inflammatory medicine Vioxx. Despite the evident emnity between the two men, Waxman, like Burton, felt that the vaccine compensation program he'd helped create was failing some of the families who were in it. He and Burton joined forces to sponsor bills that would have expanded the program. The legislation was never brought to a vote. Neither Burton nor Waxman carried much weight on the Hill in the late 1990s.

The most visible product of Burton's hearings was a 2001 report on conflicts of interest in the committees that oversee vaccines at the CDC and the FDA. At the time of his investigation, Burton's daughter, Danielle, had become one of the first parents in the country to file an autism claim in the vaccine court. As part of his freewheeling reform efforts, Burton was pushing to liberalize the vaccine court's criteria and expand its payments. He never disclosed at the hearing that he had his own conflict of interest—the fact that his grandson had a case pending before the court.[37]

As he built his insurgency against the vaccine establishment, Burton was supported by a large community of volunteer professionals, mostly the parents of autistics, furious and grief-stricken about their children and sure they had found the culprit. One well-educated, well-to-do group of parents called themselves Safe Minds—Sensible Action for Ending Mercury-Induced Neurological Disorders. The group was an Internet phenomenon, a network of parents from New Jersey to Georgia to California united by the conviction that vaccines had damaged their children. Many of the parents were pursuing alternative treatments of the sort that Rimland promoted. In October 1998, the issue of mercury poisoning was discussed for the first time at a gathering of Rimland-affiliated doctors who called their loosely knit group Defeat Autism Now!, or DAN! for short. One of the physicians, a vaccine skeptic named Stephanie Cave, of Baton Rouge, Louisiana, reported that she had "chelated" two or three boys with autism who had elevated levels of lead in their blood. In chelation, a patient gets a dose—through IV or a topical cream—of a potent sulfur-based chemical called a succimer, which binds with mercury and other heavy metals, which the patient subsequently excretes. Cave told the other doctors that the children had done better cognitively and behaviorally after the chelations. A new therapy for autism was born.

The only environmental agent that was indisputably associated with autism was the rubella virus, which frequently caused autism when it infected a fetus during the first month of pregnancy.[38] Congenital rubella, of course, had been eliminated by a vaccine—the MMR shot. In 2001, several Safe Minds members published an article in *Medical Hypothesis*—a forum for intriguing, unproven ideas—titled "Autism—a Novel Form of Mercury Poisoning." The authors were Sallie Bernard, Lyn Redwood, Albert Enayati, Heidi Roger, and Teresa Binstock. One curious fact about the authors was that at least a few of them had children who were not exposed to the larger doses of thimerosal that were being blamed for the "epidemic" of autism.[39] Bernard's autistic son Billy had been born in 1987, before Hib and hepatitis B vaccines were given to infants. Billy was a triplet, had been born five weeks premature and weighing just three pounds, and had remained hospitalized for a month—a scenario often linked to organic brain damage. Binstock was an adult with Asperger's syndrome, a mild form of autism. Redwood and Roger apparently had children who received all the thimerosal-containing shots. Another leader of the group, Liz Birt, blamed MMR for her child's illness.[40]

Over the years, the salient fact for many of these and other parents was that their children had physical symptoms that would be addressed, at least partially, by alternative treatments premised on the idea that they were overcoming damage caused by vaccination. If a gluten-free, casein-free diet improved their children, these parents believed, it confirmed the theory that vaccines had done something bad to the gut. If chelation worked, it proved that heavy metals, including thimerosal, were to blame for autism. The fact that the establishment, the American Medical Association, and the American Academy of Pediatrics, supported universal vaccination and were skeptical about gluten-free diets and chelation therapy only cemented the notion that their indifference to the "cure" for autism was a cover-up for their role in causing it. The us-against-them dynamic that had characterized the struggles of Rimland and Fisher was honed to a sharp edge in the autism battle. Whether they were scientists, physicians, or parents, the two sides lived in parallel universes. The great majority of practitioners would be convinced by published research showing no evidence of a link between vaccination and autism. The people who blamed vaccines had their own diets and physicians—even their own epidemiologists and basic scientists.

To be sure, some of the signs and symptoms of autism had rough parallels in the literature on mercury poisoning, whose symptoms can appear months after exposure and which afflicts only a minority of the

population exposed. In the first half of the twentieth century, for example, thousands of children in the United States and Europe became ill
with Pink disease, or acrodynia, which caused peeling skin on the
extremities as well as mental disturbances, pathological shyness, and
regressive behavior. In 1945, a keen-eyed Cincinnati pediatrician named
Josef Warkany noticed a common risk factor in Pink children: they had
all received teething powders containing calomel, or mercurous chloride. In Switzerland, children treated for parasites with a mercurial remedy called *Wurmschokolade* (worm chocolate) had similar symptoms.
Only about 1 in 500 children who received teething powder treatments
got Pink disease, roughly the same ratio of American children with
autism. After mercury was removed from the powders in the late 1950s,
Pink disease disappeared. Behaviorally, Pink children were sometimes
described as "walled off;" one mother said her Pink child had "behaved
like a mad dog." But while mercury might have made Pink children
"crazy" by some definitions, it had *never* made them autistic. And Pink
children had consumed upwards of a *gram* of mercury in the first year of
life, compared to one five-thousandth of a gram for the most heavily
vaccinated autistic children.

The *Medical Hypothesis* article was repeated and chewed over and spat
out all across the Internet. It was the start of a flood of literature, some
legitimate, some extremely sophomoric, linking thimerosal to vaccines. If
you downloaded enough of these articles, mercury could become something of an organizing principle. It could explain a lot of what was wrong
not only with your child but also with the culture in general. It could
explain the increase in ADHD, and Kip Kinkel and Columbine. "It colors the way you perceive things," said George Mead. "I see kids in the
airport who are behaving in classically spectrum ways, tantrumming, and
what's happened is, as a culture we've come to say, 'Oh that's just him
having a fit.' But you talk to any teacher and they say this is not the way
things were. Things are worse than ever now. There is a very large percentage of children out there affected. And yeah, I believe heavy metals
could be the cause of all of it."

Chelation was the hot new treatment for autistics in the summer of
2002 when I first visited the Meads. The process could cause liver damage, so people chelating their children used a variety of supplements to

protect them. It might have been prudent to wait for research that showed something definitive about mercury and autism, the parents admitted. But like the white rabbit in *Alice in Wonderland*, they had "No time! No time!" Their children's brains were growing and they needed to act now. So they chelated their kids, sometimes feeding them the chemicals once every week or so for a period of months. Mercury wasn't the only metal that came out in their urine—there was also arsenic, lead, cadmium, and helpful metals like zinc—the parents would feed their children supplements of the good metals after each round of chelation. Toxicologists would point out that these metals come out of anyone whose system is challenged with a powerful succimer. There was no baseline information for "normal" mercury in a child's body, or a normal response to chelation. But the phenomenon of multiple metal removal had led the Meads to shift their culprit from mercury alone to heavy metals in general.

In fact, the Meads had initially blamed the MMR shot for their child's illness, because Willie's symptoms had begun some time after his second birthday. Tory Mead first told me that "mercury in the MMR shot" had caused her son's illness. I pointed out that the MMR shot did not contain mercury. Later, the Meads would subscribe to a new theory rolled out by some of the DAN! physicians, which was that thimerosal damaged the children's immune systems, thereby rendering them unable to resist the live viruses in the MMR shot. This was wild speculation—what about the zillions of other viruses they encountered in everyday life?—but perhaps a convenient means of keeping the MMR and thimerosal camps within the movement from openly disagreeing with each other.

In August 2001, the Meads had become one of the first families in the country to sue the vaccine industry over the use of thimerosal. They filed two lawsuits, in effect hedging all bets about the impact of the chemical. A classic product liability suit sought damages for Willie's autism; a class action was filed on behalf of Eleanor, Willie's blonde, Barbie-loving older sister, along with 30 million other kids. Eleanor, said attorney Michael L. Williams, "has no deficits as far as we know." The claim filed on her behalf was a medical monitoring lawsuit, which would require vaccine companies to fund studies looking for subtle neurological and cardiovascular effects from mercury in the vaccinated kids, regardless of how normal or abnormal they seemed. The monitoring lawsuit was later dropped when the Meads filed their claim through the federal vaccine compensation program, which did not hear class-action cases.

When I met the Meads they lived in a large, handsome ranch-style house on the outskirts of Portland, with plenty of white upholstery, Audubon prints on the walls, and a big yard full of rose bushes and Douglas Fir trees. Fighting mercury—legally, politically, medically, nutritionally—was the Meads's life. They had taken out a $400,000 line of credit on their house to pay for special care for Willie, including a team of therapists who came to the house every day. The Meads estimated they were paying $8,000 a month to improve their son's chances. All autism experts agree that early intervention is key to overcoming the deficits of the disease. For the Meads, the approach was to attack on all fronts—the traditional behavioral therapies coupled with nutrition, secretin, chelation, supplements. In their long, airy kitchen, a large white board contained a schedule of the children's pills and tinctures—Eleanor was taking many of them prophylactically—and therapy sessions.

A cabinet full of supplements in blue bottles, labeled with names like Milk Thistle and Ambertose and Chemet—occupied the opposite wall, by the telephone. There were 45 bottles in this cabinet alone. It was awfully hard to know which of these substances and therapies was having any effect. But to the extent that their son's symptoms improved while he was on the DAN! regimen, the Meads felt justified in pursuing it. They felt angry, and perhaps a little smug, toward the medical establishment, which didn't figure at all in their lives anymore. Good old American pragmatism had won out long ago over evidence-based medicine in the Mead household.

George had laid out all the knowledge the Meads gleaned from Defeat Autism Now! and other Internet sites in three big flow charts full of boxes and arrows. In 2002 he was particularly avid about the theories of William Walsh, a biochemist who ran a clinic for disturbed children outside Chicago. Walsh's theory, later developed by chemist Richard Deth at Northeastern University and nutritional scientist Jill James at the University of Arizona, was that autistic children had a defect in metal metabolism that caused their brains to be damaged by toxins that other children got rid of. To listen to George's unsettling, warp-speed rap,

Walsh's theory was the basis for a worldview that explained each waking moment of young Willie's life.

I spent a day with Willie and his mother to get a feel for daily life in a DAN! family. Arriving at 8 A.M. one Tuesday, I found the boy breakfasting on quinoa-flour waffles and chopped pear, washed down with a sippy cup full of stevia-sweetened water into which his mother had stirred a dozen supplements, including high-dose minerals and vitamins, cod liver oil, antifungals, glycoproteins, amino acids, and digestive enzymes. Chelation was causing Willie to grow yeast in his stomach, so the Meads had stopped chelating for a while. The glycoproteins supposedly were killing the yeast. As he ate breakfast Willie bounced up and down while twirling his hair. "See that jumping?" George asked. "That's die off. The yeast are dying. When they die they release toxins, and that causes him to jump up and down and spin like that."

After breakfast, Willie calmed down. The supplements and the diet, the Meads were convinced, had improved their son's overall health. They were also using Applied Behavioral Analysis, a painstaking training program that is regarded as the gold standard in the care of autistics. Beth Steege, who directed Willie's behavioral treatments, arrived at 8:30 A.M. With a little prompting, Willie greeted her, and Steege, a trim, muscular woman with short brown hair, led him down to the basement playroom. His school day began with an articulation lesson. Steege showed him a card with a picture of a light switch on it. Willie, brown-haired, blue-eyed, dressed in plaid flannel shorts and a red zip-up shirt, stared at it for a moment. "On," Steege said.

"Onghhh," Willie responded.

She showed Willie a card with a ghost on it, but instead of saying "Boo," on cue, he twirled his hair and rubbed his nose. Steege took his hands and gave his arms a shake. He jumped into her lap and squealed. After a while they returned to the flash cards, but Willie kept looking away. He seemed sleepy. "Sometimes he's really there," Steege said, "Other times he's not. When I started with him he couldn't even point to something he wanted."

After a while they took a break. Willie was allowed to wander around the backyard while his mother fixed him a lunch of organic chicken sausages and carrots. "Everything we do is controversial now, we're totally controversial," Tory said jauntily. She said she had become "heavy and authentic and no fun at cocktail parties anymore," but the preppie, peppy Tory, who was slim, pretty, and wore her straight black hair in a pageboy, was still on display. Many old friends had lost interest in the Meads, she said, but the experience had deepened her understanding of the world. "We would be a lot less strident," she added, "if there was a little more compassion for the fix we're in."

Around the time that Willie started his treatments for autism, Eleanor, who was 4 at the time, threw some tantrums, was wakened with nightmares, and generally acted up. The Meads's response was to have her tested for mercury. "We did a workup and it spooked us enough that we put her on the diet and chelated her for mercury, and as you can see she's doing great," said Tory. Indeed, Eleanor was great, with her blonde pigtails and bubbly, talkative manner. Could it be, I asked, that she was "acting up" because her brother had just been diagnosed with a terrible affliction, and she was upset and needed some attention? "When we put her on the diet, her tantrums stopped," Tory said. "If you've lost one child to autism then you're very careful."

Public Health as Public Enemy

In 1943, when he coined the term "autism," Kanner noted that all of the first 11 children he saw were from "highly intelligent families."[41] Not only were the men well-established professionals but most of the women also had college educations, unusual at the time. More recently, an article in *Wired* magazine called autism the "geek disease," and a leading British autism expert hypothesized that technology had increased autism's prevalence by creating a Darwinian niche for people with the particular skills that "high-functioning" autism, at least, confers—such as absorption in detailed work and special abilities in math. The "assortative mating" patterns of the last few decades, in which more nerdy women were in the workplace where they met similarly nerdy men, meant that Kannerian, semi-autistic men were more likely to marry women something like themselves than they might have been in the past.[42]

Safe Minds activists were infuriated by these articles, which they perceived as heaping more blame on the victim.[43] But a major epidemiological study conducted in California in 2003 found that women over the age of 35 were four times more likely than women under 20 to have autistic children. Better-educated women were also more susceptible, the study found. Those with postgraduate degrees were four times more likely to have autistic children than women who hadn't been to college. When maternal age was removed as a variable, the highly educated women were still at twice the risk of having autistic kids.[44] The article offered no speculation on why this might be the case. Presumably, educated mothers were more likely to get their children diagnosed. But it could also be that "assortative mating," pairing up obsessively brainy people, was creating more autistics. It was a theory, anyway.

Among the parents of autistics I spoke with while preparing this book were the chief of pediatric immunology at a large medical school, a senior FDA research scientist, numerous high-achieving lawyers, several wealthy businessmen, an NIH geneticist, and an emergency medicine doctor-turned-thriller-writer. Whether or not the genetics of these high-achievers had any relevance to their children's sickness, their professional experience and single-mindedness certainly helped push their theories about autism into the limelight.

I met many of the parents at a Defeat Autism Now! conference in Boston attended by 1,100 people. The agenda was a mixture of motivational exercises, scientific presentations, diet advice, and railings against the establishment. Barbara Loe Fisher was introduced with the words, "What does it feel like to be vindicated?" and her talk greeted with a standing ovation. I sat with Rimland at dinner one night and asked him about his long experiences in the autism world. He was personable and emphatic and told me that he had never been wrong about anything. "I'm very careful before I make any claims," he explained. His current thesis, he said, was that vaccines were probably not responsible for the major decline in diseases like whooping cough, measles, and diphtheria. We sat with a few of Rimland's autistic friends, with whom he was quite affectionate. People came by to share a joke or a clap on the back. It was clear that Rimland had won a lot of loyalty over the decades.

I also met Mark Blaxill, a handsome, well-coiffed and slightly stiff

Boston management consultant who directed Safe Minds. Blaxill's day job was to help Fortune 500 companies think about intellectual property, marketing, and operations in ways that made them more efficient and profitable. His company's website said that Blaxill developed some of the firm's "core concepts," such as "time-based competition" and "cycle-time reduction." Blaxill has remarkable self-confidence and a facility with complicated concepts. He had made it his second job to pick apart the epidemiological studies he disagreed with—the ones that showed either that thimerosal was not linked to autism or that autism was not increasing.

"This is an area where establishment science does not tread," he told me as we passed tables where vendors sold gluten-free diet foods and supplements that were proclaimed "Autism Safe!" Blaxill's daughter, 6 at the time we met, regressed into autism in her second year but had improved thanks to the supplements and diet, he said. "This is where the risk takers are. The ethos of this is 'You can do something,' not 'sit back and accept the inevitable.' " Reflecting his marketing background, Blaxill gave grand names to his critiques of the immunization program's work. "The Governance Problem" was how he described the incestuousness of the public health world; "the Hidden Horde Hypothesis," his dismissive term for the idea that only detection of autism had changed, not levels of the disease. For Blaxill, the lack of treatment for autism reflected poor business performance, like GM producing another clunker. If assembly lines in China could make error-free motherboards, why couldn't Merck make accident-free vaccines? "The medical community is not serving this [autism] problem, it's an awful, miserable underperformance," he told me. "They don't know what's going on, they ignore the evidence, they try to explain inconvenient facts away. They don't take seriously any of the unconventional treatments yet everyone knows that the glutein-free, casein-free diet works. Nobody knows how."

Blaxill, Bernard, and other Safe Minds members had modeled themselves on the AIDS group ACT-UP, which had great success in shaking up the bureaucracy and generating new research and medical progress. Just as AIDS activists had stormed the NIH in the 1980s, antimercury activists shouted down scientists at the campus where they had been invited to discuss autism research in 2001. "We are no longer susceptible to your propaganda!" one of the mothers yelled.[45] ACT-UP's tactics were often ugly, but they got results. Much the same may be said, years hence, for Safe Minds, considering how much autism research has increased recently. "If I had an advocacy group," said one beleaguered CDC scientist, "I would do exactly what they did."

Critics of the CDC claimed that putting it in charge of vaccine safety research was "like relying on Ford and Firestone to investigate the SUV rollovers," as Rimland scoffed. Here, the antiestablishmentarians were not alone. While most people in public health doubted that the world's premier public health agency would engage in widespread fraud and suppression, some felt that the mere suspicion of such action, because of the structure of the vaccine safety system, justified taking the safety branch out of the CDC's immunization program where it had always resided.

In the congressional attacks on the vaccine system, Burton was soon joined by Dave Weldon, a Florida Republican, physician, and member of the House Appropriations Committee. Weldon had been asked to join the cause by a friend, Dr. Jeff Bradstreet, who had an autistic child and mixed alternative and traditional medicine with biblical teaching in his therapeutic practice. After leaving the Royal Free Hospital under pressure in 2002, Andrew Wakefield had for a time transferred his practice to Bradstreet's clinic in Florida, which advertised that its treatment of autistic children succeeded thanks to Bradstreet's "vast research and grasp of the Scripture."[46]

In late 2001, Safe Minds activist Liz Birt, a Chicago attorney, used the Freedom of Information Act to uncover classified study drafts, meeting notes, and e-mails that offered a window into government scientists' discussions about the thimerosal issue. Safe Minds and Burton would use these documents as weapons in their war against the vaccine establishment. What the documents plainly showed was that the government anticipated what would later come to pass—a serious public relations problem resulting from thimerosal. The documents did not show any definitive evidence of a link between thimerosal and autism or other neurological problem. However, a civil jury might see that differently. By June 2000, the Vaccine Safety Datalink study that Orenstein and Halsey had discussed at the August 1999 meeting in Washington had been largely completed, and its results were troubling.

The study, whose lead author was a Belgian epidemiologist at CDC named Thomas Verstraeten, showed a slightly increased risk of tics, speech delays, and other neurodevelopmental lags in children at two West Coast managed care organizations who had received relatively larger amounts of thimerosal. But there were many factors that made the early results shaky. Verstraeten sought to determine whether tiny differences in thimerosal ingestion—differences of 50 millionths of a gram of thimerosal, spread out over a period of months—could differentially affect the developmental outcome of babies. He took groups of babies that had received 25, 50, and 87 micrograms of mercury by age three months, and looked to see whether years later these babies had different levels of neurological problems. But a baby in the 25-microgram group at two months might have gotten 50 micrograms more a week later, which made the "stratification" of the study problematic. A lack of knowledge of how much thimerosal was in the mother's blood or breast milk made the comparisons even more meaningless, especially since the toxicokinetics of ethyl mercury were unknown. Then, too, parents who brought their babies in for shots right on schedule—and therefore had the highest thimerosal exposures in the early months—were probably the most attentive parents, and might notice their child's speech delay or tics at an earlier age. Conditions like speech delay, which might or might not be the beginnings of an autism diagnosis, were soft outcomes that weren't necessarily tracked carefully by an HMO.

All these doubts led Orenstein to call for a consultation to discuss the results. A total of 54 people were at the meeting at the Simpsonwood resort outside Atlanta. The press was not invited, and CDC officials asked those present to keep quiet about the data until they were presented to the Advisory Committee for Immunization Practices two months later. Verstraeten presented the study and two other CDC epidemiologists, Robert Davis and Phil Rhodes, gave critiques of his work. Then the consultants, who included environmental health, toxicology, infectious disease, and other specialists, were invited to comment. Some were upset by the data. "I just got called out at 8 o'clock for my daughter-in-law delivering a son by C-section. I don't want that grandson to get a thimerosal-containing vaccine until we know what is going on," said Richard Johnston, who had chaired an Institute of Medicine vaccine safety committee years earlier. Others were less worried about the actual harm than by what the results could mean. "The medical/legal findings in this study, causal or not, are horrendous," said Robert Brent, a Philadelphia pediatrician. "If an allegation was made that a child's neurobehavioral findings were caused by thimerosal-containing vaccines, you could readily find a junk scientist who would support the claim with a reasonable degree of certainty. But you will not find a scientist with any integrity who would say the opposite with the data that is available. We are in a bad position from the standpoint of defending any lawsuits if they were initiated."

At the end of the meeting, when the consultants were asked whether the data gave reason to be worried about thimerosal, on a scale of 1 to 5 (1 being strong correlation, 5 being weak) all but one believed the evidence of harm from thimerosal to be 1 or 2. William Weil, a Michigan environmental health specialist, was the outlier, giving it a 4. (Later, Weil told me that despite his concerns about thimerosal, he thought it farfetched to believe it caused autism.)

Safe Minds released the transcript of the meeting in 2001, and it quickly became the centerpiece of an attack on the National Immunization Program by Safe Minds and members of Congress to whom it indicated an effort to suppress damaging evidence about thimerosal. In the meantime, to give the study more power, the authors brought in 34,000 additional children from another HMO. These were children who were now old enough to have potentially received diagnoses for neurological problems. Safe Minds interpreted this expansion as an effort to water down the results. The new results wiped out the earlier signal of damage from thimerosal—a relief to everyone except those who had a vested interest in a bad outcome. Safe Minds leapt on the fact that Verstraeten had departed the CDC and joined GlaxoSmithKline before the final study had been finished, which the group saw as evidence his silence had been bought. In fact, Verstraeten's visa had run out. GlaxoSmithKline, which makes vaccines in the Belgian city of Rixenaart, was his best offer back home.

To be sure, there was a lot riding on the results, and it was always possible to redo a study in a way that might change them. "There are lots of people who don't want them to find this association," Neal Halsey told me in 2002. "You can imagine. The manufacturers, everyone else. . . .

The people in the vaccine safety branch are embattled and criticized from both sides." But Halsey, the Institute of Medicine, and most other scientists I spoke with did not feel that the various iterations the study went through were signs of a cover-up. It was quite normal for a difficult study of this type to be examined, pondered, prodded, and adjusted before publication.

In Atlanta, Chen's Vaccine Safety and Development Branch chased down each lead the critics presented. When Safe Minds attacked the scientists for including children who were too young in the study cohort, they had added older kids. When Safe Minds chided them for excluding children with congenital and perinatal defects, they restored kids whose "perinatal" problems were things like diaper rash. But this was never going to be a study that answered all the questions. Chen ordered a follow-up study that carefully assessed about 1,000 children who had been exposed to different quantities of thimerosal. The $2 million study, shaped using the methodology that Phillipe Grandjean had used in the Faeroes, was contracted to an outside consultant to avoid any claim of conflict of interest.

You didn't have to be of conspiratorial bent to believe some harm *could* have resulted from thimerosal. In the years after Simpsonwood, dozens of studies were conducted to examine the evidence of a link between thimerosal and autism or other developmental disorders. Epidemiological studies produced in England, Denmark, Sweden, Canada, and the United States tended to dismiss the thimerosal thesis. The Danish study, published in late 2002 in the *Journal of the American Medical Association*, was the most important. Danish children were not vaccinated against *Haemophilus influenzae* type b or hepatitis B, but virtually all Danish children were vaccinated with DTP vaccines that contained more thimerosal than the American shot. In the 1960s, a Danish child received 200 micrograms of ethyl mercury through DTP by 15 months of age. From 1970 to 1992 the children's vaccines contained 125 micrograms by 10 months. That wasn't as much as the 187.5-microgram maximum that an American child would have received by six months, but it was comparable. More importantly, *all Danish children had received thimerosal-free vaccines since 1992.* The results were clear. Autism incidence was flat until 1990 in Denmark, then steadily increased after thimerosal had been removed.[47]

Rimland and Safe Minds attacked the work because it was funded by the Statens Serum Institut, the governmental vaccine-making agency where the authors worked.[48] Their hostility was palpable. When Morton Hviid, the study's lead author, came to the Institute of Medicine in 2004, he had to be ushered in through a back door because of death threats. In the English study, and in the 1,000-child neurodevelopmental outcomes test, whose results were discreetly unveiled to U.S. scientists and mercury activists in 2005, additional thimerosal early in a child's life seemed to correlate with *improved* IQ and speech development.[49] Both findings showed a small increase in tics associated with greater thimerosal use— possibly a chance association.

Such was the data that mainstream pediatricians and infectious disease specialists were looking at. But while they were assimilating these reassuring studies, Mark and David Geier, a father-son pair who worked out of their home in Silver Spring, Maryland, put out a stream of damning studies in small journals. Mark Geier and his son had a business that specialized in offering testimony on behalf of claimants in the Vaccine Injury Compensation Program. They weren't always regarded as reliable witnesses. Special masters of the vaccine court had thrown out their testimony on at least ten occasions, on the grounds that they lacked expertise.[50] The Geiers' work on thimerosal was so full of holes, in the view of most vaccine scientists, that it was hard to interpret it as anything other than an effort to boost their own legal strategy in the vaccine cases. The Institute of Medicine panel that studied the vaccine-autism link found their results "uninterpretable."[51]

When the Geiers mined Vaccine Adverse Event Reporting System (VAERS) data, for example, their methodology was to assume that these reports were literally true, although anyone could make a VAERS report at any time. In a 2003 study, for example, the Geiers compared VAERS reports after thimerosal-containing and thimerosal-free DTaP shots from 1992 to 2000, and found a sixfold risk of autism or mental retardation following vaccination with thimerosal. But the study was simply riddled with holes. It was based entirely on unverified reports to VAERS; there was no indication which other shots the children had received, and, because of the sample size, if only one of the children with autism was misclassified, it would have changed the relative risk by half. Most importantly, the authors failed to note that most parents who reported autism as an aftereffect of vaccination to VAERS only did so after reports of a link started to appear in the news media in 1999. In other words, the Geiers were reporting an artifact of the vaccine scare.[52]

Some of the most damning evidence against the thimerosal and MMR theses was contained in a report that vaccine critics often trumpeted as proof of a vaccine-provoked epidemic. These were the data from a 1999 California Department of Developmental Services report. Like reports from other states, it showed a dramatic increase in autistic children seeking the state's services from 1987, when 2,778 autistics were on the state's rolls, to 1998, when there were 10,360.[53] An impressive diagram of this increase was projected on a screen at one of Dan Burton's hearings. "Look at that graph," Burton said. "They are having an epidemic out there."[54] But if you looked closely, you saw that the vaccines Burton was blaming for autism were unlikely to have had anything to do with the increases. While MMR vaccination rates increased 14 percent from 1980 to 1994, autism intakes to state services programs increased 373 percent. *Haemophilus influenza* type b vaccination was introduced for infants beginning in 1991; hepatitis B vaccination in 1993. The biggest increase in cases of autism occurred in the cohort of children born from 1987 to 1992.[55]

Mainstream autism experts expounded much more obvious explanations for the increase in California: diagnostic changes, new laws that expanded federal payments to care for autistics, and greater parental awareness of these resources—as well, possibly, as the greater risk of autism in children of older parents. The jumps in California autism diagnoses closely followed changes in the diagnostic manual that expanded the definition of autism spectrum disorders. These changes came in 1980, 1987, and 1994. And in 1990, Congress passed the IDEA legislation, which qualified autism disability for federal funding. Thereafter, states were obliged to report all cases of autism. A study in Minnesota provided one clear indication that greater monitoring and awareness were increasing autism statistics.[56] Diagnoses of autism disorders among Minnesotan six-year-olds increased from 13 per 10,000 to 35 per 10,000 from the 1995–96 to 1999–2000 school years. At the same time, as the program gradually took in more patients, the number of autistics counted *in the same group of kids* also tripled. For example, a prevalence of 13 autism spectrum cases per 10,000 pupils was reported in 1995 among children born in 1989. In the 2000–2001 school year, the prevalence rate for this same group of children was reported at 33 per 10,000. Between the ages of 6 and 11, these kids had suddenly "become" nearly three times as autistic. The obvious explanation was that authorities were enrolling them in programs more aggressively.[57]

In Britain, where it was common for parents of autistics to blame MMR rather than thimerosal for their children's condition, autism diagnoses increased steadily from 1979 to 1992, when they peaked. MMR vaccination had begun in Britain in the late 1980s.[58] In a study that sought to determine what trigger factors parents blamed for autism, domestic stress, seizures, or viral illness were blamed for regressive autism before 1998. After the publication of Wakefield's paper, parents were more likely to attribute their children's problems to vaccination. Chart reviews of 92 children with autism whose parents blamed MMR found that the parents of 36 of those children had been concerned about their child's mental health *prior to the MMR shot.*[59]

But if the population surveys did not support the idea of large increase in vaccine-related mental illness, what about the closer-grained biological studies? Here, two strands of research conducted between 1999 and 2005 seemed to show conflicting evidence. Two studies conducted shortly after the thimerosal study broke found that thimerosal was metabolized fairly quickly in the human body.[60] An important study in monkeys, comparing the metabolism of injected ethyl mercury and eaten methyl mercury, found that total mercury was cleared out of the monkey bodies 5.4 times faster when they were injected with ethyl rather than fed methyl mercury.[61] The ethyl mercury's half-life in the monkeys' brains was also smaller—about three or four times less of it concentrated in the brain, although more of it converted to inorganic mercury, which might be more dangerous.

In any case, these studies on small groups of animals might not pick up a genetically susceptible population. Other studies suggested that autistics might have biochemical or cellular deficiencies that rendered them more susceptible to damage from thimerosal. *In vitro* work conducted by scientists such as Boyd Haley of the University of Kentucky and David Baskin at Baylor, found that thimerosal was very damaging to brain cells. Haley, former chairman of his school's chemistry department, had been a controversial proponent of the theory that mercury from dental fillings was a major contributor to Alzheimer's disease. He called methyl and ethyl mercury "oink and oink-oink," and theorized that Gulf War syndrome was caused by thimerosal in shots the soldiers received.[62] But his *in vitro* studies of metal toxicity were generally not considered reliable predictors of health effects.[63] Other authors tried to establish that autistics had unique metabolisms that rendered them susceptible to damage by heavy metals like mercury. Mady Hornig, a Columbia University scientist, found a strain of mice that demonstrated "autistic" behavior—obsessive grooming, poor sociability, aggression—when exposed to thimerosal. Another paper published early in 2005 by Jill James, a University of Arizona nutritionist, and colleagues, found lower levels of glutathione, a protein that protects cells against damage from free radicals, in autistic children.[64] Since autistics have weird diets and glutathione is very sensitive to diet, James's study—which had large margins of error and overlap between the two groups—was intriguing but seemed to need confirmation.

In 2001, an Institute of Medicine panel on vaccine safety reported that a link between thimerosal and neurodevelopmental delays was "biologically plausible," and called for more research. In its follow-up, published in June 2004, the panel decided that enough research had been done to make a more definitive pronouncement. Brushing aside the Geiers' research for its "methodological flaws and nontransparent analytical methods," the panel said that the bulk of population studies showed no serious effects from either thimerosal or MMR on children's mental health. The controversial Verstraeten study, the panel found, had been conducted properly.[65] The panel categorically rejected Wakefield's thesis that MMR caused autism, but noted charitably that Wakefield's work shouldn't all be tossed out with the bathwater: "If some developmentally disabled children have a previously unrecognized intestinal pathology causing distress, pain, and a cycle of diarrhea and constipation, treatment could lead to a better quality of life and possibly to behaviors conducive to better learning."

As for evidence of thimerosal damage, the panel pointed out the extensive literature indicating peculiar immune patterns or genetic immunology in autistics. None of these studies, the panel said, showed that the immune problems caused autism. It was more likely that immune and behavioral problems could be part of a disorder that science suggested was 90 percent inheritable.[66] The panel found that studies of mercury in hair were also inconclusive. Some parents, such as Lynn Redwood, became convinced of the mercury factor when large amounts of mercury turned up in their children's hair.[67] Others suggested that less mercury came out in the hair of autistic children, because they were unable to excrete mercury bound to brain cells.[68]

The committee recognized that a small group of children might theoretically be genetically vulnerable to mercury, but found no corroborating data linking vaccines or vaccine components to autism based on genetic susceptibility. While it was possible that some people with autism, perhaps even a subgroup that could eventually be identified by genetic markers, had abnormal immune reactions and abnormal mercury metabolism, "vaccination of these individuals does not cause these abnormalities or autism itself," the panel concluded. "From a public health perspective the committee does not consider a significant investment in studies of the theoretical vaccine-autism connection to be useful at this time . . . the benefits of vaccination are proven and the hypothesis of susceptible populations is presently speculative. Using an unsubstantiated hypothesis to question the safety of vaccination and the ethical behavior of those governmental agencies and scientists who advocate for vaccination could lead to widespread rejection of vaccines and inevitable increases of serious infectious diseases."[69]

The foes of thimerosal and vaccination were, predictably, infuriated. The most respected, professional voice in medicine had just told them to stop wasting their time. "A grave disservice has been done to science today," intoned Safe Minds' Blaxill. The activists loathed Marie McCormick, the prim director of the panel, referring to her as "the church lady." Representative Weldon was convinced that the CDC had cooked the Verstraeten study.[70]

Even as the Safe Minds activists raged, the data continued to weaken their case. In the second quarter of 2006, more then 6,000 new autistic 3- to 5-year-olds entered the California mental health system—roughly 40 percent more than had entered the system in the last quarter of 2002. Unlike the 2002 cohort, few, if any, of the newly admitted children had received relevant amounts of thimerosal in their vaccines.[71]

If, as most vaccine experts believe, the thimerosal and MMR alarms will soon fade into history as the latest craze of the autism world, some good will have been done, and some harm. Paul Offit, who had criticized the removal of thimerosal from the start, now felt vindicated. Others such as Orenstein felt it was still a good idea to remove the stuff. "Clearly, you'd like to have had data to show there was a risk," he told me after he'd left the CDC. "But the fact that we know it's possible to produce these vaccines with no or minimal thimerosal, means it's worthwhile getting it out."

The short-term impact of the thimerosal controversy was a shortage of the DTP and tetanus vaccines. The FDA had been pressuring Wyeth to upgrade its factory in Marietta, Pennsylvania, where vaccines had been made since the 1880s. But Wyeth had barely a quarter of the market for DTP. It was inclined to get out of the market, and the request for thimerosal-free vaccine was the final push it needed. Its competitors, Aventis and GlaxoSmithKline, were coming out with products that combined DTP with hepatitis B or Hib in a single convenient shot, and Wyeth wasn't interested in regearing to make thimerosal-free shots in a factory that it believed was obsolete. Another, smaller DTP manufacturer, Baxter Hyland, also got out of the market, much to the chagrin of John Robbins, whose laboratory had invented its one-antigen pertussis-component vaccine, Certiva. Merck announced in September 1999 that its thimerosal-free hepatitis B vaccine was available, though it apparently distributed the old shot into 2002.[72]

The thimerosal switchover also complicated efforts to eradicate hepatitis B. Hospitals were supposed to continue to vaccinate children if their mothers tested positive for the hepatitis B surface-antigen. But many at-risk women had either never been tested or were tested somewhere other than the hospitals where they gave birth. Four babies were born with hepatitis B in late 1999 at the Children's Hospital of Philadelphia, which neglected to vaccinate them. "It may be 20 years before these children get their cancer," said Barbara Watson, a public health official.

On the positive side, the controversy forced mainstream neurologists to take a closer look at the treatments being pursued by the people it was supposed to be serving.[73] A survey of the evidence found that neither B6 megadoses nor the gluten-free, casein-free diet had been properly studied, though both were used by thousands of autistic children. The same was true for chelation therapy, although there were signs that chelation was starting to fade in favor of a new rage for methyl-B12 with folinic acid, to directly increase the glutathione levels of children who, per James's study, were deficient in this protein.[74] The deaths of two autistic children following chelator injections in 2005 raised new doubts about the therapy.

Despite Rimland's insistence that secretin was a godsend, science showed otherwise. Quick funding by the NIH and CDC helped generate 13 placebo-controlled studies of the potentially dangerous hormone, 12 of which showed no more benefit from secretin than saltwater.[75] Some of these studies, however, showed that a saltwater placebo was quite potent, at least in the short term. In one study in which patients were given either secretin or a placebo, and then, after a period of several weeks, switched groups, 27 families guessed their child's assignment correctly and 27 guessed wrong.[76] Adrian Sandler and James Bodfish, who conducted secretin trials at the University of North Carolina, found that 30 percent of both groups showed significant improvement after infusion. Their study was conducted in early 1999 when secretin was hot in the media.

"On the designated mornings of the infusions, families gathered in the waiting room, talking and sharing their hopes and dreams as they waited for the injections; taking a pill seems so much less definitive than inserting a needle, watching the blood flow up the tubing of the butterfly, and then infusing the mysterious liquid with a syringe. What does our study show about the mechanisms of the placebo effect? It is implausible that the children with autism actually showed improved behavior because of their own expectancies of improvement. In contrast it is clear that parents of children with autism are exquisitely attuned to variations in their children's behavior, so that improvements which occurred because of day-to-day variability might have been attributed to treatment effects." The point was, autism symptoms exhibited a natural waxing and waning.[77]

Meanwhile, the thimerosal thesis was gaining a life of its own in the public. Don Imus, the notorious New York "shock jockey," dedicated several radio shows to blasting thimerosal in 2005 and 2006. One frequent guest was freelance journalist David Kirby, who had published a book about the thimerosal story told from the perspective of Safe Minds and its antivaccine allies. Burton's inquiry into the vaccine industry showed no signs of stopping, although Burton was demoted to chair of a Government Reform health subcommittee at the end of 2002. The investigation had been successful in creating a climate of doubt around vaccines that helped produce nearly 10,000 legal claims from parents of autistics against the vaccine program.

In early 2005, Dave Weldon and Carolyn Maloney, a New York Democrat, introduced a bill to eliminate mercury from all vaccines, although truth be told there was less mercury in vaccines given to infants than there had been at any time in the past 60 years. Legislatures in California, Missouri, Iowa, and 17 other states jumped on the bandwagon by passing laws against mercury in vaccine. In Illinois, the legislature banned the distribution of vaccines containing even trace amounts of mercury, a move that vaccine producers said would force them to halt shipments of vaccine to the state. Since thimerosal was part of the process for making vaccines, the legal attack on thimerosal had expanded, in practical terms, to an attack on vaccination itself.

A chunk of the news media had bought into the thimerosal story—especially the partisan news media as typified by Imus, Chuck Scarborough, and other television and radio personalities. Thimerosal was the Pamela Anderson of news stories—dumb, maybe, but oh so sexy. Science and the knee-jerk media were headed in opposite directions on a story like thimerosal. The culture of science was to be cautious to the point of excess—never to err on the side of prematurely accepting a hypothesis and tossing out a proven principle. But proven principles were boring—if it wasn't new it wasn't news, and if it didn't bleed it didn't lead. The thimerosal story had all the elements necessary for a certain brand of pseudo-investigative, sensationalist news reporting. There was the "secret conference" of "drug company reps and government bureaucrats," the "published studies" (mostly by the Geiers) that showed harm, the compellingly sick children, several of whom "miraculously recovered" thanks to the treatments offered by "brave mavericks" like Wakefield and

Rimland. There were parents like Lisa Sykes, an Episocopalian minister in Richmond, Virginia, who held up a photograph of her child at a scientific hearing in 2004 and declared, "The Greeks have a story—Icarus. We have soared too high. . . . We are decimating an entire branch of the human genome, taking out our brightest kids. Infectious diseases were random in their victims, but thimerosal is selective."

Part of the environmentalist community also jumped onto the thimerosal bandwagon.[78] Robert F. Kennedy Jr., an attorney better known for fighting river pollution and coal-burning power plants, stated that "mounting evidence" implicated thimerosal in the autism "epidemic" and went on Jon Stewart's Comedy Central show to promote his views.[79] Parents who blamed their kids' health problems on the stuff were growing increasingly strident, sometimes to the point of know-nothingism. "What you know in your heart is that the explanation has to be simple, and it is. We have a generation of children suffering from mercury poisoning," wrote J. B. Handley, a San Francisco investment banker. "It is that simple. Research it. The truth will set you free. There is no controversy."[80] Handley's group, calling itself "Generation Rescue," ran a series of full-paged ads in *The New York Times* demanding that Bush support mercury-free vaccines.[81]

Meltdown at the CDC

While the Institute of Medicine was deliberating on thimerosal, the antivaccine lobby, led by Burton and Weldon, had leveled its guns on the National Immunization Program. Declaring that he had found a "disturbing pattern which merits a thorough, open, timely and independent review by researchers outside of the CDC, HHS and the vaccine industry," Weldon demanded that the CDC hand over the original data used in the Verstraeten study.[82] The data was owned by the HMOs who provided it, and the demand raised legal and scientific issues. First of all, the vaccine establishment mistrusted the Geiers. The HMOs balked at the requests for data because they violated patient confidentiality. Some scientists and healthcare officials feared that the Geiers would use the data to fish for customers for their partners in the legal profession. Even with names removed, it was easy to locate people through symptoms and illnesses listed in the database. The impasse was broken with a compromise that allowed the Geiers relatively free access to the data. The Geiers then produced a series of studies linking thimerosal to vaccines. Presumably they might be of use in a courtroom.

By mid-2006 there were 5,100 legal claims pending before the National Vaccine Injury Compensation Program. Another 4,700 had been filed pending the court's decision on how to proceed on the autism docket, which included an estimated 208,000 pages of documents deposed by attorneys for the parents.[83] About 300 other cases, having spent the requisite nine months in the court, had been removed and were now pending in civil courts.[84] The thimerosal outcry deeply disturbed leading scientists of the immunization program. The vaccine people at CDC felt they were doing God's work, and they couldn't understand why people were attacking them. In part, this reflected their own slightly flawed self-image. "For many people who work in immunization it has become a bit of a religion to them—the ends justify the means," one CDC scientist told me. "And it's an incestuous circle of folks. That's not true only in this field. The problem here is that people can get into their heads that they are doing good and anything threatening their doing good is a no-good."

Being a pragmatic group of people who were used to solving problems, some of the scientists started thinking about how they could improve the public's confidence in their work. They'd been punched in the nose by the congressional bully. It didn't matter that they were the smart kids and the bully was a moron—he still ran the playground. Eventually, two of the scientists created new approaches to the challenge. One was Roger Bernier, a New Hampshire native who had first gone to work at the CDC in 1969. After giving congressional testimony on one occasion, Bernier had been approached by Rick Rollens, an influential vaccine critic from California, and told that CDC studies were "dead in the water" as soon as they were even published. Bernier found it depressing that this good research was being held in such contempt. During a sabbatical in 2003 he started pondering what he could do about it. "I realize Rick Rollens doesn't speak for 280 million Americans, but I felt it deserved some attention," Bernier recalled. "I diagnosed the situation as one of lacking trust." He set himself the goal of building trust in the immunization program by increasing public participation in it.

For starters, Bernier held a retreat for "stakeholders"—the annoying catch-all term for people who give a damn about vaccinations, in this case—at a resort near Racine, Wisconsin. Vaccine and mercury activists attended the meeting, as did pharmaceutical representatives. Bernier felt it had been a success and convinced the National Vaccine Advisory Committee to break off a chunk of a problem—how to prepare the public for the pandemic flu—and give it to his public participation group to make a decision. The group would include everyone from Native American tribal leaders to the Consumers' Union.

Bernier didn't feel that the CDC's vaccination program had made any major mistakes. "I'm a happy camper," he said. But scientists and the public had grown too far apart, and direct participation in decision-making would make people feel good. Public policy involved values, and the public could help determine what values went into decisions. It was refreshing—inspiring, even—to hear Bernier, an intense, burly man with an impish smile and a Down East accent, describe his goals. "There is no mechanism for input from citizens to express the values they share and how they would weigh things," he told me. "We as scientists lack a way of deciding how to value things other than our own values."[85]

Bernier was a top scientist at the immunization program, a freewheeling right-hand man to the program director. Bob Chen, meanwhile, posited a more concrete proposal, which was to take vaccine safety out of the immunization program altogether and give it independence within the CDC, or to make it a stand-alone safety agency like the National Transportation Safety Board, which investigates plane crashes. Some felt that Chen's idea, which Bernier and Orenstein opposed, offered a more substantial way of regaining public confidence in vaccines. There was a certain irony in the situation, in that for the past two years, Chen had been forbidden from talking to the press about vaccine safety because his view wasn't the official line. At the immunization program's safety branch, the pressures of the thimerosal battle had taken a steep toll. The 25 staffmembers, intensely loyal to Chen, had been subjected to legal threats, subpoenas, and a barrage of requests for information from Weldon and Burton. Some antivaccine activists had even called the scientists at their homes and on direct lines, accusing them of falsifying data and threatening to kill them. In February 2004, the FBI briefed the staff on how to search for bombs. Security guards were stationed at the program's office in Atlanta.

It didn't help that Chen was increasingly on the outs within the vaccine program. Although he had worked closely with Orenstein for more than a decade, starting in 2002 the relationship grew more distant when Orenstein brought in a new deputy, Steve Cocchi, and Melinda Wharton became Chen's direct boss. Cocchi and Wharton were close, but Chen didn't get along with them. He started receiving reprimands for continuing to speak out in favor of his proposal. The strains were personal, but also organizational—there was an intrinsic conflict of interest in the way the National Immunization Program was structured—and the thimerosal headache only deepened them. Although they did not exactly suppress the Verstraeten study, senior immunization officials took longer than usual to clear it. There were pros and cons, some philosophical, some practical, to removing vaccine safety from the immunization program, where it had become a coveted piece of turf. Orenstein, Cocchi, and others felt that vaccine safety belonged in their division, where experts in vaccinology could quickly respond to perceived problems. The quick response to Rotashield, they felt, was the perfect example of how well the system could work, and those who urged the nation to vaccinate its children should be responsible for making sure the vaccines were safe.

In early February 2004, Orenstein left the immunization program after leading it for 15 years, and Cocchi took his place as an acting chief. The next month CDC director Julie Gerberding, whom Congressman Weldon had been bombarding with letters demanding changes in the immunization program, announced that a Blue Ribbon Panel would decide where the safety branch would be housed in the future. Although the vaccine critics didn't realize it, they were demanding a move that Chen had asked for all along. A few weeks before Gerberding's announcement, an aide held two bull sessions with the safety branch scientists, who told the aide that conflict of interest was complicating their work. While there were no major smoking guns in the form of suppressed studies, the staff complained that higher-ups scrutinized their findings excessively when they showed harm from a vaccine. One scientist said he had been criticized for a study that showed an apparent increase in Guillain-Barre syndrome following influenza immunization. He was told that the study "may hinder our ability to increase flu vaccine uptake." (In the end, the study was published.) Some staffers pointed out that all members of the immunization program had been asked to sign pledges committing themselves to increasing the national immunization rate. It was only after the safety branch members pointed out that this conflicted with their duty that they were exempted from the pledge.

After the bull session, Gerberding forbade the safety branch members from speaking directly with the Blue Ribbon Panel, which included Fisher, Blaxill, leaders of autism organizations, and other consultants. Many safety branch scientists were shocked at the process. Chen, who had originated the idea under discussion, was forced to rehearse his remarks in front of superiors before he gave a presentation to the panel. When he veered from his prepared remarks in response to questions, Cocchi became so angry that he issued Chen an official letter of reproval. Finally, Cocchi called Chen to his office and told him to look for a new job. His 16 years as chief of the CDC's vaccine safety activities were over. Chen was crestfallen, colleagues said. He felt that an intellectual debate on vaccine policy had been violated by authoritarian tactics. As a first-generation immigrant from a totalitarian state it was especially unsavory for Chen, those who knew him said. His firing came as a shock to some vaccinologists, like Halsey and Katz. "It borders on a crime," said one FDA regulator who'd been a colleague of Chen's. Others shook their heads and tut-tutted. It was a bureaucratic conflict he'd been bound to lose, some said.

In March 2005, Chen was vindicated when Gerberding took the Blue Ribbon Panel's advice and announced that the safety branch would be moved into her office, where it would answer directly to Dixie Snider, the CDC's chief scientist. Now it was time for Chen's foes in the agency to charge that crude politics were overturning good science. Just as safety branch members felt they were under attack, some veteran vaccine scientists felt that their chief had caved in to the demands of "poor parents whose minds have been poisoned," in Steve Cocchi's words. "We're the sacrificial lamb," he told me bitterly. "The director [Gerberding] is figuring, we'll throw him [Weldon] a bone and he'll maybe not go after CDC some other way. It's a cynical horsetrading type of situation. This is what the antivaccine movement wants to happen. You remove the voice of reason, the voice of the system trying to find scientific truth. The purpose is to neutralize or destroy or cripple the vaccine program."[86]

The battle with the autism community had been a bruising one for the vaccinologists. Although Chen was said to be pleased that the safety program had been moved, the more suspicious critics, like Fisher, suspected it would be no more independent under the direct control of the CDC director. The episode showed how vulnerable the vaccine program was to public criticism and a few opinionated members of Congress. Had the uproar been about anything real? Was it a real scandal, or a real witchhunt, or something confusingly in between? The autism controversy would remain an open chapter in the history of vaccines, at least for a few years. If autism rates suddenly plunged to pre-1990s level, it would be hard not to suspect that the thimerosal thesis had some basis in fact. But by mid-2006 this wasn't happening. The rates of autism remained as high as ever.

EPILOGUE

—

OUR BEST SHOTS

Every man is entitled to his own opinions, but not to his own facts.

— Daniel Patrick Moynihan

When thimerosal was long gone from vaccines people kept discovering that their children were being made autistic by it. Searching minds grasped this attractive explanation and wouldn't let go. "It was absolutely the vaccines. My husband and I have no doubt about that at all," Andrea Taube of Tucson told a reporter in 2005. Her son was normal, she said, until a few days before his first birthday, when he got four shots—"all with mercury"—and then began showing symptoms of autism.[1] His first birthday was in 2003, by which point thimerosal remained, in fact, only in one shot, against influenza.

On a humid summer morning in 2005, about 250 parents of autistic children rallied in Washington to demand that the government remove all mercury from vaccines. The writer David Kirby, along with Dan Burton and five other members of Congress, riled up the crowd. A woman handed out black rubber bracelets bearing the words "Mercury hurts. Truth heals." Activists spoke tearfully about government betrayal, and an autistic boy came shyly to the mike and shouted, "Giving mercury to kids is stupid!" They weren't against vaccines, the speakers insisted—

they only wanted to get mercury out of vaccines, and get the government to acknowledge the enormity of its error. Parents with their sick children in tow carried signs depicting the NIH, CDC, and Institute of Medicine as the three monkeys, with the words "hear, speak, see truth." Another sign plead "Vaccines are great to prevent disease, but make them safe for children please!"

Earlier that morning, as the crowd marched past the Department of Health and Human Services, Barbara Loe Fisher yelled, "HHS—Tell the truth about vaccines!" It was "something I've been wanting to do for a long time!" she told me. As she waited in the hot sun on the Capitol's West Lawn to speak to the crowd, Fisher was worrying a bit about the future. It was possible, she acknowledged, that autism rates would continue to be high years after thimerosal was gone. People would figure that thimerosal hadn't been to blame, and they'd get a false sense of security about the safety of vaccines in general. "Thimerosal is only part of the problem," she told me. Vaccines, Fisher was convinced, caused the immune system to go awry in a way that affected the mind. Perhaps microbial poisons crossed the blood-brain barrier, or vaccines overwhelmed the immune system, or the very absence of the diseases that vaccines prevented somehow threw the body off kilter.[2]

You could find some PhDs and MDs who believed any of these theories, but they could only be described as fringe science. Fisher was accustomed to brushing off the sneers of the establishment, however. To the public health authorities who dismissed her all-encompassing doubts she asked for the grand experiment, an impossible experiment, really, that antivaccinists had been demanding since the nineteenth century—give vaccines to 10,000 kids and leave 10,000 kids unvaccinated, then check their health over the next 20 years. In the meantime, the precautionary principle should apply. It was time for a moratorium on new compulsory vaccines. "How do we know," she asked me, "that the human race was not supposed to have a bit of suffering from infectious disease at the beginning of life, in order to have an immune system that would be able to meet the challenge of viruses and bacteria along the line? How do you know that?"

Like so many vaccine skeptics, Fisher preferred a do-it-yourself trial by fire, the hardy individualist approach to disease. And she wanted the intuitive spiritual values of everyday people like her allowed into the temple of science policy. Legislatures, which were representatives of the people, and not the utilitarian professional elite, should decide which vaccines to administer to children, she said. To an extent, this was already happening. Whatever they thought or said about Fisher, policymakers had cracked open the door for her. She sat on the FDA's vaccine advisory panel from 1999 to 2003 and in the week before the rally, she and Paul Offit had sat together—they argued, and Fisher cried, but wasn't that democracy in action?—on a panel discussing how to implement pandemic flu vaccination. Dan Salmon, a Johns Hopkins University vaccine expert, invited Fisher to colloquia on vaccine exemption laws. In 2003, 13 separate states were considering legislation that would have loosened vaccination requirements, no small thanks to the effort of Fisher and her organization. Salmon and Halsey countered with a model law that would allow philosophical as well as religious exemptions while tightening the exemption process to make sure people who shunned vaccines were doing it out of conviction and not convenience.[3]

One of the ironies of the Internet age, said Roger Bernier, the CDC vaccine expert, was that with scads of information at their disposal, it was easier for people who'd already made up their minds to live inside a hall of mirrors, where the only information they saw was the kind they wanted to see. "That is a serious blow," Bernier told me, "Because in a democracy we need give and take and compromise. If we're not skilled at that, it bodes ill." Bernier's goal was to improve the climate in which vaccination public policy was developed "and if enough people from different sectors are working together over time it will improve the overall climate. Barbara and I share a common vision for citizens' having a meaningful voice. And I think this is the best chance around. There ain't nothing like it. If it can be successful, there will be created a forum where different perspectives can be aired in a more productive way than has been the case the past few years."[4]

The vaccination program needed all the help it could get. While many parents worried about the unproven link to autism, a potentially far more serious crisis facing the system was a shortage of vaccines. Scientists who worked with the influenza virus believed that a flu pandemic, which could kill millions of people whose immune systems were unfamiliar with it, was inevitable. Many of them thought it possible that the lethal H5N1 strain of avian flu, which had killed dozens of Asian farmers by 2006, and was cropping up around the world, might begin spreading from person to person, which would allow it to quickly multiply across the globe. The 1918–19 pandemic, which had a mortality rate of less than 5 percent, had killed an estimated 50 million people. The new virus was killing 50 percent of those known to have been sickened by it. The math of mass death, though speculative, was frightening.

The spectre of a flu pandemic was uppermost in the minds of public health officials as they prepared for the annual flu season in 2004, when suddenly the British government announced on October 3 that it was shutting down a facility where Chiron, one of three U.S.-licensed flu vaccine manufacturers, made its shots. The Food and Drug Administration immediately sent inspectors to England, and they confirmed that the plant was contaminated by *Serratia marcescens*, a rather quotidian bacteria. The closure immediately cut the U.S. flu vaccine supply in half, and set off several months of inquisitive media coverage and hearings before Congress. Embarrassingly, the CDC had just added middle-aged adults and 6- to 23-month-old babies to the list of people for whom it advised annual flu shots. Now it had to tell Americans to forgo vaccination unless they were elderly or chronically ill.

Because the previous year's Fujian flu strain had seemed particularly virulent—the 2003 season had started early and 152 children died during it (no one could say whether that was an unusually high number, since this was the first year child flu deaths had been tracked)—people panicked when they couldn't get vaccine in 2004. There were long lines at the supermarkets and pharmacies that offered it, and sick and infirm people sometimes had to go home unvaccinated. "You can't do this to people!" a frail old man shouted, as a crowd of angry people were turned away from a social services center on the Upper West Side. "It's criminal!"[5]

Eventually the public shrugged its shoulders and gave up on flu vaccination, and we ended up with surplus vaccine. As in past years, millions of doses were thrown out. But in the meantime, the FDA, underfunded whipping boys who couldn't afford to serve coffee at their own meetings, were lambasted for failing to take earlier notice of the problems at the Chiron plant. They defended themselves as best they could against the requisite venting of congressional spleen. Flu vaccines were grown in chicken eggs, they explained, and each year a new strain was chosen for the vaccine. Each strain grew better or worse than the last and each year's vaccine was more- or less-effective. Depending on how scary the headlines were in a given year, people either ignored flu vaccination or hollered for it. The manufacturers might sell 80 million doses, or 20 million. The rest they threw away. For those reasons, and others, it wasn't a good business.

"Inherently we undervalue what vaccines do. When vaccines work well what happens? Nothing," Dr. Bruce Gellin, a senior Health and Human Services vaccine official, told a reporter. "And that's hard for people to value."[6] The way it worked, people only appreciated vaccines when they couldn't get them. And the well ran dry in the fall of 2004.

The flu vaccine was not the only one we didn't have enough of. From 2000 to 2002, there had been shortages of MMR, DTP, chickenpox, and pneumococcal vaccines. Each had a different explanation, but they all had to do with the tiny size of the vaccine industry. MMR and chickenpox production went down while Merck refurbished part of its West Point plant in response to FDA citations. Stronger-than-anticipated demand for Wyeth's new Prevnar pneumococcal vaccine caused the company to run short. Wyeth dropped out of DTP production in 2000 rather than refit its factory to make thimerosal-free vaccine.

It wasn't as if we hadn't had warnings of a crisis before. In 1976 and 1977, following the swine flu fiasco, many were convinced that the vaccine-making establishment in the United States was collapsing. The authoritative journal *Science*, in an editorial titled "Our last vaccine?" bemoaned the "moribund" state of the biologics industry. Difficult work, tough regulations, and low profits were driving companies out of the field. "Plants and research facilities lie empty, or are sold to foreign firms; research and development efforts are at a standstill." The only solution, *Science* editorialized, was to have the government take legal responsibility for vaccines, or even take over manufacturing and distribution itself.[7] Salk called for a "mini-Manhattan project."[8] Neither happened.[9]

In 2004, recriminations over vaccine supply had a more partisan tone. Where had all the vaccine companies gone? Republicans blamed the trial lawyers—a handy surrogate for the Democratic Party. One of the accusers, ironically, was Representative John L. Mica of Florida, who had organized a hearing to air criticism of the hepatitis B shot. "You know we don't even make ladders in the United States. Ladders and vaccines. Why? Because of the field day we've created for trial attorneys," Mica said. "I'm pretty bitter about it."[10] The Republicans pointed out that because the 1986 vaccine compensation law had a loophole (which Democratic leader Waxman had insisted upon) allowing parents to take their cases to civil court if they weren't satisfied with the federal vaccine court's ruling, thousands of autism lawsuits were poised to flood the courts. Starting in 2002, with the support of Senate leader Bill Frist, the Bush administration repeatedly sought legislation to plug the loophole. Just as insistently, public interest groups and Democrats fought the bill, calling it an example of the Republican administration's coziness with the pharmaceutical industry.

Liability was no doubt a problem and an expense, and it was easy to trash the trial lawyers, unless you happened to be defended by one. But lawsuits were not, in fact, the main force that had winnowed out vaccine makers. The trouble with the American vaccine system was that the safe, effective shots we relied upon to protect us from the scourges of infectious disease were expensive and difficult to make—yet they had to be cheap enough to be widely used or they would not protect the community. Vaccines were square pegs that didn't fit into the triangular holes of market capitalism.

In 1967 there were 26 companies making vaccines in the United States.[11] In 2006 only four major firms remained on the market—Merck, Sanofi (formerly Aventis-Pasteur), GlaxoSmithKline, and Wyeth. Two other firms (Chiron, Medimmune) made flu vaccine for the U.S. market, while ID Biomedical, a Canadian flu-vaccine maker, was poised to enter the market. The lavishing of billions of taxpayer dollars for bioterrorism vaccines had drawn in other companies, such as Acambis and Vaxgen— the failed AIDS vaccine maker.

The companies that bailed out of the vaccine market over the past four decades had done so for reasons of profitability. Of the 26 producers from 1967, seven made single specialty vaccines of little strategic significance and three were state laboratories that made small quantities of DTP.[12] Of the remaining 16 firms, 11 gradually consolidated to form the current big four. Five others—Eli Lilly, Pitman-Moore (a division of Dow Chemicals), Pfizer, Cutter, and Parke-Davis—simply stopped making vaccines. The first to go was Cutter, which bombed out of polio vaccine and eased out of the DTP market because of low profits. Pfizer was next. It produced vaccines at a former biological weapons plant in Terre Haute, Indiana, leased from the Department of Defense. The market for Pfizer's inactivated polio disappeared in the early 1960s, and it got out of human vaccines altogether in 1967 after withdrawing its dangerous inactivated measles vaccine. Pfizer is still a market leader in pet and livestock vaccines.

The next to go were Lilly and Dow. Lilly had built its Greenville, Indiana, biologics plant in 1914, originally to make smallpox vaccine, and decided—after hearing Salk's advice—that it would be unprofitable to build a new one. Dow still had a market share for its measles vaccine when it backed out of the business. June Osborn, who oversaw the measles portion of the FDA's major review of vaccines in the early 1970s, recalled that when she started the review there were five licensed measles vaccine makers. When she finished, only Merck was left. "If they have a fire at Merck we're in terrible trouble," she remembered a colleague saying.[13]

Parke-Davis, the very first vaccine maker to be licensed by the Treasury Department, in 1902, was the last major company to leave the business. Parke-Davis had suffered liability losses because of its defective DTP-polio combination in the early 1960s, and in 1976 it had to junk a quantity of swine flu vaccine that contained antigens for the wrong strain. Warner-Lambert, which purchased Parke-Davis a few years later, cut loose the vaccine division, which became King Pharmaceuticals. Under the name Parkedale Pharmaceuticals the company was still making flu vaccine near Detroit in 2002, when the FDA demanded one too many changes in its production line and it finally threw in the towel.

In no case was liability the main reason a company quit the vaccine business. "It was sort of tragic. You could see what was happening, that all these firms were getting out of the business because it wasn't profitable," recalled Don Hill, a licensing officer and later chief of licensing at Biologics from 1964 to 1990. Hill allowed the companies to keep their licenses for as long as possible with the hope that they would reenter the market. But "let's face it, preventive medicine isn't too popular, because after you vaccinate people that's it, right?" Added Paul Parkman:

"Biologics are bad news. You could develop the tenth cholesterol drug and make a zillion dollars even if you have a lot of competition. If you make a vaccine all you're doing is asking for trouble. There are a few dedicated people who will do it but it's not a born winner."[14]

The transformation of the vaccine industry reflected trends in the overall transformation of late-twentieth-century American industry in general. Pharmaceutical companies expanded, merged, consolidated, and cast off less profitable ventures, including "loss leaders" like vaccines.[15] Biologicals barely qualified as footnotes in the official histories of these firms. Vaccines were rarely blockbusters;[16] in 2005, they made up 10 percent or less of the sales of the four big companies.[17] These remaining firms had long historic commitments to vaccines. Merck, at times dragged kicking and screaming by Maurice Hilleman, had stayed in the business. Yet even respectable Merck came close to closing its vaccine division at least twice after 1980.[18] Sanofi grew out of two government enterprises: Connaught Laboratories, owned by the province of Ontario, and Pasteur-Merieux, founded by Charles Merieux, an early Pasteurian, and at times partly owned by the French government. Wyeth had been making vaccines in Pennsylvania for well over a century. The subsidiary of GlaxoSmithKline that made vaccines at Rixensart, Belgium, had been doing so since 1956.[19]

There were many complex reasons why vaccine making incurred so many risks and made relatively meager profits. The organisms were fickle. It was expensive and time-consuming to test vaccines on people because such efficacy trials were "hostage to the timeline of nature," as Alan Shaw, a Merck vice president, put it.[20] To test a cholesterol pill, you gave it to someone with high cholesterol. When you conducted a placebo-controlled trial of a vaccine, you had to wait for an outbreak of the disease to test your product's powers. You couldn't intentionally expose masses of people to dangerous pathogens.

A particular paradox of flu vaccines was that you couldn't be sure they would work from one year to the next. Public health officials were inevitably crying wolf about approaching flu seasons that might not end up worse than the last, and whose strains might or might not be prevented effectively by the current vaccine. Even in the best of years, flu vaccine probably does little to reduce the morbidity and mortality of the disease in the elderly. The shrunken thymuses of old people—the thymus being the place where T-cells learn to stimulate antibodies—can't generate as robust a response to the vaccine as younger thymuses do. A 2005 study that compared the flu mortality rate to the percentage of vaccinated during a series of flu seasons found little difference, despite the increasing percentage of the elderly who got the shots.[21] "We're not saying the vaccine is inherently bad," one of the study's authors hastened to add. "But it's certainly not as good as we thought it could be."[22]

Flu vaccine's efficacy, on a population-wide scale, could be boosted by vaccinating children, and indeed, the ACIP and Red Book committees in 2002 started encouraging the vaccination of children 6 to 23 months old. If you had a feeble old father living in a nursing home, the best vaccine to protect him was one for your visiting toddler (and for the woman who cleaned his room) although the toddlers themselves might not get that much direct benefit.[23] And with thimerosal in most of the flu vaccines, it was a hard sell.[24] Another reason for expanding flu vaccination, obviously, was that it expanded the market for flu vaccine, which meant that companies in the business would make more money and have more vaccine available in years when it was especially needed. Manufacturers were pushing for the CDC to recommend that everyone be vaccinated, which would "provide the impetus on the part of vaccine manufacturers to increase their production capacity to meet routine demand," James Young of Medimmune told a congressional hearing.[25]

Price was another obvious vexation for the vaccine business. Inactivated flu vaccines sold for $8 a dose in 2004. Considering that from year to year the producers had no idea how much vaccine people would buy, it was not surprising that Wyeth had chosen to get out of the market in 2003, switching to a live nasal spray flu vaccine for which it charged $50 a dose. (When demand for that product was less than expected, Wyeth abandoned the flu vaccine business entirely, leaving the nasal spray vaccine to its partner, Medimmune.) Merck departed the flu market in 1982. Hilleman had long been skeptical of flu shots, and the company was losing money hand over fist making them. Drug companies at the time allowed doctors and pharmacists to return unused flu vaccine for reimbursement. One year, nearly all of Merck's vaccine was returned.[26]

The other price constraint was the trap of a single buyer. Under the Vaccines for Children program, roughly 55 percent of vaccines are now purchased by the federal government, which guarantees a market for producers, but also puts pressure on the price. The vaccine producers support the program because it guarantees a market for their product, but they want the government to pay more. "While it is an obligation of government to be a prudent purchaser, it is also an obligation of the government to protect the public's health," said Peter Paradiso, a vice president for scientific affairs at Wyeth. "By overemphasizing the former, one risks jeopardizing the latter."[27]

When they do see promise in vaccines, most pharmaceutical companies seem to see it in novel vaccines against previously untargeted germs. The newer vaccines can be priced higher, but there is a catch there as well. Such vaccines are aimed at illnesses that are either more rare—like meningococcal meningitis—or less serious, like rotavirus. Because their benefit is more limited than the older vaccines, they require larger clinical trials to assess the risks before they can be recommended to large populations. "With the major diseases—measles, polio, pertussis—you were protecting everyone at risk, and the burden you were reducing was very, very big," says David Salisbury, who has headed Britain's vaccination programs since 1989. "But now we're moving down against diseases that affect fewer and fewer people, but you still have to vaccinate everyone to get the same return."[28] More regulatory hurdles meant that companies justifiably charged higher prices, but higher prices made the vaccines less attractive to the government or the insurance companies that paid for them. It was a vicious circle, though not always insurmountable.

On a technological level it was a time of great promise for the vaccine enterprise. Inhalable flu, measles, and rubella vaccines—safer, easier to use, and more immunogenic—were coming on line. Genetically engineered typhoid vaccines were being tested for use in places like New Delhi, India, where 2 percent of the population still fell ill with typhoid fever each year.[29] Scientists were growing antigens in potatoes and tobacco plants that could easily be harvested, made into powders, and eaten. Recombinant, antigen-bearing bacteria were being put into candy and bananas, although such vaccines were years from the market and might never really work. DNA vaccines could program the coding of useful antigens directly within the body's cells, to create a powerful immune response. The remarkable speed with which biologists could now read the entire genomes of microbes gave vaccine makers a wealth of new information about these pathogens and a host of new molecules around which to design vaccines.[30] Vaccines against two common infectious ailments, respiratory syncytial virus and group A streptococcus bacteria, seemed to be on the near horizon, after being shelved following disastrous trials in the 1960s.[31]

For the immediate future, most of the new vaccines were designed to be given, initially at least, to teenagers or preteens. The CDC and Academy of Pediatrics advisory committees in 2005 recommended that a new vaccine against four strains of meningococcus be given routinely to 11 and 12 year olds as well as teens going off to college if they hadn't received a shot previously. *Neisseria meningitis*, the targeted bacteria, caused a horribly fast-moving infection that killed 10 percent of those it infected. Sad reports about the sudden deaths of students in college dormitories, where the germ colonized easily, had ensured that it would be a much sought-after vaccine. Another important vaccine that was approaching the market was the highly effective, yet controversial, vaccine against several strains of human papilloma virus, or HPV. Although pap smears, which detected early changes in cervical cells caused by the virus, had dramatically reduced the rates of cervical cancer, 4,000 women a year still died of the disease in the United States, and hundreds of thousands died in poor countries.

The politics of HPV were similar to those of hepatitis B, because the papilloma virus lived in the nether parts and spread through sexual intercourse. Opponents of the vaccine had mobilized quickly through the Internet, although the rationale for their opposition was no different from the one that preachers had expressed 200 years ago. No doubt most of the children born to religious conservatives would some day have sex with at least one person; some portion of them would also probably inject illegal drugs. Yet cultural conservatives had taken issue with mandatory hepatitis B vaccination because they felt that the protection of their daughters and sons encouraged them to commit ungodly acts. Christians whose opposition to hepatitis B vaccine had only slowly taken shape were mobilized against compulsory HPV vaccination fully a year before the vaccine even got to the market. "If a 10- or 12-year-old is given a vaccine to protect against sexually transmitted disease, then it's implied they'd be engaging in risky sexual behavior," said Pia de Solenni, director of "life and women's issues" for the Family Research Council.[32] The group opposed giving the vaccine to children. "It shouldn't be mandatory," said Bridget Maher, another spokeswoman. "Parents should decide."[33]

But if the vaccine was not compulsory, it would be unlikely to have much impact on cervical cancer, which had largely become a disease of poverty in the United States. Women who got regular medical care were screened for cervical cancer. As a group, poor women didn't see doctors as often and were unlikely to shell out $50 to $75 for the vaccine. They would remain unprotected unless they were vaccinated while still in school.

Too Much Regulation?

Whether they were attempting to bring new vaccines to market or to keep old facilities going, vaccine makers complained of a rapid growth of regulations. Both company and FDA officials mentioned two major factors: the incorporation of Good Manufacturing Procedures, a Talmud-like collection of rules applied to each process in a production line, and less cordial relations between industry and the FDA's regulatory staff. In the 1970s, officials from Biologics tested the final product for sterility. In 2005, they tested each step in the process to make sure the process was sterile. This seemed like a good precaution, but no one could cite an instance in which it had saved a life, and it cost money and time. Vaccine license applications that used to be contained in a few binders now filled the back of FedEx trucks. And it wasn't always easy to characterize each stage of the process when you were growing stuff in eggs. "You have more people sampling, more people watching to make sure you are sampling, lots of documents to sign off," Paradiso told me. "It's hard to put your finger on how this has happened. But it's a little scary to see where it's ended up."

Paradiso and others say that the culture at Biologics has gradually changed in the decades since the agency left NIH. Men and women like John Robbins, Rachel Schneerson, and Paul Parkman, who were vaccine inventors as well as vaccine regulators, have moved on. "At NIH we were a research organization with regulatory authority. After switching to FDA we became a regulatory agency with a small research agenda," said Mike Williams, who worked on flu vaccine at Biologics from 1968 to

1997. Some date the tightening of the screws to the mid-1990s, when FDA field inspectors, who were accustomed to examining drug factories, started to run vaccine plant inspections as well. The new group, which called itself "Team Biologics," was led by field inspectors, and they were tough. More and more warning letters were issued to manufacturers, and in private the manufacturers started calling them "inspection Nazis."[34]

Mike Williams mentioned an episode in the 1980s when a problem in a filling line led an inspector to shut down Merck's entire West Point, Pennsylvania, factory. Biologics director Hank Meyer intervened with senior FDA officials to keep the country's only MMR supply from drying up. "This was a classic example of benefit/risk thinking in vaccines," says Williams. "It's one thing to shut down a drug that may have a shelf life of five or ten years. But to shut down the sole provider of a childhood vaccine—you better have some serious problems before you make that decision."

The FDA's regulatory process could certainly stand improvement. But loosening a few regulations would not do much to alleviate the crisis in American vaccine readiness. The bottom line was that American corporations in the twenty-first century were expected to post rapidly growing profits, which they weren't going to harvest from cheap vaccines like the ones against flu, measles, and pertussis. America's vaccination program faced what one expert called a "toxic mixture" of a corporate culture that had no patience for vaccines and a political culture unwilling to accept government responsibility. Budget cuts and outsourcing had crippled our ability to respond to crises in infectious disease. Only when the government feared a crisis did it intervene. No price was too high to pay for anything that had the magic word "terrorism" attached to it. Congress was willing to authorize $1.9 billon to build and maintain a stockpile of smallpox vaccine, and $1.4 billion to create and stockpile a new anthrax vaccine. From 2002 to 2006 it spent $33 billion on biodefense.[35] Yet in 2003, the NIH invested less than $70 million on influenza vaccine research. Presumably, politicians felt that while they'd be blamed for failing to anticipate a bioterror attack, germs were no one's fault. It took Hurricane Katrina to make the Bush administration recognize that a natural disaster like, say, pandemic flu could be every bit as deadly—and as politically costly—as a terrorist attack. And of course, it was a whole lot likelier.[36]

Even after signing off on a $7 billion pandemic flu preparedness program, based on plans that had circulated but had been ignored for years, the government tried to eliminate the right to sue or collect compensation for vaccine injuries.

After the 2004 flu vaccine crisis, the NIH agreed to subsidize a new $150 million Sanofi-Pasteur flu vaccine plant. In March 2006, the ACIP recommended annual vaccination of 2- to 5-year-olds, a measure that would bolster the bottom line of flu vaccine makers.[37] Other proposals called for the government to centralize the purchase of vaccines or to follow the lead of other countries, including Britain, and become the sole distributor of vaccines.[38] In Britain, the government maintains a steady supply of all necessary vaccines, purchasing them through contracts that are renewed each year. It stockpiles the vaccines and dispenses them in response to orders from general practitioners, who administer them to children. The government also has established a rational system for vaccinating its citizens.

"When your baby is six or seven weeks old a letter comes through the door that says, 'Please take your baby in at 3 P.M. a week from Wednesday to this clinic to have his vaccination.' Four weeks later another one comes. Your general practitioner has received an identical card saying that you are coming," says David Salisbury, Britain's vaccination chief. Brits apparently do not resent "Big Brother" for letting them know it's time to vaccinate. Although vaccination is not compulsory there, and Britons are generally more skeptical of vaccination than are Americans, they report for their child's "jab" with surprising alacrity. Fully 94 percent of British babies receive four DTP shots by age 2—considerably higher than the U.S. rate. Even at the height of the MMR scare, the rate of British one-year-olds vaccinated against measles fell only as low as 79 percent (from 92 percent).

Salisbury and his cohorts felt that hepatitis B was too uncommon a disease in Britain to merit blanket vaccination of infants—mothers are screened, and the babies of hepatitis B–positive mothers are vaccinated. As for chickenpox, Britain preferred to stiff-upper-lip the disease. Salisbury and other specialists were concerned that eliminating chickenpox among small children would increase the rates in older adults, and also increase the incidence of shingles. Salisbury concedes, too, that

given British skepticism, "we have to move very carefully or else it's counterproductive." His agency conducts twice-yearly surveys to get the public mood on vaccination—to find out what their sources of information
and worries are about vaccines and disease. The government responded
promptly to public dismay over a series of meningitis deaths in the late
1990s by recommending vaccination of infants against type C meningococcal bacteria, thus eliminating a disease that the United States had yet
to tackle.

Enter the Robber Baron

Vaccine supply is difficult enough in the United States. But in the poorest
countries of the tropical world, the unavailability of vaccines means that
vaccine-preventable diseases kill millions of children each year. Millions
more die of AIDS, malaria, and tuberculosis, effective vaccines for which
have been slow to come. In 1998, however, a new hope arose in the battle
to bring effective vaccines to the third world. Like John D. Rockefeller a
century earlier, Bill Gates is a wealthy man who was known for his
cutthroat business practices but then suddenly devoted a considerable
portion of his immense fortune to combating infectious disease.
Rockefeller's money created the world's foremost biomedical research
institute and funded drives to eliminate tropical diseases like yellow fever,
yaws and hookworm. Much of Gates's largesse was more narrowly fun-
neled into a single tool of biotechnology: vaccines.

The Bill and Melinda Gates Foundation announced on February 6,
1999, that it would provide $100 million for vaccine research to a Seattle
group that funds third world health initiatives. Over the following five
years the foundation provided nearly $2 billion in grants for vaccines and
vaccinology, and promised billions more. Some of the money would go to
purchase existing vaccines, beginning with *Haemophilus influenzae* type b,
which poor countries have been unable to afford. By July 2003, the Gates
fund had helped provide more than 4 million children with Hib vaccines.
A pilot project in Gambia wiped out invasive Hib disease by 2005.[39]
Another portion of Gates's money went to efforts to develop vaccines
against malaria, AIDS, TB, and other diseases, and to improve vaccine
needles, often a source of disease themselves. Despite the money, an

effective AIDS vaccine seemed as much as a decade away in 2005. HIV attacks the very cells that a vaccine stimulates to produce immunity, creating unique challenges.

Different but equally mind-boggling problems have slowed development of a malaria vaccine. The malaria parasite passes through various stages in its human and mosquito hosts, making it an elusive foe. Scientists at the Naval Medical Research Hospital in Bethesda, Maryland, had worked on malaria vaccines for decades, and their research program was kept alive by the military's interest in protecting its troops. GlaxoSmithKline had been ready to drop its experimental malaria vaccine in 1999 before Gates came through with support.[40] The Gates Foundation's Malaria Vaccine Initiative hired Regina Rabinovich, chief of the microbiology branch at the National Institute for Allergic and Infectious Diseases, to lead a multimillion-dollar program to fund malaria vaccines. It announced its first major success in 2005 when the delightfully named Rip Ballou, a former navy researcher now with GlaxoSmithKline, announced that an experimental vaccine against the mosquito-borne parasite had protected 58 percent of recipients against severe disease in a trial in Mozambique.

The Gates Foundation hired Richard Klausner, former director of the National Cancer Institute, as its medical director and began working closely with the new Global Alliance for Vaccine Initiatives. By 2005, the Global Alliance had received commitments of $1.3 billion and spent more than $400 million on vaccine programs. Gates had promised another $750 million for the following decade, and the Norwegian government chipped in $180 million. Britain and France established an International Financing Facility for Immunization. Governments would commit long-term money to the fund, which would sell the commitments on bond markets to raise cash for its programs. Although the Bush administration declined to support the facility, it was expected to contribute as much as $8 billion over 10 years to vaccine initiatives.[41] All in all, donors had promised $35 billion to fight infectious disease in the developing world since Gates first entered the fray.[42]

All this "big dough," in Paul de Kruif's memorable formulation, would increase demand and thus keep multinationals in the vaccine game, it was hoped, while also stimulating third world vaccine manufacturers, who could provide affordable vaccines to those who needed them most. By 2005, the company that sold the most vaccines was neither Merck nor Sanofi-Pasteur, but a Pune, India-based firm called the Serum Institute of India. Its vaccines were considerably cheaper than most of the multinational companies' products. Latin American countries bought most of their measles and measles-rubella shots from the Indians, and hepatitis B vaccine from a South Korean firm. Companies in Indonesia and Brazil were also entering the global market. These companies generally lacked the technology or intellectual property rights to make the most elaborate conjugate or reassortant vaccines, but the Gates and its partners were working on technology transfer agreements.

The Gates Foundation came on the scene as the earlier Children's Vaccine Initiative was winding down. When the eradication of smallpox was officially confirmed in 1980, something less than 20 percent of the children in the world's poor countries were being vaccinated against other killers of childhood. In 1984, public health officials and activists began regular meetings under the Rockefeller Foundation's auspices, and in 1990, UNICEF announced the Children's Vaccine Initiative. This coordinated effort brought vaccines against measles, diphtheria, pertussis, rubella, polio, and tetanus to more than 75 percent of the world's population by the time it ended in 1999.[43]

The twenty-first-century push to expand vaccination in the third world had the potential to resolve a decades-old debate about the value of technological fixes for the world's poor. In the 1950s, the philosopher of science C. P. Snow argued that science and technology could alleviate the poverty of the third world and that these tools must be brought to bear lest we "sit comfortably in our sitting rooms and watch people die on our television screens." The literary critic F. R. Leavis countered that the scientific revolution required an infrastructure before it could do its work. Improvements in third world health, he believed, would only come about through development of the countries' political and economic systems.[44] Leavis was echoing the claims of Thomas McKeown, a British sociologist who had argued that nutrition and sanitation, rather than specific medical interventions, were responsible for the improved health of the West in the twentieth century. Many leftist advocates of third world development argued that vaccination campaigns per se were of little lasting value to societies plagued by a multiplicity of health problems rooted in poverty and underdevelopment.

Others, such as D. A. Henderson, Harvard economist Barry Bloom, and Bill Foege, the former CDC director, argued just the opposite. "From my perspective, immunization is a foundation stone for all modern public health because it's inexpensive and democratic in that it provides social justice to everyone," Foege told me. "It often will protect a child for a lifetime with one or two or three visits, you don't have threat of resistance as you do with antibiotics and pesticides, and it involves every aspect of public health. You have to do surveillance, you have logistics problems, you need teams of people and evaluation. And so it becomes training wheels for other public health. And if you can't run a vaccine program it's not likely you can do other things."

In some countries, warfare and corruption have rendered immunization campaigns fruitless as a means of lowering childhood mortality and misery. But in fact, in most of the world, top-down organized vaccination campaigns have proved successful, at least at lowering child mortality rates. Latin America, where Ciro de Quadros of the Pan American Health Organization organized polio and measles eradication campaigns beginning in 1985, is a case in point. De Quadros, a wiry Brazilian who cut his teeth as the chief epidemiologist of smallpox eradication in Ethiopia, wielded tenacity, brains, and charm to make the campaigns work. Inspired partly by Cuba's successful eradication of the two diseases, the campaigns triumphed in a region plagued by overcrowded slums, corruption, dictatorship, and war. For El Salvador's vaccination effort, for example, de Quadros met with leftist guerrilla representatives in Washington, DC, to arrange "Days of Peace"—cease-fires for vaccination. The last case of polio in the Western Hemisphere lamed Luis Fermin Tenorio Cortez, a Peruvian two-year-old, in August 1991. The year 2002 marked the last locally generated measles outbreak in the region— in Venezuela. Since then, measles has been introduced into the region several times by visitors from Korea and Japan, which have rather slipshod measles vaccination programs, but each imported outbreak has been quickly snuffed out.

While real progress is being made fighting vaccine-preventable diseases, the sweeping triumph of the smallpox eradication campaign has thus far not been duplicated. In 1988, Unicef began a drive to eradicate polio by 2000, but by late 2006 the campaign was still ongoing with no guarantee of success. To be sure, the global burden of polio had declined tremendously, from about 350,000 cases when the campaign began to 784 in 2003.[45] But it had cost a bundle—around $3 billion by mid-2005— and an upsurge of cases in the new millenium had been discouraging to all involved. By comparison, the smallpox eradication cost about $100 million, took a little over 10 years to complete, and finished only nine months behind schedule. Polio was a trickier disease to fight, and the growth of the world's slums has compounded the difficulties of surveillance and eradication.

Scientists discovered in the early 1970s that polio vaccine did not work as well in the tropics, where the gut flora is more plentiful and interferes with the vaccine's ability to protect. If it took three doses of oral polio vaccine to immunize American children, it might take eight in the Congo, for example. And it was awfully difficult to organize eight campaigns to get viable polio vaccine to children in the Congolese bush. There were other unforeseen obstacles to polio eradication. First was the danger of mutated vaccine virus. In 2003, Haiti and the Dominican Republic experienced 23 cases of polio associated with a vaccine strain that had circulated in the population. Two years later, oral polio virus strain turned up mysteriously in three Amish children in Minnesota, though it hadn't been used in the United States in many years. It seemed that a tiny minority of individuals continued to excrete polio virus up to 20 years after they'd been vaccinated and were potential "Polio Marys" throughout that time. Tests were finding poliovirus in sewage. None of this meant that wild polio transmission couldn't be stopped. But it might take years or decades before we could stop vaccinating billions of people. Meanwhile, vaccine-associated polio paralyzed from 250 to 500 people each year.[46]

Still, by 2003 the campaign seemed to be making progress. Polio remained endemic only to Nigeria, Pakistan, India, Afghanistan, and Egypt, and the World Health Organization decided to focus its National Immunization Days on those five countries. This turned out to be a big mistake. In May 2003, rumors began to circulate in northern Nigeria's mostly Muslim Kano state that polio was contaminated with AIDS virus or with hormones that would reduce fertility. The most vocal proponent of the theory was Ibrahim Datti Ahmed, a Kano physician and president of Nigeria's Supreme Council for Sharia Law, who concluded that the United States was using polio vaccine for the purpose of genocide. He got his information from the Internet.[47] All that year, vaccination stopped in areas of Nigeria with sharia law, while WHO officials scrambled to convince the clerics that the vaccine was not a Western conspiracy. Tests showed the vaccine contained no extraneous agents, but the clerics were only mollified when WHO agreed to import vaccine made in Indonesia, a Muslim country.

Such problems were not new to global immunization campaigns. "I learned 40 years ago," Foege said, "that if you tangle with culture, culture will always win." But critics of the polio campaign said that its organizers had failed to anticipate the impact of such problems. "It was a chronicle of an outbreak announced," de Quadro told me. "The problem was not Nigeria. The problem was to stop it there at the border. If they had continued the National Immunization Days in the border countries, Nigeria couldn't have exported it. They would be the only country in the world with polio, and they'd have to do something about it." Instead, the virus quickly spread into seven neighboring countries—Niger, Burkina Faso, Chad, Ghana, Togo, Benin, and Cameroon—and before long polio was reported in an additional 11 countries, from Ethiopia to Indonesia. South Asia continued to make inroads against the disease, but people were tiring of the seemingly endless campaign in areas where other health problems remained severe.[48]

It was an extremely trying time for the polio fighters. And for measles, which was far deadlier than polio, eradication was only a pipe dream. While biologically speaking it might even be more feasible than eliminating polio, "another eradication campaign is so politically loaded that I don't see it in the near future," Walter Orenstein told me.

The Road Not Taken

One of the Rockefeller Institute's great biologists of the mid-twentieth century was Rene Dubos, a microbiologist whose research with bacteria from a New Jersey cranberry led to the creation of the first antibiotics. Dubos had an extremely plastic concept of the role of the environment in biological events, and in his later years he grew increasingly skeptical of modern medicine.[49] Born in France, Dubos was also one of Pasteur's biographers and was particularly fascinated by Pasteur's intellectual agility as he pondered questions of pure science while devising means of answering the practical concerns of French industry, the French state, and eventually the world.[50] In 1865 Pasteur's studies of silkworm diseases saved the silk industry, but he often regretted that he had been unable to return to the silkworm work. "If I were to undertake new studies on the silkworm diseases," he wrote years later, "I would direct my efforts to the environmental conditions that increase their vigor and resistance." The complex relationship of culture and environment and their impact on human health was not a subject Pasteur was able to delve into; he is remembered for the brilliance of his reductionist work—the discovery that one could attribute an illness to a specific germ.

At the start of the twenty-first century there were still people asking whether there was something inherently counterfeit—and therefore, inevitably doomed—about solving our problems with the reductionist, technical fix of vaccination. Was a holistic approach better? Just as a polio eradication campaign might divert resources from other, more crucial though less dramatic health goals, was it possible that a vaccination diverted the immune cells from their highly evolved system of dealing with life's trials? I asked Orenstein whether the possibility of unknown consequences of introducing new vaccines ever gave him sleepless nights. "Well, you never know when you do anything what the long-term consequences will be," he said. "It would be nice if life were a placebo-controlled trial and you could just go back and redo it based on what you learn. But we always have to take the best decisions, setting up mechanisms to understand what's happening, sometimes making adjustments. You always fear, 'Could you be doing something that over the long term will make things worse?' But to take that to the logical extreme it could paralyze you into total inaction."

Because vaccinologists can see the targets they are aiming for and the means of conquering them, while so much else of science is so fabulously nebulous, they feel an urgency in moving forward to combat new causes of sickness and death. This is a noble enterprise with an asterisk, like most such enterprises. Certainly, policymakers can advance more or less cautiously in changing the microbiological environment through vaccination. But the alternative to action is helplessness. "Should we have never used Hib vaccine, because we didn't know for certain that older kids wouldn't get the disease?" Orenstein asked. "For those who trained in pediatrics in the 1970s, Hib was a horror—there were 12,000 meningitis cases a year, 25,000 cases of invasive disease. To not have used that vaccine because we didn't have definitive answers about whether the ecology would be changed with other kinds of *Haemophilus*—those are things that are only knowable in the long run. We have to take our chances and then follow up."

Reductionism remains the basis of science, although we all recognize that systems are interrelated and changing one element may affect others. The history of vaccines has shown that unexpected results are the rule rather than the exception. Vaccinologists would do well to remember this as they move toward new discoveries and recommendations in an enterprise that has accomplished tremendous good.

At the same time, the public and politicians need to understand the threat that thoughtless actions pose to the great immunological commons we have built by decades of vaccination of children. If vaccination is truly important, we need to support it, by welcoming interested parties in discussions of its true risks and benefits. Elected officials must play their part by removing ideological blinders. The history of our vaccination programs has shown that the immunological commons must be carefully tended, with subsidies for vaccines when called for, legal protection from lawsuits for firms that are doing their honest best to create safe vaccines, and the bottom-line assurance that those who are accidentally damaged by vaccination will be properly compensated. Only then will vaccination fulfill its promise as our collective commitment to protect children from infectious disease.

ACKNOWLEDGMENTS

I AM GRATEFUL to many friends, colleagues, and fellow vaccine obsessives who contributed ideas, advice, and facts to this book, which germinated from a series of magazine assignments. I owe a debt to Chuck Lane, who saw fit to put my first vaccine article on the cover of *The New Republic*, and to Steve Coll, who allowed me to portray the vaccine court in *The Washington Post Magazine*. Amy Meeks and the late Mike Kelly at *The Atlantic Monthly* and Daniel Zalewski, then at *The New York Times Magazine* helpfully guided my investigations of vaccine refuseniks and the autism controversy. Carol Lloyd, Jennifer Sweeney, Joan Walsh, and Karen Croft at *Salon* ran many of my stories, as did Emily Bazelon at *Slate*.

Sarah Chalfant has been the perfect agent, doggedly shaking sense out of me while insisting that without passion for the work there was no sense in it at all. My editor, Angela Vonderlippe, offered a valuable critique; Lydia Fitzpatrick has been a helpful guide; Carol Rose an excellent copy editor. Peter Lewis, Tamara Razi, Hanna Rosin, David Plotz, Max Kelly, Liza Mundy, David and Steven Talbot and Camille Peri, Nick and Emily Allen, Ben Coates and Julian Allen were attentive listeners and enthusiasts, as were colleagues Dawn McKeen and Matt Davis. I benefited from the historical perspective of Mark Bradley and the medical and public health grounding of my sister Susie Allen, my late, great sister Katie Allen, my sisters-in-law Cindy Talbot, Martha Kowalick, and Pippa Gordon, and brother-in-law Richard DiCarlo. Ann Hulbert, a role model to writers of narrative nonfiction, was amazingly encouraging and gave invaluable book-writing and editing advice. The following people also read all or portions of the book and provided vital corrections, advice, and support: Ben Coates, Bruce Weniger, Bob Chen, Sarah Depres, Eric Colman, Jeffrey Schwartz, Neal Halsey, Vincent Fulginiti, Leonard Seeff, James Whorley, Edward Morman, Andrea Rusnock, and Kendall Hoyt. My parents, Dick and Barbara Allen, provided sharp copyediting and steadfast encouragement. Paul Offit, another reader, was open, knowledgeable, and passionate about vaccines.

The Rockefeller Archive provided a stipend that allowed me to review its fastidiously kept records. Thanks to Darwin Stapleton and Tom Rosenbaum for their cordiality and guidance. Ed Morman of the Wood Institute at the College of Physicians of Philadelphia provided a research fellowship, advice, and humor. Nava Hall of the Wood Institute and David Rose of the March of Dimes dug up interesting illustrations. I am also indebted to the staffs of the National Archives II, the American Philosophical Society, the Medical Heritage Library in Cincinnati, the Library of Congress' Prints and Photographs Division, and the Harry Ransom Humanities Research Center at the University of Texas, all of whom helped me locate piquant literature and illustrations. Stephen Greenberg and Crystal Smith of the History of Medicine Division at the National Library of Medicine were infallibly patient and helpful. In Philadelphia, Tom Kellogg helped me investigate the family background of Porter Cope.

While not interviewed for this book, Cliff Shoemaker, Jane el-Dahr, Geoff Evans, Bruce Gellin, Jeffrey Baker, and Robert D. Johnston were helpful reality checkers. Victor Harding generously shared his memories and files from the DTP trials. Al Mehl has been a warm, inspiring, and honest friend. The late Maurice Hilleman was particularly charming, patient, and entertaining. I won't soon forget the two days spent in his unique company. Barbara Loe Fisher and Neal Halsey have been reluctant but important resources in separate ways. I thank Dr. Halsey for allowing me to audit the vaccinology course he taught at Johns Hopkins University in the winter of 2002 and Lyn Redwood for inviting me to a NIEHS-sponsored conference in 2005.

My son Isaac and daughter Lucy were interested in the less scary parts of this book and even drew on them for their own imaginative projects.

Finally, I can't properly express the gratitude and admiration I have for my wife, Margaret Talbot, a kind person who has never thought to discourage my dreams. She is a great editor, a fabulous thinker, my soulmate and copilot.

NOTES

INTRODUCTION

1. *The Smallpox Vaccination Program: Public Health in an Age of Terrorism* (Washington: Institute of Medicine, 2005): 26.
2. Cf. Ken Alibek and Stephen Handelman, *Biohazard: The Chilling True Story of the Largest Covert Biological Weapons Program in the World, Told from the Inside by the Man Who Ran It* (New York: Random House, 1999); Judith Miller, Stephen Engelberg, and William J. Broad, *Germs: Biological Weapons and America's Secret War* (New York: Simon & Schuster, 2001).
3. John Modlin, interview with author, Washington, June 2003.
4. Marvin Olasky, "Premature Obituary," *The Washington Times*, 12 November 2001, A17.
5. Bruce Gellin, Edgar Marcuse, et al. "Do Parents Understand Immunizations? A National Telephone Survey," *Pediatrics* 106 (November 2000): 1097–102.

CHAPTER 1

1. Linda Pollock, *A Lasting Relationship: Parents and Children over Three Centuries* (Boston: University Press of New England, 1987): 102.
2. The synopsis of Mather's shortcomings owes much to Kenneth Silverman, *The Life and Times of Cotton Mather* (New York: Harper & Row, 1984).
3. Cotton Mather, *The Angel of Bethesda* (Worcester: American Antiquarian Society, 1972): 96.
4. Ola Elizabeth Winslow, *A Destroying Angel: The Conquest of Smallpox in Colonial Boston* (New York: Houghton-Mifflin, 1974): 48; and Donald R. Hopkins, *The Greatest Killer: Smallpox in History* (University of Chicago Press, 2002): 249.
5. Cotton Mather, *Account of the Method and Success of Inoculating the Smallpox, In Boston and New England* (Boston: J. Peele, 1722): 3. In my quotations from this and other eighteenth-century documents, I have taken the liberty of using modern spellings.
6. Mather, *Angel of Bethesda*, 50.
7. Mather, *Account*, 17.
8. Alexander Hamilton, and Carl Bridenbaugh, *Gentleman's Progress: The*

Itinerarium of Dr. Alexander Hamilton, 1744 (Chapel Hill: University of North Carolina Press, 1948): 116–17.

9. Winslow, *Destroying Angel*, 34.

10. William Douglass, *Dissertation Concerning Inoculation of the Small-pox: Giving Some Account of the Rise, Progress, Success, Advantages, and Disadvantages of Receiving the Small Pox by Incisions; Illustrated by Sundry Cases of the Inoculated* (London: D. Henchman, 1730): 5.

11. Ibid., 4; and Dennis Melchert, *Experimenting on the Neighbors: Inoculation of Smallpox in Boston in the Context of Eighteenth Century Medicine*, doctoral dissertation, University of Iowa, 1974, 153–54.

12. William Douglass, *The Abuses and Scandals of Some Late Pamphlets in Favour of Inoculation of the Small Pox, Modestly Obviated, and Inoculation Further Consider'd in a letter to A—— S—— M. D. & F. R. S. in London* (Boston: J. Franklin, 1722): 2.

13. Melchert, *Experimenting*, 200.

14. Mather, *Angel of Bethesda*, introduction.

15. Melchert, *Experimenting*, 226–27.

16. Douglass quoted in John B. Blake, *Public Health in the Town of Boston, 1630–1822* (Cambridge: Harvard University Press, 1959): 65.

17. Douglass, *Abuses*, 8–9.

18. Ibid., 28.

19. Benjamin Franklin and William Herberden, *Some Account of the Success of Inoculation for the Small-pox in England and America: Together with Plain Instructions, by Which Any Person May Be Enabled to Perform the Operation, and Conduct the Patient through the Distemper* (Philadelphia: W. Strahan, 1759): 6.

20. Melchert, *Experimenting*, 218–19.

21. William Cooper, *A Reply to the Objections Made against Taking the Small Pox in the Way of Inoculation from Principles of Conscience: in a Letter to a Friend in the country / By a Minister in Boston* (Boston: S. Gerrish, 1730).

22. Louise A. Breen, "Cotton Mather, the 'Angelical Ministry,' and Inoculation," *Journal of the History of Mediciine and Allied Science* 46 (3) (July 1991): 335–37.

23. Melchert, *Experimenting*, 160.

24. Ibid., 159–60.

25. Breen, "Cotton Mather," 338.

26. Mather, *An Account of the Incident and Method*.

27. Thomas H. Brown, "The African Connection. Cotton Mather and the Boston Smallpox Epidemic of 1721–1722," *Journal of the American Medical Association* 15 (21 October 1988): 2247–49.

28. Ibid.

29. Isobel Grundy, *Lady Marty Wortley Montagu* (Oxford: Clarendon Press, 1999).

30. Ibid., 98–104.

31. Ibid., 211.

32. Ibid., 217.

33. Zabdiel Boylston, *An Historical Account of the Small-Pox Inoculated in New England, upon All Sorts of Persons, Whites, Blacks, and of All Ages and Constitutions* (London: S. Chandler, 1726); Mead cited in Arnold C. Klebs, "Historic Evolution of Variolation," *Bulletin of the Johns Hopkins Hospital* (1) (1913): 73.

34. Henry Lee Smith, "Dr. Adam Thomson, the Originator of the American Method of Inoculation for Smallpox," *Johns Hopkins Hospital Bulletin* 20 (February 1909): 50.

35. D. Carroll, "Medical Practice and Practitioners in Eighteenth Century Maryland," *Maryland State Medical Journal* 21 (1972): p. 57–58.

36. Thomas Ruston, *Essay on Inoculation* (London: J. Payne, 1767) 10.

37. David Van Zwanenenberg, "The Suttons and the Business of Inoculation," *Medical History* 22 (1) (1978).

38. Reginald Fitz, "The Treatment for Inocluated Smallpox in 1764 and How It Actually Felt," *Annals of the History of Medicine* (1942): 110–13.

39. John Adams autobiography, *Adams Family Papers: An Electronic Archive.* Part 1, 1764–1765, sheet 9 of 53, retrieved from www.masshist.org/digital adams/aea/browse/autobio1.html.

40. Patrick Henderson, "Smallpox and Patriotism, The Norfolk Riots, 1768–69," *The Virginian Magazine* 73 (1965): 413–24; Frank Dewey, *Thomas Jefferson, Lawyer* (Charlottesville: University Press of Virginia, 1986): 45–56; Willard Sterne Randall, *Thomas Jefferson, a Life* (New York: Henry Holt, 1993): 130–35.

41. Elizabeth A. Fenn, *Pox Americana: The Great Smallpox Epidemic of 1775–82* (New York: Hill and Wang, 2001): 84–85.

42. Letter from John Adams to Abigail Adams, 29 March 1776, [electronic edition] *Adams Family Papers: An Electronic Archive*, Massachusetts Historical Society, retrieved from www.masshist.org/digitaladams.

43. Natalie S. Bober, *Abigail Adams, Witness to a Revolution* (New York: Simon & Schuster, 1995): 78.

44. Sarah B. Dine, "Inoculation, Patients, and Physicians: The Transformation of Medical Practice in Philadelphia, 1730–1810," *Transactions and Studies of the College of Physicians of Philadelphia* 19 (1997): 67–94.

45. Blake, *Public Health in Boston*, 76–78.

46. Ibid., 94.

47. Franklin and Heuberden, "Some account," 7.

48. Ibid., 140.

49. Benjamin Waterhouse, *A Prospect of Exterminating the Smallpox. Part II* (Cambridge: Harvard University, 1802): 6. John Blake has argued that smallpox hazards were worse in Philadelphia than in Boston because the expense of variolation kept Philadelphia's poor from gaining its protec-

tion. But Sarah Dine found in her inspection of the account books of late-eighteenth-century Philadelphia doctors that they variolated carpenters and laborers as well as landowners. By the 1790s, more than a third of some city doctors' fees during the spring months derived from variolations. Rush himself variolated thousands.

50. Dine, "Inoculation, Patients," 71.

51. Roy Porter, *English Society in the 18th Century* (London: Allen Lane, 1982): 273; Statistical Table

52. See www.census.gov/population/censusdata/table-2.pdf.

53. John Haygarth, *An Inquiry How to Prevent the Small-pox: And Proceedings of a Society for Promoting General Inoculation at Stated Periods, and Preventing the Natural Small-pox, in Chester* (London: J. Monk, 1784): 157–60.

54. John Haygarth, *Sketch of a Plan to Exterminate the Smallpox from Great Britain* (London: Johnson, 1793).

CHAPTER 2

1. Albert Marrin, *Edward Jenner and the Speckled Monster* (New York: Dutton, 2003): 64.

2. John Baron, *The Life of Edward Jenner, MD*, Vol. 1 (London: Henry Colburn, 1827): 6–7.

3. A. S. MacNalty, "The Prevention of Smallpox: From Edward Jenner to Monckton Copeman," *Medical History* 12 (1) (1968): 5.

4. Thomas Dudley Fosbroke, *The Berkeley Manuscripts* (London: Nichols, 1821): 165–66.

5. Edward Jenner, *An Inquiry into the Causes and Effects of the Variolae Vaccinae* ... (London: Sampson Law, 1798): 37.

6. Donald R. Hopkins, *The Greatest Killer: Smallpox in History* (Chicago: University of Chicago Press, 2002): 81–82.

7. Ibid.

8. John Blake, *Benjamin Waterhouse and the Introduction of Vaccination: A Reappraisal*. (Philadelphia: University of Pennsylvania Press, 1957): 25.

9. Hopkins, *Greatest Killer*, 262–67.

10. Jenner, *An Inquiry*, 30–31.

11. Peter Razzell, *The Conquest of Smallpox: The Impact of Inoculation on Smallpox Mortality in Eighteenth Century England* (London: Caliban, 1977) and *Edward Jenner's Cowpox Vaccine: The History of a Medical Myth* (London: Caliban, 1977).

12. Derrick Baxby, *Edward Jenner's Cowpox: The Riddle of Vaccinia Virus and Its Origins* (London: Heinemann, 1981); A. Herrlich et al., "Experimental Studies on Transformation of the Variola Virus into the Vaccinia Virus, *Artchiv fuer die Gesamte Virusforschung* 12 (1963): 579.

13. Herrlich, "Experimental Studies," 579.

14. Joseph J. Esposito, personal communication, June 2005.

15. Thomas Malthus, *An Essay on the Principle of Population* (London: Murray, 1817).

16. E. Blanche Sterling, "Child Hygiene in Human Ecology," in *A Decade of Progress in Eugenics: Proceedings of the Third International Congress of Eugenics* (Baltimore: Williams and Wilkins, 1934), 343–49. Sterling, a U.S. Public Health Service officer, told the gathered eugenicists that while "I have no doubt we do save some that are unfit . . . it is quite impracticable to separate the fit from the unfit in any public health program. We cannot limit our diphtheria immunization to the children of college professors or refuse to vaccinate the feeble-minded son of the day laborer."

17. Gillray's drawing can be viewed at www.bl.uk/onlinegallery/features/pictures .html. The monstrous description of Jenner is mentioned in Genevieve Miller, *Letters of Edward Jenner* (Baltimore: Johns Hopkins Press, 1983): 38.

18. Benjamin Moseley, *Commentaries on the Lues Bovilla or Cow Pox* (London: Longman, Hurst, Rees, and Orme, 1806).

19. Genevieve Miller, *Letters of Edward Jenner* (Baltimore: Johns Hopkins Press, 1983): 28, 62.

20. Miller, *Letters of Edward Jenner*, 41.

21. Henry J. Parish, *A History of Immunisation* (London: Livingston, 1965): 26–29.

22. Carlos Franco-Paredes et al., "The Spanish Royal Philanthropic Expedition to Bring Smallpox Vaccination to the New World and Asia in the 19th Century," *Clinical Infectious Diseases* 41 (2005): 1285–89.

23. Joseph C. Hutchison, *Vaccination and the Causes of the Prevalence of Smallpox in New York in 1853–4* (Brooklyn 1854).

24. Hopkins, *Greatest Killer*, 85.

25. John Duffy, *The Sanitarians* (Champaign: University of Illinois Press, 1990): 56.

26. Hopkins, *Greatest Killer*, 89–91.

27. John Duffy, "School Vaccination: The Precursor to School Medical Inspection," *Journal of History of Medicine and the Allied Medical Sciences* 33 (3) (1978): 344–55.

28. Frank Foster, *A Report on Animal Vaccination at the New York Dispensary in the Year 1872* (New York: J. Amerman, 1872): 1–16.

29. Ibid., 11.

30. John J. Buder, "Letters of Henry Austin Martin: The Vaccination Correspondence to Thomas Fanning Wood, 1877–1883," master's thesis, University of Texas, 1991, 1–12.

31. S. Monckton Copeman, *Vaccination: Its Natural History and Pathology* (New York: Macmillan, 1899): 183–84.

32. Jean-Charles Sournia, *The Illustrated History of Medicine* (London: Harold Starke, 1992): 431.

33. Wade Oliver, *The Man Who Lived for Tomorrow; a Biography of William Hallock Park* (New York: E.P. Dutton, 1941): 132–33. The establishment was later moved to 916 Second Avenune, then to 326 East 44th Street and eventually to Otisville, Long Island.

34. See, for example, Tom Rivers, *Reflections on a Life in Medicine and Science* (Cambridge: MIT Press, 1967).

35. According to retired Wyeth vaccine maker Alan Bernstein, the practice continued—down to the full-time poop-scooper—through 1975, when Wyeth stopped making smallpox vaccine.

36. Copeman, *Vaccination*, 198, 217.

37. Nadja Durbach, "They Might as Well Brand Us," *Social History of Medicine* 13 (2000): 47–57.

38. H. Rider Haggard, *Doctor Therne* (London: George Newnes, 1902): 123–25.

39. Alfred Russel Wallace, *Vaccination a Delusion; Its Penal Enforcement a Crime* (London: Anti-Vaccination League, 1901).

40. Ibid, 267–68.

41. Ibid, 223.

42. Ibid., 242–74.

43. Scott Edward Roney, *Trial and Error in the Pursuit of Public Health: Leicester, 1849–1891*, doctoral dissertation at the National Library of Medicine, 2002.

44. Michael Shermer, *The Borderlands of Science* (Oxford University Press, 2001): 162–64.

45. This comes from Steven Lehrer, *Explorers of the Body* (New York: Doubleday, 1979).

46. William G. Rothstein, "When Did a Random Patient Benefit from a Random Physician: Introduction and Historical Background," *Caduceus* 12 (3) (1996): 3, cited in Bert Hansen, "New Images of a New Medicine: Visual Evidence for the Widespread Popularity of Therapeutic Discoveries in America after 1885," *Bulletin of the History of Medicine* 73 (1999): 629–78.

47. George Bernard Shaw, *The Doctor's Dilemma, with a Preface on Doctors* (New York: Brentano's, 1913): vi–xc.

48. Cyril M. Dixon, *Smallpox* (London: Churchill, 1962): 452–69. According to Dixon's data, from 1953 to 1957 there were 34 cases of smallpox and 10 deaths in England and Wales. From 1951–1958 there were 243 serious reactions, including 42 deaths, attributed to vaccine.

CHAPTER 3

1. Alfred Russel Wallace, *Vaccination a Delusion; Its Penal Enforcement a Crime* (London: Anti-Vaccination League, 1901): 1.

2. Edmond Esquerre, "Safeguarding Virus and Antitoxins," letter in *The New York Times*, 24 November 1901, 5.

3. Robert D. Johnston, *Radical Middle Class: Popular Democracy and the Question of Capitalism in Progressive Era Portland, Oregon* (Princeton University Press, 2003): 353.

4. Genetic microbiology has revealed various forms of low-mortality small-pox, which in the early twentieth century went under names such as Amass and alastrim. For simplicity's sake I am calling all mild forms of the disease *V. minor*.

5. See www.fda.gov/oc/history/makinghistory/100yearsofbiologics.html. It is important to note that the vaccine and sera regulation law was partly a response to the Philadelphia vaccine tragedy, partly to another disaster. In November 1901, the bacteriological laboratory of St. Louis revealed that 13 children had died after contracting tetanus from its diphtheria anti-toxin, grown by Dr. Amand Ravold in a horse named Joe that was later found to be infected with tetanus.

6. Ibid.

7. John F. Anderson, "Federal Control of Vaccine Virus," *Journal of the American Medical Association* 44 (1905): 1838–40.

8. Cf. Lincoln Steffens, *The Shame of the Cities* (New York: Hill & Wang, 1904).

9. Edward T. Morman, "Scientific Medicine Comes to Philadelphia: Public Health Transformed," doctoral thesis, 1986, 100–12.

10. Sam Alewitz, *Filthy Dirty: A Social History of Unsanitary Philadelphia in the 19th Century* (New York: Garland, 1989): 6, 50–57.

11. Walter Reed, "What Credence Should Be Given to the Statements of Those Who Claim to Furnish Vaccine Lymph Free of Bacteria," *Journal of Practical Medicine* 5 (1895): 532–34.

12. C. Probst, "Smallpox in Ohio," *Journal of the American Medical Association* 33 (23 December 1899): 1589–90.

13. It was eventually determined (Joseph Esposito, personal communication) that *V. minor* and *V. alastrim* were distinct entities. Both were considerably milder than *V. major*.

14. Charles Chapin, "Variation in Type of Infectious Disease as Shown by Changes in Smallpox in the United States, 1895–1912," *Journal of Infectious Diseases* 13 (2) (1913):172.

15. "Isolation, Its Value and Limitations," 1921 speech in *Papers of Charles V. Chapin*, (New York: Commonwealth Fund, 1934); Chapin and Joseph Smith, "Permanency of the Mild Type of Smallpox," *Journal of Preventive Medicine* (1932): 273–320.

16. Chapin, "Variation," Chapin and Smith, "Permanency," and "Changes in Type of Contagious Disease," *Journal of Infectious Diseases* 1 (1) (1926–1927): 1–29.

17. "Smallpox Spreads Despite Vaccination," *Philadelphia North American*, 15 December 1901, 1.

18. "Law Requiring Vaccination of School Children Sustained," *Pennsylvania Board of Health, Annual Report* 16 (1899–1900): 892–93.

19. Pennsylvania Board of Health, "Proceedings and Papers of the 8th Annual Meeting of the Associated Health Authorities of Pennsylvania," *Annual Report* 17 (1900–1901): 298.

20. "More Cases of Lockjaw," *Philadelphia North American*, November 11, 1901.

21. "Vaccination Boy, Tetanus-Struck, May Recover," *Philadelphia North American*, 11 November 1901; "Three More Are Killed by Lockjaw," 12 November 1901.

22. Pennsylvania Board of Health, *Annual Report* 17 (1901): 20.

23. "Forty Doctors to Fight Smallpox," *Philadelphia North American*, 6 December 1901.

24. Pennsylvania Board of Health, *Annual Report* 18 (1902): 25.

25. "Smallpox Shows No Abatement," *Philadelphia North American*, 17 December 1901.

26. "Camden's Smallpox Scare," *The New York Times*, 5 December 1901, 5; "Health Board to Take City in War on Smallpox," *Philadelphia North American*, 4 December 1901.

27. Statistics from Chapin and Smith, "Permanency," 280; Joel G. Breman, et al., "The Last Smallpox Epidemic in Boston and the Vaccination Controversy, 1901–1903," *New England Journal of Medicine* 344 (5) (2001): 375–79.

28. Joel G. Breman et al., "The Last Smallpox Epidemic in Boston and the Vaccination Controversy, 1901–1903," *New England Journal of Medicine* 344 (5) (2001): 375–79.

29. Most of this account is drawn from Patsy Gerstner, "Smallpox, the Chamber of Commerce, and the Reshaping of the City's Public Health Department in the Early 20th Century," paper presented at the 16th annual Western Reserve Studies Symposium (2001); and from "How We Rid Cleveland of Smallpox," *Cleveland Medical Journal* (February 1902), which reports on a speech by Friedrich, and a spirited debate that followed among various city physicians, at the December 1901 meeting of the Cleveland Medical Society.

30. See, for example, Judith Walzer Leavitt, *The Healthiest City: Milwaukee and the Politics of Health Reform* (Madison: University of Wisconsin Press, 1996); and J. J. Morclay, "Variola in Buffalo," *Buffalo Medical and Surgical Journal* 21 (8) (1882): 345.

31. Martin Friedrich, "How We Rid Cleveland of Smallpox," *Cleveland Medical Journal* (February 1902): 77–78.

32. Ibid., 79–81.

33. "Tom Johnson's Man Defeats Smallpox," *Philadelphia North American*, 15 December 1901, 1.

34. Friedrich, "How We Rid," 81–89.

35. Charles-Edward A. Winslow, "Man and the Microbe," *Popular Science Monthly* (July 1914): 19.

36. The account of the founding of the U.S. chapter is in a typewritten 1916 history by Porter F. Cope in the Pitcairn Archive, Bryn Athyn, PA. It includes a copy of a letter from Wilder to Cope recalling the early days of the movement. Other elements are contained in a logbook titled "Anti-Vaccination Society of America, 1895–1898," Wood Institute of the College of Physicians and Surgeons, Philadelphia.

37. Henry Bergh, "The Lancet and the Law," *North American Review* 134 (February 1882): 161–80; and Zulma Steele, *Angel in Top Hat* (New York: Harper and Bros., 1942): 137–40, 273.

38. Susan Lederer, *Subjected to Science; Human Experimentation in America Before the Second World War* (Baltimore: Johns Hopkins Press, 1995): 90–107.

39. Quoted in Annie Riley Hale, *These Cults* (New York: National Health Foundation, 1926), retrieved from www.soilandhealth.org/02sov/0303critic/03031cults/cults-toc.htm.

40. F. L. Oswald, cited in W. J. Furnival, *Professional Opinions Against Vaccination* (London: Stone, 1906).

41. Barbara Loe Fisher, author interview, June 2005; In addition to the Cope account, letters and diary entries on the early history of the antivaccination movement, including a list of early members, are contained in two logbooks filed as Z10c 17 [antivaccinationist list, Philadelphia? 1900], Wood Institute for the History of Medicine, College of Physicians of Philadelphia.

42. Reuben Swinburne Clymer, *Vaccination Brought Home to You* (Terre Haute: Hebb, 1904): 30.

43. Ibid., 53.

44. Clymer, who is worshipped as an immortal prophet by followers of his Rosicrucian sect, brought a variety of seemingly unconnected elements into his health philosophy.

45. Correspondence in logbook, filed under Z10c17, Wood Institute for the History of Medicine, College of Physicians of Philadelphia.

46. Cope, unpublished memoir of the American antivaccinationist movement, 10, at Pitcairn Library, Bryn Athyn, Pennsylvania.

47. "Porter F. Cope, 81, Author and Editor," *The New York Times*, 21 December 1950, 29.

48. Pennsylvania Board of Health, *Annual Report* 18 (1901–2), 25.

49. Ibid, 266.

50. Ibid., 7, 482.

51. Pennsylvania Board of Health, *Annual Report* 19 (1902–3): 27.

52. Ibid., 201–4.

53. Proceedings of State Public Health Conference, 9 May 1902, in Pennsylvania Board of Health, *Annual Report* 19, 316–17.

54. "Vaccination Is the Only Remedy," *Cleveland Plain Dealer*, 20 June 1902.

55. Philadelphia Bureau of Health, *Annual Report* (1902): 114–17.

56. Ibid., 90–91.

57. Philadelphia Bureau of Health, *Annual Report* (1903): 41.

58. Ibid., 30–31.

59. Joseph McFarland, "Tetanus and Vaccination: An Analytical Study of 95 Cases of This Rare Complication," *Proceedings of the Philadelphia County Medical Society* 2 (1902): 166–78.

60. Milton Rosenau, "The Bacteriological Impurities of Vaccine Virus," *U.S. Hygienic Laboratory Bulletin* 12 (March 1903): 5–49.

61. H. M. Alexander to McFarland, 31 January 1902, Cage Z10/204/Box 1, Correspondence, McFarland Papers, College of Physicians of Philadelphia.

62. John P. Anderson, "Federal Control of Vaccine Virus," *Journal of the American Medical Association* (10 June 1905): 1840.

63. Anderson to SGO, 21 June 1906, Folder 3655-1901, Box 345, Central File 1897-1923, Records of the PHS, Record Group 90, National Archive II.

64. This account comes from Alan Bernstein, who was in charge of Wyeth's Marietta plant from 1975 to 1987.

65. Alexander to McFarland, 31 January 1902, McFarland Papers, Box 1 Medico-Legal, Wood Institute for the History of Medicine, College of Physicians of Philadelphia.

66. "Exposure of a Disreputable Proceeding," Mulford pamphlet dated 2 February 1902 in McFarland Papers, Box 1 Medico-Legal, Wood Institute for the History of Medicine.

67. Parke-Davis letter to McFarland, CageZ10/204/Box 3, correspondence, Wood Institute for the History of Medicine.

68. Clymer, *Vaccination Brought*, 46.

69. Rosenau to SGO, 2 February 1906, Folder 3655-1906, Box 342, RG 90, National Archives.

70. Joan Retsinas, "Smallpox Vaccination; A Leap of Faith," *Rhode Island History* (November 1979): 113–28.

71. "Justifiable Measures for the Control of Infectious Disease," in *Papers of Charles V. Chapin*, 78.

72. Barbara Guttmann Rosenkrantz, *Public Health and the State: Changing Views in Massachusetts, 1842–1936* (Cambridge: Harvard University Press, 1972): 124.

73. Lora Little, "Mass Meeting in Philadelphia," *The Liberator* (June 1906): 82.

74. Ralph C. Williams, *The US Public Health Service, 1798–1950* (Landover: Commissioned Officers Association of the USPHS, 1951).

75. Charles M. Higgins, *Vaccination and Lockjaw: Assassins of the Blood* (Brooklyn, 1916): 26.

76. Cf. Cotterrill to Wyman, 8 February 1907, Folder 3655-1907, Box 343, RG 90, National Archives; and Rosenau to Wyman, 16 March 1908, Folder 3655-1908, Box 344, RG 90, National Archives.

77. Ibid., numerous letters in Folder 3655-1907, Box 343; also W. C. Hassler to Wyman, 3 July 1906, Folder 3655-1906, Box 342, RG 90, National Archives.

78. The state ordered strict cleanliness and record-keeping, oversight by a physician or veterinarian, the hiring of a full-time worker devoted to prompt removal of cow droppings, and twice-annual suspension of operations for thorough cleanings.

79. Ravold was disgraced in 1901 when 13 children died of tetanus poisoning from a diphtheria serum produced at his laboratory.

80. John R. McKiernan, "Fevered Measures: Race, Communicable Disease and Community Information on the Texas Mexico Border, 1880–1923," doctoral thesis, University of Michigan, 2002, 221–23, 257–96.

81. C. S. Carr quoted in Clymer, *Vaccination Brought*, 77–79.

82. From Martin S. Pernick, "Public Health Then and Now: Eugenics and Public Health in American History," *American Journal of Public Health* 87 (11) (1997): 1768–70.

83. Johnston, *Radical Middle Class*, 205.

84. Josepha Sherman, *Johnny Gruelle and the Story of Raggedy Ann and Andy* (Hockessin: Mitchell Lane Publishers, 2005): 87–92. A framed picture of Raggedy Ann hangs over the receptionist's desk at Barbara Loe Fisher's National Vaccine Information Center.

85. "A Prayer," *Life* 1268 (14 February 1907): 239; Lederer, *Subjected*, 42.

86. Cope to Pitcairn, 10 December 1915, Anti-Vaccination File, Pitcairn Archive.

87. Richard R. Gladish, *John Pitcairn, Uncommon Entrepreneur* (Bryn Athyn: Academy of the New Church, 1989): 331.

88. Quoted in Pitcairn's obituary, *Vaccination Inquirer*, 1 September 1916.

89. J. F. Potts, *Vaccination* (London, National Anti-Vaccination League, 1883).

90. Naomi Rogers, *An Alternative Path: The Making and Remaking of Hahneman Medical College and Hospital of Philadelphia* (Piscataway: Rutgers University Press, 1998): 9.

91. John Pitcairn, *Both Sides of the Vaccination Question* (Philadelphia: Anti-Vaccination League of America, 1911): 1.

92. Gladish, *Pitcairn*, 335.

93. Little, "Mass Meeting," 82.

94. Cope, report to Miss Cyrial Odhner, 12 December 1916, 21–22. Anti-Vaccination File, Pitcairn Archive.

95. Clymer, *Vaccination Brought*, introduction.

96. Little has been ably memorialized by Robert D. Johnston. See "The Myth of the Harmonious City," *Oregon Historical Quarterly* (Fall 1998); *The*

Radical Middle Class; Popular Democracy and the Question of Capitalism in Progressive Era Portland, Oregon (Princeton University Press, 2003).

97. "Harmonious City," 275.

98. Ibid., 273.

99. *Vaccination Inquirer*, 2 August 1909.

100. Charles V. Chapin, "A Report on State Public Health Work Based on a Survey of State Boards of Health," 1915, in Folder 191, Box 3, Series 908, RG 3, Rockefeller Foundation, RAC.

101. Frank Schamberg, "What Vaccination Has Really Done," *Ladies' Home Journal*, June 1910.

102. Correspondence in Vaccination file, Pitcairn Archive.

103. "Anti-Vaccination Reports on Alleged Injuries from Compulsory Vaccination in the Philadelphia Area," 183-paged report, Philadelphia, 1912 at Wood Institute for the History of Medicine, College of Physicians of Philadelphia.

104. J. M. Hodge, "A Physician's Reasons for Having Renounced Vaccination," *Medical Century* (February 1914): 136.

105. *Buffalo News*, 24 January 1914, 1; "Conditions at the Falls," *Buffalo Express*, 29 January 1914, 1.

106. Charles-Edward A. Winslow, *The Life of Herman M. Biggs* (Philadelphia: Lea & Febiger, 1929): 264–66.

107. "Biggs Threatens the City Again," *Niagara Falls Gazette*, 13 February 1914.

108. "City Will Not Be Quarantined," *Niagara Falls Gazette*, 12 February 1914.

109. "History and Frequency of Smallpox Vaccinations and Cases in 9,000 Families," *Public Health Reports* 51 (16) (1936): 445–48.

110. Chapin, "Permanency," 275.

111. Johnston, *Middle Class*, 212.

112. "State again Defied by Delaware town," *Philadelphia Bulletin*, 15 January 1926; "Freed in Vaccination Row," *Philadelphia Bulletin*, 31 January 1926.

113. The most important step was widespread introduction of a new vaccination method, developed by the Public Health Service's James Leake, in which a drop of vaccine was put on the skin and the virus injected by repeated, gentle poking with a sterilized needle.

114. "Deliberately Endanger the Public," *The New York Times*, 6 May 1914, 10.

CHAPTER 4

1. Deborah Dwork, *War Is Good for Babies and Other Young Children: A History of the Infant and Child Welfare Movement in England, 1898–1918* (London: Tavistock Publications, 1987): 209.

2. Richard Krugman, author interview, October 2004.

3. Henry Mustard and Philip W. Hendrick, "Generalized Vaccinia; A Study of 15 Cases," *Journal of Pediatrics* (1948): 281–94.

4. Michael Pollak, "A Distant, Troubling Echo from an Earlier Smallpox War," *The New York Times*, 17 December 2002, F1.

5. Tom Rivers, *Reflections on a Life in Medicine and Science* (Cambridge: MIT Press, 1967): 385–89.

6. *U.S. News and World Report*, 7 May 1947, 26; James S. Simmons, "Mr. LeBar and World Health," *Public Health in the World Today* (1949).

7. Albert Cowdrey, *War and Healing: Stanhope Bayne-Jones and the Maturing of American Medicine* (Baton Rouge: Louisiana State University Press, 1992).

8. Gerald L. Geison, *The Private Science of Louis Pasteur* (Princeton University Press, 1995): 125–31.

9. Most Pasteur material comes from Geison, *Private Science*; Rene Dubos, *Louis Pasteur, Free Lance of Science* (New York: Little, Brown, 1950); and Arthur M. Silverstein, "Pasteur, Pastorians and the Dawn of Immunology: The Importance of Specificity," *History and Philosophy of the Life Sciences* 22 (2000): 29–41.

10. Burt Hansen, "New Images of a New Medicine: Visual Evidence for Widespread Popularity of Therapeutic Discoveries in America after 1885," *Bulletin of the History of Medicine* (December 1999): 647.

11. Zulma Steele, *Angel in Top Hat* (New York: Harper and Bros., 1942): 137–40.

12. Thomas H. Huxley, known as "Darwin's bulldog" for his fervent defense of evolution, came eloquently to Pasteur's defense. The opposition, he wrote, "proceeds partly from the fanatics of laissez-faire, who think it better to rot and die than to be kept whole and lively by State interference, partly from the blind opponents of properly conducted physiological experimentation, who prefer that men should suffer rather than rabbits or dogs, and partly from those who for other but not less powerful motives hate everything which contributes to prove the value of strictly scientific methods of inquiry in all those questions which affect the welfare of society."

13. Geison, *Private Science*, 218–20.

14. Wade Oliver, *The Man Who Lived for Tomorrow, a Biography of William Hallock Park* (New York: E. P. Dutton, 1941): 103; also Paul de Kruif, *Microbe Hunters* (New York: Harcourt Brace, 1926): 197–200.

15. Oliver, *Man Who*, 70.

16. Evelynn Maxine Hammonds, *Childhood's Deadly Scourge: The Campaign to Control Diphtheria in New York City, 1880–1930* (Baltimore: Johns Hopkins University Press, 1999): 169, 202–3; Oliver, *Man Who*, 459.

17. Oliver, *Man Who*, 106–25.

18. De Kruif, *Microbe Hunters*, 199; Hammonds, *Deadly Scourge*, 124–25.

19. Oliver, *Man Who*, 36.

20. George Bernard Shaw, "Quintessence of Ibsenism Now Completed" (Boston: B. R. Tucker, 1891).

21. Von Pirquet and Schick, *Serum Sickness*, 1905; quoted in Antoni Gronowicz, *Bela Schick and the World of Children* (New York: Abelard-Schuman, 1954): 64–69.

22. E. W. Goodall, "A Clinical Address on Serum Sickness," *The Lancet* (2 March 1918), and *The Lancet* (9 March 1918).

23. Barbara Guttman Rosenkrantz, *Public Health and the State: Changing Views in Massachusetts, 1842–1936* (Cambridge: Harvard University Press, 1972): 115–18.

24. Hammonds, *Deadly Scourge*, 189.

25. John Duffy, *The Sanitarians* (Champaign: University of Illinois Press, 1990): 196.

26. Cf. Anne B. Newton, "Must Children Have Children's Diseases?" *Ladies' Home Journal*, April 1910.

27. "A Plan to Make All Children Immune to All Acute Infections," *Current Opinion* (November 1916), 328.

28. Elaine Goodale Eastman, "The Waste of Life," *Popular Science Monthly* (August 1915): 187–94.

29. Duffy, *Sanitarians*, 249.

30. *The Rockefeller University, Achievements* (New York: Rockefeller University Press, 2000): 39, 148–49.

31. *Siberian Husky: A Brief History of the Breed in America*, retrieved from www.shca.org/shcahp2d.htm.

32. Hammonds, *Deadly Scourge*, 213.

33. Joseph McFarland, "The Beginning of Bacteriology in Philadelphia," *Bulletin of the History of Medicine* 5 (1935): 148–72.

34. On display at the New Orleans Historical Pharmaceutical Museum.

35. Paul de Kruif, *The Sweeping Wind, a Memoir* (New York: Harcourt, Brace & World, 1962): 30.

36. Rivers, *Reflections*, 170.

37. Duffy, *Sanitarians*, 264.

38. Charles F. McKhann, "The Prevention and Modification of Measles," *Journal of the American Medical Association* 109 (25) (1937): 2034–37.

39. Peter Aaby, "Severe Measles: A Reappraisal of the Role of Nutrition, Overcrowding and Virus Dose," *Medical Hypothesis* 18 (2) (1985): 93–112.

40. George Edgar Vincent, "Standards and Authority," *Journal of the American Medical Association* 56 (12) (1911): 894–96.

41. Susan Lederer, *Subjected to Science: Human Experimentation in America Before the Second World War* (Baltimore: Johns Hopkins Press, 1995): 126–31.

42. Paul Starr, *The Social Transformation of American Medicine* (New York: Basic Books, 1982): 160.

43. Quoted in Hammonds, *Deadly Scourge*, 15.

44. "Heath and National Security," speech at the Harvard School of Public

Health on 24 November 1944, in *Collected Papers of James Simmons*, National Library of Medicine collection.

45. Elizabeth Fenn, *Pox Americana: The Great Smallpox Epidemic of 1775–82* (New York: Hill & Wang, 2001): 35–50.

46. U.S. Department of the Army, Office of Surgeon General, *Medical and Surgical History of the Civil War* 6: 625–38.

47. Ibid., 5: 67; 6: 638–48.

48. William A. Green, "Vaccination and its Results," *Atlanta Medical and Surgical Journal* 8 (6) (1867): 241–46.

49. Mary Gillett, *The Army Medical Department, 1865–1917* (Washington: Center for Military History, 1995): 237–38.

50. Ibid., 258, 273, 301, 395.

51. Walter Reed, Victor C. Vaughan, and Edward O. Shakespeare, *Report on the Origin and Spread of Typhoid Fever in US Military Camps During the Spanish War of 1898* (Washington: GPO, 1900).

52. Most of this account comes from Leonard Colebrook, *Almroth Wright, Provocative Doctor and Thinker* (London: Heinemann, 1954); Anne Hardy, "Straight Back to Barbarism; Anti-typhoid Inoculation and the Great War, 1914," *Bulletin of the History of Medicine* 74 (2000): 265–90, and an exchange of letters between Wright and Pearson in the *British Medical Journal* (5 November 1904): 1243–46; (12 November): 1343–45; (19 November): 1432; (26 November): 1489–91; (3 December): 1542; (10 December): 1614; (24 December): 1727; (31 December): 1775–76).

53. Myron M. Levine, "Typhoid Fever Vaccines," *Vaccines*, 3rd ed. (Philadelphia: W. B. Saunders Co., 1999): 788.

54. Churchill, *London to Ladysmith via Pretoria*, 1900, quoted in Hardy, "Straight Back," 274–75.

55. Gillett, *Medical Department*, 348.

56. M.W. Ireland, "Communicable and Other Diseases," *History of the Medical Department of the U.S. Army in the World War*, IX (Washington: GPO, 1928): 45.

57. Allan Chase, *Magic Shots, a Human and Scientific Account of the Long and Continuing Struggle to Eradicate Infectious Diseases by Vaccination* (New York: Morrow, 1982): 68–69.

58. William Osler, *Bacilli and Bullets* (London: University of Oxford Press, 1914): 1–8.

59. At a November 1918 Fargo, North Dakota, gathering of the American Medical Freedom League, the group called "public attention to the usurpations of State Medicine, holding the same to be un-American, despotic and intolerable." *Vaccine Inquirer*, 11 November 1918.

60. Derek Linton, "Was Typhoid Inoculation Safe and Effective During Word War I?" *Journal of the History of Medicine* 55 (2000), 123–31.

61. Ireland, *History of the Medical*, 15–45, 314.

62. Victor Vaughan, *A Doctor's Memories* (Indianapolis: Bobbs-Merrill, 1926): 397–99.

63. In Willa Cather's novel of World War I, a shipboard physician is stunned that "vigorous, clean-blooded young fellows of nineteen and twenty turned over and died," and wonders whether "all these inoculations they've been having, against typhoid and smallpox and whatnot, haven't lowered their vitality." *One of Ours* (New York: Vintage, 1922): 264.

64. James H. Cassedy, *Charles Chapin and the Public Health Movement* (Cambridge: Harvard University Press, 1962): 191.

65. Jeffrey P. Baker and Samuel L. Katz, "Childhood Vaccine Development: An Overview," *Pediatric Research* 55 (2) (2004): 394; G. Feldberg, *Disease and Class: Tuberculosis and the Shaping of North American Society* (Piscataway: Rutgers University Press, 1995).

66. "The Present Status of Antityphoid Vaccination," *Journal of the American Medical Association* 56 (25) (1911); F. W. Hachtel and H. W. Stoner, "Inoculation Against Typhoid," *Journal of the American Medical Association* 59 (15) (1912): 1364–69.

67. Selwyn Collins, "History and Frequency of Typhoid Fever Immunizations and Cases in 9,000 Families," *Public Health Reports* 51 (1928): 897–926; "Typhoid in Large American Cities," *Journal of the American Medical Association* 74 (10) (1920): 673.

68. Michigan Department of Health, "Typhoid Fever Eradication, 1934," mimeographed report in National Library of Medicine collections, 5.

69. "General Simmons of Harvard dead," *The New York Times*, 2 August 1954.

70. J. S. Simmons, "Health, the Number One Freedom," *Southern Medicine and Surgery* 107 (1945): 4–11.

71. Interview with Monroe Eaton, May 1951, Lewis W. Hackett Notes for International Health Division history, V. III, 1117–18, in Folder 191, Box 3, RG 3, International Health Division, Rockefeller Archive Center.

72. Excerpt from Trustee Bulletin, December 1940; Folder 100, Box 11, Series 100, RG 1.1, International Health Division, Rockefeller Archive Center. For examples of correspondence see Influenza Vaccine—Distribution, January 20-June 18, 1941, Folder 164, Box 15, RG 5/4, and Influenza Vaccine—Inqiuiries 1936–1947, same folder.

73. Arthur P. Long, "The Army Immunization Program," *Preventive Medicine in World War II*, vol. 3, *Personal Health Measures and Immunization* (Washington: U.S. Army Medical Dept., 1955): 288–97.

74. Ibid., 328.

75. Ibid., 285.

76. Fred Soper, *Ventures in World Health: The Memoirs of Fred Soper*, ed. John Duffy (Washington: Pan American Health Organization, 1977): 260.

77. Albert Cowdry, *Fighting for Life: American Military Medicine in World War II* (New York: Free Press, 1994): 123.

78. Paul Weindling, "La Victoire par les Vaccins," in Anne-Marie Moulin, *L'Aventure de la Vaccination* (Paris: Librairie Atheme Fayard, 1996): 229–47.

79. Ibid., 246.

80. Herald R. Cox and John E. Bell, "Epidemic and Endemic Typhus," *Public Health Reports* 55 (110) (1940): 110–15.

81. Lewis Hackett note from interview with George Strode, Folder 86.101, Box 7h, Series 908, Rockefeller Foundation Archives, Rockefeller Archive Center.

82. Allen Raymond, "Now We Can Lick Typhus," *Saturday Evening Post*, 22 April 1944; Fred Soper et al., "Typhus Fever Control in Italy With Louse Powder," *American Journal of Hygiene* 45 (1947): 305–44.

83. Typewritten speech in folder "Talks, 1944," USA Typhoid Commission Papers, History of Medicine Division, National Library of Medicine.

84. P. O. Williams, "Brief Reviews," *Medical History* (1983).

85. Thomas Norton, mimeographed report, "The Foundation's Research Center in New York," 1946, in Folder 92, Box 11, Series 100, Record Group 1, Rockefeller Foundation, Rockefeller Archive Center.

86. Wilbur Sawyer, "Discourse on Receiving Richard Pearson Strong Medal, 7 April 1949, in folder 1939-51: "Articles, Speeches and Bibliography," Box 1, Wilbur Sawyer Collection, History of Medicine Division, National Library of Medicine.

87. Karl F. Meyer, Johannes Bauer, Wilbur A. Sawyer, et al., "The Recent Outbreak of Jaundice Among Soldiers in the Western United States," *American Journal of Hygiene* 40 (1944): 35–104.

88. Interview with Harry Burruss, 17 April 1986, Harry Burruss Collection, Folder 3, p. 5, History of Medicine Division, National Library of Medicine.

89. John P. Fox et al., "Observations on the Occurrence of Icterus in Brazil Following Vaccination Against Yellow Fever," *American Journal of Hygiene* 36 (2) (1942): 104.

90. BJ to Bauer, 7 April 1943, Folder 142, Box 17, Series 100, RG 1, Rockefeller Foundation Archive, Rockefeller Archive Center.

91. Notes on Wilbur A. Sawyer, History, Greer Williams, Box 7h, Series 908, RG3; Greer Williams Notes, Infectious Hepatitis-Jaundice in Army. Etc., folder 86.18, Box 7b Series 908, RG 3, Rockefeller Foundation, Rockefeller Archive Center.

92. Hacket to Bayne-Jones; Hackett to Soper, Folder 18.2, Box 2, Series 908, RG 3, Rockefeller Foundation Archive, Rockefeller Archive Center.

93. George Merck, "Bioterrorism," *Military Surgeon* (March 1946): 237–43.

94. Fox in Ed Regis, *Biology of Doom* (New York: Henry Holt, 1999): 9–11.

95. Statement of J. H. Bauer, Folder 397, Box 36, Series 4, International Health Bureau, RG 5, Rockefeller Institute Virus Laboratory, Rockefeller Archive Center.

96. Bauer to Sawyer, 11 April 1942, Folder 131, Box 16, Series 100, RG 1, Rockefeller Foundation Archives, Rockefeller Archive Center.

97. Cowdrey, *War and Healing*, 144–45.

98. John R. Paul and Horace T. Gardner, "Viral Hepatitis" in *Medical History of World War II*, vol. 5, ed. E. C. and P. M. Huff et al. (Washington: GPO, 1960): 422–23.

99. "Timeline" Folder Jaundice—IHD, 1941-2, Box 1120 HD 710, RG 90, National Archives.

100. Turner to Bauer, 30 July 1940, Folder HD 710 (Virus Disease) Jaundice. Blood Serum in YF Vaccine. Correspondence and Russian Manuscript, WWII Administrative Records—entry 31 (ZI), Box 1132, RG 112, National Archives.

101. Personal journal, January–December 1941, Box 10, Fred Soper Papers. History of Medicine Division, National Library of Medicine.

102. Kenneth Goodner, the New York lab chief, had tried to substitute chick embryo juice for human serum but hadn't been able to produce good vaccine. Available military records lack evidence of whether Simmons and Bayne-Jones were fully aware of the risks of the serum-containing vaccine. One assumes they were; Bayne-Jones had been dean of Yale Medical School.

103. Sawyer to Soper, 6 December 1941, Folder 129, Box 16, Series 100, RG 1, Rockefeller Foundation Archive, Rockefeller Archive Center.

104. John Farley, *To Cast Out Disease: A History of the International Health Division of the Rockefeller Foundation* (Oxford University Press, 2004): 173.

105. "Greer Williams Notes, Infectious Hepatitis," Folder 86.18, Box 7b, Series 908, RG 3, Rockefeller Foundation Archives, Rockefeller Archive Center.

106. Lawrence K. Altman, "Karl Meyer, Viral Scientist, Dies," *The New York Times*, 29 April 1974.

107. Karl Meyer oral history, HMD Reading Room, National Library of Medicine.

108. Farley, *Cast Out*, 178.

109. Leonard Seeff, author interview, June 2005.

110. "Hawley" to Maj. Gen. James C. Magee, Surgeon General, U.S. Army, 17 July 1942, Folder HD 710 (Virus Disease) Jaundice. Infectious Hepatitis or Icterus, Incidence in ETO in 1942 to 1944. Entry 31 (ZI), Box 1118, RG 112, National Archives.

111. Cowley, *War and Healing*, 145.

112. Philip Minor, remarks at FDA conference, "Evolving Perspectives in Cell Substrates for Vaccine Development," 10 September 1999. Also see Rudi Schmid, "History of Viral Hepatitis: A Tale of Dogma and Misinterpre-

tation," *Journal of Gastroenterology and Hepatology* 16 (7) (2001): 718. Minor states that he heard the story from a late Wellcome-Burroughs executive.

113. "Diary of Dr. J. H. Bauer," Folder 146, Jaundice-Reports, 1942, Box 17, Record Group 1, Rockefeller Archive Center.

114. Meyer to Hackett, 18 March 1960, Folder 86, series 908; J. H. Bauer diary entry, Folder 146, "Jaundice-Reports, 1942," Box 17, Series 100, RG 1. Rockefeller Foundation Archive, Rockefeller Archive Center.

115. Sawyer to Strode, 23 March 1942, Folder 130, Box 16, Series 100, RG 1. Rockefeller Foundation Archive, Rockefeller Archive Center.

116. Bauer to Sawyer, 11 April 1942, Folder 104, Box 12, Series 100, RG 1.1. Rockefeller Foundation Archives, Rockefeller Archive Center.

117. Porter Crawford to Sawyer, 9 September 1942, Folder 138, "Jaundice, Sept–Oct 1942," Box 16, Series 100, RG 1, Rockefeller Foundation Records, Rockefeller Archive Center.

118. Bayne-Jones handwritten notes from a 17 June 1942 meeting, entry 31(ZI) Box 1120, HD 710 (Virus Disease) Jaundice. Infectious Hepatitis or Icterus, IHD 1941 and 1942; Box 1118, HD710, World War II Administrative Records, RG 112, National Archives.

119. Interview with Mason Hargett by Victoria Harden, 1985, Folder 2, Box 1, Mason Hargett Collection. History of Medicine Division, National Library of Medicine, 6–14.

120. Harry Burruss, *Oral History*, 16–17.

121. An estimate of 330,000 people infected by the vaccine comes from James E. Norman and Leonard B. Seeff, et al., "Mortality Follow-Up of the 1942 Epidemic of Hepatitis B in the U.S. Army," *Hepatology* 18 (4) (1993): 790–97.

122. Bayne-Jones to Sawyer, 25 February 1943, Folder 141, "Jaundice," Box 17, Series 100, RG 1, and Memo, J. H. Bauer, 30 October 1942, Folder 138, "Jaundice Sept–Oct 1942," Box 16, Series 100, RG 1, Rockefeller Foundation Archives, Rockefeller Archive Center. Also "Analysis of Yellow Fever Serum Donors," Folder HD 710 (Virus Disease etc. . . .), Box 1118, and "Analysis of Yellow Fever Serum Donors, 1941–43," Box 1120. Entry 31 (ZI), RG 112, WWII Administrative Records, National Archives.

123. Stanhope Bayne Jones, "Vaccination Against Yellow Fever," in *Preventive Medicine in World War II*, vol. 3, 361; and Wilbur Sawyer et al., "Jaundice in Army Personnel in the Western Region of the U.S.," *American Journal of Hygiene* 40 (1944): 37–104.

124. Paul and Gardner, "Viral Hepatitis," 452.

125. John A. Rogers, "The Outbreak of Jaundice in the Army," SGO Circular #95, 31 August 1942, Folder HD 710 (Virus Disease) Jaundice. Infectious Hepatitis or Icterus, Incidences and Control 1946-47, Box 1118, entry 31 (ZI) RG 112, WWII Administrative Records, National Archives.

126. Morris Fischbein, "Jaundice in the Armed Forces," *Journal of the American Medical Association* 120 (1) (1942).

127. Soper to Hackett, 23 March 1960, Folder 86.18, Box 7b Series 908, RG 3, Rockefeller Foundation, Rockefeller Archive Center.

128. Greer Williams, *Plague Fighters* (New York: Scribners, 1969): 322. In his 5 January 1948 letter to the Nobel Committee for Physiology and Medicine, Albert Sabin nominated Theiler for his 1930–1937 work on the yellow fever vaccine in particular and for furthering viral research in general with his work with tissue cultures and his development of the mouse as a viral test animal. Sabin Archive, Medical Heritage Library, University of Cincinnati.

129. Leonard Seeff, et al., "A Serologic Followup of the 1942 Epidemic of Post-Vaccination Hepatitis in the U.S. Army," *New England Journal of Medicine* 316 (16) (1987): 965–70.

130. Ross L. Gauld, "Infectious Hepatitis in Germany," 15 December 1946 report, Folder HD 710 (Virus Disease) Jaundice. Infectious Hepatitis or Icterus, Incidences and Control 1946-47. Box 1118, entry 31(ZI), RG 112, WWII Administrative Records, National Archives.

131. Tom Rivers, *Reflections on a Life in Medicine and Science* (Cambridge: MIT Press, 1967): 498–99. Paul A. Lombardo, "'Of the Utmost National Urgency': The Lynchburg Colony Hepatitis Study, 1942," *In the Wake of Terror: Medicine and Morality in a Time of Crisis* (Cambridge: MIT Press, 2003): 3–15.

132. Baruch S. Blumberg, *Hepatitis B: The Hunt for a Killer Virus* (Princeton: Princeton University Press, 2002): 129.

133. See Hargett and Harry Burruss oral histories; Jeff Widmer, *Spirit of Swiftwater, 100 Years at the Pocono Labs* (Swiftwater, PA: Connaught Laboratories, 1997): 43.

134. J. S. Simmons, "Public Health as a Weapon for National Defense," *Transactions and Studies of the College of Physicians of Philadelphia* 19 (1) (April 1951): 1–13.

135. Simmons, "Health, the Number One Freedom," 7.

CHAPTER 5

1. John Duffy, *The Sanitarians* (Champaign: University of Illinois Press, 1990): 270–79.

2. A. S. Goldman et al., "What Was the Cause of Franklin Delano Roosevelt's Paralytic Illness?" *Journal of Medical Biography* 11 (4) (November 2003): 232–40.

3. Hugh Gallagher, author interview, 5 May 2004.

4. Jane S. Smith, *Patenting the Sun: Polio and the Salk Vaccine* (New York: W. Morrow, 1990): 59.

5. Gallagher interview.

6. Franklin D. Roosevelt, *Nothing to Fear: The Selected Addresses of Franklin Delano Roosevelt, 1932–45* (Boston: Houghton-Mifflin, 1946): 68.

7. Paul de Kruif, *The Sweeping Wind, a Memoir* (New York: Harcourt Brace and World, 1962): 177–79; see also Correspondence, 1936, De Kruif–Stokes, Joseph P. Stokes Collection, American Philosophical Society.

8. Gallagher interview.

9. Natalie Rogers, *Dirt and Disease: Polio before FDR* (Piscataway: Rutgers University Press, 1992): 13.

10. Ibid., 20–64.

11. Ibid., 106–37.

12. Ibid., 123–32.

13. John R. Wilson, *Margin of Safety: The Story of the Poliomyelitis Vaccine* (London: Collins, 1963): 42–43.

14. Kathryn Black, *In the Shadow of Polio: A Personal and Social History* (Boston: Addison-Wesley, 1996): 63–64.

15. Ibid., 60–65.

16. "What Ever Happened to Polio?" Exhibit at American Museum of Natural History, Washington, DC, 2005.

17. Albert Weihl, author interview, February 2005.

18. Tom Rivers, *Reflections on a Life in Medicine and Science* (Cambridge: MIT Press, 1967): 237.

19. "New Weapon Against Infantile Paralysis," *Literary Digest*, 1 September 1934; "New Vaccine Promises Conquest of Polio, *Literary Digest*, 21 July 1934.

20. At least according to Rivers, *Reflections*, 189–90.

21. "Dr. Brodie, Expert in Serum Research," *The New York Times*, 12 May 1939. This obituary gives no cause of death and does not mention the vaccine fiasco.

22. De Kruif to Stokes, 18 May 1936, Correspondence—de Kruif, Joseph Stokes Collection, American Philosophical Society.

23. Rivers, *Reflections*.

24. Greer Williams, *Virus Hunters* (New York: Knopf, 1959): 268.

25. Wilson, *Margin of Safety*, 147.

26. Maurice Hilleman, author interview, January 2004.

27. Albert Sabin, "St. Louis and Japanese B Types of Epidemic Encephalitis: Development of Non-Infective Vaccines," *Journal of the American Medical Association* 122 (1943): 477–86.

28. Smadel's laboratory later modified the Waring blender to avoid the escape of viral particles, see Smadel, "The Hazards of Acquiring Virus and Rickettsial Diseases in the Laboratory," *American Journal of Public Health* 41 (1951): 788–95.

29. Allan B. Brandt, "Polio, Politics, Publicity and Duplicity: Ethical Aspects in the Development of the Salk Vaccine," *International Journal of Health Services* 8 (2) (1978): 257–70.

30. Black, *In the Shadow*, 77.

31. Richard Carter, *Breakthrough: The Saga of Jonas Salk* (New York: Trident Press, 1966): 102.
32. Williams, *Virus Hunters* 239–43.
33. The description of Enders comes from author interviews with Maurice Hilleman, Sam Katz, and Gilbert Schiff, as well as contemporary literature.
34. Williams, *Virus Hunters*, 252–53.
35. Sam Katz, author interview, February 2004.
36. Williams, *Virus Hunters*, 268.
37. David M. Oshinsky, *Polio: An American Story* (Oxford University Press, 2004): 189.
38. Carter, *Breakthrough*, 69.
39. Julius Youngner, author interview, September 2004.
40. Carter, *Breakthrough*, 115–19.
41. Oshinsky, *American Story*, 100–4, 146–50.
42. Carter, *Breakthrough*, 118–19.
43. Oshinsky, *American Story*, 139.
44. John R. Paul, *History of Poliomyelitis* (New Haven: Yale University Press, 1971): 445.
45. Wilson, *Margin*, 135.
46. The accounts of Sabin's personal difficulties come from author interviews with Katz, Hilleman, Schiff, and two neighbors in Cincinnati.
47. Edwin Lennette, Oral History, History of Medicine Division, National Library of Medicine, 90.
48. Ibid., 285–86.
49. Robert M. Chanock, "Reminiscences of Albert Sabin and his Successful Strategy for the Development of the Live Oral Poliovirus Vaccine," *Proceedings of the Association of American Physicians* 108 (March 1996): 118.
50. Carter, *Breakthrough*, 171.
51. Rivers, *Reflections*, 466–67.
52. "Poliomyelitis: A New Approach," *Lancet* 259(6707) (1953): 552.
53. See Edward Hooper, *The River* (New York: Little Brown, 1999), an enormous book that examines Hooper's theory that Koprowski's 1957 oral polio vaccine trial in Congo spread HIV-AIDS from chimpanzees to humans. Other controversies are outlined in Koprowski's authorized biography, Roger Vaughan, *Listen to the Music: the Life of Hilary Koprowski* (New York: Springer Verlag, 2000).
54. Rivers, *Reflections*, 234.
55. Smith, *Patenting*, 123.
56. Robert Hull, author interview, August 2004.
57. Carter, *Breakthrough*, 131.
58. Ibid., 135.

59. O'Connor liked Hershey, and it was a good midway point for the polio scientists.
60. Paul, *History of Polio*, 418.
61. Carter, *Breakthrough*, 194.
62. Ibid., 154.
63. Paul Offit, *The Cutter Incident* (New Haven: Yale University Press, 2005): 43.
64. Karl Meyer, Oral History, History of Medicine Division, National Library of Medicine, 190–96.
65. Williams, *Virus Hunters*, 292–96.
66. Carter, *Breakthrough*, 105.
67. Thomas H. Weller, *Growing Pathogens in Tissue Cultures* (Sagamore Beach, MA: Science History Publications, 2004): 81–83.
68. Smith, *Patenting*, 243–45.
69. See "Bradford Hill and the Randomized Controlled Trial," *Pharmaceutical Medicine* 6 (1992): 23–66, for a complete account of the 1948 trial of streptomycin against tuberculosis. Bradford Hill also used randomizing methods for a pertussis vaccine trial during World War II.
70. James Shannon, Oral History, History of Medicine Division, National Library of Medicine, 94–95.
71. Some (including Wyeth's Alan Bernstein) have hypothesized that the addition of thimerosal may, ironically, have saved Salk's vaccine. Thimerosal reduced the immunogenicity of the polio antigens in the vaccine trial lots—*it may also have killed live polio contained in the lots.*
72. Carter, *Breakthrough*, 223.
73. Warren Winkelstein, author interview, April 2003.
74. "Activities of the Department of Health, Education, and Welfare relating to polio vaccine," *15th Report of the Committee on Government Operations* (18 August 1957): 7–19. Eli Lilly, which produced two-thirds of the Salk vaccine, cleared $20 million in profit on it in the first six months of 1956.
75. Offit, *Cutter*, 60.
76. Carter, *Breakthrough*, 271; Rivers, *Reflections* 549–50.
77. Oshinsky, *American Story*, 211–12.
78. Offit, *Cutter*, 63–65.
79. Winkelstein interview.
80. "Miracle of Modern Medicine Failed a Little Girl," *Seattle Post-Intelligencer*, 22 August 1955, 1.
81. Neal Nathanson and Alex Langmuir, "The Cutter Incident. Poliomyelitis following Formaldehyde-Inactivated Poliovirus Vaccination in the United States during the Spring of 1955. *American Journal of Epidemiology* 142 (2) (1995) (reprint): 108–40.
82. Offit, *Cutter*, 89.

83. "Parents Reaffirm Faith in Vaccine After Girl Dies," *Philadelphia Inquirer*, 28 May 1955, 1.

84. Author interviews with Neal Nathanson, Paul Offit, and Donald A. Henderson.

85. *Technical Report on the Salk Poliomyelitis Vaccine* (Washington: U.S. Public Health Service, 1955): 23–46.

86. Maryn McKenna, *Beating Back the Devil* (New York: Free Press, 2004): 11–20.

87. Neal Nathanson, author interview, February 2004.

88. Julius Youngner, author interview, September 2004; *Technical Report*, 45–66; *Report on Poliomyelitis Vaccine Produced by the Cutter Laboratory* (U.S. Public Health Service: August 25, 1955): 6.

89. Shannon, Oral History, 94–95.

90. Weller, *Growing Pathogens*, 84–89.

91. "17,000 Pupils Drop Out of Vaccine Program," *Philadelphia Inquirer*, 9 June 1955.

92. William Foege, author interview, June 2004.

93. Carter, *Breakthrough*, 311.

94. Graham Wilson, *Hazards of Immunization* (London: Athlone Press, 1967): 124–25.

95. J. S. Wilson, *Margin*, 131–32.

96. Foege interview.

97. Wilson, *Margin*, 314–15.

98. Salk noted during the March 1961 hearing (U.S. House of Representatives, Subcommittee of the Committee on Interstate and Foreign Commerce, "Polio Vaccines," 16–17 March 1961, 303) that a large pocket of French Canadians around Montreal had resisted vaccination, as they had during the smallpox epidemic of 1888 (see Michael Bliss, *Plague: A Story of Smallpox in Montreal* [New York: HarperCollins, 1991]). As a result, Montreal suffered a major polio outbreak in 1959.

99. *Public Acceptance for the Salk Vaccine Program*. American Institute of Public Opinion. February 1957. In collection of the National Library of Medicine.

100. Oshinsky, *American Story*, 250–52.

101. Wilson, *Margin*, 211–19.

102. For example, see Geor ge Dick, "The Whooping Cough Vaccine Controversies," in *New Trends and Developments in Vaccines*, ed. A. Voller and H. Friedman (Baltimore: University Park Press, 1978): 184–85.

103. Hilary Koprowski, "Poliomyelitis: Visit to Ancient History," in *Microbe Hunters, Then and Now*, ed. Hilary Kropowski and Michael B. A. Oldstone (Lansing, MI: Medi-Ed Press, 1996): 141.

104. Hilleman to Stokes, folder "Polio, 1955-56 #1," Joseph Stokes Collection, American Philosophical Society.

105. House subcommittee hearing, March 1961, 263–66.

106. Ibid., 279–83, 304, 315.

107. Gilbert Schiff, author interview, September 2004.

108. House subcommittee hearing, March 1961, 255.

109. Ibid., 307.

110. Carter, *Breakthrough*, 356; Salk vaccine was not the only loser, so were Koprowski and Cox. Though the three oral vaccines were similar, tests at the NIH and by Joseph Melnick's laboratory in Houston found the Cox and Koprowski vaccines slightly more likely to revert to virulence after passing through children's intestinal tracts.

111. Luther Terry, *The Association of Cases of Poliomyelitis with the Use of Type III Oral Poliomyelitis Vaccines* (Washington: Public Health Service, 20 September 1962).

112. D. A. Henderson, author interview, June 2005.

113. Wilson, *Margin*, 26.

114. Bernice Eddy interview with Edward Shorter, 1986, on cassette at History of Medicine Division, National Library of Medicine.

115. Maurice Hilleman, author interview, January 2004.

116. Maurice Hilleman, "Discovery of Simian Virus 40 and Its Relationship to Poliomyelitis Virus Vaccines," *Developments in Biological Standardization* 94 (1998): 183–90.

117. Hilleman interview with Edward Shorter, 1987, on cassette at History of Medicine Division, National Library of Medicine.

118. House subcommittee hearing, March 1961, 311–12. At Koprowski's Wistar Institute, Leonard Hayflick had developed an eternal culture of human cells, derived from an aborted fetus, the Wi-38 cell line. After years of resistance by federal vaccine regulators, Plotkin's rubella vaccine, which used the cell line, was licensed in 1979 and incorporated into Merck's MMR shot.

119. Louis Galambos and Jane E. Sewell, *Networks of Innovation: Vaccine Development at Merck, Sharpe & Dohme, and Mulford, 1895–1995* (New York: Cambridge University Press, 1995): 60–61.

120. Keerti Shah, "Simian Virus 40 and Human Disease," *Journal of Infectious Diseases* 190 (15 December 2004): 2061–63.

121. SV-40 is the subject of Deborah Bookchin and Jim Schumacher, *The Virus and the Vaccine* (New York: St. Martin's Press, 2004).

122. "Salk Foresees Cooperation Era," *The New York Times*, 3 November 1970, 17.

123. "Prospects for Peace between the Arab Countries and Israel," Sabin Archive, Medical Heritage Library, University of Cincinnati.

124. Black, *Shadow of Polio*, 371.

CHAPTER 6

1. Proceedings of the conference were published in the *American Journal of Diseases of Childhood* 103 (March 1962). References here are from pages 219–392.

2. *New England Journal of Medicine* 263 (4): 153–84.

3. Cheap and effective antibiotics led pharmaceutical companies to abandon antibacterial vaccines in the 1960s and 1970s. A classic example is the orphaned pneumococcal vaccine that Robert Austrian of the University of Pennsylvania had worked on for decades.

4. Neal Halsey, Mark Papania, and Peter Strebel, "Measles Vaccines," in *Vaccines*, 4th ed. (Philadelphia: Saunders, 2004): 389–441.

5. Alexander Langmuir, "Medical Importance of Measles," *American Journal of Diseases of Childhood* 103 (1962): 224–26.

6. Tom Rivers, *Reflections on a Life in Medicine and Science* (Cambridge: MIT Press, 1967): 396.

7. Greer Williams, *Virus Hunters* (New York: Knopf, 1959): 399.

8. See Joel Warren, "Industrial Production of Primary Tissue Cultures," in *Cell Cultures for Virus Vaccine Production*, NCI Monograph 29 (1968): 35–43.

9. Sam Katz, author interview, February 2004.

10. "The Dr. Maurice Hilleman Symposium August 30, 1999 in Honor of his 80th Birthday," *Journal of Human Virology* 3 (2) (2000): 60–61.

11. Paul Offit, author interview, April 2005.

12. U.S. Dept of the Army Bureau of Public Relations. "Biological Warfare: Report to the Secretary of War by George Merck," 237–43.

13. Maurice Hilleman, author interview, January 2004.

14. Notes, "George Merck, June 1955," Vannevar Bush Collection, Library of Congress.

15. Hilleman, author interview. Louis Galambos and Jane E. Sewell, *Networks of Innovation: Vaccine Development at Merck, Sharpe & Dohme, and Mulford, 1895–1995* (New York: Cambridge University Press, 1995): 57–64; Dickinson Richards to Max Peters, 28 August 1969, Sabin Archive, Medical Heritage Library, University of Cincinnati.

16. Galambos and Sewell, *Networks of Innovation*, 62

17. I interviewed Hilleman twice in 2004.

18. Gilbert Schiff, author interview. September 2004.

19. Hilleman enjoyed telling a story about the smallpox division at Squibb, which consisted of a veterinarian, an assistant, and many calves. At one point the company sent a time-and-motion engineer to see if the company could dispense with the assistant, who among other things scraped up the cow manure. With relish, Hilleman described the Taylorist's comeuppance: "The guy was picking his nose and got a primary take," he laughed, holding his hand six inches from his face. "His nose was out to here!"

20. See http://poisonevercure.150m.com/diseases/inside_scoop_on_the.htm.

21. Offit interview.

22. Press release in "Measles #4" folder, Joseph Stokes Collection, American Philosophical Society.

23. Robert Chanock, author interview, February 2004. Hilleman claimed that Schwarz and other scientists in Sabin's laboratory were brought there from Germany as part of Operation Paperclip, which put Nazi scientists in America. I could not corroborate Hilleman's statement. German AIDS researcher Manreid Koch, who worked for Sabin as a young man and later researched the activities of Nazi biologists, says Hilleman's claim is untrue.

24. Schiff interview.

25. Susan Dandoy, "Measles Epidemiology and Vaccine Use in Los Angeles County, 1963 to 1966," *Public Health Reports* 82 (8) (1967): 659–66.

26. "Cuba Denies Role," *The New York Times*, 6 May 1960, 3.

27. Cf. Hilleman, confidential 1 July 1959 summary of a WHO conference on live polio vaccine; in "Polio #1, 1955–56" folder, Joseph Stokes Collection, American Philosophical Society.

28. U.S. House, *Intensive Immunizations Programs: Hearing before Committee on Interstate and Foreign Commerce on HR 10541*, 15–16 May 1962, 45–52.

29. Ibid., 77–78.

30. Ibid., 74.

31. Although the NIH established multiyear contracts with firms such as Lilly, Pfizer, and Cutter Laboratories, little came of these efforts.

32. "U.S. Health Goal: 75-Year Life Span," *The New York Times*, 6 August 1965, 13.

33. "U.S. Plans Drive to End Measles," *The New York Times*, 6 March 1967, 43.

34. U.S. House, *Intensive Immunization Programs*, 80–87.

35. Edward A. Mortimer et al. "Long-term Follow-up of Persons Inadvertently Inoculated with SV40 as Neonates," *New England Journal of Medicine* 305 (25) (1981): 1517–18.

36. Schiff interview.

37. *American Journal of the Diseases of Childhood* 109 (1965): 232–37.

38. C. Henry Kempe and Vincent Fulginiti, "Altered Reactivity to Measles Virus," *Journal of the American Medical Association* 202 (12) (1967): 1075–80; Richard G. Lennon and Warren Winkelstein Jr., "Skin Tests with Measles and Poliomyelitis Vaccines in Recipients of Inactivated Measles Virus Vaccine," *Journal of the American Medical Association* 200 (4) (1967): 275–80.

39. Franklin White, "Prior use of Killed Vaccine as a Factor in Measles Incidence in Canada, *International Journal of Epidemiology* 28 (1999): 1185.

40. Jeffrey Koplan and Alan Hinman, "Public Health Policy Toward Atypical Measles Syndrome in the United States," *Medical Decision Making* 2 (1) (1982): 71–77.

41. E. M. Nichols, "Atypical Measles, Syndrome, a Continuing Problem," *American Journal of Public Health* 69 (2) (1979).

42. D. Annunziato et al., "Atypical Measles Syndrome, Pathologic and Serologic Findings," *Pediatrics* 70 (2) (1982). An even graver disaster was narrowly avoided in the late 1960s, when Fulginiti and others pulled the

plug on an inactivated respiratory synctial virus vaccine. Children vaccinated with the experimental FI-RSV vaccine remained susceptible to wild
virus, and once infected suffered severe inflammations and dangerous
constriction of the lungs.

43. Vincent Fulginiti, author interview, May 2004.

44. Warren to Sabin, File 14, Box 22, Sabin Archive, Medical Heritage
Library, University of Cincinnati.

45. Huntley Collins, "The Man Who Saved Your Life," *Philadelphia Inquirer
Magazine*, 29 August 1999.

46. Galambos and Sewell, *Networks of Innovation*, 101–3.

47. In an ironic twist to the Merck-Schwarz confrontation, the tides were
turned on Hilleman when SmithKline used some clever microbiological
parsing to create its own vaccine from the Jeryl Lynn virus. The
SmithKline scientists cultured a vial of Merck's vaccine and separated out
two strains of mumps that had been present in Ms. Hilleman's throat.
SmithKline then took one of the strains and patented it.

48. Lou Cooper, author interview, May 2004.

49. Louis Z. Cooper, "The History and Medical Consequences of Rubella,"
Review of Infectious Diseases 7 (supplement) (1985): S2–10.

50. Thomas H. Weller, *Growing Pathogens in Tissue Cultures* (Sagamore Beach,
MA: Science History Publications, 2004): 139–49.

51. Sabin played a key role by suggesting to Weller that he add rabbit serum
to his neutralizing mixture, which produced antibody complexes that confirmed the presence of the virus.

52. Paul Parkman, author interview, May 2004.

53. "German Measles Epidemic," *Time*, 24 April 1964, 42.

54. Stanley Plotkin, "Birth and Death of Congenital Rubella Syndrome,"
Journal of the American Medical Association 251 (15) (1984): 2003–4.

55. Walter Orenstein and Barry Sirotkin, "The Opportunity and Obligation
to Eliminate Rubella from the Unitd States." *Journal of the American
Medical Association* 251 (15): 1988—94.

56. Stella S. Chess, "Autism in Children with Congenital Rubella," *Journal of
Autism and Childhood Schizophrenia* 1 (1) (1971): 33–47.

57. Louis Z. Cooper, author interview, March 2004.

58. Cf. Galambos and Sewell, *Networks of Innovation*, 105–7.

59. Cf. Jon Nordheimer, "Measles Outbreaks Fuel Vaccine Debate," *The New
York Times*, 16 May 1977.

60. As we will learn in chapter 9, pertussis vaccination seems to wear off after a
few years, leaving older children and adults susceptible. The disease is less
serious in older people, but they can still spread the germ to infants, whom
pertussis can kill or maim.

61. Alan Hinman, author interview, May 2004; Jane E. Brody, "City's Rubella

Drive on TV 'Sells' Children on Need for Shots," *The New York Times*, 14 June 1970; Paul Parkman, "Making Vaccination Policy, the Experience with Rubella," *Clinical Infectious Disease* 28 (suppl 2) (1999): S140–46.

62. Paul Parkman, "Making Vaccination Policy, the Experience of Rubella," *Clinical Infectious Diseases* 28 (suppl 2) (1999): S140–6.

63. Janet Englund, "In Search of a Vaccine for Respiratory Syncytial Virus: The Saga Continues," *Journal of Infectious Diseases* 191 (7) (2005): 1036–39.

64. Harold M. Schmeck, "In Health, the Accent Switches to Prevention," *The New York Times*, 12 January 1970; "Measles Have Just About Had It," *The New York Times*, 26 March 1967.

65. Ruth Kempe, H. K. Silver, V. Fulginiti et al., "C. Henry Kempe, MD-Physician, Investigator, Mentor and Humanist," *American Journal of the Diseases of Childhood* 138 (3) (1984): 223–32.

66. Kempe's curriculum vitae provided by Anne Kempe.

67. C. Henry Kempe, "Acceptance of the Howland Award," *Pediatric Research* 14 (1980): 1155–61.

68. "Notes & Discussion," *Proceedings of the American Pediatrics Society* 5 pt. 2 (1967): 1017–22.

69. Maurice Hilleman, "Toward Control of Viral Infections of Man," *Science* 164 (879) (1969): 506–14.

70. William Foege, author interview, June 2004.

71. Cyril M. Dixon, *Smallpox* (London: Churchill, 1962): 402.

72. Thomas J. Halpin et al., "Measles in Ohio—Update," *Ohio State Medical Journal* 73 (3) (1977): 147–49.

73. Discussed in Walter Orenstein, Alan Hinman, and Lance Rodewald, "Public Health Considerations in the United States," in *Vaccines*, 4th ed. 1011.

74. "Back to Quarantine?" *Newsweek*, 28 October 1974, 107.

75. Charles L. Jackson, "State Laws on Comuplsory Immmunization in the United States," *Public Health Reports* 84 (9) (1969): 787–95.

76. "Report and Recommendations of the National Immunization Work Group on Consent," Annex A, 1977, in collection of the National Institutes of Health Library; Nordheimer, "Measles Outbreaks."

77. Dandoy, "Measles Epidemiology," 659–66.

78. Peter Isaacson, "Allergic Reactions Associated with Viral Vaccines," *Progress in Medical Virology* 13 (1974): 240.

79. James D. Cherry et al., "Urban Measles in the Vaccine Era: A Clinical, Epidemiologic, and Serologic Study," *Journal of Pediatrics* 81 (2) (1972): 217–30.

80. Alan Hinman and Walter Orenstein, "The Immunization System in the United States—the Role of School Immunization Laws," *Vaccine* 29 (17, Supplement 3) (1999): S19–24.

81. Alan Hinman, "The New U.S. Initiative in Childhood Immunization," *Bulletin of the Pan-American Health Organization* 13(2) (1979): 169–76.

82. Joseph Califano, *Governing America* (New York: Simon & Schuster, 1981): 172–81.

83. Hinman, "U.S. Initiative," 171.

84. Rudy Johnson, "Patterson Fights Rise in Measles," *The New York Times*, 27 December 1973, 78.

85. Foege interview; also Plotkin, *Vaccines*, 1012.

86. Orenstein and Hinman, "Immunization System," S19–24.

87. Califano, *Governing*, 180.

CHAPTER 7

1. Jeffrey Shwartz, author interview April 2004.

2. Barbara Loe Fisher and Kathi Williams, author interview, June 2004.

3. Schwartz, author interview, September 2004.

4. Vincent Fulginiti, "Controversies in Current Immunization Policy and Practices: One Physician's Viewpoint," *Current Problems in Pediatrics* 6(6) (1976): 4.

5. U.S. Senate, Committee on Government Operations, *Consumer Safety Act of 1972: Hearing before the Subcommittee on Executive Reorganization and Government Research*, 92nd Congress, 20, 21 April and 3, 4 May 1972 (Washington: GPO, 1972): 487.

6. Nicholas Wade, "Division of Biologics Standards: In the Matter of J. Anthony Morris," *Science* 175 (24) (1972): 861–66; "Division of Biologic Standards: Scientific Management Questioned" *Science* 175 (25) (1972): 966–69; "DBS: The Boat that Never Rocked," *Science* 175 (27) (1972): 1225–30; "DBS: Agency Contravenes Its Own Regulations," *Science* 176 (1) (1972): 34–35.

7. Wade, "DBS Agency Contravenes," 34; U.S. Senate, Comptroller General of the United States, *Problems Involving the Effectiveness of Vaccines: Report to the Subcommittee on Executive Reorganization of Committee on Government Relations* (Washington: GPO: 1972); U.S. Senate Hearing, *Consumer Safety*, 435, 469–71.

8. *The Virus and the Vaccine* portrays Morris as a heroic figure; Parkman, Hilleman, Chanock, and others I spoke with were sympathetic during Morris's confrontation with Murray but later concluded that he was unmanageable and performed unacceptable work. Morris did not respond to interview requests.

9. Morris was reinstated under Harry Meyer, only to turn on his new boss in 1975 with another grievance procedure. Scientists who took part in the second review of Morris's work determined that it was subpar.

10. M. Mitka, "1918 Killer Flu Virus Reconstructed, May Help Prevent Future Outbreaks," *Journal of the American Medical Association* 294 (19) (2005): 2416–19.

11. Richard E. Neustadt and Harvey Fineberg, *The Swine Flu Affair* (Washington: Department of Health, Education and Welfare, 1978): 57.

12. Therman E. Evans, "Blacks' Fear of the Swine-Flu Shots," *The Washington Post*, 26 November 1976.

13. Richard Cohen, "Science as Fiction Makes Skeptical Fan," *The Washington Post*, 14 November 1976.

14. Cf. Victor Cohn, "All Vaccine Drives Falter," *The Washington Post*, 13 November 1976; "Washington and Business: Vaccine Puts Focus on Liability Problem," *The New York Times*, 14 October 1976; Stuart Auerbach, "States Low in Polio Vaccine," *The Washington Post*, 14 September 1976.

15. Philip M. Boffey, "Soft Evidence and Hard Sell," *The New York Times Magazine*, 15 September 1976; Bayard Webster, "Man in the Middle of Flu Vaccine Program," *The New York Times*, 16 October 1976; Lawrence K. Altman, "Disease Unit Plans Big Re-Evaluation," *The New York Times*, 17 April 1977.

16. David S. Gillmor, "How Much for the Patient, How Much for Medical Science," *Modern Medicine* (7 January 1974): 30–33.

17. My reading of the Willowbrook case is based largely on *The Willowbrook Wars*, by Sheila M. and David J. Rothman (New York: Harper and Row, 1984): 262–77, which casts Krugman's experiments as unscrupulous, and on my interviews with Richard Krugman and Louis Cooper, who maintain that the scientist was unfairly villified.

18. Richard Krugman, "Editorial: Immunization 'Dyspractice': The Need for 'No Fault' Insurance," *Pediatrics* 56 (2) (1975): 159–60.

19. "Compensation for Vaccine-Related Injury," *AAP Notes and Comments* (March 1979).

20. Paul Offit, *The Cutter Incident* (New Haven: Yale University Press, 2005): 148–53.

21. Barbara Loe Fisher and Harris Coulter, *DPT: A Shot in the Dark* (New York: Warner Books, 1984).

22. Paul Parkman, author interview, June 2004.

23. Irving Ladimer, "Mass Immunizations: Legal Problems and a Proposed Solution." *Journal of Community Health* 2(3) (1977): 189–208.

24. Margaret Pittman, "History of the Development of Pertussis Vaccine," *Devlopments in Biological Standardization* 73 (1991): 11–29.

25. James Cherry, "Pertussis in the Preantibiotic and Prevaccine Era," *Clinical Infectious Diseases* 28 (suppl 2) (1999): S107–11.

26. Cf. L. A. Lurie and S. Levy, "Personality Changes and Behavior Disorders

of Children Following Pertussis," *Journal of the American Medical Association* 120 (12) (1942): 890–94.

27. "Pertussis vaccines omitted from NNR," *Journal of the American Medical Association* (February 1931): 613.

28. Grace Eldering, "Symposium on Pertussis Immunization in Honor of Dr. Pearl L. Kendrick in her 80th Year: Historical Notes on Pertussis Immunization," *Health Laboratory Science* 8 (4) (1971): 200–19.

29. Pittman, "History," 11–29; John Robbins and Rachel Schneerson, author interview, June 2004; Charles Manclark, author interview, August 2004.

30. J. Cameron, "Some Problems—and Their Solutions—in the Control of DTP Vaccines," *Developments in Biological Standardization* 41 (1978): 45–53; "Problems Associated with the Control Testing of Pertussis Vaccine, *Advances in Applied Microbiology* 20 (1976): 57–80.

31. In a 1985 deposition in *Nelson v. American Cyanamid*, Henry Piersma, Lederle's director of biologics (1944–54) and quality control (1954–72) described how Lederle switched from acellular to whole-cell pertussis vaccine after the Kendrick/Eldering tests were developed.

32. Thorvald Madsen, "Vaccination against Whooping Cough," *Journal of the American Medical Association* 101 (3) (1933): 187–88.

33. R. K. Byers and F. C. Moll, "Encephalopathies Following Prophylactic Pertussis Vaccine," *Pediatrics* 1 (4) (1948): 437–57.

34. John A. Toomey, "Reactions to Pertussis Vaccine," *Journal of the American Medical Association* 139 (7): 448–50.

35. Reimert Ravenholt, Astrid Ravenholt, et al., "Immunizable Disease Occurrence and Prevention in Seattle," *Public Health Reports* 80 (11) (1965).

36. Eldering, "Symposium on Pertussis," 200–19; and A. A. Miles and G. S. Wilson, eds., *Topley and Wilson's Principles of Bacteriology and Immunity* (Baltimore: Williams and Wilkins, 1964), 2014–16.

37. Justus Stroem, "Is Universal Vaccination Agianst Pertussis Always Justified?" *British Medical Journal* (October 22, 1960): 1184–86, B. D. Bower and P. M. Jeavons, "Complications of Immunization," *British Medical Journal* (12 November 1960): 1453; Stroem, "Further Experience of Reactions" *British Medical Journal* (11 November 1967): 320–23.

38. Joseph Lapin and Carl Weihl, "Extracted Pertussis Antigen, a Clinical Appraisal," *American Journal of the Diseases of Childhood* 106 (August 1963): 124–29.

39. Charles Manclark, "The Current Status of Pertussis Vaccine: An Overview," *Advances in Applied Microbiology* 20 (1976): 1–7.

40. Ibid., 7.

41. Ibid, 7.

42. "DPT Vaccine under Scrutiny in Wake of Huge Award," UPI, 19 October 1987; Amy Tarr, "DTP Vaccine Injuries; Who Should Pay?" *National Law Journal* (1 April 1985).

43. Kay Johnson and Barbara Richards, "Federal Immunization Policy and Funding; A History of Responding to Crises," *American Journal of Preventive Medicine* 19 (suppl 3) (2000): 99–112.

44. Howard Tint to C. J. Kern, Wyeth Labs, 20 May 1975 (Exhibit 65, *Miller v. Connaught*), courtesy of Victor Harding.

45. C. N. Christensen, "Pertussis Vaccine Encephalopathy," *Eli Lilly Report* (1962): 10.

46. Memo from M. V. Quarry, 29 April 1966 (Exhibit 24, *Miller v. Connaught*).

47. According to Alan Bernstein, who was Wyeth's production manager, Lilly claimed it had 20 percent of the market. Wyeth sales representatives claimed it was as high as 50 percent.

48. Internal Lederle memo, 17 April 1967 (Exhibit 18, *Miller v. Connaught*); Lederle memo, R. Meschke to S. A. Flaum, 26 April 1977 (Exhibit 12, *Miller v. Connaught*).

49. Lloyd Colio, scientist with Richardson-Merrill (Connaught), deposition 21 March 1986 (*Terry Lynn Hall v. Connaught*).

50. Meschke to Flaum.

51. Wyeth scientists Alan Bernstein, Frank J McCarthy, memos, 1977 (Exhibits 14–16, *Miller v. Connaught*).

52. Yuji Sato and Hideo Arai, "Leucoctysosis-Promoting Factor of Bordetella Pertusiss, I Purification and Characterization." *Infection and Immunity* 6 (6) (1972): 899–904; Sato, Arai, and Kenji Suzuki, "Leucoctysosis-Promoting Factor of Bordetella pertusiss, II. Biological Properties, *Infection and Immunity* 7(6) (1973): 992–99.

53. Yuji Sato, Charles Manclark, et al., "Role of Antibody to LPF Hemagglutinin in Immunity to Pertussis," *Infection and Immunity* 31 (3) (1981): 1223–31.

54. Sato, Manclark, et al., "Separation and Purification of the Hemagglutinins from *Bordetella pertussis*," *Infection and Immunity* 41 (1) (1983): 313–20.

55. M. Kulenkampff et al., "Neurological Complications of Pertussis Inoculation," *Archives of the Diseases of Children* 49 (1) (1974): 469.

56. E. Gangarosa, R. Chen, et al., "Impact of Anti-Vaccination Movements on Pertussis Control: The Untold Story," *The Lancet* 351 (9099) (1998): 356–61.

57. J. Isacson et al., "How Common Is Whooping Cough in a Nonvaccinating Country?" *Pediatric Infectious Disease Journal* 12 (1993): 284–88.

58. Alan Hinman et al., "Acellular and Whole Cell Pertussis Vacines in Japan: Report of a Visit by U.S. Scientists," *Journal of the American Medical Association* 257 (10) (1987): 1351–56.

59. Gangarosa, "Untold Story," 357–59; Hinman, "Vaccines in Japan," 1355; and internal memo, Alan Bernstein to Larry Hewlett, Wyeth, 27 August 1979 (Exhibit 68, *Miller v. Connaught*).

60. Yuji Sato, author interview, July 2004; Charles Manclark, author interview, June 2004.

61. Roger Bernier et al., "Diphtheria-Tetanus Toxoid-Pertussis Vaccine and Sudden Infant Death in Tennessee, *Journal of Pediatrics* 101 (3) (1982): 419–21.

62. A. V. Jonville-Bern et al., "Sudden Unexpected Death in Infants Under 3 Months of Age and Vaccination Status-Case-Control Study," *British Journal of Clinical Pharmacology* 51 (3) (2001), 271–76; M. R. Griffin et al., "Risk of Sudden Infant Death Syndrome After Immunization with Diphtheria-Tetanus-Pertussis Vaccine," *New Engalnd Journal of Medicine* 319 (10) (1988): 618–23. According to Wyeth internal correspondence, one FDA official suggested asking for a recall of the lot but was rebuffed by superiors, though Wyeth eventually withdrew it voluntarily. Company officials wanted to include a predisposition to SIDS as a contraindication to DTP vaccination, to cover their litigation risks. But Harry Meyer, who had taken over as the vaccine division's chief from Roderick Murray in 1972, strongly opposed such a move, saying that expert reviews had found no link between SIDS and DTP vaccine. See also Geier and Geier, "The True Story of Pertussis Vaccine: A Sordid Legacy?" *Journal of the History of Medicine and Allied Sciences* 57 (3) (2002): 249–84.

63. Bernstein to Hewlett (Exhibit 68, *Miller v. Connaught*).

64. Alan Hinman, author interview, May 2004.

65. U.S. Senate Committee on Labor and Human Resources, *Hearing: Vaccination Injury Compensation*, 3–4 May 1984, 37–43.

66. Denis J. Hauptly and Mary Mason, "The National Childhood Vaccine Injury Act: The Federal No-Fault Compensation Program That Gives a Booster for Tort Reform," *Federal Bar News & Journal* 37 (8) (1990): 452–58.

67. U.S. Senate Committee on Labor, *Hearing*, 3–4 May 1984, 8–11.

68. Tim Westmoreland, author interview, April 2004.

69. Donna Middlehurst, "An Open Letter to Parents," *DPT News* 1 (2) (1984): 8–9.

70. U.S. House Committee on Energy and Commerce, *Vaccination Injury Compensation*, 9–10 September 1984, 82.

71. Martin Smith, "National Childhood Vaccine Injury Compensation Act," *Pediatrics* 82 (2) (1988): 264–69.

72. U.S. House, *Vaccination Injury Compensation*, 270–75.

73. Tim Westmoreland, author interview.

74. Mortimer deposition (Exhibit 52, *Cossette Krause et al. v. F. K. Abbousy, MD, et al.*, Case No. 82-1232, State of Ohio, Stark County).

75. J. D. Cherry and E. A. Mortimer, "An 'Old' Bacterial Vaccine with New Problems," *Pediatric Immunization Today: A Symposium* (Swiftwater: Connaught, 1979); and "Damage Lawsuits Settled for Millions," *Fresno Bee*, 5 December 1984.

76. James Cherry, "Pertussis Vaccine Encephalopathy: It Is Time to Recognize

It as the Myth That It Is," *Journal of the American Medical Association* 263 (12) (1990): 1679–80 (Erratum in *JAMA* 263 [16] [1990]: 2182).

77. Victor Harding, author interview, May 2004; Janny Scott, "Researcher to Clarify Ties to Drug Company," *Los Angeles Times*, 24 March 1990, B3; U.S. House of Representatives, *Are Scientific Misconduct and Conflicts of Interest Hazardous to Our Health?* (Washington: GPO, 1990): 65.

78. "Damage Lawsuits Settled."

79. *Hearing on Vaccine Development before the Subcommittee on Oversight and Investigations of the Senate Committee on Energy and Commerce*, 13 March 1985, 241–57.

80. James Cherry et al., "Pertuss and Its Prevention: A Family Affair," *Journal of Infectious Diseases* 161 (March 1990): 473–79.

81. David Klein, author interview, September 2005; William Jordan, author interview, August 2005; and K. Edwards and M. Decker, "Acellular Pertuss Vaccines for Infants," *New England Journal of Medicine* 334 (6) (1996): 391–92.

82. Stanley Plotkin and Michel Cadoz, "The Acellular Pertussis Vaccine Trials: An Interpretation," *Pediatric Infectious Disease Journal* 16 (5) (1997): 508–17.

83. It would turn out that whole-cell vaccines in use in the United States starting in the mid-1980s were already waning in effectiveness. In the mid-1980s, Lederle had found a way to produce a less-reactive whole-cell vaccine by reducing the number of organisms in the brew. The vaccine caused fewer fevers, but it wasn't as effective in preventing whooping cough. Cherry believes that Connaught did something similar with its vaccine. When Connaught's vaccine was used as a control in the European trials, it worked less well than any of the acellular versions.

84. David Klein, "From Pertussis to Tuberculosis: What Can Be Learned?" *Clinical Infectious Diseases* 30 (suppl 3) (2000): S302–8.

85. These were the three-component vaccine, Infanrix, produced by GlaxoSmithKline, and the five-component Daptacel, made by Connaught Canada.

86. James Cherry and Patrick Olin, "The Science and Fiction of Pertussis Vaccines," *Pediatrics* 104 (6) (1999): 1381–84.

87. Tim Westmoreland, author interview, June 2004.

88. Martin Smith, author interview, March 1998.

89. Jeffrey Schwartz, author interview, October 2004.

90. Cliff Shoemaker and Mark Horn, author interviews, January 1998.

91. Ibid.

92. Fulginiti, "Controversies in Current Immunization Policy," 3–35.

93. Vincent Fulginiti, "Pertussis Disease, Vaccine and Controversy," *Journal of the American Medical Association* 251 (2) (1984): 251.

94. "Protective Efficacy of the Takeda Acellular Pertussis Vaccine Following

Household Exposure of Japanese Children," *American Journal of the Diseases of Children* 144 (8) (1990): 899–901.

95. Britain's DTP controversy had begun in 1974 and led to a drastic decline in pertussis vaccination. Gordon Stewart, an outspoken physician, published his pithy views frequently in the British medical and lay press. Jeffrey P. Baker's article, "The Pertussis Vaccine Controversy in Great Britain, 1974–1986," in *Vaccine* 21 (25–26) (2003): 4003–10, contains an excellent review.

96. R. J. Robinson, "The Whooping Cough Immunization Controversy," *Archives of Disease in Childhood* 56 (1981): 577–80.

97. Christopher L. Cody et al., "Nature and Rates of Adverse Reactions Associated with DTP and DT Immunization in Infants and Children," *Pediatrics* 68 (5) (1981): 650–60.

98. Larry Baraff et al., "Infants and Children with Convulsions and Hypotonic-Hyporesponsive Episodes Following DTP Immunization: Follow-up Evaluation," *Pediatrics* 81 (6) (1988): 79–94.

99. Walter Orenstein and John Livengood, "Family History of Convulsions and Use of Pertussis Vaccine," *Journal of Pediatrics* 115 (4) (1989): 527–31.

100. See, for example, Mortimer, "Pertussis Vaccines," in *Vaccines*, 2nd ed. (Philadelphia: Sanders, 1994): 102.

101. Institute of Medicine: Committee to Review the Adverse Consequences of Pertussis and Rubella Vaccines, *Adverse Effects of Pertussis and Rubella Vaccines*, Christopher Howson et al., eds. (Washington: National Academy Press, 1991): 86–124.

102. Robert Chen et al., "Risk of Seizures after Receipt of Whole-cell Pertussis or MMR Vaccine," *New England Journal of Medicine* 345 (9) (2001): 656–61.

103. "Hypotonic-Hyporesponsive Episodes Reported in VAERS, 1996-1998," *Pediatrics* 106 (4) (2000): e52.

104. David Klein, author interview. September 2004.

105. Transcript, 12 December 1991 meeting of the Advisory Committee on Childhood Vaccines, copy made available to author by Health Resources and Services Administration (HRSA).

CHAPTER 8

1. Karen De Witt, "Faith over Medicine in Philadelphia's Measles Outbreak," *The New York Times*, 23 February 1991.

2. Mike Owen, "Pastor Urges Congregation to Fight Fear after Four Die in Measles Epidemic," Associated Press, 17 February 1991.

3. Larry Tye, "Bigtime Measles Outbreak Hits US," *Boston Globe*, 26 March 1990.

4. George Curry, "Rights Case Grows over Measles Vaccinations," *Chicago Tribune*, 25 March 1991.

5. Susan Okie, "Vaccination Record in US Falls Sharply," *The Washington Post*, 24 March 1992.

6. Robert Byrd, "Most 2-Year-Olds Haven't Had Their Shots, Federal Report Says," Associated Press, 13 February 1992; "Nearly Half the Children Entering New Orleans Public Schools Unvaccinated," Associated Press, 16 March 1991.

7. Philip Hilts, "US Vaccine Plan Uses Welfare Offices," *The New York Times*, 17 March 1991.

8. Spencer Rich, "Child Vaccine Shortage Seen at Health Centers," *The Washington Post*, 20 May 1991.

9. Loring Dales, "The Measles Epidemic," 1990 videotaped presentation available at National Library of Medicine.

10. Okie, "Vaccination Record."

11. Holcomb Noble, "Incentive Program Raises Immunization Rates," *The New York Times*, 7 October 1998, A14. Chicago for many years required welfare recipients to report for their benefit packages monthly, rather than quarterly, until they showed their children had been immunized.

12. Paul Bedard, "Bush Focuses on Childhood Disease," *Washington Times*, 14 June 1991; Robert Pear, "Bush Defers an Emergency Plan to Provide Vaccines for Children," *The New York Times*, 22 June 1991.

13. Robert Pear, "The 1992 Campaign: White House, Bush Announces New Effort to Immunize Children," *The New York Times*, 13 May 1992.

14. "President Assails 'Shocking' Prices of Drug Industry: Plans a Vaccine Program," *The New York Times*, 13 February 1993, A1; "Clinton, in Compromise, Will Cut Parts of Childhood Vaccine Plan," *The New York Times*, 4 May 1993, A1.

15. This discussion relies heavily on my interview with Alan Hinman, May 2004.

16. The Nixon and Reagan administrations proposed eliminating PL 317 and including the money in block grants, but Congress successfully fought the cuts.

17. Walter Orenstein, author interview, February 2005.

18. Neal Halsey et al., in *Vaccines*, 4th ed. (Philadelphia: Saunders, 2004): 395.

19. Sam Katz, "The History of Measles Vaccine and Attempts to Control Measles," in *Microbe Hunters Then and Now*, ed. Hilary Kropowski and Michael B. A. Oldstone (Lansing: Med.-Ed Press, 1996): 69–76.

20. National Vaccine Advisory Committee, "The Measles Epidemic: The Problems, Barriers and Recommendations," *Journal of the American Medical Association* 266 (11) (1991): 1547–52.

21. "Clinton Seeks 20-Cents-a-Pack Cigarette Levy Hike," *Los Angeles Times*, 24 February 1997; "Child Immunizations Rise Sharply, U.S. Says," *The Washington Post*, 24 July 1997.

22. Rachel Schneerson et al., "Haemophilus Influenzae Type B Polysaccharide

Protein Conjugates: Model for a New Generation of Capsular Poly-saccharide Vaccines," *Progress in Clinical Biological Research* 47 (1980): 77–94.

23. "Whitestone's Story Speaks Volumes on Courage, Faith," *Chicago Tribune*, 20 July 1995; Paul Bonner, "Miss America Urges State to Invest in Disabled, Encourages Students," *Durham Herald-Sun*, 20 April 1995.

24. J. C. Butler, "Nature Abhors a Vacuum, but Public Health Is Loving it: The Sustained Decrease in the Rate of Invasive Haemophilus Influenzae Disease;" *Clinical Infectious Diseases* 40 (6) (2005): 831–32.

25. S. I. Pelton et al., "Seven-Valent Pneumococcal Conjugates Vaccine Immunization in Two Boston Communities;" *Pediatric Infectious Disease Journal* 23 (11) (2004): 1015–22.

26. "Direct and Indirect Effects of Routine Vaccination of Children with 7-Valent Pneumococcal Conjugate Vaccine," *Morbidity & Mortality Weekly Report* 54 (36) (2005): 893.

27. J. Wahlberg, "Vaccinations May Induce Diabetes-related Autoantibodies in One-Year-Old Children," *Annals of the New York Academy of Sciences* 1005 (2003): 404–8; for record of 26 September 2003 court ruling see www.uscfc .uscourts.gov/Opinions/Specmast/Millman/MILLMAN.baker.pdf.

28. "Proceedings of the Symposium on Ethical Issues in Human Experimentation: The Case of Willowbrook State School Research," NYU School of Medicine Student Council, 24 March 1972, in collection of the National Library of Medicine.

29. Ibid.

30. Neal Halsey and Caroline Breese Hall, "Control of Hepatitis: To Be nor Not to Be" *Pediatrics* 90 (2 pt 1) (1992): 274–77.

31. Immunization Practices Advisory Committee, "Prevention of Perinatal Transmission of Hepatitis B Virus: Prenatal Screening of All Pregnant Women for Hepatitis B Surface Antigen," *Morbidity and Mortality Weekly Report* 39 (1990): 8–19.

32. Leonard Seeff, author interview, June 2005.

33. Josh Sharfstein, *The American Prospect*, 8 May 2000, 15–18.

34. Sally Squires, "Hepatitis B Vaccinations: Pediatricians Begin Immmunizing Infants to Curb the Spread of the Disease," *The Washington Post*, 5 May 1992.

35. "Experience and Reason," *Pediatrics* 95 (5) (1995): 764–65.

36. Gary L. Freed et al., "Family Physician Acceptance of Universal Hepatitis Immunization of Infants," *American Journal of Family Practice* 36 (2): 153–57.

37. "Hepatitits B Today: New Guidelines for the Pediatrician," *Pediatric Infectious Disease Journal* 12 (5) (1993): 427–53.

38. Halsey and Hall, "Control of Hepatitis," 274–77.

39. David L. Wood et al., "California Pediatricians' Knowledge of and Response to Recommendations for Universal Infant Hepatitis B Immunization," *Archives of Pediatric and Adolescent Medicine* 149 (7) (1995): 769–73.

40. "Achievements in Public Health: Hepatitis B Vaccination—United States, 1982–2002," *Morbidity and Mortality Weekly Report*, 28 June 2002.

41. "Incidence of Acute Hepatitis B, 1991–2002, U.S.," *Morbidity and Mortality Weekly Report*, 1 January 2004.

42. "Questions About Varicella Vaccine," *Baltimore Sun*, 18 March 1995.

43. M. Takahashi and A. Gershon, "Varicella Vaccine," in *Vaccines*, 2nd ed. (Philadelphia: Saunders, 1994): 395–96.

44. Philip Brunnell to Albert Sabin, 28 June 1977, Sabin Archive, Medical Heritage Library, University of Cincinnati.

45. Susan Ellenberg, author interview, June 2005.

46. Arthur Lavin, "Questions about Varicella Vaccine," *Pediatrics* 98 (6 pt 1) (1996): 1225.

47. Stanley Plotkin, "Varicella Vaccine," *Pediatrics* 97 (2) (1996): 251–53.

48. Lynn Grossman, "Chickenpox Vaccine Slows Rash of Outbreaks," *USA Today*, 18 January 1999, 4D.

49. Jane Seward et al., "Varicella Disease after the Introduction of Varicella Vaccine in the United States, 1995–2000," *Journal of the American Medical Association* 287 (5) (2002): 606–11.

50. Stanley J. Schaffer and Sandra Bruno, "Varicella Immunization Practices and the Factors That Influence Them," *Archives of Pediatric and Adolescent Medicine* 153 (April 1999): 357–62.

51. Kathleen Stratton et al., "Adversen Events Associated with Childhood Vaccines Other than Pertussis and Rubella; Summary of a Report from the Institute of Medicine," *Journal of the American Medical Association* 271 (20) (1994): 27–34.

52. Orenstein interview.

53. Penelope H. Dennehy and Joseph S. Breese, "Rotavirus Vaccine and Intussusception: Where Do We Go from Here?" *Infectious Disease Clinics of North America* 15 (1) (2001): 187–209.

54. Margaret B. Rennels, "The Rotavirus Vaccine Story: A Clinical Investigator's View," *Pediatrics* 106 (1) (2000): 123–25.

55. Details of this are contained in the transcript of the 12 December 1997 meeting of the FDA Vaccines and Related Biologics Advisory Committee and were published in Roger Glass et al., "Lack of an Apparent Association between Intussusception and Wild or Vaccine Rotavirus Infection," *Pediatric Infectious Disease Journal* 17 (10) (1998): 924–25.

56. Roger Glass, "Rotavirus Vaccine: When Is a Vaccine Safe Enough?" Symposium, 2003 Biotech Industry Organization Annual Meeting, 23 June 2003, Washington, DC.

57. Roger Glass, "Cost-effectiveness Analysis of a Rotavirus Immunization Program for the United States," *Journal of the American Medical Association* 279 (17) (1998): 1371–76.

58. American Academy of Pediatrics, Committee on Infectious Disease, "Prevention of Rotavirus Disease: Guidelines for Use of Rotavirus Vaccine," *Pediatrics* 103 (6) (1998): 1483–91.

59. Paul Offit, author interview, March 2005.

60. Lauran Neergaard, "FDA Approves World's First Vaccine Against Children's Diarrhea," Associated Press, 1 September 1998.

61. "Suspension of Rotavirus Vaccine after Reports of Intussusception—United States, 1999," *Morbidity and Mortality Weekly Report*, 53 (34) (3 September 2004): 786-89.

62. Mark Benjamin, "UPI Investigates: The Vaccine Conflict," UPI, 20 July 2003.

63. U.S. House, Committee on Government Reform, *Conflicts of Interest in Vaccine Policy Making: Majority Staff Report*, 15 June 2000.

64. Lone Simonsen, "Effect of Rotavirus Vaccination Programme on Trends in Admission of Infants to Hospital for Intussusception," *The Lancet* 359 (9289) (2001): 1224–29.

65. Letter, *The Lancet* (23 March 2002): 359

66. Martha Iwamoto et al., "A Survey of Pediatricians on the Reintroduction of a Rotavirus Vaccine," *Pediatrics* 112 (1) (2003): e6–10.

67. Albert Mehl, personal communication to author, June 2005.

68. Glass, "Rotavirus Vaccine."

69. Jon Cohen, "Rethinking Vaccine's Risk," *Science* 293 (5535) (2001): 1576–77.

70. Trudy V. Murphy et al., "The First Rotavirus Vaccine and Intussusception: Epidemiological Studies and Policy Decisions," *Journal of Infectious Diseases* 187 (8) (2003): 1309–13.

71. Margaret Rennels, "The Rotavirus Vaccine Story: A Clinical Investigator's View," *Pediatrics* 106 (1) (2000): 123–25; with critical response from Guy Lonergan, a Montreal physician, on electronic edition.

72. Lance Gordon, remarks at "Second National Vaccine Advisory Committee Workshop on Strengthening the Supply of Vaccines in the United States," 25 January 2005, Washington, DC; Orenstein interview.

73. Orenstein interview.

74. Paul Offit, author interview, March 2005.

75. Justin Gillis, "Rotavirus Vaccine Urged for Babies," *The Washington Post*, 22 February 2006, A8.

CHAPTER 9

1. See, for example, "An Emotional Debate on Childhood Vaccination," *Chicago Tribune*, 10 May 1994, A1; "Parents Weight Decision to Vaccinate in Low-Rate Boulder," *Boston Globe*, 9 September 2002, A3; "Values, Health

Concerns Clash," *Albany Times-Union*, 13 March 1977, B1; "To Immunize or Not; Alt Medicine Stokes the Debate," *Seattle Post-Intelligencer*, 31 August 1995, C1; Katherine Seligman, "Vaccine Backlash," *San Francisco Chronicle Magazine*, 8; A. Siedler et al., "Progress toward Measles Elimination in Germany," *Journal of Infectious Diseases* 187 (suppl 1) (2003): S208–16; B. Hanratty et al., "United Kingdom Measles Outbreaks in Non-Immune Anthroposophic Communities: Implications for the Elimination of Measles from Europe," *Epidemiology and Infection* 125 (2000): 377–83; R. Klein et al., "Durchimpfund der Schlanfaenger in Deutschland, Date aus der KJGD-Kampagne," *Kinderarztliche Praxis* 2 (suppl) (1999): 46–51.

2. John Hickenlooper, author interview, November 2004.

3. George Orwell, *Collected Essays, Journalism and Letters*, vol. 4: *In Front of Your Nose* (New York: Harcourt Brace Jovanovich, 1968): 259–60. The essay in question, written in 1946, was a response to a comment by a character in G. B. Shaw's *Saint Joan*, to the effect that modern-day people, paradoxically, were far more credulous than their ancestors.

4. Cf. Arthur Allen, "Vaccination Rejection," *The New Republic*, 25 March 1998.

5. Fawn Vrazo et al., "Prescriptions for Painkillers Drop Sharply in Wake of Safety Concerns," *Philadelphia Inquirer*, 15 February 2005.

6. Barbara Loe Fisher, author interview, May 2005.

7. Robert D. Johnston, *The Politics of Healing: Essays in the 20th Century History of North American Alternative Medicine* (London: Routledge, 2001): 271.

8. Steve Robison, "Parental Attitudes toward Immunization in Ashland, Oregon," presented at 38th National Immunization Conference, 11–14 May 2004.

9. Allison Kempe, author interview, December 2004.

10. D. Feikin et al., "Individual and Community Risks of Measles and Pertussis Associated with Personal Exemption to Immunization," *Journal of the American Medical Association* 284 (24) (2000): 3145–50.

11. Thanks to Robert T. Johnson, ed., *Politics of Healing*, 259, for this gem.

12. Michael Palmer, *Fatal* (New York: Bantam, 2002). Palmer, at a 2003 conference of Defeat Autism Now! said that vaccines had caused his own son to become autistic.

13. Phyllis Schlafly, "A Society of Snoops?" *Washington Times*, 1 December 1999.

14. Matthew Schneirov and Jonathan David Geczik, "Beyond the Culture Wars: The Politics of Alternative Health" in Johnston, ed., *Politics of Healing*, 250–51. These academics found a working-class Christian group in Pittsburgh working closely with left-wing alternative practitioners.

15. Leonard G. Horowitz, *Emerging Viruses: Aids and Ebola: Nature, Accident or Intentional?* (Rockport, MA: Tetrahedron, 1996). Horowitz frequently

addresses UFO conferences, where he gives talks such as "DNA and the Alien Threat to Humanity" (see www.tetrahedron.org/speak.htm).

16. Allen, "Vaccination Rejection."

17. U.S. House, *The Autism Epidemic—Is the NIH and CDC Response Adequate?: Hearing before Committee on Government Reform*, 18 April 2002, 141–42.

18. Ivan Illich, *Medical Nemesis: The Expropriation of Health* (New York: Pelican Books, 1976).

19. In Peggy O'Mara, *Vaccination: The Issue of Our Times* (Santa Fe: Mothering Magazine, 1997): 33, 174–75.

20. *Children's Healthcare Is a Legal Duty, Inc.*, newsletter #5 (CHILD Inc., 2004); also see "Importation-Associated Measles Outbreak—Indiana, May–June 2005," *Morbidity and Mortality Weekly Report* 54 (28 October 2005).

21. Charles Rosenberg, "Alernative to What? Complementary to Whom? On Some Aspects of Medicine's Scientific Identity," 18 July 2002 lecture at the National Institutes of Health.

22. Harris Coulter, *Homeopathic Science and Modern Medicine: The Physics of Healing with Microdoses* (Richmond, CA: North Atlantic Books, 1980): 11.

23. "Triumph over the Most Terrible of the Ministers of Death," *Annals of Internal Medicine* 128 (9) (1997).

24. Coulter, *Homeopathic*, 93.

25. Ibid., 122.

26. Ibid., xii–xiv.

27. "Two-Month-Old Baby Dies—Child Abuse Possible," *Orlando Sentinel*, 30 November 1997.

28. Harold Buttram and John Chriss Hoffman, *The Immune Trio* (Quakerstown: Humanitarian Press, 1993): 28–29.

29. Ibid., 60–74.

30. See www.veracity.org/Americans%20Invasion.html.

31. Ibid.

32. Reuben Swinburne Clymer, *The Age of Treason: The Carefully and Deliberately Planned Methods Developed by the Vicious Element of Humanity for the Mental Deterioration and Moral Debasement of the Mass as a Means to Their Enslavement. Based on Their Own Writings and Means Already Confessedly Employed* (Quakerstown: Humanitarian Press, 1957).

33. Citations from Clymer, *The Age*, 97, 108, 207–8, 221.

34. Cf. Anthony Storr, *Feet of Clay: Saints, Sinners, and Gurus* (New York: Free Press, 1996), for his observations about Steiner.

35. Franz Kafka, *The Diaries, 1910–23* (New York: Shocken, 1975): 45–49.

36. Rudolf Steiner, *The Occult Significance of Blood* (London: Rudolf Steiner Press, 1967).

37. Rudolf Steiner, *Health and Illness*, vol. 1 (Spring Valley, NY: Anthroposophic Press, 1981): 83–86.

38. Rudolf Steiner, *Theosophy of the Rosicrucian* (London: Rudolf Steiner Press, 1981): 68–69.

39. Rudolf Steiner, "Illness Ocurring in Different Periods of Life," in *Health and Illness*, vol. 1, 34; "Why Do We Become Sick?" 27 December 1922 lecture, in *Health and Illness*, vol. 2, 148–49; "Diphtheria and Influenza, Cross Eyes," 20 January 1923 lecture, vol. 2, 101–2; 247, "Breathing, Circulation—Jaundice, Smallpox; Rabies," 27 January 1923 lecture, in *Health and Illness*, vol. 2, 115.

40. Rudolf Steiner, *Manifestations of Karma* (London: Rudolf Steiner Press, 1995): 140, 146.

41. Francis X. King, *Rudolf Steiner and Holistic Medicine* (York Beach, ME: Nicolas-Hays, 1987): 90–91.

42. Rudolf Steiner, "The Crumbling of the Earth and the Souls and Bodies of Man," Dornbach lectures, 7 October 1917, in an article by Philip Incao at www.waldorflibrary.org/Journal_Articles/GW3414.pdf.

43. Dan Dugan, e-mail communication, December 2004.

44. Scott Olmstead, personal communication, October 2001.

45. Philip Incao, "Childhood Illness, Vaccination and Child Health," *Renewal: A Journal of Waldorf Education* 7 (1) (Spring 1998).

46. *Alternative Medicine Digest* 19 (1999); www.industryinet.com/~ruby/vac_child_health.html.

47. T. Shirakawa, J. Hopkin, et al., "The Inverse Association between Tuberculin Responses and Atopic Disorder, *Science* 275 (5296) (1997): 77–79.

48. P. A. McKinney et al., "Early Social Mixing and Childhood Type 1 Diabetes Melitus," *Diabetes Medicine* 17 (3) (2000): 236–42; "Is Atopy a Protective or a Risk Factor for Cancer? A Review of Epidemiological Studies," *Allergy* 60 (9) (2005): 1098–111.

49. A. Ponson et al., "Exposure to Infant Siblings During Early Life and Risk of Multiple Sclerosis," *Journal of the American Medical Association* 293 (4) (2005): 463–69.

50. Michael Odent, "Pertussis Vaccine and Asthma—Is There a Link?" *Journal of the American Medical Association* 272 (8) (1994): 592–93. For example of negative studies, see C. Gruber et al., "Do Early Childhood Immunizations Influence the Development of Atopy and Do They Cause Allergic Reactions?" *Pediatric Allergy and Immunology* 12 (6) (2001): 296–311; L. Nilsson et al., "Allergic Disease at the Age of 7 Years after Pertussis Vaccination in Infancy," *Archives of Pediatric and Adolescent Medicine* 157 (2) (2003): 1184–87; C. Gruber et al., "Transient Suppression of Atopy in Early Childhood is Associated with High Vaccination Coverage" *Pediatrics* 111 (3) (2003): e282–88; H. R. Anderson et al., "Immunological Symptoms of Atopic Disease in Children: Results from the International Study of Asthma and Allergies in Childhood," *American Journal of Public Health* 91 (7) (2001): 1126–29, G. Martignon et al., "Does Childhood Immunization against

Infectious Disease Protect from Development of Atopic Disease?" *Pediatric Allergy and Immunology* 16 (3) (2003): 193–200. In the latter study, the rate of asthma was doubled in unvaccinated children. For a good discussion of the surprising complexity of asthma risk factors, see M. T. Salam et al. "Early-Life Environmental Risk Factors for Asthma: Findings from the Children's Health Studies," *Environmental Health Perspectives* 112 (6) (2004).

51. For example, see A. M. Krieg et al., "CpG DNA: Trigger of Sepsis, Mediator of Protection, or Both?" *Scandinavian Journal of Infectious Disease* 35 (9) (2003): 653–59; Krieg, "Enhancing Vaccines with Immune Stimulation CpG DNA," *Current Opinions in Molecular Therapy* 3 (1) (2001): 15–24.

52. Stephen Smith, "The Asthma Riddle," *Boston Globe*, 13 April 2004.

53. Alexis Carrel, *Man the Unknown* (New York: Harper and Brothers, 1935): 114–15.

54. Murphy, *Vaccination Dilemma*.

55. Ibid., 10.

56. Ibid., 9.

57. Dan Salmon, unpublished data.

58. Feikin et al., "Individual and Community."

59. Daniel Salmon, Neal Halsey, et al., "Factors Associated with Refusal of Childhood Vaccines among Parents of School-Aged Children," *Archives of Pediatric and Adolescent Medicine* 195(5) (2005): 470–76.

60. Daniel Salmon, Neal Halsey, et al., "Knowledge, Attitudes and Belief of School Nurses and Personnel and Associations with Nonmedical Immunization Exemptions," *Pediatrics* 113 (6) (2004): 552–59.

61. Kumanan Wilson et al., "Changing Attitudes toward Polio Vaccination: A Randomized Trial of an Evidence-Based Presentation versus a Presentation from a Polio Survivor," *Vaccine* 23 (23) (2005): 3010–15.

62. Lorrie Quick of the Colorado Department of Public Health, author interview, November 2004.

63. See www.hapihealth.com.

64. J. G. Wheeler et al., "Barriers to Public Health Management of a Pertussis Outbreak in Arkansas," *Archives of Pediatric and Adolescent Medicine* 158 (February 2004): 146–52.

65. K. Edwards and N. Halasa, "Are Pertussis Fatalities in Infants on the Rise?" *Journal of Pediatrics* 43 (5) (2003): 552–53.

66. K. Edwards, "Is Pertussis a Frequent Cause of Cough in Adolescents and Adults? Should Routine Pertussis Immunization Be Recommended?" *Clinical Infectious Diseases* 32 (12) (2001): 1698–99; and L. D. Senzilet et al., "Pertussis Is a Frequent Cause of Prolonged Cough Illness in Adults and Adolescents," 1691–97 in same issue.

67. John Robbins, author interview, March 2004; James Cherry, author interview, June 2004

68. These were Connaught's two-component Tripedia and Lederle-Wyeth's low-antigen, four-component Acel-Immune, based respectively on the Japanese Biken and Takeda vaccines. James Cherry et al., "Comparative Effects of Lederle/Takeda Acellular Pertussis Component DTP Vaccine and Lederle Whole Cell DTP Vaccine in German Children after Household Exposures," *Pediatrics* 102 (3) (1998): 551.

69. P. Olin et al., "Declining Pertussis Incidence in Sweden Following the Introduction of Acellular Pertussis Vaccine," *Vaccine* 21 (17–18) (2003): 2015–21; A. E. Tozzi et al., "Clinical Presentation of Pertussis in Unvaccinated and Vaccinated Children in the First Six Months of Life," *Pediatrics* 112 (5) (2003): 1069–75. In Sweden, whose children were unvaccinated from 1979 to 1996, the rates of pertussis among two-year-olds fell from 1,466 per 100,000 in 1994 to 40 per 100,000 in 2000. Of 145 hospital admissions for pertussis during a three-year study period, 116 were unvaccinated. Out of a total of 529 German children hospitalized with pertussis from June 1996 to 1998, 423 were unvaccinated, and only 2 had gotten three or more doses of vaccine. See Karen R. Broder, "Pediatric Burden of Pertussis: Infants and Children Aged under 10—United States," 2 September 2004 presentation, Pertussis Working Group; P. Jurertzko et al., "Effectiveness of Acellular Pertussis Vaccine Assessed by Hospital-Based Active Surveillance in Germany," *Clinical Infectious Diseases* 35 (2) (2002): 163–67.

70. C. R. Vitek, "Increase in Deaths from Pertussis among Young Infants in the United States in the 1990s," *Pediatric Infectious Disease Journal* 22 (7) (2003): 628–34.

71. U. Heininger et al., "A Controlled Study of the Relationship Between Pertussis Infections and Sudden Unexpected Deaths among German Infants," *Pediatrics* 114 (1) (2004): e9–15.

CHAPTER 10

1. Leo Kanner, "Autistic Disturbances of Affective Contact," *The Nervous Child* 2 (3) (1943): 217–50.

2. Neal Halsey, author interview, May 2002.

3. Neal Halsey, author interview, April 2002.

4. The origins of the thimerosal story are described in *Hepatitis Control Report*, Summer 1999.

5. European Agency for the Evaluation of Medicinal Products, "Statement on Thiomersal Containing Medicinal Products," 8 July 1999, at www.emea .eu.int/pdfs/human/press/pus/2096299EN.pdf.

6. Neal Halsey, "Increased Mortality after High Titer Measles Vaccines: Too Much of a Good Thing," *Pediatric Infectious Disease Journal* 12 (6) (1993): 462–65.

7. U.S. Environmental Protection Agency, *Mercury Study: Report to Congress,*

vol. 4, *An Assessment of Exposure to Mercury in the United States*, 1997, retrieved from www.epa.govmercury/report.htm.

8. M. Bigham and R. Copes, "Thiomersal in Vaccines," *Drug Safety* 28 (2) (2005): 89–101.

9. Robert Chen, author interview, May 2002.

10. Tom Verstraeten e-mail to Phillipe Grandjean, 14 July 2000, at www.no mercury.org/science/documents/FOIA_emails_11-03.pdfmemo.

11. Peter Paradiso, speaking at National Vaccine Advisory Committee-Sponsored Workshop on Thimerosal in Vaccines, Bethesda, Maryland, 12 August 1999, 217.

12. Myron Levine, "Merck Knew about Thimerosal, Memo Shows," *Los Angeles Times*, 8 February 2005.

13. National Vaccine Advisory Committee, "Workshop on Thimerosal in Vaccines," Bethesda, Maryland, 12 August 1999.

14. Susan Levy and Susan Hyman, "Use of Complementary and Alternative Treatments," *Pediatric Annals* 32 (10) (2003): 685–91.

15. Eric Colman, author interview, June 2002.

16. Leo Kanner, "Early Infantile Autism Revisited," *Psychiatry Digest* 29 (2) (1968): 17–28.

17. I am in debt to Edward Dolnick's book *Madness on the Couch* (New York: Simon & Schuster, 1998) for much of the material on Rimland and Kanner.

18. "Problems of Nosology and Psychodynamics," in *Childhood Psychosis: Initial Studies and New Insights* (Washington: Winston, 1973).

19. Dolnick, *Madness*, 167–233.

20. Bernard Rimland and Stephen M. Edelson, *Treating Autism* (San Diego: Autism Research Institute, 2003): 14–18.

21. Richard Noll, "A Blood Test for Madness? Serology, Psychiatry and Dementia Praecox, 1912–1917," paper presented at American Association for the History of Medicine conference, Madison, Wisconsin, April 2004.

22. Mary and T. Campbell Goodwin, "Malabsorption and Cerebral Dysfunction: A Multivariate and Comparative Study of Autistic Children," *Journal of Autism and Childhood Schizophrenia* 1 (1) (1971): 48–62.

23. Bernard Rimland, author interview, May 2002.

24. Andrew J. Wakefield et al., "Ileal-lymphoid-nodular Hyperplasia, Non-Specific Colitis, and Pervasive Developmental Disorder in Children," *The Lancet* 351 (9103) (1998): 637–41; Simon H. Murch et al., "Retraction of an Interpretation," *The Lancet* 363 (9411) (2004): 750.

25. A detailed and convincing version of the MMR controversy is Michael Fitzpatrick's *MMR and Autism: What Parents Need to Know* (London: Routledge, 2004); *MMR: Science and Fiction*, by *Lancet* editor Richard Horton (London: Granta, 2004), is a rushed but thought-provoking account; Murch et al., "Retraction."

26. "Parent Groups and Vaccine Policymakers Clash over Research into Vaccines, Autism and Intestinal Disorders," at www.909shot.com/Press Release/pr030398autism.htm.

27. Eliot Marshall, "A Shadow Falls on Hepatitis B Vaccination Effort; Autoimmune and Nervous Disorders Resulting from Vaccine," *Science* 281 (5377) (1998): 630–31.

28. *Children's Healthcare Is a Legal Duty, Inc.*, newsletter #3, 2002.

29. See, for example, "Who Decides What Drugs Are Forced on Children?" *The Phyllis Schlafly Report* 34 (7) at www.eagleforum.org/psr/2001/feb01/psrfeb01.shtml; "Is Hillary Really for the Children?" www.townhall.com/opinion/column/phyllisschlafly/2000/11/08/167571.htm; Andrew Schlafly, 12 May 2003 testimony against chickenpox vaccine, at www.aapsonline.org/stateis/njvac.htm.

30. Kathleen Stratton, Marie C. McCormick, et al., *Immunization Safety Review: Hepatitis B and Demyelinating Neurological Disorders* (Washington: National Academies Press, 2002).

31. U.S. House, *Hearings before the Committee on Government Reform*, 107th Congress, 1st session, 26 April 2001, 308–9.

32. "No Link to Autism Proved at Hearing," *Indianapolis Star*, 12 January 2003; Evans Witt, "Birch Society Funds Movement for Laetrile," Associated Press, 10 October 1977.

33. Beth Clay, author interview, May 2005; Maurice Possley, "North Chicago Med School Sued on Cancer Cure," *Chicago Tribune*, 29 July 2004, NS1.

34. See www.Crp.org/politicians/summary.asp?CID=N0000001000cycle%2000.

35. See www.whale.to/v/sarkine.html; by "nine vaccines," Burton-Sarkine presumably was describing each type of antigen as a separate vaccine.

36. OPV and MMR do not contain thimerosal, and the only licensed Hib-Hep B combination, Merck's Comvax, contains no preservative. See www.merck.com/product/usa/pi_circulars/c/comvax/comvax_pi.pdf.

37. The Burton vaccine court case was disclosed to me by Laura Millman, a special master.

38. Stella S. Chess, "Autism in Children with Congenital Rubella," *Journal of Autism and Child Schizophrenia* 1 (1) (1971): 33–47.

39. Many of the parents' experiences with vaccines and autism are described in David Kirby, *Evidence of Harm* (New York: St. Martin's Press, 2005): 18–34, 60.

40. Liz Birt, author interview, May 2002.

41. Kanner, "Autistic Disturbances," 217–18.

42. Simon Baron-Cohen, "Have the Airplane and the Computer Changed the Architecture of the Mind? And Is That Why Autism Is on the Increase?" www.edge.org/3rd_culture/bios/baroncohen.html; Simon Baron-Cohen, "Two New Theories of Authism: Hyper-Systematization and Assortative Mating," *Archives of the Diseases of Children* 91 (1) (2006): 2–5.

43. Mark Blaxill, author interview, May 2002.
44. Baron-Cohen, "Two New," 4; and Lisa A. Croen et al., "Descriptive Epidemiology of Autism in a California Population: Who Is at Risk?" *Journal of Autism and Developmental Disorders* 32(3) (2002): 217–24.
45. Kirby, *Evidence of Harm*, 105.
46. See www.gnd.org/dr_jeff/dr_jeff.htm.
47. A. Hviid et al., "Association between Thimerosal-Containing Vaccine and Autism," *Journal of the American Medical Association* 290 (2003): 1763–66.
48. Bernard Rimland and A. Hviid, correspondence, "Association between Thimerosal-Containing Vaccine and Autism," *Journal of the American Medical Association* 291 (20) (2004): 180–81.
49. J. Heron et al., "Thimerosal Exposure in Infants and Developmental Disorder," *Pediatrics* 114 (3) (2004): 577–83; and Paul Offit and a member of the Vaccine Safety Branch, author interviews.
50. Memorandum opinion, *John and Jane Doe 2 v. Ortho-Clinical Diagnostics, Inc.*, U.S. District Court for the Middle District of North Carolina, July 6, 2006, case 1:03-cv-00669, p. 9.
51. Kathleen Stratton, Marie C. McCormick, et al., *Immunization Safety Review: Vaccines and Autism* (Washington: National Academies Press, 2004): 65.
52. Sarah Parker et al., "Thimerosal-Containing Vaccines and Autism Spectrum Disorder: A Critical Review of Published Original Data," *Pediatrics* 114 (3) (2004): 793–804; and Mark R. Geier and David A. Geier, "Neurodevelopmental Disorders after Thimerosal-Containing Vaccines: A Brief Communication," *Experimental Biology and Medicine* 3 (228) (2003): 660–64.
53. California Department of Developmental Services, *Changes in the Populations of Persons with Autism and Pervasive Developmental Disorders in California's Developmental Services System 1987–1998* (Sacramento: DDS, 1999), available at www.dds.cahwnet.gov/autism/pdf/autism_report_1999.pdf.
54. U.S. House, *Autism—Why the Increased Rates? A One-Year Update: Hearings before the Committee on Government Reform*, 107th Congress, 1st Session, 25 and 26 April 2001, 1–2, 8.
55. Loring Dales, "Time Trends in Autism and in MMR Immunization Coverage in California," *Journal of the American Medical Association* 285 (9) (2001): 1183–85; AAP, Committee on Infectious Diseases, "Haemophilus Influenzae Type b Conjugate Vaccines: Recommendations for Immunization of Infants and Children 2 Months of Age and Older," *Pediatrics* 88 (1) (1991): 169–72.
56. J. G. Gurney et al., "Analysis of Prevalence Trends of Autism Spectrum Disorder in Minnesota," *Archives of Pediatric and Adolescent Medecine* 157 (July 2003): 622–27.
57. Craig J. Newschaffer et al., "National Autism Prevalence Trends from U.S.

Special Education Data," *Pediatrics* 115 (3) (2005): 277–82. Newschaffer, a Hopkins researcher who has an autistic son, used U.S. census data and Department of Education statistics to examine autism spectrum diagnoses in children born from 1975 to 1995. The biggest increase occurred for cohorts born from 1987 to 1992, with a somewhat smaller increase in successive years.

58. R. Lingam et al., "Prevalence of Autism and Parentally Reported Triggers in a North East London Population," *Archives of the Diseases of Children* 88 (2003): 666–70.

59. Reported rates of autism changed dramatically depending on the definition of the disease and the size of the study, with smaller studies generally finding higher rates. For 16 surveys conducted between 1966 and 1991, the median prevalence rate was 1 in 2,500, while for 1992–2001 the median rate was about 1 in 800. Three more recent studies showed rates of about 1 in 150 for all forms of the autism spectrum. But all the more recent studies searched for children more thoroughly. Studies that relied upon administrative records, such as the figures from California, found lower rates. This convinced one author, Eric Fombonne, that study methodology "could account for most of the variability in published prevalence estimates." Other epidemiologists disagreed. Newschaffer, for one, felt that some of the increase was unrelated to diagnostic shift.

60. G. Stajich et al., "Iagrogenic Exposures to Mercury after Hepatitis B Vaccination in Preterm Infants," *Journal of Pediatrics* 136 (5) (2000: 679–81; M. Pichichero et al., "Mercury Concentrations and Metabolism in Infants Receiving Vaccines Containing Thimerosal," *The Lancet* 360 (9347) (2002): 1137–41.

61. G. J. Harry et al., "Mercury Concentrations in Brain and Kidney Following Ethylmercury, Methlmercury and Thimerosal Administration to Neonatal Mice," *Toxicology Letters* 154 (2004): 183–89; Tom M. Burbacher et al., "Comparison of Blood and Brain Mercury Levels in Infant Monkeys Exposed to Methylmercury or Vaccines Containing Thimerosal," *Environmental Health Perspectives*, published online 21 April 2005.

62. Boyd Haley, author interview, June 2002.

63. Thomas W. Clarkson et al., "The Toxicology of Mercury—Current Exposures and Clinical Manifestations," *New England Journal of Medicine* 349 (18) (2003): 731–37.

64. Jill James, "Thimerosal Neurotoxicity Is Associated with Glutathione Depletion, Protection with Glutatione Precursors," *Neurotoxicology* 26 (1) (2005); "Metabolic Markers of Oxidative Stress and Impaired Methylation Capacity in Children with Autism," *American Journal of Clinical Nutrition* 80 (6) (2004): 1611–17.

65. Stratton, *Autism*, for example 52–62, 75–77, 143–47. The report contains an avalanche of evidence refuting the vaccine-autism link.

66. Michael Rutter, "Genetic Studies of Autism: From the 1970s into the Millenium," *Journal of Abnormal Child Psychology* 28 (1) (2000): 3–14.

67. Kirby, *Evidence of Harm*, 66.

68. Amy Holmes, author interview, May 2002.

69. Stratton and McCormick, *Vaccines and Autism*, 128–45.

70. David Weldon, "Are Autism, Vaccines and Mercury Related?" *The Hill*, 9 February 2005.

71. See www.dds.cahwnet.gov/FactsStats/pdf/Jun06_Quarterly.pdf and www.dds.cahwnet.gov/FactsStats/pdf/Dec02_Quarterly.pdf.

72. Myron Levin, "Merck Misled on Vaccines, Some Say," *Los Angeles Times*, 7 March 2005.

73. Levy and Hyman, "Use of Complementary," 685–91; and "A Radical Approach to Autism," Amy Dockser Marcus, *The Wall Street Journal*, 15 February 2005, D1.

74. Discussion with a leading DAN! practitioner who wished to remain anonymous.

75. B. E. Esch and J. B. Carr, "Secretin as a Treatment for Autism: A Review of the Evidence," *Journal of Autism and Developmental Disorders* 34 (5) (2003): 543–56.

76. J. Coplan et al., "Children with Autistic Spectrum Disorders. II: Parents are Unable to Distinguish Secretin from Placebo under Double-Blind Conditions," *Archives of Diseases of Childhood* 88 (2003): 737–39.

77. A. D. Sandler and J. W. Bodfish, "Placebo Effects in Autism: Lessons from Secretin," *Developmental and Behavioral Pediatrics* 21 (5) (2000): 347–49.

78. Environmental Working Group, "Overloaded? New Science, New Insights about Mercury and Autism in Susceptible Children," at www.weg.org/reports/autism/execsumm.php.

79. Robert F. Kennedy, "Playing Politics at Kids' Expense," *The Press-Enterprise*, 10 April 2005.

80. "An Open Letter to the American Public," 22 October 2004, at www.autismone.com. For further scrutiny of the Handley family's unusual perspective on autism, see Angela Valdez, "Curing Jamie Handley" *Willamette Week*, 12 October 2005; and discussion in *A Photon in the Darkness*, at http://photoninthedarkness.blogspot.com, a blog by critics of the thimerosal-causes-autism theory. Handley's son was born in late 2002, when mercury had been removed from most vaccines, but in interviews Handley has insisted that his son was mercury-poisoned.

81. *The New York Times* ads appeared on 8 June and 14 November 2005.

82. David Weldon, "Open Letter to Julie Gerberding," 31 October 2003, at www.momsonamission.org/Autism_Center_Dr_Weldon_Respond.shtml.

83. Office of Special Masters, U.S. Court of Federal Claims, "Autism Update, December 22, 2005," at www.uscfc.uscourts.gov/OSM/Autism/Autism_Update_12_22_05.pdf.

84. Peter Paradiso, author interview, March 2005.

85. Roger Bernier, author interview, April 2004.

86. Stephen Cocchi, author interview, February 2005.

EPILOGUE

1. Carla McClain, "Vaccine Tie to Autism Gains New Supporters," *Arizona Daily Star*, 3 July 2005.

2. Barbara Loe Fisher, author interview, July 2005.

3. Daniel Salmon and Andrew Wiegel, "Religious and Philosophical Exemptions from Vaccination Requirements and Lessons Learned from Conscientious Objectors from Conscription," *Public Health Reports* 116 (2001): 289–95; Daniel Salmon, Neal Halsey, et al., "Public Health and the Politics of School Immunization Requirements," *American Journal of Public Health* 95 (5) (2005): 778–83.

4. Roger Bernier, author interview, July 2005.

5. Robert D. McFadden, "Frustration and Fear Reign over Flu Shots," *The New York Times*, 16 October 2004, A1.

6. Stephen Smith, "Flu Shot Shortage Shows System Flaws," *Boston Globe*, 10 October 2004.

7. Alfred M. Prince, "Our Last Vaccine?" *Science* 195(4284) (1977): 1287.

8. Jonas Salk, "Wanted: A Single Flue Vaccine," *The Washington Post*, 21 November 1976, 36.

9. Cf. Philip K. Russell, "Heading Off a Crisis in Vaccine Development," *Issues in Science and Technology* 11 (1995): 28.

10. David Brown, "Fixing Vaccine Supply System; Task Will Not Be Easy," *The Washington Post*, 9 October 2004.

11. Cf. Paul Offit, "Why Are Pharmaceutical Companies Gradually Abandoning Vaccines?" *Health Affairs* 24 (3) (2005): 622–30.

12. The 26 figure comes from "Historical Record of Vaccine Product License Holders in the United States," Annex H in *The Children's Vaccine Initiative, Achieving the Vision* (Washington: National Academy Press, 1993).

13. Speech at "A Scientific Colloquium Honoring Maurice Hilleman," American Philosophical Society, Philadelphia, 26 January 2005.

14. Paul Parkman, author interview, May 2004.

15. Cf. Jean Kozubowski, "Connaught Takes Risks, Shows Swift Growth," *Northeast Pennsylvania Business Journal*, September 1988, 20.

16. Inactivated polio vaccine was Lilly's topselling product in 1955, according to *Eli Lilly and Company: Innovation, Diversification, and Globalization* (Mountain View, CA: Market Intelligence, 1993).

17. Offit, "Why Are," 624.

18. Roy Vagelos, remarks at "Hilleman Symposium"; Fran Hawthorne, *The Merck Druggernaut* (Hoboken, NJ: Wiley & Sons, 2003): 53–54.

19. See www.gsk-bio.com/webapp/CM/CM_Histoire_Developpement.jsp.
20. "Second NVAC Workshop on Strengthening the Supply of Vaccines in the United States," Washington, DC, 24 January 2005.
21. Lone Simonsen, "Impact of Influenza Vaccine on Seasonal Mortality in the U.S. Elderly Population," *Archives of Internal Medicine* 165 (3) (2005): 265–72.
22. David Wahlberg, "Two Public Agencies Disagree on Value of Flu Shots," *Atlanta Constitution*, 14 February 2005; Lone Simonsen et al., "Mortality Due to Influenza in the United States—an Annualized Regression Approach Using Multiple-Cause Mortality Data," *American Journal of Epidemiology* 163 (January 15, 2006): 181–87.
23. Patricia Reaney "No Evidence Flu Vaccine Works in Kids, Study Finds," Reuters, 25 February 2005.
24. Jon Cohen, "Vaccine Policy: Immunizing Kids against Flu May Prevent Deaths among the Elderly," *Science* 306 (5699) (2004): 1123.
25. Susan Heavey, "Flu Vaccine Makers Say Government Must Increase Demand," Reuters, 12 February 2004.
26. Mike Williams, author interview, June 2005.
27. U.S. House, *Hearing of the Health, Oversight and Investigations Subcommittees of the House Energy and Commerce Committee on the Flu Vaccine,*" 18 November 2004.
28. David Salisbury, author interview, January 2005.
29. L. Von Seidlein, "The Need for Another Typhoid Fever Vaccine," *Journal of Infectious Diseases* 192 (3) (2005): 357–59.
30. Cf. B. L. Laube, "The Expanding Role of Aerosols in Systemic Drug Delivery, Gene Therapy and Vaccination, *Respiratory Care* 50 (9) (2005): 111–76; H. O. Alpar et al. "Strategies for DNA Vaccine Delivery," *Expert Opinion on Drug Delivery* 2 (5) (2005): 829–42. For a more sober view, see Maurice Hilleman, "A Simplified Vaccinologist's Vaccinology for the Pursuit of a Vaccine Against AIDS," *Vaccine* 16 (8) (1998): 778–93.
31. Cf. K. L. Kotloff and J. B. Dale, "Safety and Immunogenicity of a Recombinant Multivalent Group A Streptococcal Vaccine in Healthy Adults: Phase 1 Trial," *Journal of the American Medical Association* 292 (11 August 2004): 709–15; K. L. Kotloff and J. B. Dale, "Progress in Group A Streptococcal Vaccine Development," *Pediatric Infectious Diseases Journal* 23 (August 2004): 765–66; and A. P. Durbin and R. A. Karron, "Progress in the Development of Respiratory Syncytial Virus and Parainfluenza Virus Vaccines," *Journal of Infectious Diseases* 191 (1 April 2005): 1093–94.
32. Fawn Vrazo, "Promising New Vaccines Could Wipe Out Cervical Cancer," *Philadelphia Inquirer*, 4 July 2005.
33. Meghan Meyer, "Cancer Funds a Morality Issue?" *Miami Herald*, 11 July 2005.

34. Mike Williams, author interview, July 2005; D. A. Henderson, author interview, July 2005.

35. Milton Leitenberg, "Bioterror, Hyped," *Los Angeles Times*, 17 February 2006.

36. "An Interview with David S. Fedson, MD," *Biosecurity and Bioterrorsm* 3 (1) (2005): 9–15.

37. "Industry Needs Shot in the Arm," *Forbes*, 10 May 2005.

38. Susan Levy, "The Face That Helped Alter Nation's Flu Policy," *The Washington Post*, 7 March 2006, B1.

39. S. Dominguez and R. S. Daum, "Toward Global Haemophilus influenzae Type B Immunization," *Clinical Infectious Diseases* 37 (12) (2003): 1600–2; "Elimination of HiB Disease from the Gambia after the Introduction of Routine Immunization," *The Lancet* 366 (9480) (2005): 144–50.

40. "Malaria: The Sting of Death," *Los Angeles Times*, 30 June 2005.

41. Stephen Hadler, author interview, February 2005.

42. Jon Cohen, "The New World of Global Health," *Science*, 13 January 2006, 162.

43. Ernst R. Berndt and John A. Hurvitz, "Vaccine Advance Purchase Agreements for Low-Income Countries: Practical Issues," *Health Affairs* 24 (3) (2005): 653.

44. June Goodfield, *A Chance to Live* (New York: Macmillan, 1991): 171.

45. Bruce Aylward et al., "OPV Cessation—the Final Step to a 'Polio-Free' World," *Science* 310 (2005): 625–26.

46. Ibid., 625.

47. John Murphy, "Distrust of U.S. Foils Effort to Stop Crippling Disease," *Baltimore Sun*, 4 January 2004.

48. Donald G. McNeil Jr. and Celia Dugger, "On The Brink: Rumor, Fear and Fatigue Hinder Final Push to End Polio, *The New York Times*, 21 March 2006, A1.

49. See, for example, Rene Dubos, *Mirage of Health* (New York Harper and Row, 1959): 25, 75, 94, 190–91.

50. Rene Dubos, "Pasteur's Dilemma: The Road Not Taken," in *ASM News* 40 (1975): 703–9.

INDEX

ABC, 263, 388
abortions, 235–36, 390–91
  deaths caused by, 237
  illegal, 235–36, 237–38
  legalization of, 217, 237
  rubella exposure and, 235–36, 237–38
  scientific use of fetuses from, 238, 390–91
  therapeutic, 235–36
Abramson, Jon, 381
Acosta, Carmen, 116
Acosta, Ismael, 116
acrodynia (Pink disease), 396
ACT-UP, 402–3
acupuncture, 361
Adams, Abigail Smith, 42, 43
Adams, John, 41–42, 43
Adehlardt, Frances, 230
adenovirus, 234
adenovirus vaccine, 223
Adkin, Thomas F., 78
Aedes aegypti mosquitoes, 144
Aetna insurance, 178
AFL-CIO, 229
Africa, 71, 149, 207
  smallpox in, 12, 37, 243
African Americans, 76, 81, 97, 205, 244, 294
Agency for Evaluation of Medicinal Products, 377
Age of Treason, The (Clymer), 341
Agriculture Department, U.S., 96, 141
Ahmed, Ibrahim Datti, 440
AIDS, 16, 53, 284, 297, 316, 332, 426, 435, 440
  activism and, 402–3
  global pandemic of, 207, 209, 225
  origins of, 225
  research in, 201, 213, 225, 302, 426
air, 65
  pollution of, 67, 334
Albert Einstein Medical School, 302
Alexander, H. M., 93–94
Ali, Muhammad, 245
Alice in Wonderland (Carroll), 397
allergic responses, 125–26, 143, 146, 241, 253, 275, 317, 347, 348, 386

Alliance for Progress, 229
Alpert, Joel, 243
alternative medicine, 18, 86, 103–4, 108, 230, 330, 331, 334–39, 345–47, 364, 373, 383, 386, 403
  see also homeopathy; naturopathy
Alzheimer's disease, 410
American Academy of Pediatrics (AAP), 243, 278, 279, 283, 287, 292, 299, 311, 315, 317, 324, 326, 373, 381–82, 395, 431
  Infectious Disease Committee of, 375
  Red Book Committee of, 288, 376, 429
American Anti-Slavery Society, 103
American Association of Physicians and Surgeons, 340
American Cancer Society, 210
American Friends Service Committee, 225–26
American Journal of Hygiene, 268
American Legion, 260
American Medical Association (AMA), 61, 72–73, 97, 207, 270, 286, 395
  Council on Defense of Research in, 106
  experimental ethics concerns of, 158, 189
  House of Delegates of, 204
  New and Nonofficial Remedies of, 268
American Pediatrics Society, 242
American Public Health Association, 173
American Revolution, 41, 42, 131
Amundsen, Roald, 127
anaphylaxis, 125–26, 175
Anderson, Christine, 351
Anderson, H. B., 136
Anderson, John F., 73
Anderson, Monika, 351
Anderson, Porter, 307
Angell, Marcia, 330
Angel of Bethesda, The (Mather), 27
animals, 90, 224
  experiments on, 68, 81, 85, 86, 121–22, 123, 124–25, 126, 145–47, 172, 174–75, 176, 180, 185, 186, 191, 200, 202, 207, 209–10, 211, 212, 220–21, 225, 268–69

animals (*continued*)
organized protection of, 85–86
wild, 355
Anopheles mosquitoes, 142
anthrax, 121, 222, 331
2001 mailing of, 13, 324, 368–69
anthrax vaccine, 13–14, 433
anthroposophy, 342–50
antibiotics, 120, 129, 161, 162, 169, 216, 219,
223, 307–8, 340, 352, 358, 359, 363
antibodies, 14–15, 19, 36, 122, 177, 178, 180,
186, 218, 306–7, 368
harvesting of, 19, 72
maternal, 167, 169, 244, 297, 304
anti-bubonic plague serum, 105
anticancer vaccine, 310
antidepressants, 330, 384
antidiuretics, 223
Antietam, Battle of, 132
antigenic drift and shift principle, 223
antimony, 33, 39, 40, 42
anti-Semitism, 183, 341
antitoxin, 19, 105, 123, 125
see also tetanus antitoxin
Anti-Vaccination League (England), 64, 69,
85–86, 135
Anti-Vaccination League (Pennsylvania),
102–3
Anti-Vaccination League of America, 103–4,
106
Anti-Vaccination Society, 86, 87, 339
antivaccine activism, 17–18, 30–34, 56–57,
61, 64–69, 73, 82, 85–88, 89, 95,
97–111, 134, 135–36, 139, 328–70,
388
animal rights and, 85–86, 103
artists and intellectuals in, 99, 342–47
books in support of, 337–38
cultural pessimism in, 348–49
entrenched attitudes in, 353–54
funding of, 100–102, 357
homeopathy and, 82, 87, 101, 103–4, 330,
335–38, 345
individual freedom and, 103–4, 105–6
medical professionals' embrace of, 353–54,
356
politics and, 102–4, 136, 334–35
propaganda of, 142, 334–35, 337–38, 341,
354, 356–58
Raggedy Ann doll symbol in, 99
religious and philosophical views in,
100–101, 230, 294–97, 328, 332–34,
335–37, 339–50
rising disease rates caused by, 354, 366, 369
websites of, 354, 356, 357, 375
antivivisection, 86, 103, 124–25

Anti-Vivisection Society, 86
Anzio, Battle of, 165
Appel, Maddie, 236
Applied Behavioral Analysis, 399
Arab/Israeli conflict, 213
Arizona, University of, 398, 410
Armed Forces Epidemiological Board,
137–38, 174
Armstrong, Louis, 166
Army, U.S., 133, 136, 159
malaria control agency of, 162
Medical School Laboratory of, 116
101st Airborne Division of, 278
reformed medical practices in, 75, 133, 135
Army Medical Museum, 134
Army of Northern Virginia, 132
arsenic, 397
Artenstein, Mal, 234
asbestos poisoning, 213
Ashbridge, John, 73, 91
Asperger's syndrome, 395
Aspinwall, William, 44, 50, 51, 175
Association of Waldorf Schools of North
America, 346
asthma, 15, 240, 258, 338, 347, 348, 349,
362
Atlantic Monthly, 352
attention deficit hyperactivity disorder
(ADHD), 15, 287, 292, 329, 338, 349,
396
autism, 15, 18, 329, 335, 338, 349, 371–420
alternative approaches to, 383, 386–87
biological origin of, 387
diagnosis of, 373, 374, 380, 390, 408
educating children with, 373, 399
enhanced abilities with, 372, 384
parents of children with, 384–86, 388, 391
prevalence of, 372, 373, 387, 400, 408–9,
420
rubella exposure and, 234, 236, 394
symptoms of, 371–72, 373, 374–75, 383,
384, 386–87, 395–96
vaccine link to, 362, 372–75, 387–88,
390–91, 394–400, 403–20,
421–23
"Autism—A Novel Form of Mercury
Poisoning" (Bernard, Redwood,
Enayati, Roger and Binstock), 395,
396
autoimmune diseases, 13, 16, 163, 329,
389–90, 391
Aventis-Pasteur, 96, 285, 321, 412, 426
aversion therapy, 386
Avery, Oswald, 307
avian flu, 16, 260
H5N1 strain of, 424

babies:
  bundling of, 328
  developmental delay in, 256, 359, 404
  premature births of, 160
  rubella, 233–34, 236–38
  sudden death of, 256, 277, 287, 288, 356, 357, 369
Babies Milk Fund, 171, 231
Baby and the Medical Machine, The (Little), 105
baby boomers, 138, 379
Bacille Calmette-Guerin (BCG) vaccine, 20, 136–37
bacteria, 19, 53, 65, 75, 123, 260, 266
  killed, 129
  live attenuated, 348
  modified, 20–21, 136
  pus, 94
  recombinant, antigen-bearing, 430
  saprophytic, 94
bacterins, 258, 259
bacteriology, 65, 73, 81, 90, 92–95, 98, 121, 123, 133, 149, 166, 180, 344
Baker, Sara Josephine, 115, 127
Ball, Leslie, 375–76
Ball, Robert, 376
Ballou, Rip, 436
Baraff, Larry, 290
Bartlett, Ethan, 361
Baruch, Hermann, 127
Baskin, David, 410
Batt, Gail Adams, 193
battered child syndrome, 218, 241, 339, 341–42
Bauer, Johannes, 146, 151, 154–55
Baxby, Derrick, 55
Baylor, University, 214, 389, 410
Bayne-Jones, Stanhope, 138, 150, 153, 157
Bay of Pigs invasion, 229
Beasley, Oscar, 88, 102
Beck, Parker, 388–89
Beck, Victoria, 388–89
behavioral problems, 253, 268
Behring, Emil von, 123
Bellevue Hospital, 235
Belli, Melvin, 263
Belt, J. H., 84
Bennet, Byron, 180
benzethonium chloride, 264
Bergen-Belsen concentration camp, 140, 141
Bergh, Henry, 85
Berlin, 60, 63–64, 122, 123, 177
Bernard, Billy, 395
Bernard, Sallie, 395
Bernier, Roger, 416–17, 423
Bettleheim, Bruno, 385
Biemiller, Andrew, 229

Biggs, Herman M., 74, 110, 111, 123, 126, 130, 131
Bill and Melinda Gates Foundation, 435–36, 437
  Malaria Vaccine Initiative of, 436
Binstock, Teresa, 395
biochemistry, 127
Biologics, 427–28, 432–33
biometrics, 133
bioterrorism, 149–51, 330, 368–69, 426
biowarfare, 143, 222
Birch, John, 56
Birt, Liz, 395, 403
birth control, 341
Birthday Ball Commission, 166, 172, 173
birth defects, 35, 98, 233–34, 235–38
birth rates, 129
Bishop, Ruth, 318
Black, Kathryn, 170–71, 214
Blake, Aki, 360–61
Blake, Finn, 360, 361
Blake, Serena, 361
Blavatsky, Helena Petrovna, 343
Blaxill, Mark, 401–2, 411, 419
blindness, 237, 238, 303, 338
blood, 48, 140, 218
  clotting of, 231, 317, 325
  hemorrhaging of, 144
  infection spread through, 98, 140, 147–48
  mingling of, 65, 151, 343
  pollution of, 56, 87, 100, 101, 133, 139, 140, 158, 343
  red cells in, 125
  testing of, 151, 153, 223
  umbilical cord, 380
  vomiting of, 144
  white cells in, 133, 267, 347
bloodletting, 27, 33, 38, 39, 40, 48, 52–53, 125, 337
Bloom, Barry, 438
Bloomsberg School of Public Health, 378
Bodfish, James, 413
Bodian, David, 174, 177–78, 196–97
Boerhaave, Hermann, 40
Boer War, 133
Boettjer, Joanne, 171
Boettjer, Joseph, 171
Bogart, Humphrey, 166
Boone, Cynthia, 391
Bordet, Jules, 268
Bordetella pertussis, 268
Boscawen, Frances, 25
Boston, Battle of, 131
Boston, Mass., 25–36, 45, 61, 74
  Board of Health in, 51, 82
  Brigham and Women's Hospital in, 220

Boston, Mass. (*continued*)
　Children's Hospital in, 226, 243, 270
　Lying-In Hospital in, 220
　1702 smallpox epidemic in, 28, 35
　1721 smallpox epidemic in, 25–26, 27–36, 38
　1730 smallpox epidemic in, 31
　1753 smallpox epidemic in, 33, 43–44
　vaccination campaigns in, 81–82
Boston University, 380
botulism, 152
Boylston, Tommy, 29
Boylston, Zabdiel, 29–33, 34, 41–42
Bradstreet, Jeff, 403
brain, 174–75, 197
　damage to, 41, 147, 217, 234, 251–52, 268, 288–91, 298, 306, 338, 339, 342, 398, 410
　frontal cortex of, 234
　inflammation of, 80, 117, 146, 338
Brandt, Alan, 177
Brandt, Edward, 281–82
Brazil, 71, 75, 147–48, 151, 152, 155, 156, 183
Brebner, William, 186
Brent, Robert, 405
British Parliament, 57, 70, 69
Brock, Susan, 391
Brodie, Maurice, 172–74, 184, 187
bronchiolitis, 240
bronchitis, 258
Brooklyn City Dispensary, 59–60
Brower, William, 79
Brown, Charles W., 107
Bryn Athyn Academy, 100, 101
Buchenwald concentration camp, 385
Buck v. Bell, 98–99
Buescher, Ed, 184
Bugher, John, 148
Bumpers, Betty, 245
Bumpers, Dale, 245
Bureau of Biologics (1972–82), 258, 271, 276
Bureau of Land Management (BLM), 355
Burnett, McFarlane, 178
Burney, Leroy E., 207
Burroughs Wellcome Institute, 147
Burton, Dan, 322, 326, 329–30, 374, 391–94, 408, 414, 415, 417, 421
Burton-Sarkine, Christopher, 392–93
Burton-Sarkine, Danielle, 392–93, 394
Bush, George H. W., 299–300, 310
Bush, George W., 17, 415, 426, 433, 436
　"compassionate conservatism" of, 13
　2002 smallpox vaccination of, 11–12
Bush, Vannevar, 138, 216, 222–23
Butler, Paul, 204

Buttram, Harold, 338–42

cadmium, 27, 397
Cadoz, Michel, 285
Cagney, James, 166
Califano, Joseph A., 245–46
California, University of, 194
　in Los Angeles (UCLA), 283, 290, 368, 369
　in San Francisco, 192, 219
California Board of Health, 162
California Department of Developmental Services, 408
California Department of Health, 198
calomel (mercurous chloride), 41, 396
Cambridge Board of Health, 98
Camden, N.J., 202
　Board of Health in, 80–81, 94
　smallpox in, 77, 92–93
　tetanus death in, 79–81, 92–93
Cameron, Jack, 269
Canada, 101, 131, 231, 369
cancer, 15, 19, 258
　breast, 18, 330, 346
　cervical, 431–32
　childhood, 348, 349, 384
　deaths from, 18, 231, 349
　liver, 148, 153, 158, 312
　research on, 211, 212, 213
　"war" on, 240
Cantor, Eddie, 166
Capitol University of Integrative Medicine, 392
Carnegie, Andrew, 100
Caroline, Princess of England, 38
Carpenter, Arthur, 166
Carrel, Alexis, 349
Carson, Paul, 82
Carson, Rachel, 142
Carter, Jimmy, 243, 245, 247, 258, 262
Carter, Rosalynn, 245, 247
Carver, George Washington, 162
Case Western Reserve Medical College, 83, 283
Castro, Fidel, 132, 208, 228, 392
cataracts, 233
Cave, Stephanie, 394
Cawley, Morris, 78
CBS, 187
celiac disease, 386
cell biology, 127
Center for Biologics Evaluation and Regulation, 259
Centers for Disease Control (CDC), 11, 13, 155, 204, 217, 243, 245, 246–47, 259, 260, 265, 277, 281, 302–4, 309, 316, 321–26, 351, 369, 375, 377–78, 380,

381, 383, 403–4, 405, 413, 415–20, 422, 431
Advisory Committee for Immunization Practices (ACIP) of, 12, 239, 300, 301, 310–12, 318, 321, 323, 324, 326, 376, 404, 429
  crisis of credibility in, 17, 415–20
  Epidemic Intelligence Service of, 199, 200–201, 316
  National Immunization Program at, 316–17, 403
  origin of, 162
  Polio Surveillance Unit at, 202–3
Central Intelligence Agency (CIA), 12, 193, 335
Certiva, 412
Chalabi, Ahmed, 150
Chambron, Dr., 62, 63
Chancellorsville, Battle of, 132
Chanock, Robert, 184, 240
Chapin, Charles, 95
Chautauqua movement, 108
chelation, 394, 396–97, 399, 400, 413
Chelsea Hospital, 56
chemotherapy, 352
Chen, Bob, 316–18, 321, 323, 326, 379, 380, 406, 417–20
Cheney, Richard, 13
Cherry, James, 283, 288, 290, 354, 396
Chesterton, G. K., 86
Chiang Kai-Shek, 316
Chicago, University of, 223, 224, 316
Chicago Tribune, 157, 286–87
Chicago World's Fair of 1893, 94
chicken cholera, 62, 121
chickenpox, 71, 179, 218, 232, 301, 308, 312–16, 360
chickenpox vaccine, 17, 239, 312–16, 351, 425
chickens, 140–41, 221
chicken soup, 351
child abuse, 218, 241, 339, 341–42
child labor, 90
Children's Defense Fund, 299
Children's Vaccine Initiative, 437
China, 191, 200, 364
Chiron, 286, 310, 424, 426
cholera, 12, 62, 68, 74, 118, 168
cholera vaccine, 20, 269
cholesterol drugs, 428
Christensen, C. N., 273
Christianity, 343
  evangelical, 336
  fundamentalist, 362
  mystical, 100–101, 342
Christian Science, 230, 333–34, 337

Churchill, Winston, 134, 154
Churchill Hospital (England), 347
Church of Illumination, 339
Cincinnati, University of, 210
Cincinnati Bible Seminary, 392
Citizens Medical Reference Bureau, 102, 136
City College of New York, 179, 302
Civilian Defense Volunteer Organization, 116–17
civil rights movement, 216, 225
Civil War, U.S., 58, 60, 61, 75, 131–32
Clark, Fred, 319, 320–21, 326
Clarkson, Thomas, 380
Clay, Beth, 393
Clean Air Act, 255, 334
Cleveland, Ohio, 82–85, 90, 92
  disinfection and isolation smallpox program in, 82–83, 84, 85, 90
Cleveland Medical Association, 83
Cline, Henry, 49–50
Clinton, Bill, 300, 301, 305, 316
Clinton, Hillary Rodham, 305, 334
Clymer, Reuben Swinbourne, 87, 339, 340–41, 345
Cocchi, Steve, 418, 419–20
Cohen, Richard, 261
cold war, 12, 177, 228–29, 262
Colio, Lloyd, 273
College of Physicians, 93
Colman, Eric, 383
colonics, 44, 103
Colorado, University of, 218, 242, 262
Columbine shootings, 396
coma, 268
Comedy Central, 415
common cold, 221, 223, 350
communicable diseases:
  campaigns against, 11–14, 15, 16–17
  decline in, 15, 16, 129–30, 136, 162, 169, 328
  fear of, 12–13, 17, 25, 30, 34, 118
  intentional spread of, 218
  modern increases in, 243–44, 354, 366, 369
  research on, 127, 128, 130, 137–38
  spirituality and, 328, 344
Communists, 122, 177, 206, 341
Comprehensive Childhood Immunization Act (1993), 300
concentration camps, 140, 141, 143, 150, 385
Confederate Army, 132
Congregationalist Church, 31
Congress, U.S., 59, 72, 98, 135–36, 194, 205, 208, 237, 238, 260–61, 286–87, 301, 324, 405, 420
  see also House of Representatives, U.S., Senate, U.S.

conjugate pneumococcal vaccine, 307
Connaught Laboratories, 96, 260, 266, 269, 274, 283, 286, 428
conscientious objectors, 158
Consumers' Union, 417
Continental Congress, 42–43
Converse, Judy Lafler, 390
convulsions, 52, 79, 268, 270, 279, 290
Cook, James, 47
Coolidge, Calvin, 173
Cooper, Louis Z., 235–39
Cooper, William, 34–35
Cope, Porter F., 88, 100, 101, 102–3, 107–8
cortisone, 223
coughing, 217, 267–68, 360, 361, 364, 368
Coulter, Harris, 279–80, 337, 338, 340, 387
Country Gentleman, 166
Court of Appeals, U.S., 261, 287
"Cow Pock—or—the Wonderful Effects of the New Inoculation, The" (Gillray), 57
cowpox, 46–55
    origins of, 54–55
    symptoms of, 49, 50, 51, 53, 59
cowpox lymph, 49, 51, 52, 54, 57–58, 61, 64, 65, 83, 95, 132
cowpox virus, 49–53
    harvesting of, 49, 51, 52, 54, 55, 57–58, 59, 60, 61, 62–63, 65
    mutations of, 52, 54–57, 58
    supply and distribution of, 50–51, 52, 57–64, 65
    testing of, 50
cows, 20, 54–55, 61–64, 65
    foot-and-mouth disease in, 65
    vaccinia virus in, 12, 27, 46, 117, 120
Cox, Herald, 141, 181, 182–83, 206
Cox, James, 163
Coxe, John Redmond, 51
Cox-2 inhibitors, 330
Coyle family, 77
"creation science," 336
Crick, Francis, 197
Crippler, The, 166
Criss Prize, 197
Crohn's disease, 388
Crowley, Dorothy, 198
Cuba, 75, 132, 134, 145, 208, 213, 228–29, 239, 305, 392, 438
Culver, Bettyan, 182
Cutter Analytic Laboratories, 96, 196, 197–99, 201–2, 203, 205, 210, 218, 258, 263, 273, 427
cytokines, 347

Dachau concentration camp, 141, 385

Dalhonde, Laurence, 30
Dartmouth College, 12
Darwin, Charles, 68, 400
Davis, H. H., 80
Davis, Robert, 404
daycare centers, 216, 298, 334
DDT, 118, 141–42, 359
deafness, 217, 233, 234, 237, 238, 298, 306, 307
Declaration of Independence, 43
Defeat Autism Now! (DAN!), 394, 397, 398–99, 401
Defense Department, U.S., 427
dehydration, 296, 320, 387
de Kruif, Paul, 46, 129, 166, 172, 193, 436–37
delirium, 217, 219
democracy, 130, 165, 423
Democratic National Committee, 203–4
Democratic Party, 82, 103, 163, 165, 205, 229, 244, 300, 425–26
dengue fever, 183
Denver Children's Hospital, 332
depression, 384
Depression, Great, 161, 163, 165, 168, 268
de Quadros, Ciro, 305, 438, 440
de Solenni, Pia, 431–32
Deth, Richard, 398
diabetes, 111, 329
    juvenile, 15, 308, 348
diarrhea, 33, 35, 39, 78, 109, 167, 169, 217, 323, 338, 387
Dick, George, 276
diet, 39, 40, 48, 359, 400
    gluten-free, 386, 387, 395, 402, 413
    high-protein, 154
diethylene glycol, 377–78
Dingell, John, 284
diphtheria, 56, 72, 109, 120, 123–28, 229–30, 340, 344
    deaths from, 78, 91, 105, 124, 136, 162, 169, 269
    successful campaigns against, 74, 111, 124–28, 129, 162, 247, 331, 349
    swab test kits for, 123, 124
diphtheria, tetanus, and pertussis (DTP) vaccine, 251–57, 268, 273, 298, 307, 311, 325, 369, 376, 379, 389, 406, 412, 426–27, 434
    adverse effects and death from, 251–55, 270, 277–80, 287–93, 323, 328, 396, 373, 374, 387
    cost of, 273, 283
    see also diphtheria antitoxin; pertussis vaccine; tetanus antitoxin
diphtheria antitoxin, 19, 92, 123–26, 129, 136, 175, 216, 229, 264

anaphylactic response to, 125–26
diphtheria toxin mixed with, 126, 127, 128
intraspinal injection of, 124
Diphtheria Prevention Commission, 128
diphtheria toxoid, 128, 139
disease:
  animal transmission of, 53, 54–55
  childhood, 15, 120, 127
  chronic, 16
  germ theory of, 28, 32, 36, 53, 58, 98, 104, 120
  hereditary, 98–99
  insect-borne, 142
  passive acceptance of, 28
  prevention of, 27
  as sign of sin, 27, 28, 36, 41
  treatment of, 27, 33
  tropical, 56
  see also communicable diseases; *specific diseases*
Dissatisfied Parents Together (DPT), 254–55, 256, 331
DNA, 55, 175, 207, 430
Doctor's Dilemma, The (Shaw), 68
Doctor Therne (Haggard), 66
Doden, Melinda, 361, 367
dogs, 84, 125, 127
  rabid, 65, 68, 121–22
  sled, 127, 130
Dohan, F. C., 386–87
Dorgan, Byron, 366
Dougherty, Edward, 80
Douglass, William, 31–33, 37, 40, 43
Dow Chemicals, 227, 427
Dowdle, Walt, 299
DPT: A Shot in the Dark (Fisher and Coulter), 281, 338
DPT: Vaccine Roulette, 251–55, 256, 277, 278
drugs, 116, 161–62, 215
  addiction to, 308, 340
  marketing of, 330
  medication errors with, 330
  mood-altering, 330, 341
  painkilling, 167, 330
  recall of, 330
  regulation of, 72, 330
D.T. Watson Home for Crippled Children, 186
Dubos, Rene, 440–41
Duchez, Alphonse, 138
Duffy, John, 126, 162
Dunbar, Bonnie, 389–90
Dunning, Richard, 27
Durgin, Samuel Holmes, 82

ear infections, 351
Ebola virus, 334–35
Economic Security Act (1935), 136
eczema, 241, 243
eczema vaccinata, 71
Eddy, Bernice, 202, 209–11
Edison, Thomas Alva, 120
Edmondsen, Ron, 362–63, 365
Edmonston, David, 216, 219, 247
Egars, Elise, 352
Egars, Johnnie, 351–52
eggs, 140–41, 176, 220, 224–25, 238, 425
Egypt, ancient, 167
Ehrlich, Paul, 72
Eisenhower, Dwight D., 189, 190, 192, 197, 198, 202, 203, 204, 205, 301
Elderling, Grace, 268
elections, U.S.:
  of 1920, 163
  of 1932, 165, 166
  of 1976, 260
  of 1996, 305
  of 2004, 366
electrical fields, 169
Elgin, William L., 103
Eli Lilly, 172, 186, 195, 203, 209, 231, 232, 271, 273, 274, 275, 427
Ellenberg, Susan, 314, 325, 326
El Salvador, 16, 305
Emory University, 302
Empty Fortress (Bettleheim), 385
Enayati, Albert, 395
encephalitis, 71, 117, 152, 242, 292
  measles-linked, 217, 218
  vaccine-linked, 270, 287, 288–89
Endeavor, HMS, 47
Enders, John, 174, 181, 185, 187, 188, 196, 203, 224
  measles research of, 216, 219–21, 226, 227–28
  polio research of, 178–79, 180, 216
endometriosis, 356
endotoxin, 348
Environmental Protection Agency (EPA), 376, 378
Ephedra, 392
epidemiology, 53, 198–99, 217, 231, 239, 259, 292, 314, 355, 369, 406
epilepsy, 276
Erlichman, I. Fulton, 199
Erlichman, Pamela, 199
E. R. Squibb, 176, 223
Espionage Act, 105
eugenics, 97–99, 133
evolution, 66, 97, 336
Eyler, Bill, 359–60, 362

Family Research Council, 432
Fanning, Shirley, 246
Farrakhan, Louis, 335
Faucci, Anthony, 221–22
Fay School, 219–20
FDA Modernization Act (1997), 377
Federal Bureau of Investigation (FBI), 182, 418
federal disability programs, 237
Federal Register, 377
Fenichel, Gerald, 292–93
fever, 29, 35, 41, 44, 48, 52, 59, 79, 115, 126, 144, 217, 219, 221, 226, 231, 233, 234, 350, 367
Field, Charles J., 87–88, 102
Field, Victoria, 87–88
filamentous hemagglutinin antigen (FHA), 275
Findlay, G. M., 147, 153
Fishbein, Morris, 157
Fisher, Barbara Loe, 86, 253–54, 279–80, 287, 292, 330–31, 334, 338, 346, 349, 354, 357, 374, 388, 393, 395, 401, 419, 422–23
Fisher, Christian, 253, 280, 292
flesh-eating disease, 240
Flexner, Abraham, 74
Flexner, Simon, 74, 168, 172
flies, 168
Flower Hospital, 146
Foege, William H., 204, 243, 245, 262, 265, 438
food:
    additives in, 90, 341
    chemicals in, 142–43
    denatured, 340
    intolerance for, 387
    organic, 399
    pollution of, 142, 147, 356
    regulation of, 74, 96, 126–27
    trace chemicals in, 141, 142
    whole, 104, 364
Food and Drug Administration(FDA), 98, 257, 258, 259, 262, 279, 281, 285, 290, 319, 321–22, 325, 383, 401, 424, 425
    Center for Biological Evaluation and Research of, 314, 375–76
    products regulated by, 273–74, 330, 380–81, 392
    Team Biologics of, 433
    vaccine advisory panel of, 423
food and drug regulation system, 72–73
foot-and-mouth disease, 65
Ford, Gerald R., 260
formaldehyde, 20, 84, 85, 90, 185, 188, 201, 232, 264

Forrestal, James, 165
Fort Dix, 234, 259–60
Foster, Frank P., 61
Foster, Vince, 392
fowl cholera, 12, 62
Fox, John, 174, 183
Fox, Leon A., 115, 142, 151–52
Fraenkel, Karl, 123
France, 61, 121, 130
Francis, Don, 302
Francis, Thomas, 138, 174, 179, 180, 185, 192, 194–95, 196, 203
Franco-Prussian War, 60, 61
Franklin, Benjamin, 30, 33, 34, 42, 43–44
Franklin, Francis, 42
Franklin, James, 30
Franklin, John, 156
Franklin, Sarah, 42
Frederick Stearns, 96
free radicals, 342
French army, 60, 134, 139
Freud, Sigmund, 327, 384
Friedberger, Ernst, 135
Friedman, Lorraine, 180
Friedrich, Martin, 82–85, 90
Fries, Stephen M., 370
Frist, Bill, 426
Fulginiti, Vincent, 232, 257, 288
fumigation, 90, 168

Gaines, William, 384
Gallagher, Hugh Gregory, 164, 165, 167
Gallo, Robert, 335
gamma globulin, 19, 153
gangrene, 132
Gard, Sven, 179, 181, 188
Garfield, James A., 96
Garland, Judy, 166
Garrison, William Lloyd, 103
gastrointestinal disorders, 132, 386, 387–88
Gates, Bill, 325, 435, 436
Gauld, Ross, 158
Geier, David, 407, 410, 414, 415–16
Geier, Mark, 407, 410, 414, 415–16
Geiger, Jack, 195
Geison, Gerald, 122
Gellin, Bruce, 425
Generation Rescue, 415
genes, 20, 267, 386, 431
George Williams Hooper Institute, 152
Gerberding, Julie, 418–19
German measles, see rubella
Germany, Nazi, 139, 140, 141, 149, 158, 159, 163, 165, 183, 219, 241
germs, 15, 18, 19, 53–54, 71, 84, 120–21, 344
    disease theory based on, 28, 32, 36, 53, 58

germ warfare, 16, 143
Gershon, Michael D., 371
Gillick, Ed, 109–10
Gillray, James, 57
Glass, Roger, 294, 319–20, 321, 323–24
GlaxoSmithKline, 319, 324–25, 405, 412, 426, 436
Global Alliance for Vaccine Initiatives, 325, 436
glycerine, 64, 92, 93, 117
Goethe, Johann Wolfgang von, 342
Goffe, Alan, 219
Goldie, Joseph, 80
Golkiewicz, Gary, 293
Goodall, E. W., 126
Good Manufacturing Procedures, 432
Goodpasture, Ernest William, 140
Gore, Al, 286
Gorgas, William Crawford, 145
Gottesdanker, Anne, 263, 266
Grandjean, Philippe, 380, 406
Grant, Jim, 279
Grant, Marge, 279
Grant, Scott, 279
grease (horse disease), 54, 55
Gregg, Norman McAllister, 233
Groff, George, 89–90
Gruelle, Johnny, 99
Gruelle, Marcella, 99
Guillain-Barre Syndrome, 163, 261, 262, 317, 419
guinea pigs, 81, 123, 124–25, 126, 226
Gulf War syndrome, 410

Haagen, Eugen, 149–50
Haemophilus influenzae (Hib) bacterial vaccines, 20, 269, 306, 307, 311, 381, 393, 395, 408, 435, 442
Haemophilus influenzae type a, 307
Haemophilus influenzae type b (Hib), 306, 307, 308, 328, 331, 376, 406, 412, 435, 442
Haggard, H. Rider, 66
Hahnemann, Samuel, 345
Hahnemann Hospital, 80
Haiselden, Harry, 98
Halberstadt, A. H., 88
Haldol, 386
Haley, Boyd, 409–10
Hall, Carolyne Breese, 312
Halsey, Neal, 292, 311–12, 313, 375–79, 381–83, 403, 405–6, 419, 423
Halsey, William F., 375
Halstead, Lauro, 214
Hammond, William, 177, 203
Handley, J. B., 357, 415
Hardegree, Carolyn, 319

Harding, Victory, 266–67, 272, 275, 283
Harding, Warren, 163
Hargett, Mason V., 156, 158
Harmon, Heath, 367, 369
Harris, Oren, 229
Harvard University, 26, 29, 35, 50, 82, 145, 181, 185, 220, 221, 336
School of Public Health of, 138, 157, 178
Hatch, Orrin, 286
Haverford College, 167
Hawkins, Paula, 254, 278, 281
Hayes, Helen, 167
Haygarth, John, 45
Hazards of Vaccination, The (Wilson), 203
Hazelton, Laura, 80
Hazelton, Thomas, 80
H-bomb tests, 191
headache, 115, 167
healers, 56, 78, 87
Health, Education and Welfare Department (HEW), U.S., 196, 204, 229, 247, 252, 258, 262
Education Office of, 245–46
Health and Human Services Department (HHS), U.S., 260, 281, 284, 293, 422, 425
health insurance, 18, 298, 300, 305
health maintenance organizations (HMOs), 305, 317–18, 404, 405, 415
heart disease, 99, 173, 234, 393
Heidelberg, University of, 150
Heisenberg uncertainty principle, 189
Helms-Burton Act, 397
hemiparesis, 280
hemophilia, 308
Hempil, Betty Lee, 212
Henderson, D. A., 209, 299, 438
Henderson, Jessica, 86
hepatitis, 17, 148, 153, 156–58, 235, 263, 302
hepatitis A, 147, 153
hepatitis B, 12, 16, 143, 153, 156, 157, 158, 308–13, 389–91, 412
stigma of, 390
transmission of, 147, 148, 311
hepatitis B vaccine, 175, 235, 309–13, 314–15, 317, 324, 376, 378, 381, 389–90, 392, 395, 426
hepatitis C, 143, 312
herbs, 39–40, 361, 374
Hering, Constantine, 349–50
Herlihy, Walter, 389
Hickenlooper, John, 328–29, 367
Hickenlooper, Teddy, 367
Higgins, Charles, 106
Hill, Don, 427
Hillary, Edmund, 217–18

Hilleman, Jeryl Lynn, 232–33
Hilleman, Maurice, 176–77, 184, 188, 189,
    221–28, 314, 335, 381, 428
  character and personality of, 175, 221,
    223–24
  death of, 221
  dozens of vaccines developed by, 221–22,
    223, 225, 310
  education and early scientific work of, 223,
    224–25
  influenza research of, 221, 223, 224–25
  measles research of, 225–26, 227–28,
    232–33
  polio research of, 208, 209–10, 211–13
  rubella research of, 238–40
Hinman, Alan, 16, 155, 246–47, 265, 277,
    301, 316, 317
Hinman, E. Harold, 16, 155
Hippocratic oath, 27
HIV, 207, 225, 335, 436
H. K. Mulford Company, 65, 92, 93–94, 95,
    99–100, 103
H. M. Alexander, 93–94, 95
Hobby, Oveta Culp, 196, 197, 204
Hodge, John, 103
Hodge, J. W., 103, 104, 108–9, 110–11
Hogan and Hartson, 278
hog fever, 183
holistic therapies, 334, 336–37, 345, 352, 441
Holmes, Oliver Wendell, 98–99
homeopathy, 85, 279–80, 358
  antivaccine sentiment in, 82, 87, 101,
    103–4, 330, 335–38, 345
  principles of, 345
Homo sapiens, 53
hookworm, 144
Hooper, Edward, 207
Hoover, Herbert, 127
Hopkins, Julian, 347–48
hormone replacement therapy, 330
hormones, 341, 440
Hornig, Mary, 410
Horowitz, Leonard, 334–35
horsepox, 55
horses, 54–55, 143
  antibodies from, 72, 79, 123, 125–26
Horsfall, Frank, 138
Horstmann, Dorothy, 174, 178, 197
hospice care, 316
hospitals:
  intensive care units (ICUs) in, 353, 358
  neonatal wards in, 160, 353, 366–67
  unsanitary conditions in, 74–75
  variolation, 44, 45
Houlihan, Ralph, 197–98
Houlton, Robert, 41

House of Representatives, U.S.:
  Appropriations Committee of, 403
  Commerce Health Subcommittee of,
    255
  Government Reform Committee of, 326,
    389, 414
  Health and Environment Subcommittee
    of, 299
  Interstate and Foreign Commerce
    Committee of, 203
Howe, Howard, 185
Hull, Robert, 186, 203
human papilloma virus (HPV), 431
Hume, Alexander Hope, 70
Hunter, John, 47
Hurricane Katrina, 13, 433
Hurricane Rita, 13
Hussein, Saddam, 11–12
Hutchison, Joseph C., 59–60
Hviid, Morton, 407
hydrocephalus, 98, 220
hygiene hypothesis, 347–50
Hyland, Baxter, 412
hypertension, 40
hypoxia, 356

ID Biomedical, 426
Illich, Ivan, 335
immune gamma globulin (IGG), 19, 177,
    218, 219, 226, 227, 310
immune system, 52, 125–26, 133, 167,
    174–75, 259, 347
  antibodies created in, 14–15, 19, 36
  hygiene hypothesis and, 347–50
  manipulation of, 14–15, 19, 36, 348
  weakening of, 52, 104, 297, 314, 340, 352,
    354, 356, 358, 422
immunity, 19, 49, 54
  genetic, 168
immunization records, 81, 246
immunology, 72, 125–27
Imus, Don, 414
Incao, Philip, 345–47, 349–50, 367
incubators, 160
India, 133, 186, 209, 336
  smallpox in, 12, 241, 246, 302–4
  State Serum Institute of, 324–25
industrial chemicals, 142–43
Infantile Autism (Rimland), 385, 386
infantile paralysis, see polio
inflammatory bowel disease, 388
influenza, 423–25
  avian, 16, 260, 424
  deaths from, 136, 138, 169, 424, 429
  Fujian strain of, 424
  transmission of, 260, 344

influenza pandemic of 1918–19, 13
  bird-to-human transmission of, 260
  deaths from, 136, 138, 178, 224, 424
influenza pandemic of 1957, 223, 224–25
influenza vaccine, 13, 20, 137–39, 140,
    428–29, 434
  contamination of, 424
  demand for, 138–39, 257–58, 354, 425
  research and development of, 137, 138,
    224–25
  shortages of, 354, 424
  substandard batches of, 257–58
influenza virus, 423–24
  constant mutation of, 137, 259
  transmission of, 260
informed consent, 71, 161, 190, 202, 261–62
Ingalls, Theodore, 218
*Inquiry Into the Causes and Effects of the Variolae
    Vaccine . . .* (Jenner), 49, 53, 54, 57–58,
    88
Institute of Medicine (IOM), 231, 291, 317,
    354, 378, 390, 404–5, 406, 407, 415,
    422
Institute of Physicians and Surgeons, 78
Institutional Review Boards, 48
insulin, 15
International Chiropractors' Association, 340
International Conference on Measles
    Immunization (1961), 215–16, 217,
    218–19, 226–27
International Financing Facility for
    Immunization, 436
International News Service, 195
International Symposium on Pertussis (1963),
    273
Internet, 329, 358, 372
  antivaccine websites on, 354, 356, 357, 375
intussusception, 319, 321, 322–24
ipecac, 33
Iraq, 11, 12, 17, 297
Iraq War, 17
iron lungs, 160, 170, 171, 191, 195, 214
Irons, Jeremy, 343
Isaacs, Ellen, 309
Iverson, Portia, 388

Jacobson, Henning, 82, 98, 103
Jacobson v. Massachusetts, 98–99, 106, 108
James, Jill, 398, 410, 413
Japan, 65, 121, 129, 140, 150, 163, 165,
    274–77, 284–85, 347–48
Japanese encephalitis B virus, 139
  vaccine for, 174, 176, 183
jaundice, 147, 148, 152–57, 175
Jefferson, Thomas, 42, 51
Jenner, Edward, 46–50, 62, 142

  animal and nature studies of, 47, 48, 54–55,
    61
  character and personality of, 47, 48
  family background of, 47
  as Fellow of Royal Society, 48
  opposition to, 56, 57, 85, 105
  publications of, 49, 53, 54, 57–58, 88
  smallpox vaccine developed by, 12, 27, 44,
    46–50, 52–58, 62, 71, 120, 175
  variolation of, 48
Jenner, Robert, 49
Jesty, Benjamin, 53–54
John Birch Society, 230, 392
Johns Hopkins University, 74, 138, 151, 155,
    156, 178, 181, 185, 189, 200, 201, 292,
    353, 371, 377, 379, 423
  Institute for Vaccine Safety at, 375
Johnson, Lyndon B., 230
Johnson, Tom, 82
Johnston, Richard, 404–5
joint pains, 126
Jones, Mother, 104
Journal of Autism and Child Schizophrenia, 372
Journal of the American Medical Association
    (*JAMA*), 15, 129, 157, 176, 187, 270,
    283, 406
J. R. Geigy, 141
Jungle, The (Sinclair), 72

Kafka, Franz, 343
Kanner, Leo, 371–72, 383–85, 388, 400
Kapikian, Albert, 318–19, 321, 323
Kaposi's syndrome, 116
karma, 344–45
Kass, Leon, 316
Katz, Samuel, 178, 220–221, 227–28, 232,
    262, 265, 300, 314, 383, 419
Keenan, Blaise, 360
Kelchner, William I., 79
Keller, Helen, 237
Kempe, Alison, 332
Kempe, C. Henry, 192, 196, 218–19, 232,
    241–43, 265, 332, 339, 375
Kendrick, Pearl, 268–71
Kennedy, John F., 198, 208, 215, 258
  vaccination promotion of, 228–30
Kennedy, Robert F., 415
Kentucky, University of, 409
Kerr, Randy, 193
Khrushchev, Nikita, 206
kidneys, 219, 220, 231
  Willms tumor of, 352
King, Gloria, 117
King, Martin Luther, 216
King Pharmaceuticals, 427
Kinkel, Kip, 396

Kinnear, Johnnie, 290
Kirby, David, 414, 421
Kirkman Labs, 386
Kirkpatrick, William F., 186
Kissinger, Henry, 335
Kitasato, Shibasaburo, 123
Klausner, Richard, 436
*Klinische Studien über Vakzination und vakzi-
    nale Allergie* (von Pirquet), 125
Koch, Robert, 64, 65, 122, 123, 344
Koehler, Bonnie, 365–66
Kolmer, John, 173–74, 184, 187
Koplan, Jeffrey, 322
Koprowski, Hilary, 146, 181, 183, 184–86,
    205, 206–7, 209, 212, 213, 276
Kora Indians, 358
Korean War, 190, 200
Krugman, Richard, 241, 262, 263, 272
Krugman, Saul, 116, 232, 235, 241, 242,
    262–63, 309–10
Kuomintang Party, 316

LaBar, Eugene, 115–16
Ladies' Home Journal, 106
laetrile, 392
Lal, Ashanti Priori, 303
Lamaze classes, 280
Lancet, 158, 185, 271, 387, 388
Landsteiner, Karl, 168, 306–7
Langmuir, Alexander, 199, 200, 217–18
Lasker, Albert, 238
Lasker, Mary, 238
Lasker Award, 197
Laughlin, William, 110
Lawton, Steve, 278
Lazear, Jesse, 130, 144
lead, 379, 394, 397
Leake, James P., 173
Leavis, F. R., 437
Lederle Pharmaceuticals, 176, 183, 206, 260,
    261–62, 266, 270, 272, 273, 274, 283
Lee, Benjamin, 61–62, 75, 77, 78, 80, 81, 89,
    92, 296
Lee, Robert E., 132
leeches, 337
Legal Aid Board, 388
Legionnaire's disease, 260–61
leptospirosis, 152
Letter to a Friend in the Country . . . (Cooper),
    34–35
Lettsom, John, 50
leukemia, acute lymphoblastic, 348
Leverson, Montague, 86–87
Lewis, James, 180
Libby, I. Lewis "Scooter," 13
liberalism, 13

lice, 139–40, 141
Life, 99, 235
life expectancy, 130, 162, 230
Liffrig, Mike, 366
lime, 168
Lincoln, Abraham, 100
Lirugen, 227
Little, Kenneth, 105
Little, Lora C., 99, 104–5, 130
Little House on the Prairie, 350–51
Little Turtle (Indian chief), 42
liver, 154, 396
    bacterial infection of, 152
    cancer of, 148, 153, 158, 312
    cirrhosis of, 148, 312
Lloyd, Wray, 146
lockjaw, *see* tetanus
Loeffler, Friedrich, 123, 124
London, 26, 45, 49–50, 67, 147, 153–54
    1721 smallpox epidemic in, 38
Longworth, Alice Roosevelt, 165–66
Los Angeles, Children's Hospital in, 318
Luddites, 105, 333, 349
Ludlow, Daniel, 47
lungs, 240, 269, 352
Lutaud, August, 122
Lynchburg State Colony, 158

Maas, Clara, 144
MacArthur, Mary, 167
Macbeth (Shakespeare), 220
MacCalum, A. O., 147
McCarthyism, 197
McCormick, Marie, 411
Macfadden, Bernarr, 103, 104
McFarland, John, 108
McFarland, Joseph, 92–95, 102, 125, 128
Machado Ventura, Jose, 228
McKeown, Thomas, 437
McKinley, William, 96
McLean, Va., Franklin Sherman Elementary
    School, 160–61, 193
McNeill, William, 53
Mad, 384
Madras Hospital, 241
magnesium, 386
Maharishi cult, 335–36
Maharishi University, 336
Maher, Bridget, 432
Maitland, Charles, 38
Maitlin, Hugh, 178–79
Maitlin, Mary, 178–79
malaria, 12, 118, 122, 132, 144, 162, 435
Maloney, Carolyn, 414
Maltese Falcon, The, 324
Malthus, Thomas Robert, 56

Manclark, Charles, 271, 272, 276–77, 290
Manhattan Project, 138, 162, 180, 222
Mann, William, 211
Marcet, Alexander, 58
March of Dimes, 160–61, 165, 166, 193, 207, 237
March on Washington, 216
Marek's disease, 221
Marietta College, 210
Marine Hospital, 155
Martin, Edward, 91
Martin, Henry Austin, 61
Marx, Groucho, 183
Maryland, University of, 319
    Medical School of, 320
Mather, Abigail, 35
Mather, Cotton, 25–32, 34–37, 39, 43, 44, 46, 104
    character and personality of, 26, 30, 32
    criticism of, 29, 30–32, 40
    erudition and scholarship of, 26, 27, 36–37
    fatherhood of, 26, 28, 35–36
    political bent of, 26, 29, 35, 38
    religious conviction and practice of, 26–27, 30, 32
    variolation introduced in America by, 25–32, 35–37
Mather, Hannah, 35
Mather, Increase (father of C. Mather), 26, 28, 35–36
Mather, Increase (son of C. Mather), 35–36
Mather, Richard, 26
Mather, Samuel, 35–36
Mead, Eleanor, 373, 397–98, 400
Mead, George, 17–18, 371, 372–75, 396–400
Mead, Richard, 40
Mead, Tory, 371, 373–74, 384, 397–400
Mead, William, 371, 372–75, 397–400
measles, 35, 53, 109, 174, 215–21, 225–28, 336, 347
    adverse effects and deaths from, 35, 102, 105, 129–30, 169, 217, 219, 269, 298, 303
    atypical, 231–32, 258
    decline of, 129–30, 169, 240, 247, 297, 331–32
    epidemics of, 102, 219, 231, 244, 294–97, 354
    as ghetto disease, 244
    misdiagnosis of, 232
    symptoms of, 217, 219, 231, 349–50
    transmission of, 217, 246, 299, 305–6
measles, mumps, and rubella (MMR) vaccine, 233, 240, 247, 252, 286, 317, 325, 373, 388, 392–93, 394, 408–10, 412, 433
measles vaccine, 12, 17, 20, 140, 191, 216–28,
    239–40, 244–47, 388
    Jeryl Lynn strain of, 233
    killed-virus, 231–32, 244, 298
    live-virus, 231, 244, 298
    research and development of, 218–28, 231–33, 375
    testing and trials of, 219, 221, 225, 226, 231, 235, 304, 377
measles virus, 232
    Edmonston strain of, 219–20, 226, 247
    wild, 220, 226
Medicaid, 238, 293, 300, 332
Medical Hypothesis, 395, 396
Medical Nemesis (Illich), 335
medical science:
    education in, 40, 74, 82, 83
    growth of confidence in, 28, 111, 119, 120, 130, 136, 190
    humors theory of, 39, 133, 345
    innovation and progress in, 27–28, 65, 71, 74, 111, 118, 120, 161–62
    place of surgeons in hierarchy of, 32
    professional competition in, 30, 122
    sanitary reforms in, 65, 67, 68, 71, 72, 74, 117, 120
    secular vs. sacred authority and, 27, 28, 31–32, 34–35, 41, 56, 100–101
Medical Society of Washington, D.C., 193
Medicare, 293
medicines, 38
    patent, 78, 86, 103
    see also drugs; vaccines
Medico-Chirurgical College, 92, 102
Medimmune, 426, 429
Meese, Ed, 287
Mehl, Al, 323
Meister, Joseph, 121
Melnick, Joseph, 174, 205
Menard, St. Yves, 62, 63
meningitis:
    aseptic, 317
    bacterial, 306
    Hib, 21
    meningococcal, 430
meningitis vaccine, 17, 139, 306–7
mental illness, 386
mental retardation, 234, 235, 238, 407
Merchant's Gargling Oil, 103
Merck, George W., 149, 222–23
Merck, Sharpe, and Dohme Institute for Therapeutic Research, 223
Merck and Co., 75, 95, 209, 213, 221–23, 226–28, 233, 238, 260, 286, 307, 310, 314, 320, 321, 325, 326, 381, 402, 412, 425, 426, 427, 428, 429, 433, 437

Merck Institute for Therapeutic Vaccines, 175
mercurous chloride (calomel), 41, 399
mercury, 39, 40–41, 42, 52, 337, 357, 373–82, 397, 414–15
  acceptable limits of, 376, 378, 381
  ethyl, 373, 374, 376, 378–79, 382, 406, 409
  in food chain, 356, 380, 382
  methyl, 382, 409
  in thimerosal, 356, 373–74, 375–79, 381, 396, 406, 407
  toxicity of, 379–80, 395–96, 411
  treatment with, 27, 33, 379, 396
Merieux, Charles, 428
Merrill-National, 95
Metabolife, 392
Metropolitan Life Insurance Co., 128
Mexican Army, 134
Mexico, 16, 97, 115, 116, 139, 358
Meyer, Hank, 433
Meyer, Harry, 238–39, 278
Meyer, Karl Friedrich, 152–54, 189–90
Mica, Dan, 254–55, 278, 281, 389
Mica, John L., 254, 278, 389, 390, 391, 426
mice, 146, 174, 176, 268–69, 275, 410
Michigan, University of, 180, 195
Michigan State Health Department, 268
Michigan State Public Health Laboratory, 284
Microbe Hunters (de Kruif), 46, 166
microbiology, 120, 128, 180, 189, 223
microorganisms, 129
Middlehurst, Donna, 252–53, 281, 282
Middlehurst-Schwartz, Julie, 280–81, 282, 288
Midway, Battle of, 153
Milbank Fund, 128
Military Air Transport Service, 177
milk, 137, 169
  adulteration of, 78, 90
  pasteurization of, 74, 126–27, 129, 169
milkmaids, 47, 48, 49, 53, 54
Millar, Don, 245, 246
Millbank, Jeremiah, 166
Miller, Clinton R., 230
Miller, Neil Z., 338
Minnesota, University of, 130, 252
Miracle Worker, The, 237
miscarriage, 236
Mitchell, John, 99
MMWR, 246, 320
Modlin, John, 12
Mohican Indians, 51
Molacek, Gingy, 363–64, 365
Monitoring System for Adverse Events Following Immunization, 259

monkeys:
  Cercopithecus, 211, 225
  experimentation on, 145–46, 147, 172, 180, 185, 186, 191, 200, 209–10, 212, 220–21, 225
  rhesus, 211, 225
Montagu, Edward, 37
Montagu, Lady Mary Wortley, 37–40, 46
Montagu, Mary, 38
Montaigne, John, 324
Montana State College, 224
moon shot, 162, 217
Moran, Evelina, 353
Moran, Helena, 353
Moran, Marty, 353
Moraten, 227–28
Mormon Church, 106
Morris, J. Anthony, 262
Morrison, J. H., 156
Morse, Wayne, 204
Mortimer, Edward, 283–84, 288
Moseley, Benjamin, 56, 57
Moskowitz, Richard, 335
mosquitoes:
  Aedes aegypti, 144
  Anopheles, 142
  eradicatioin of, 148
  malaria, 132
Mountain, Isabel Morgan, 185
Moynihan, Daniel Patrick, 421
Mulford, H. K., 94
multiple sclerosis, 389
Mulvaney, Richard, 161
mumps, 218, 232–33, 247, 347
  research on, 174, 233
  Urabe strain of, 317
mumps vaccine, 17, 20, 216, 225, 233, 240
Murray, Roderick, 257, 258
Murrow, Edward R., 196

Nader's Raiders, 257, 258
Naito, Ryoichi, 150
Napoleon I, Emperor of France, 50
Nashville Children's Hospital, 366
Nathanson, Neal, 199–200, 201
National Academy of Sciences, 197, 378
National Anti-Vaccination League, 86
National Cancer Institute, 436
National Childhood Encephalopathy Study (NCES) (UK), 289, 290, 291
National Drug Co., 158
National Foundation for Infantile Paralysis, 162–63, 170, 179–82, 186, 188, 189, 191–92, 194, 195, 196, 198, 199, 204, 228
  fund-raising of, 163, 164–67, 181–82

Hershey conference of, 184–85, 187
Immunization Committee of, 181, 184–85, 187, 188
Vaccine Advisory Committee of, 202
National Health Federation, 230
National Health Service, 45
National Immunization Program, 229, 304, 405, 415, 418
National Institute for Allergy and Infectious Diseases (NIAID), 221–22, 322, 324, 436
National Institutes of Health (Japan), 274, 275
National Institutes of Health (NIH), 157, 158, 189, 192, 198, 200, 202–4, 215, 222, 230, 235, 238–39, 241, 268, 269, 285, 292, 329–30, 348, 308, 401, 422
   Center for Alternative and Complimentary Medicine, 393
   Division of Biologics Standards, 199, 210, 231, 238, 242, 257, 258, 270
   Laboratory of Biologics Control, 195–96, 199
National Library of Medicine, 30
National Public Radio (NPR), 326
National Research Council, 149
National Transportation Safety Board, 417
National Vaccine Advisory Committee, 281, 304–5, 382, 417
National Vaccine Establishment, 96
National Vaccine Information Center, 254, 331
National Vaccine Injury Compensation Program, 73, 393, 407, 416
Nation of Islam, 335
Native Americans, 417
natural disasters, 13, 433
Natural Hygiene Society, 230
naturopathy, 103, 104, 230, 337–38
Naval Medical Research Hospital, 436
Navy, U.S., 136
NBC-TV, 251–52
Neisseria meningitis, 431
Nelmes, Sarah, 49, 53
neurological disorders, 13, 122, 147, 240
Neustaedter, Randall, 338
New Age, 334
New Deal, 136
New England Courant, 30
New England Gazette, 30, 31
New England Journal of Medicine, 216, 330
New Republic, 14
News American, 117
Newton, Sir Isaac, 26
New York, N.Y., 60, 63, 302
   Division of Child Hygiene in, 127

hospitals in, 116, 117, 146, 235
1947 smallpox outbreak in, 115–19, 219
vaccination campaigns in, 60, 115–19, 242
New York Board of Health, 63, 116, 242
New York Central Railroad, 110
New York City Dispensary, 61
New York *Daily News*, 117
New York Department of Public Health, 239
New York Division of Health, 218
New York Herald-Tribune, 117, 118–19
New York *Medical Record*, 127
New York State Board of Health, 109
New York Times, 70, 111, 117, 142, 207, 223, 240, 286–87, 415
New York Times Magazine, 14, 378
New York University, 116, 235, 309, 373
   Medical School of, 128, 179, 183
Niagara Falls, N.Y., 103, 109–11, 131, 353
Nietzche, Friedrich Wilhelm, 343
Nixon, Richard M., 240, 244, 258
Nobel prizes, 127, 157, 168, 178, 179, 181, 197, 216, 220, 231, 306, 349
Noguchi, Hideyo, 130
North Africa, 140, 141, 153
North American, 79
North American Review, 85
North Carolina, University of, 413
Northeastern University, 398
Northwest Fever Hospital, 126
Norton, Thomas, 146, 184
Norwalk virus, 184
nuclear power, 219, 229
numerology, 345
Nuremberg trials, 158
nutrition, 129, 134, 161, 340, 386–87

obesity, 338, 383
O'Connor, Basil, 164–66, 173, 177, 181–82, 184, 187, 188, 194, 196, 202, 207, 266
O'Dwyer, William, 116
Office of Biologics Research and Review (1983–88), 258
Office of Defense Mobilization, 177
Office of Scientific Research and Development, 138, 222
Offit, Bonnie, 320
Offit, Paul, 13, 18, 319, 320–21, 326, 378, 412, 423
Oklahoma, University of, Medical School, 340
Olasky, Marvin, 13
Olitsky, Peter, 172, 183
Olympic Games of 1932, 97, 212
O'Neill, Thomas P. "Tip," 286
Onesimus (slave), 37
Orederu, Omotayo, 342

Orenstein, Walter, 245, 246–47, 300, 302–4,
    305–6, 316, 318, 322, 325–26, 377,
    383, 403, 404, 412, 417, 418, 440,
    441–42
Orient, Jane, 340
Orr, William J., 195
Orwell, George, 329
Osaka University, 314
Osborn, June, 427
Osler, William, 134–35, 313
Oslo Pediatric Hospital, 387
Ottoman Turks, 38
Owosso Sugar Co., 107
Oxford University, 135

Pacific Ocean, 47, 143, 153, 191, 197
Pallone, Frank, 377
Panama, 132, 149, 261
Panama Canal, 145
Pan American Health Organization (PAHO),
    305, 438
pap smears, 431
Paradiso, Peter, 430, 432
parasites, 19, 20, 396
Paris, 45, 62, 63, 122
    Necker Hospital in, 123
Park, William H., 74, 98, 123–24, 126, 128,
    130, 172, 173, 183, 186
Parkedale Pharmaceuticals, 427
Parke-Davis, 65, 92, 94–95, 96, 176, 203,
    260, 264, 273, 427
Parkman, Paul, 234, 238–39, 265, 285,
    427–28, 432
Pasteur, Louis, 62, 71, 85, 120–23, 124, 344
    criticism of, 85, 122
    experiments and demonstrations of,
        120–21
    rabies vaccine developed by, 12, 65, 68,
        121–22, 175–76
Pasteur Institute, 79, 140
pasteurization, 74
Pasteur-Merieux, 428
patriotism, 108, 111, 118–19, 122–23, 134–35
Paul, John, 174, 177, 181, 203
Pearl Harbor, Japanese attack on, 143, 153
Pearson, Karl, 133
Pederson, Griffin, 358
Pederson, Sharalee, 358
Pediatric Infectious Disease Journal, 319
Pediatrics, 263, 270, 320
Peebles, Thomas C., 219–20
penicillin, 118, 130, 179
Pennsylvania, University of, 13, 40, 81, 138,
    201, 218, 225, 319, 320
Pennsylvania Board of Health, 61, 81, 89,
    96–97, 101

Pennsylvania State Vaccination Commission,
    107
Pentagon Papers, 265
peptides, 348, 389
Pernick, Martin, 98
pertussis, 230, 332, 348, 350
    see also whooping cough
pertussis vaccine, 14, 20, 172, 216, 264,
    266–67, 268–93, 357, 367–68
    adverse affects of, 251–56, 265, 266–67,
        270–73, 275, 276, 277–78, 325
    booster shots of, 369
    "extracted" acellular, 270, 271–75, 284,
        285, 286, 317, 325, 352, 368, 369
    research and development of, 268–77,
        284–86, 352
    standardizing of, 268–69, 270
    whole-cell, 14, 255, 270–73, 277, 284, 289,
        290, 291, 317, 325, 338, 367
    see also diphtheria, tetanus, and pertussis
        (DPT) vaccine
petit journal, 62–63
Pfeiffer, Immanuel, 82
Pfeiffer, Richard, 133
Pfizer, 222, 231, 232, 427
pharmaceutical industry, 73, 75, 103, 116–17,
    218, 222–23, 260, 300, 330, 426–28
    mergers and acquisitions in, 95–96, 223,
        426–27, 428
    no-fault system proposed for, 263
    profits of, 331, 427
    regulation and inspection of, 95–96
    unscrupulous practices of, 128–29
pharmacies, 52, 73
Philadelphia, 30, 63, 71–72, 76–81, 86,
    87–88, 90–94, 128–29, 199, 202,
    294–98
    Children's Hospital of, 177, 320, 412
    1880–81 smallpox epidemic in, 62
    epidemics in, 40, 43, 62, 75, 77–81, 88,
        90–91, 92, 93, 102, 107–8
    Faith Tabernacle Congregation in, 295–97
    Keystone Public School in, 87–88
    Municipal Hospital in, 62, 74–75, 77,
        91–92, 106
    1901–4 smallpox epidemic in, 77–81, 88,
        90–91, 93, 296
    1989–91 measles epidemic in, 294–98
    pharmaceutical industry in, 73, 75, 92,
        93–95, 129
    Philadelphia Hospital in, 92
    political corruption in, 73–74
    population growth in, 74
    poverty and disease in, 73–75, 77–78,
        107–8
    1793 yellow fever plague in, 40, 144

smallpox eradication programs in, 41, 42–43, 44, 73–75, 77–79, 81, 87–88, 97, 101–2, 119
variolation system in, 41, 42–43, 44
Philadelphia and Reading Railways, 77
Philadelphia Medical Times, 62
Philadelphia News American, 78
philanthropy, 163, 164–67, 238, 435–47
Philippines, 75, 132, 186
Philosophical Transactions, 31
Philosophical Treatises, 37
Phipps, James, 49
Physical Culture, 103
physicians, 16, 29–33
    apprenticeship to, 31, 47
    British, 12, 27, 38–40, 45, 46–50, 56, 62
    military, 131
    mistrust of, 38, 57, 104, 105
Picasso, Pablo, 213
picric acid, 172
Piehn, I. H., 87
Pink disease (acrodynia), 396
Pitcairn, Harold, 102
Pitcairn, John, 100, 101, 102–3, 104, 105, 106–8
Pitcairn, Raymond, 100, 102
Pitfield, Robert L., 81
Pitman-Moore, 196, 227, 427
Pittman, Margaret, 268–69, 270
Pittsburgh, University of, 177, 179, 180–81
Pittsburgh Municipal Hospital, 191
Pittsburgh Plate Glass Co., 100
Pittsburgh Press, 190
Pius XII, Pope, 158, 189
placebo effect, 226, 337
placentas, 169, 220, 244, 297, 304, 379
plague, 68, 118, 139
Plagues and Peoples (McNeill), 53
plague vaccine, 139, 296
Pliny, 57
Plotkin, Stan, 238–239, 285, 314, 315, 319, 320–21, 324, 326, 382
pneumococcal bacteria, 139, 307
pneumococcal vaccine, 129, 139, 425
pneumonia, 116, 129, 130, 231, 296
Pocono Laboratories, 95–96
polio, 159, 439–40
    deaths from, 130, 168, 170, 214
    decline of, 14, 160, 161, 208, 213–14, 216, 240, 266, 331, 338, 359
    1894 Vermont outbreak of, 167
    epidemics of, 167, 168–70, 228–29
    fear of, 162, 170, 191, 214
    historic civilian mobilization against, 162–63, 165–67, 172, 173, 193, 207, 266

paralysis and suffering from, 160, 161, 163–65, 167, 168, 170–71, 182, 186, 191, 214
    resurgence of, 214
    symptoms of, 161, 163–65, 167
    transmission of, 169, 172, 191
polio vaccine, 12, 17, 216, 228–29, 342
    adverse effects and deaths from, 173, 195, 198–201
    contamination of, 198–201, 209–13
    field trials of, 160–61, 189–90, 191–95, 202, 206–7, 208, 210, 375
    manufacture of, 194, 197–203, 205, 206, 207, 211, 212
    oral, 20, 146, 172, 184, 204, 205–9, 210–12, 264, 276, 317
    research and development of, 130, 162–63, 164–67, 172–97, 201, 203
    see also Sabin polio vaccine; Salk polio vaccine
polioviruses, 167, 168, 172
Polyclinic Hospital, 117
polymerase chain reaction (PCR) testing, 368, 369
polysaccharide molecules, 21, 306
polysaccharide vaccine, 20
Population (Malthus), 56
Portier, Paul, 125
postal workers, 13, 14
potassium bichromate, 121
Potts, J. F., 100–101
Pouchet, Felix, 121
Powers, Grover, 241
poxviruses, 55
Pratt, Douglas, 376
pregnancy, 382
    miscarriage in, 236
    rubella dangers in, 218, 233–34, 235–36
    termination of, *see* abortions
Prevnar, 425
Price, David, 189
Priest, Percy, 203
Prince v. Commonwealth of Massachusetts, 295
prisoners of war (POWs), 140, 227
Progressive Party, 72, 73, 99, 127
prostitution, 218
protein-polysaccharide conjugate vaccine, 306
proteins, 126, 143, 259, 285, 390
    egg, 261
    F, 232
    inflammatory, 347
    islet-activating, 275
    myelin basic, 174
    pertussis toxin, 275
    recombinant bacterial, 348

Prudden, William, 63, 74
Public Citizen, 330
Public Health and Marine Hospital Service,
　91
　Hygiene Laboratory of, 72, 73, 95–96, 125
Public Health Laboratory Service, 203, 219
public health movement, 12
　death of "health soldiers" in, 130
　egalitarian systems of, 45, 162
　eugenicist element in, 97–99
　evolution and change in, 14, 65, 69, 72,
　　161–62
　funding of, 162–63
　health goals of, 21, 98
　health inspectors in, 78, 184
　leaders and administrators of, 12, 13, 15,
　　16–17, 21, 51, 66, 68, 82–85, 90, 111,
　　177, 301, 375
　patriotism linked to, 108, 111, 118–19,
　　122–23, 134–35
　public sympathy and respect for, 108, 120,
　　375
　resistance to authority of, 45, 56, 68
　sanitation reform in, 90–91, 161, 167, 235
　shaping of public opinion by, 15, 69
　social contract inherent in, 17, 266
Public Health Service, U.S., 98, 130, 136,
　　155, 173, 194, 202–3, 208, 228, 231,
　　265, 284, 285, 302, 329
　Commissioned Corps of, 200
　Rocky Mountain Laboratories of, 141, 156
Public Ledger, 102
Puck, 122
Pure Food and Drug Act (1906), 72
purgatives, 379
purges, 39, 40, 48
　mercury-induced, 33, 41
Puritanism, 26, 32, 43, 297
Purivax, 212

Q fever, 183
Quadrigen, 264, 279
Quakers, 43, 225–26
quarantine, 25, 28, 32, 43, 51–52, 67, 69, 76,
　　88–89, 97, 105, 110–11, 117, 169, 243,
　　294, 296, 304, 336
　national security and, 220
　police enforcement of, 230
Queen Mary, 182

rabbits, 121, 122, 172, 174, 175, 176
rabies, 121, 122
rabies vaccine, 12, 65, 68, 121–22, 175–76,
　　183
Rabinovich, Regina, 436
Race Betterment Foundation, 98

racism, 168, 341, 343–44
rashes, 115, 117, 126, 217, 219, 231, 242, 256,
　　349
Rats, Lice and History (Zinsser), 140, 178
Rauh, Louise, 231
Ravold, Armand N., 97, 451, 455
Razzell, Peter, 55
Reader's Digest, 166
Reagan, Nancy Davis, 166
Reagan, Ronald, 166, 278, 283, 284, 286, 287,
　　298, 299, 310
Red Cross, 156
Redding, Josephine, 86
Redwood, Lyn, 395, 411
Reed, Walter, 75, 130, 133, 134, 144, 177
Reichelt, Karl, 387
Reinart, Charles, 295
Rennels, Margaret, 319, 321, 325
Republican Party, 73, 102, 229, 286, 300,
　　425–26
Resciniti, Anthony, 254
Resciniti, Leo, 254
respiratory disorders, 132, 167, 214, 258, 259,
　　267, 363, 369
respiratory syncytial virus (RSV), 184, 240,
　　431
Reyes, Anita, 264–65
Reye's syndrome, 290
Reyes v. Wyeth Laboratories, 264–65, 266
Rhodes, Phil, 404
Rhogam, 378–79
Ribicoff, Abraham, 229, 230, 257, 258, 262
Richardson-Merrill, 273
Richet, Charles, 125
Rickettsia genus, 140
Rimland, Bernard, 384, 385–86, 387, 388,
　　389, 394, 395, 401, 403, 406–7, 413,
　　415
Rimland, Gloria, 384
Rimland, Mark, 384
River, The (Hooper), 207
Rivera, Geraldo, 263
Rivers, Tom, 116, 117–18, 129, 172, 173, 174,
　　175, 180, 181, 184, 185, 186–87, 188,
　　196, 202, 203, 218
Robbins, Frederick, 178, 231
Robbins, John B., 273–74, 284, 307, 369, 412,
　　432
Rochester University, 380
Rockefeller, John D., 100, 435
Rockefeller Foundation, 130, 141, 151–52,
　　157, 183, 437
　International Health Board of, 127, 144,
　　145
Rockefeller Hospital, 116, 172
Rockefeller Institute for Medical Research,

116, 127, 130, 138–39, 140, 144–50,
151–52, 154–56, 166, 168, 169, 171,
174, 175, 183, 223, 306–7, 349,
440–41
International Health Division of, 138–39,
144, 147
Rockwell, Norman, 252
Rocky Mountain Spotted Fever, 140, 183,
232
Rodee, Marie, 292–93
Roe v. Wade, 237
Roger, Heidi, 395
Rogers, Paul, 255, 278
Rollens, Rick, 416
Roman Catholic Church, 235–36, 390–91
Rommel, Erwin, 153
Rooney, Mickey, 166
Roosevelt, Anna Eleanor, 160, 161
Roosevelt, Franklin Delano, 160–66, 167,
222, 266
death of, 194
hand-controlled car of, 164
1920 vice presidential campaign of, 163
polio and paralysis of, 161, 163–65
political vision of, 164
presidential campaigns and elections of,
165
radio addresses of, 165
Roosevelt, Theodore, 72, 166
Rosenau, Milton, 125
Rosenberg, Charles, 336
Rosicrucian Fraternity, 87, 333–34, 339,
340–41, 342, 345
Ross, Robert, 294–97
Rotashield, 318–19, 321–25, 418
Rotateq, 326
rotaviruses, 318–20, 323, 324, 430
rotavirus vaccine, 224, 318–26
Rothschild, Kathleen, 390
Roux, Emile, 123, 124
Royal Academy, 50
Royal Army College, 133
Royal Commission on Vaccination, 65, 69
Royal Free Hospital, 388, 403
Royal Jennerian Society, 50, 66
Royal Society, 26, 31, 36, 37, 38, 48, 49
RSV vaccine, 240
rubella, 218, 233–40, 297, 320, 338
birth defects caused by, 233–34, 235–38
decline of, 235, 247, 331–32, 394
epidemics of, 235, 236, 237, 239
intentional exposure to, 218, 235
lifelong immunity provided by, 239
rubella vaccine, 20, 216, 225, 233, 234–35,
238–40
clinical trials of, 236, 238, 239

debate over timing of, 239, 240
Parkman/Meyer, 238–39
Plotkin, 238–39
rubella virus, 238
isolation of, 234–35
RW strain of, 234–35
Rush, Benjamin, 40, 105
Russell, Frederick, 134
Russia, 11, 50, 121
Rutter, William, 310

Sabin, Albert 116, 120, 166, 192, 196, 203,
234, 260, 266
character and personality of, 180, 183–84,
208, 213, 227
polio research of, 172, 175, 176, 181,
183–85, 186, 187, 204, 205–9, 213
Sabin polio vaccine, 20, 146, 172, 184, 204,
205–9, 210–12, 228–29, 264, 317, 375
Safe Minds (Sensible Action for Ending
Mercury-Induced Neurological
Disorders), 394, 395, 400–401, 402–3,
405, 406, 407, 411, 414
Sagan, Carl, 213
Salamone, John, 317
Salem witch trials (1692), 26, 32, 34, 35
Salisbury, David, 430, 434–35
Salk, Daniel, 162
Salk, Darrell, 188, 197, 198
Salk, Donna, 188
Salk, Dora, 162
Salk, Jonas, 14, 179–82, 215, 260, 266, 274,
427
awards of, 197
celebrity of, 187–88, 190, 197
character and personality of, 163, 179–81,
182, 187, 190, 196–97, 213, 224
competitors of, 181, 182–89
education of, 179, 302
family background and youth of, 162
influenza research of, 174, 180
polio research of, 175, 179–82, 185–88,
190–97, 201, 203, 252
Salk, Jonathan, 188
Salk, Peter, 188
Salk Institute for Biological Sciences, 95, 208
Salk polio vaccine, 14, 20, 160–63, 189–206,
208, 211, 212–13, 222, 228, 229, 301,
305, 317, 375
adverse effects and deaths from, 198–201
contamination of, 198–202, 205, 210
development of, 175, 179–82, 185–88,
190–97, 201, 203
manufacture of, 194, 197–203, 207
1954 field trials of, 160–61, 189–95, 202,
208, 210, 212, 213, 223

Salk polio vaccine (*continued*)
  phasing out of, 264
Salmon, Dan, 423
Salon, 14
Sandler, Adrian, 413
Sanofi-Pasteur 75, 96, 426, 434, 437
San Quentin prison, 189–90
SARS, 16, 213, 294, 320
Satcher, David, 335
Sato, Hiroko, 274, 275–77, 284
Sato, Yuji, 274, 275–77, 284
Saturday Evening Post, 142
Sawyer, Marjorie, 157
Sawyer, Wilbur A., 145–46, 148–49, 150,
  151–52, 154–57
Scarborough, Chuck, 414
scarlet fever, 78, 91, 120, 169, 240, 269
Schamberg, Jay Frank, 106–7
Scheele, Leonard A., 198, 203, 204
Schiappacasse, Robert, 350
Schick, Bela, 126
Schiff, Gilbert, 235, 239
schizophrenia, 340, 386
Schlafly, Andrew, 391
Schlafly, Phyllis, 334, 391
Schmidt, Rosemarie, 231
Schneerson, Rachel, 307, 432
schools, 71, 78, 119, 237
  children withdrawn from, 102
  unvaccinated students excluded from, 69,
    71, 74, 87–88, 106, 245, 246, 334
  vaccinations in, 83, 99
Schuylkill River, 73, 91
Schwartz, Jeffrey H., 252–53, 254–55, 278,
  279, 280, 281–82, 287
Schwarz, Anton, 184, 227, 235
Science, 347, 425
Seahorse, HMS, 25
Seattle Post-Intelligencer, 198
secretin, 388–89, 413
See It Now, 196
seizures, 52, 79, 217, 226, 252–53, 255, 256,
  270, 280–81, 289, 291, 323, 325, 356,
  390
Selbstvergiftungen, 386
Senate, U.S., 103, 204, 368–69
  Labor and Human Relations Investigations
    Subcommittee of, 278–80, 281–83
Sencer, David, 262
Serratia marcescens, 424
Serum Institute of India, 437
Serumkrankheit, 126
serums, 19, 72, 122, 129, 149
Sever, John, 235
sexually transmitted diseases (STD), 147–48,
  308

shaken baby syndrome, 342
Shakespeare, Nicholas, 133
Shannon, James, 192, 202, 203
Sharpe and Dohme, 95, 129, 223
Shattuck, Paul, 387
Shaw, Alan, 428
Shaw, Edward, 218
Shaw, George Bernard, 68, 124–25, 133
Shays, Chris, 335
Sheppard-Towner Act (1921), 127
Shermer, Michael, 68
shingles, 314, 315, 434
Shining Mountain Waldorf School, 327–28,
  332–33, 342, 350–52, 367
Shirakawa Taro, 347–48
shock, 52, 253, 255–56, 279, 290, 292, 356
Silent Spring (Carson), 142
Simmons, James, 131, 137–38, 139, 149,
  150–51, 152–53, 154, 155, 158, 159,
  216
Simonsen, Lone, 322–23, 324
Simpsons, The, 334
Sinclair, Upton, 72
Slater, Moni, 362–63
slaughterhouses, 72
slaves, 25, 29
  variation of, 37, 42
Slee, Richard, 95
Sloane, Hans, 38
Smadel, Joseph, 116, 174, 177, 180, 181, 187,
  188, 196, 202, 203, 210, 219, 221, 223,
  230–31, 238, 258
smallpox, 12, 15 20, 25–45, 46–111, 131–32,
  302
  death from, 30, 33, 35, 38, 42, 43, 44, 45,
    59, 60, 67, 69, 74, 75, 76, 77, 81, 88,
    91, 107, 111, 115, 116, 117, 132, 338
  decline of, 115, 118, 120, 349
  diagnosis of, 116
  disinfection and isolation treatment of,
    82–83, 84, 85, 89, 90
  epidemic spread of, 16, 25–26, 27–36, 60,
    66, 71–72, 74, 75–85, 88–90, 109–11
  eradication of, 12, 45, 46, 243
  hemorrhagic, 303
  homeopathic treatment of, 345
  immunity to, 28, 99, 111
  nicknames for, 75, 76
  quarantine, flight, and prior exposure as
    protection against, 28, 29, 32, 43,
    51–52, 67
  scars from, 37, 38, 39, 52, 66, 116
  symptoms of, 27, 29, 35, 41, 44, 52, 75, 76,
    79, 82, 115, 117, 349
  transmission of, 85
  treatment of, *see* vaccinations; variation

Variola major strain of, 71, 76, 77, 81–82, 88, 111
Variola minor strain of, 71, 75–77, 101, 102, 106, 107, 109, 111
smallpox keratitis, 303
smallpox vaccination campaign (1947), 16, 242
smallpox vaccination campaign (2003), 11–12, 13, 16–17
   politicizing of, 11–12, 13, 14, 17
   resistance to, 13, 14, 17
smallpox vaccination campaigns, 51, 56–57, 60–61, 79–82
smallpox vaccine, 12, 15, 20, 129, 174, 205, 427
   adverse reactions to, 241–43, 340, 356
   campaign to end routine use of, 241–43, 265
   children as warehouses of, 58–59
   contaminated, inactive, and fake supplies of, 50–51, 52–53, 58–60, 61, 64, 66, 71–72, 74, 75, 83–84, 89, 92–95, 96
   contribution of country legend to, 46–47, 48, 54
   CVI-78, 242, 243
   Jenner's development of, 12, 27, 44, 46–50, 52–58, 62, 71, 120, 175
   production of, 18, 21, 58–64, 72, 73, 75, 92–97, 99–100, 116
   regulation of, 59–62, 72–73, 91, 95–97
   "ring vaccination" with, 243, 304
   vaccinia as active ingredient in, 11, 12, 46, 52, 54, 175
smallpox virus, 7, 27, 29, 36
   mutations of, 52, 54–55
   Soviet production of, 11
   substituting cowpox virus for, 46–50
   tracking down and destroying of, 12
   weaponized forms of, 12
Smith, Charles S., 101, 102
Smith, David, 307
Smith, Hugh, 146
Smith, James, 59
Smith, Margaret, 242
Smith, Martin, 278, 287
Smith, Marty, 160
Smith, Stuart, 290
Smith, Theobold, 63, 95, 126
Snider, Dixie, 382, 419
Snow, C. P., 437
Snyder, Jack, 141
socialist medicine, 204
Social Security Act (1935), 127
Social Security Act (1965), 238
Society for General Inoculation, 45
Society for the Prevention of Cruelty to

Animals, 85
Soper, Fred, 141, 148, 151–52, 157
South Africa, 75, 133, 134
South Fork Elementary School, 359–60
South Fork Montessori School, 360
Soviet Union, 203, 206, 207–8, 209
   biowarfare capability of, 11
   collapse of, 11
Spalding, Lyman, 50
Spanish-American War, 75, 87, 131, 132–33, 135, 136, 144, 159
spas, 103, 108, 163–64, 361
special education programs, 237
spina bifida, 98
spiritualism, 68, 344
Springer, Georg, 392
Sputnik, 228
Stafford, Robert T., 282
Stanley, Wendell, 181
staphylococcal bacteria, 52, 379
Starr, Paul, 104, 130–31
state health departments, 17, 105
Statens Serum Institut, 407
steam baths, 44, 87
Steege, Beth, 399
Steffens, Lincoln, 73
Steiner, Rudolf, 327–28, 340, 342–46
sterilization, 98–99, 152, 157
Sternberg, George, 73
steroids, 161
Stewart, Jimmy, 130, 166
Stewart, Jon, 415
Stilwell, Joseph, 153
Stimson, Henry, 157
stockyards, 63–64
Stokes, J. Buroughs, 230
Stokes, Joseph, Jr., 174, 177, 181, 225–26
Stone, Noah, 360, 363–64
Stone, Valerie, 360, 363–64
Stonebreaker, Billy, 236
Stonebreaker, Dolores, 235–36
Stovy, Elisha, 51
Strain, James, 299
Straube, Rudolph, 103
strep throat, 124, 240
streptococcal bacteria, 52
Streptococcus pneumoniae, 21, 232, 307–8
Streptococcus pyogenes, 18, 240
streptomycin, 179, 192
strep vaccine, 240
Stuart, Edwin, 102, 103
sudden infant death syndrome (SIDS), 256, 277, 287, 288, 356, 357, 369
sulfur, 27, 39, 41, 44, 337
Summers, William, 49
Sunderland, Earl of, 38

Supreme Court, U.S., 82, 237
  compulsory vaccination upheld in, 82,
      98–99, 106, 295
Sutton, Daniel, 41, 175
Sutton, Robert, 41, 55, 175
SV-40 virus, 184, 209–10, 211–13, 225,
      230–31, 258
sweats, 41, 344
  antimony, 33, 39
Swedenborgian church, 100–101, 333–34,
      345
Swedish Evangelical Lutheran Church, 98
Sweet, Ben, 184
swine flu, 16, 259, 260, 261, 262, 425
Sydenham, Thomas, 28, 104, 169
Sykes, Lisa, 415
syphilis, 52, 58, 61, 86, 87, 151

Takahashi, Michiaki, 314
Tammany Hall, 128, 166
Taube, Andrea, 421
Tebb, William, 85
Temple University, 173
Temple University Hospital, 366–67
Tennessee State Department of Public
      Health, 277
Tenorio Cortez, Luis Fermin, 438
terrorism, 19, 433
  in 9/11 attacks, 341
  germ warfare, 11–12, 16
tetanus, 17, 52, 70, 71, 117, 118, 230, 351
  deaths from, 79–80, 83, 92–93, 94
  decline of, 111, 118, 230, 247
  incubation period for, 93
  vaccine-related, 79–80, 83–84, 92–95, 102,
      108
tetanus antitoxin, 79, 80, 92, 129, 139, 229,
      264, 379
tetanus bacilli, 72, 83–84, 92–95, 96
tetanus toxoid, 139
tetanus vaccine, 17, 20, 351, 412
Thatcher, Peter, 41
Theiler, Arnold, 147
Theiler, Max, 146–48, 151, 156, 157, 174,
      183
theosophy, 342
therapeutic nihilism, 313, 351
thimerosal, 192, 196, 264, 357, 383, 384, 396,
      403–12, 413–14, 416, 417, 420, 421
  mercury in, 356, 373–74, 375–79, 381, 396,
      406, 407
Thomas Jefferson University, 213
Thompson, Adam, 40–41
Thompson, Lea, 254
Thorazine, 386
Thorpe, Helen, 328–29

thrombocytopenia, 317, 325
Thurmond, Strom, 286
ticks, 140
Tierney, Gene, 233–34
Time, 235
Times (London), 134
Tinnerholm, Eric, 264
Tint, Howard, 272–73
Tishler, Max, 223
Togo (dog), 127
Tom, Melanie, 272
Toner, Kevin, 272
Toomey, John A., 270
Topping, Norman, 189
Townsend Harris High School,
      179
toxic minerals, 27, 33, 39–41
toxic shock, 71, 240
Treasury Department, U.S., 96, 427
Trisolgen, 273, 274
Troan, John, 190
tropical diseases, 56
truancy laws, 71
Truman, Harry, 204
tuberculosis, 20, 52, 53, 68, 73, 74, 87, 109,
      122, 136–37, 144, 168, 347–48, 435
  deaths from, 78, 91, 169
  decline of, 111, 129, 162
  increase in, 297
  mycobacterium, 348
tuberculosis vaccine, 348
tularemia, 130, 330
Turner, James, 258
Turner, Thomas, 151, 189
typhoid fever, 73, 74, 97, 118, 131, 168
  deaths from, 78, 91, 102, 134, 137, 430
  decline of, 126–27, 129, 135, 137, 169, 349
Typhoid Mary, 359–60
typhoid vaccine, 20, 68, 120, 133–36, 137
typhus, 68, 118, 130, 139–42, 144, 174, 176
Typhus Commission, U.S., 141, 142
typhus vaccine, 140–42, 183

umbilical cord, 380
Uncle Tom's Cabin Show, 75–76
UNICEF, 437
Union Army, 132
United Kingdom (UK):
  rubella cases in, 240
  vaccinations in, 95, 117, 271
  variolation in, 37–40, 44, 45, 48–49, 55
United States:
  health linked to leadership role of, 131
  immigration laws in, 97, 98
  imperial military missions of, 75, 132
  military-industrial technology of, 162, 163

political corruption in, 73–74
population growth in, 45
Universitaets Kindkerklinik, 125
uranium, 180
U.S. Biologics Control Act (1902), 72
U.S. News and World Report, 362

Vaccination, Social Violence, and Criminality (Coulter)
Vaccination Act of 1871, 60, 64
vaccination laws, 60–61, 89, 106–7, 230, 263
    enforcement of, 60, 64–65, 69, 71, 74, 77, 82, 87–88, 97–100, 101–2, 106, 108, 110, 119, 244, 245
    relaxing of, 69, 84–85, 95, 106
vaccinations:
    adverse effects and risks of, 13, 14, 16–17, 18–19, 33, 52–53, 58–59, 69, 70–73, 78–82, 83–85, 86–87, 90, 92–95, 99–100, 101, 107–8, 117–18, 139, 143, 241–43, 251–55
    in armed services, 13, 42, 50, 60, 111, 118, 119–20, 131–37, 177
    changing rates of, 111
    compensation for injuries from, 263–66, 275–76, 278–84, 286–87, 293, 393, 398, 426
    compulsory, 14, 56, 60–61, 64–65, 66, 68, 69, 70–71, 78–82, 88, 95, 97–102, 106–7, 109–10, 111, 131, 134, 136, 230, 244–46
    decline in rates of, 354, 366, 369
    doubts about morality of, 27, 34–35, 56, 100–101
    early history of, 12, 15–16, 46–69
    evolving technology of, 11, 14, 15, 16, 19–20, 26, 27, 49, 52, 53, 55, 57, 58, 64–65, 83, 92, 131
    expanding of, 50, 228–32, 240, 300
    free, 78, 109, 202, 246
    as key weapon against viruses, 159
    legal liability issues of, 13, 16, 73, 80, 82, 86–87, 199, 256, 260, 261–66, 272, 274, 275–76, 278–87, 290, 293, 426, 427
    mass, 16, 45, 60, 64, 85, 115–20, 131–37, 147, 219
    national security and, 159
    1913 commission reviewing safety and efficacy of, 101
    open air, 62–63
    opposition to, 13–14, 15, 16–19, 27, 29–36, 51, 56–57, 61, 64–69, 70–71, 73, 78–88, 89, 95, 97–111, 122–23, 230, 247, 256, 294–97, 328–70

promotion of, 15, 61–62, 110, 116–17, 131, 228–30, 239, 245–47, 258, 259, 354
public confidence in, 16, 21, 228, 240, 300–301
religious of philosophical exemption from, 230, 332–33, 353, 354, 367
ring, 243, 304
as spiritual trick, 100–101, 345
Vaccine Adverse Events Reporting System (VAERS), 317–18, 321–22, 323, 407–8
Vaccine Advisory Committee, 188, 189
Vaccine Appropriations Act (1961), 229
Vaccine Court, 391
Vaccine Development Board, 230
vaccine farms, 64, 73, 75, 92, 96–97, 132
Vaccine Guide, The: Making an Informed Choice (Neustaedter), 338
vaccines:
    adenovirus, 211
    bacterial subunit, 20–21
    belief in lifesaving imperative of, 18, 29, 79
    booster shots of, 89, 108, 396
    combined, 233, 240, 247, 251–57, 264, 412
    contamination of, 16, 50–53, 58–61, 64, 66, 71–72, 74, 83–84, 89, 92–95, 96, 198–201, 231, 379, 414
    costs of, 18, 96, 194, 283, 298, 432
    definition of, 14
    edible, 213
    federal grants for development of, 230, 245–47
    history of, 12, 15–16
    inhalable, 430
    killed-bacteria, 20
    killed-virus, 20
    live-bacteria, 20
    live-virus, 20
    origins of, 12, 104
    pathogenic micro-organisms in, 14–15
    polysaccharide, 20
    recombinant, 310
    regulation of, 59–62, 72–73, 91, 95–97, 222, 256–57, 262, 273, 380, 432–35
    safety of, 51, 52, 65, 69, 72–73, 88, 117, 251–93, 316–18, 375
    shelf-life of, 96, 173
    shortages of, 92, 261–62, 299, 301, 354, 424, 425
    supply and distribution of, 75, 92–95, 96, 116–17, 244, 260
    testing of, 20, 50, 73, 81, 94, 95, 240
    therapeutic, 19
    upgrading of, 14, 19
    viral vs. bacterial, 174
    warnings required with, 264–65
    withdrawal of, 264

vaccines (*continued*)
  World War II development of, 137–59,
    183, 268–69
  see also specific vaccines
Vaccine Safety Datalink, 318, 326, 383, 403
vaccine-safety monitoring systems, 16
Vaccines: Are They Really Safe and Effective?
  (Miller), 338
vaccine schedules, 18, 19, 308, 312, 353, 361,
  376
Vaccines for Children program, 301, 429–30
vaccine stables, 48, 63
vaccinia virus, 11, 12, 46, 52, 54, 55, 62, 65,
  81, 96, 117, 132, 175, 265
Vanderbilt University, 138, 140, 292, 366
van Riper, Hart, 188
varicella vaccine, *see* chickenpox vaccine
variolation, 25–45, 175
  condemnation of, 30–34, 38, 39, 42, 43
  defense of, 34–35, 40
  definition of, 27
  early case histories of, 33
  in England, 37–40, 44, 45, 48–49, 55
  first use of, 29–30
  gradual acceptance of, 31, 36, 37–38, 43–45
  history of, 25–45
  long-term health problems suspected with,
    33–34
  mass, 43–44, 45, 131
  "preparation" for, 39–41, 48, 52–53
  risks of, 33–34, 35–36, 38, 42, 45, 52–53
  techniques of, 27, 29, 38, 39–41, 43
  vaccination compared with, 27, 46
Vaughan, Victor, 133, 134, 135
Vaxgen, 302
Vedic medicine, 335–36
Verstraeten, Thomas, 404, 405, 410, 411,
  415, 418
Veterans Administration, 157
veterinary practice, 48, 55
Victorville Army Air Field, 152
Vietnam War, 261, 278
Vincent, George Edgar, 130
Vioxx, 330, 393
Virginia Historical Society, 58
virology, 116, 145, 213, 215, 222–25
viruses:
  extinction of, 55
  "filtrable," 168
  mutation of, 52, 54, 78
  oncogenic, 210
  simian, 209, 211, 212
vitamin(s):
  B, 223, 386, 413
  C, 340
  deficiency in, 339, 342

  megadoses of, 386, 413
  soluble forms of, 386
Vogue, 86
vomiting, 33, 41, 42, 126, 217, 268, 361
von Pirquet, Clemens, 125–26

Wagner, Philip, 109
Wagstaffe, William, 39
Wakefield, Andrew, 387–88, 403, 409, 410,
  414
Wake Forest University, 381
Waldorf Schools, 327–28, 332–33, 342,
  345–46
Wallace, Lord Alfred Russel, 66–68, 104
Walsh, William, 398–99
Walter Reed Army Institute of Research, 222,
  223–24, 234, 238
Walters, Leroy B., 251
war criminals, 149–50
Ward, Elsie, 180
Ward, Joel, 368
Ward, Samuel, 43
War Department, U.S., 152
Waring, Fred, 176
Warkany, Josef, 396
Warm Springs, Ga., 163–64, 166, 214
Warner-Lambert, 427
Warren, Joel, 184
War Research Service, 222
Warshafsky, Ted, 272
Washington, D.C., 72, 73, 211
Washington, George, 42, 131
Washington, Martha, 42
Washington, University of, 310
Washington Post, 12, 161, 193, 261
Washington Post Magazine, 14, 293
Washington Senators, 194
Washington Times, 13
water:
  clean, 65, 137, 167
  contamination of, 67, 73
  filtration of, 97, 126, 129, 167, 169
  fluoridation of, 230
Watergate scandal, 261
Waterhouse, Benjamin, 44, 50–52
Waterhouse, Daniel, 50
Watson, Barbara, 367, 412
Waxman, Henry, 260, 282, 284, 286, 299,
  392, 393–94, 426
Weaver, Harry, 181, 188
Weaver, John, 91
Weibel, Robert, 226
Weihl, Albert, 171–72
Weihl, Carl, 171–72, 214, 231, 271, 273
Weil, William, 405
Weinstein, Israel, 116

Welch, William H., 74–75, 76, 77, 81, 90, 91, 106, 128
Welch, William M., 62, 74
Weldon, Dave, 403, 411, 414, 417, 418, 419–20
welfare state, 217, 299
Wellcome Institute, 219
Weller, Robert, 234
Weller, Thomas, 178, 179, 181, 191–92, 196, 200, 202, 234–35
Westmoreland, Tim, 284–85
West Nile Virus, 16, 184, 213
Wharton, Melinda, 418
wheat, 168–69
Whitestone, Heather, 307
whooping cough, 78, 120, 122, 255, 256, 331, 332, 347, 352–53, 354–55, 359–70
    after effects of, 268, 269, 358
    deaths from, 78, 162, 169, 267, 268, 269, 352, 353, 365, 366, 369, 370
    decline of, 169, 255
    rising rates of, 297, 354, 355, 359–61, 366–70
    stages of, 267–68
    testing for, 368–69
    transmission of, 353, 361, 367, 369
    underdiagnosis of, 368–69
    see also pertussis
whooping cough bacteria, 267, 268
whooping cough vaccine, 14, 17, 20, 205, 240, 251
    see also pertussis vaccine
Wiley, Harvey, 98
Willard Parker Hospital for Infectious Diseases, 116
Williams, Kathi, 253, 254, 279–80, 287, 330, 331, 346, 357
Williams, Linsly R., 109
Williams, Michael L., 397
Williams, Mike, 432–33
Williams, Nathan, 253, 254, 280
Willowbrook State School, 235, 263, 309–10
Wilson, Graham S., 203, 219
Wilson, John Rowan, 160, 170, 183, 204
Winchell, Walter, 193
Winkelstein, Warren, 194, 198
Winkler, Dawn, 354–59, 367
Winkler, Haley, 356, 359
Winkler, Levi, 355, 357, 358
Winslow, W.-E.A, 85
Winthrop, Adam, 34
Wired, 400
Wistar Institute, 213, 239
Wolfe, Sidney, 330
Wood, Leonard, 132
Woodruff, Alice, 140

Woodruff, Eugene, 140
Woodruff, Roy O., 136
Workman, William G., 195–96, 198, 204
World Health Organization (WHO), 12, 13, 46, 184, 207, 242–43, 303, 320, 323, 324
    National Immunization Days of, 439–40
World War I, 108, 119, 134–36, 138, 139, 159, 178
World War II, 16, 117, 119–20, 131, 162, 163, 168, 227, 358, 366, 375
    allied bombing raids in, 141, 143
    liberation of Europe in, 154
    Normandy invasion in, 139, 174
    Pearl Harbor attack in, 143, 153
    vaccines for, 137–59, 183, 268–69
worms, 35
Worrall, Carol, 357–58
Wright, Almroth, 68, 133–34
Wurmschokolade (worm chocolate), 396
Wyeth Laboratories, 75, 95, 196, 199–200, 201, 261–62, 264–65, 266, 273, 274, 277, 283, 307, 318–19, 320, 321, 323, 412, 425, 426, 428, 429
Wynn, Mrs. John, 77–78

Yale University, 85, 138, 157, 177, 178, 181, 235
Yale University Hospital, 171, 241
yeast, 310
yellow fever, 75, 134, 143–51, 174, 176, 261
    Asibi strain of, 146
    cause and spread of, 75, 144
    deaths from, 130, 144–45, 150
    1793 Philadelphia plague of, 40, 144
    symptoms of, 144
yellow fever vaccine, 143–58, 174, 189
    adverse effects and deaths, 16, 143–44, 146–48, 152–57
    contamination of, 16, 143–44, 146–48, 153–58
    creation of, 145–47
    human serum in, 146, 148, 151, 156–58, 175
    17 D strain of, 146–47, 148
Yellow Jack (de Kruif and Howard), 130
Yersin, Alexandre, 123
yoga, 18
Young, James, 429
Youngner, Julius, 180–81, 186, 197–98
Yurko, Alan, 339–40
Yurko, Alan, Jr., 339–40

zinc, 397
zinc sulfate, 172
Zinsser, Hans, 140, 178